Springer Series in
OPTICAL SCIENCES 111

founded by H.K.V. Lotsch

Springer Series in
OPTICAL SCIENCES

The Springer Series in Optical Sciences, under the leadership of Editor-in-Chief *William T. Rhodes*, Georgia Institute of Technology, USA, provides an expanding selection of research monographs in all major areas of optics: lasers and quantum optics, ultrafast phenomena, optical spectroscopy techniques, optoelectronics, quantum information, information optics, applied laser technology, industrial applications, and other topics of contemporary interest.

With this broad coverage of topics, the series is of use to all research scientists and engineers who need up-to-date reference books.

The editors encourage prospective authors to correspond with them in advance of submitting a manuscript. Submission of manuscripts should be made to the Editor-in-Chief or one of the Editors. See also http://www.springer.com/series/624

Junji Ohtsubo

Semiconductor Lasers

Stability, Instability and Chaos

Second, Enlarged Edition

With 169 Figures and 5 Tables

 Springer

phys.

Professor Junji Ohtsubo

Faculty of Engineering
Shizuoka University
Johoku Chome 3-5-1
432-8561 Hamamatsu, Shizuika
Japan
Email: tajohts@ipc.shizuoka.ac.jp

Library of Congress Control Number: 2007935442

ISSN 0342-4111

ISBN 978-3-540-72647-0 2nd Edition Springer Berlin Heidelberg New York
ISBN 978-3-540-23675-7 1st Edition Springer Berlin Heidelberg New York

Springer is a part of Springer Science+Business Media
springer.com

Typesetting and production: LE-TEX Jelonek, Schmidt & Vöckler GbR, Leipzig
Cover design: eStudio Calamar S.L., F. Steinen-Broo, Girona, Spain

SPIN 12066056   57/3180/YL - 5 4 3 2 1 0   Printed on acid-free paper

Preface

Preface for the second edition

Chaos research in laser physics, especially in semiconductor lasers, has developed further even after completion of the first edition of this book in the late summer of 2004, and it is still growing rapidly. For example, various forms of chaotic dynamics have been applied in newly developed semiconductor lasers, such as in vertical-cavity surface-emitting semiconductor lasers and broad-area semiconductor lasers. Chaotic dynamics plays an important role in these new lasers, even for their solitary oscillations, and control of the dynamics is currently an important issue for practical applications. Another significant advance has been made in the area of chaotic optical secure communications. Chaotic secure communications using existing public optical communications links have been tested, and successful results have been obtained. In this second edition, I have filled in the gaps in the explanation of chaotic laser dynamics in the previous edition, and I have also added several important topics that have been developed recently. In particular, a new chapter on laser stabilizations has been added, and a number of misprints in the first edition have been corrected. I believe this book will be of interest not only to researchers in the field of laser chaos, but also to those working in nonlinear science and technology.

Hamamatsu, Spring 2007 *Junji Ohtsubo*

Preface

The aim of this book is the description of the state of the art of chaos research in semiconductor lasers and their applications, and the future perspective of this field. However, for the beginner, including graduates who intend to participate newly in this field, the book starts with an introduction and explanation of chaos in laser systems and the derivation of semiconductor laser rate equations assuming two-level systems. I discuss stabilities, instabilities, and various chaotic dynamics in semiconductor lasers induced by optical and optoelectronic feedback, optical injection, and injection current modulation. As optical feedback, the effects of the conventional reflector, the grating feedback mirror, and the phase-conjugate mirror are considered. Recent results both for theoretical and experimental investigations are presented. Instabilities and chaotic dynamics for novel laser structures (self-pulsating semiconductor lasers, vertical-cavity surface-emitting semiconductor lasers (VCSELs), broad area semiconductor lasers, and semiconductor laser arrays) are also discussed not only for solitary operations but also in the presence of external perturbations.

As applications of semiconductor laser chaos, control and noise suppression of lasers based on chaos control algorithm are presented. Externally controlled lasers are also interesting for applications of new laser systems with high coherent light sources or tunable light sources. The self-mixing interferometer in semiconductor lasers is an attractive application based on dynamic properties using bistable states in optical feedback effects. I also discuss these subjects. As another application of chaos, several methods of data encryption into the chaotic carrier and its decryption are introduced for secure data transmissions and communications based on chaos synchronization in semiconductor laser systems. This book is focused on the dynamic characteristics of semiconductor lasers and their applications. Therefore, the detailed descriptions for materials and structures of semiconductor lasers are beyond the scope of this book. Of course, such characteristics are closely related to chaotic phenomena in semiconductor lasers. The interested reader is referred to the related books. For those who are interested in optics but not familiar with nonlinear systems and chaos, I have attached an appendix to describe the phenomena of chaotic dynamics and to accustom the reader to the common tools for chaos analyses in nonlinear systems. Chaos research, especially in semiconductor laser systems, is still developing rapidly and is expected to produce fruitful results not only for the fundamental research of chaos but also for applications as dynamic engineering.

Chapters 1 to 4 are devoted to the basics and the introduction of laser chaos and chaotic dynamics in semiconductor lasers, so that readers who want to know what laser chaos is and how it behaves in semiconductor lasers can follow them. Chapters 5 to 12 discuss the topics of chaos in semiconductor lasers and readers may skip to each topic according to their interest. Expected readers of this book are as follows; first, I assume those researchers who have already been involved in this field to gain an overview of the state of the art of their research. The next group is the graduate students and researchers who intend to participate in this field. For them, I have derived and explained most equations in the text from first principles as far as possible. Those readers who are familiar with electromagnetic theory and have some fundamental knowledge of optics and lasers will be able to follow the book. Finally, this book is devoted to all other researchers and engineers who are interested in dynamics in nonlinear systems and laser instabilities and applications. Since the laser is a very excellent model of a nonlinear system that shows chaotic dynamics, I believe that this book will provide useful information for readers not only in the field of optics but also in other related areas. Moreover, I hope that the ideas and techniques discussed here will give rise to a new paradigm of nonlinear systems such as chaos engineering or dynamic engineering.

For the publication of this book, I am indebted to many people. Here, I will not be able to express thanks to all those people, but, at first, I would like to thank colleagues and some previous students in my laboratory, Drs. Yun Liu, Atsushi Murakami, Keizo Nakayama, Yoshiro Takiguchi, Shuying Ye, Hong Yu, for their many discussions and support. I also extend my thanks to many other researchers at various institutions and universities who gave me fruitful discussions and advice. Those are Prof. Wolfgang Eläßer, Dr. Peter Davis, Dr. Ingo Fischer, Prof. Jia-Ming Liu, Dr. Cristina Masoller, Dr. Claudio Mirasso, Prof. Rajarshi Roy, Prof. Kevin Alan Shore, and Dr. Atsuhi Uchida. I also owe thanks to many other people with whom I had useful discussions. Finally, I express sincere thanks to Prof. Toshimitsu Asakura who gave me the opportunity to write this book and also encouraged me in various stages of the research.

Hamamatsu, April 2005   *Junji Ohtsubo*

Contents

1 Introduction

Irregularity induced by chaotic dynamics is essentially different from random fluctuation based on a stochastic process, since the chaotic system can be described by a set of rigorous equations, namely deterministic equations. Lasers are essentially chaotic systems described by nonlinear differential equations with three variables and they show a rich variety of chaotic dynamics. In this chapter, we briefly discuss laser chaos in relation to ordinary nonlinear systems and present a historical perspective of chaos research in semiconductor lasers. Then, the outline of this book will be presented.

1.1 Chaos and Lasers

It was in 1963 that Lorenz (1963) investigated the behaviors of convective fluids as a model for the atmospheric flow and showed that nonlinear systems described by three variables could exhibit chaotic dynamics. Of course, many researchers were aware of the existence of complex dynamics in well-defined systems from the beginning of early 1900s. Henri Poincaré (1913), a prominent French mathematician, at first noted the "sensitivity to initial condition." In his book, he wrote that "it may happen that small differences in the initial conditions produce very great ones in the final phenomena. A small error in the former will produce an enormous error in the latter. Prediction becomes impossible, and we have the fortuitous phenomenon."

However, modern research of chaos started from the study for irregular and complex dynamics of a nonlinear system developed by Lorenz. Chaos is not only a description of a different viewpoint of nonlinear phenomena but also itself a new physics. Chaos is a phenomenon of irregular variations of systems' outputs derived from models that are described by a set of deterministic equations. We must distinguish "chaos" from the observation of "random" events, such as the flipping of a coin, since chaos is generated in accordance with the deterministic order, namely, chaotic dynamics refers to deterministic development with chaotic outcome (Appendix A.1). The system evolves in a deterministic way and the current state of the system depends on the previous state in a rigidly deterministic way, although the systems' output shows random variations. This is in contrast to a random system where the present observation has no causal connection to the previous one. In spite of the deterministic models, we cannot foresee the future of the output, since chaos is very sensitive to the initial conditions, as Poincaré pointed out, and each system behaves completely different from each other even if the difference of the initial state is very small.

Chaos is always accompanied by nonlinearity. Nonlinearity in a system simply means that the measured values of the properties in the system depend in a complicated way on the conditions in the earlier state. Nonlinear property in a system does not always guarantee the occurrence of chaos, but some form of nonlinearity is required for the realization of chaotic dynamics. Chaos can be observed in various fields of engineering, physics, chemistry, biology, and even in economics. Though the fields are different, some of the chaotic systems can be characterized by similar differential equations. They show similar chaotic dynamics and the same mathematical tools can be applied for the analysis of their chaotic dynamics.

Nonlinear systems can be also found in optics. Many optical materials and devices show nonlinear response to the optical field and, therefore, they are candidates for nonlinear elements in chaotic systems. One such device is the laser. Since lasers themselves are nonlinear systems and are typically characterized by three variables: field, polarization of matter, and population inversion, they are also candidates for chaotic systems. Indeed, it was proved in the mid 70's by Haken (1975) that lasers are nonlinear systems similar to the Lorenz model and show chaotic dynamics in their output powers. He assumed a ring laser model and considered two-level atoms in the laser medium. Though lasers are not always described by his model, the approximations are reasonable for most lasers. Thereafter, the laser rate equations that are described by the nonlinear equations with three variables, are called Lorenz-Haken equations after their contributor (Haken 1985). However, ordinary lasers do not exhibit chaotic behavior and only a few of the lasers with bad cavity conditions show chaotic dynamics. In the meantime, chaotic behaviors were theoretically demonstrated in a ring laser system (Ikeda 1979). First, Weiss and Brocke (1986) observed Lorenz-Haken chaos in infrared NH_3 lasers.

Contrary to the prediction of Haken, ordinary lasers are stable systems and only a few systems of infrared lasers show chaotic behaviors in their output powers. Arecchi et al. (1984a) investigated laser systems from the viewpoint of the characteristic relaxation times of the three variables and categorized lasers into three classes. According to their classifications, one or two of the relaxation times are in general very fast compared with the other time scales and most lasers are described by the rate equations with one or two variables. Therefore, they are stable systems that are categorized in class A and B lasers. Only class C lasers have the full description of the rate equations with three variables and can show chaotic dynamics. However, class A and B lasers can show chaotic dynamics when one or more degrees of freedom are introduced to the laser systems.

Class B lasers are characterized by the rate equations for field and population inversion, and they are easily destabilized by an additional degree of freedom as an external perturbation. For example, solid state lasers, fiber lasers, and CO_2 lasers that are categorized as class B lasers, show unstable oscillations by external optical injection or modulation for accessible laser parameters. Semiconductor lasers, which are also classified into class B lasers and are the main topic of this book, are also very sensitive to self-induced optical feedback, optical injection from different lasers, optoelectronic feedback, and injection current modulation. A review of the earlier study of laser instabilities and chaos has been given by Abraham et al. (1988).

1.2 Historical Perspectives of Chaos in Semiconductor Lasers

Semiconductor lasers (edge-emitting and narrow stripe types), which are the main topics of this book, are intrinsically stable lasers. However, semiconductor lasers, which are described by the field and the carrier density (equivalently the population inversion) equations, can be easily destabilized by the introduction of external perturbations such as external optical feedback, optical injection, or modulation for accessible laser parameters. Since the early 80's, feedback induced instablities and chaos in semiconductor lasers have been extensively examined (Lang and Kobayashi 1980). In a semiconductor laser, the laser oscillation is affected considerably when the light reflected back from an external reflector couples with the original field in the laser cavity. A variety of dynamics can be observed in semiconductor lasers with optical feedback and they have been investigated by many researchers for the past two decades.

One of the main differences between semiconductor lasers and other lasers is the low reflectivity of the internal mirrors in the laser cavity. It ranges typically only from 10 to 30% of the intensity in edge-emitting semiconductor lasers. This makes the feedback effects significant in semiconductor lasers. Another difference is a large absolute value of the linewidth enhancement factor α of the laser media. The value of the linewidth enhancement factor $\alpha = 2 \sim 7$ was reported depending on the laser materials, while this value is almost zero for other lasers. Then, the coupling between the phase and the carrier density is encountered in the laser dynamics. These factors lead to a variety of dynamics quite different from any other lasers. At weak to moderate external optical feedback reflectivity, the laser output shows interesting dynamical behaviors such as stable state, periodic and quasi-periodic oscillations, and chaos for the variations of the system parameters. These ranges of external reflectivity are not only interesting from the viewpoint of fundamental physics, but also very important in practical applications of semiconductor lasers, such as in optical data storage systems and optical communications. Extensive lists of the recent literature for the dynamic characteristics in semiconductor lasers with optical feedback can be found in the following references (van Tartwijk and Agrawal 1998, Ohtusbo 1999, Otsuka 1999, and Ohtsubo 2002a).

Injection locking phenomena are a universal feature in lasers. Since the internal reflectivity of the facet in a semiconductor laser is very low compared with other lasers, one can easily realize injection locking from a different laser. Moreover, the effects of injection locking stand out due to the non-zero value of the linewidth enhancement factor α and one can observe not only stable injection locking but also various dynamics of unstable optical injection phenomena depending on the injection parameters. Semiconductor lasers usually have different laser oscillations characteristics for the same product number or even for the same wafer, but the oscillation frequency can be tuned on the order of GHz by changing the injection current. Therefore, a light source for injection locking to different lasers with appropriate frequency detuning is easily available. Thus, injection locking phenomena have been extensively studied in semiconductor lasers. However, earlier work was limited to stable injection locking phenomena for amplification of signals and laser stabilization. In these applications, the laser is locked to the external laser, which means that it almost copies the spectrum of the injected light. On the other hand, unstable injection locking, instabilities, and mixing of detuned frequencies occur outside the region of stable injection locking in the phase space of the frequency detuning and the injection ratio. From the viewpoint of chaos, optical injection is an addition of an extra degree of freedom to semiconductor lasers and it may induce chaotic oscillations in the laser output. It was numerically predicted (Sacher et al. 1992 and Annovazzi-Lodi et al. 1994) and experimentally demonstrated (Simpson et al. 1994 and 1995a) that an optically injected semiconductor laser follows a period-doubling route to chaos.

Direct modulation for accessible parameters is not an easy task in most lasers, however the output power of a semiconductor laser is easily controlled through the injection current and, at the same time, the laser frequency can be changed by the injection current modulation. Small amplitude injection current modulation or even large modulation under appropriate conditions for laser oscillations may produce faithful copies of the modulation amplitude for the output power in a semiconductor laser. However, modulation for the injection current is a perturbation to the laser and also the introduction of an extra degree of freedom to it. Indeed, instablities and chaotic oscillations have been observed by the injection current modulation in semiconductor lasers under the conditions of high frequency modulation with a large modulation index close to the relaxation oscillation frequency of the laser (Hori et al. 1988).

Optoelectronic feedback systems in which the emitted light from a semiconductor laser is once detected and fed back through the injection current are also studied to stabilize the laser oscillations. For a certain range of optoelectronic feedback, the laser may indeed be stabilized and the method is applied to obtain an ultra-high coherent light source. However, optoelectronic feedback has a similar effect to the above perturbations on the dynamics in semiconductor lasers. We can also observe unstable pulsation oscillations in the output of a semiconductor laser for certain conditions (Olesen et al. 1986). With the availability of high speed electronic circuits, optoelectronic feedback systems having a time response of the same order as the relaxation oscillation frequency (on the order of GHz) have been studied and useful applications of chaos dynamics have been proposed based on high-speed optoelectronic feedback (Tang and Liu 2001a, b).

Recently, a variety of novel semiconductor laser devices with different structures has been proposed and fabricated beside edge-emitting narrow-stripe semiconductor lasers, for example, self-pulsating semiconductor lasers, vertical-cavity surface-emitting semiconductor lasers (VSCELs), broad-area semiconductor lasers, and semiconductor laser arrays. These lasers themselves have extra degrees of freedom in addition to the characteristics of ordinary edge-emitting semiconductor lasers. For example, space-dependent differential terms due to a wide stripe width are introduced in the rate equations for broad-area lasers and these terms play an important role in the laser dynamics. Therefore, these newly developed lasers themselves are unstable and exhibit chaotic dynamics without any external perturbations (Yamada 1993, Law et al. 1997, and Gehrig and Hess 2000). The studies of chaotic dynamics in semiconductor lasers including new structure devices are excellent models for nonlinear chaotic systems and are very interesting from the viewpoint of basic chaotic research. Instabilities and chaotic behaviors are also greatly enhanced by additional external perturbations in the same manner as edge-emitting semiconductor lasers.

In the case of the vertical-cavity surface-emitting lasers (VCSELs), the reflectivity of the internal mirrors is very high at more than 99%, however they are also sensitive to external optical feedback due to a small number of photons in the internal cavity. Therefore, semiconductor lasers of all types are essentially very sensitive to external optical feedback. In spite of the differences of device structures, the dynamics of semiconductor lasers are the same as long as the laser rate equations are written in the same or similar forms. The dynamics of edge-emitting single mode semiconductor lasers have been extensively studied for a long time and a lot of fruitful results have been obtained. However, they are still important issues for the fundamental physics of optical chaos and also practical applications. On the other hand, little investigation into the dynamics of newly developed laser structures has been carried out.

Through external perturbations, semiconductor lasers are either stabilized or destabilized. The effects of such perturbations on laser dynamics, stability and instability, are two sides of the same coin. Examples include optical feedback, optical injection, and optoelectronic feedback. To stabilize laser oscillations, the disturbances may be weak or strong. The lasers can then be strongly stabilized under appropriate conditions of external parameters and operating conditions of the lasers. Stabilization of semiconductor lasers is very important with regard to their application. For example, frequency stabilization, linewidth narrowing, power stabilization, polarization fixing, and beam shaping are very important in optical communications, optical data storage systems, and optical measurements. In particular, ultrastabilized semiconductor lasers are expected in broad-band optical communications, high precision optical measurements, and standard light sources. Semiconductor lasers are rather unstable compared with other lasers, and their stabilization has been an important issue from the beginning of their development. For newly developed semiconductor lasers, such as VCSELs and broad-area semiconductor lasers, we can apply the same techniques of laser stabilization as those for edge-emitting semiconductor lasers. However, these lasers themselves demonstrate instabilities in their solitary oscillations. In VCSELs, controls for polarization and spatial mode instabilities are essential in applications. In broad-area semiconductor lasers, filament suppression of the oscillation pattern can greatly improve beam quality, producing a high-density beam. Such unstable oscillations are also stabilized using similar techniques through external controls, as discussed above. Semiconductor lasers are still developing, and stabilization both by the device structure and through external controls is currently an important research area.

In the meantime, important breakthroughs for applications of chaos were made in the early 90's. The ideas of chaos control and chaos synchronization were proposed and developed in this decade as common interests in various fields of nonlinear research. The ideas of chaos control (Ott et al. 1990) and chaos synchronization (Pecora and Carroll 1990) were proposed and developed in this decade. Noise suppression of feedback induced chaotic oscillations in semiconductor lasers has been proposed based on chaos control (Liu et al. 1995). Also fixed point or periodic oscillations, precursor to the onset of chaos in semiconductor lasers with optical feedback, can be used for laser control and optical measurements (Donati et al. 1995). The possibility of chaotic communications has been discussed based on chaos synchronization in two chaotic solid state laser systems (Colet and Roy 1994). After their pioneering work, the study of secure data transmissions and communications has also been discussed based on synchronization in two chaotic semiconductor laser systems (Special Issue IEEE Tans. Circuits Syst. I 2001 and Future Section IEEE J. Quantum Electron. 2002). Harnessing chaotic lasers is very attractive from the viewpoint of applications, since optics is very fast and contains parallelism as a nature of light. Applications of chaotic lasers are still growing and developing. Thus, chaotic lasers are not only important for basic research but also for engineering applications.

1.3 Outline of This Book

In this book, we focus on the dynamics and applications in semiconductor lasers subjected to external perturbations. In Chap. 2, we first introduce general forms of laser rate equations, which are equivalent to the Lorenz equations, and the classifications of lasers are given. The instabilities intrinsically involved in the rate equations are studied. Next, semiconductor lasers as class B lasers are described. The possibility of unstable oscillations in semiconductor lasers by the introduction of external perturbations is discussed. A solitary semiconductor laser is characterized by two equations for the field and the carrier density (population inversion). We then derive the forms of the rate equations for edge-emitting semiconductor lasers with a narrow stripe width in Chap. 3. Linear stability analysis used as a common tool for investigating the dynamics of nonlinear chaotic systems is introduced and the laser relaxation oscillation frequency, which plays an important role in chaotic dynamics, is derived. Several fundamental characteristics of semiconductor lasers are also introduced.

In Chap. 4, the theory of optical feedback in semiconductor lasers is presented. The effects of feedback in various external reflectors, including grating mirrors and phase-conjugate mirrors, are taken into account and the formulations of their systems are presented by introducing rate equations with optical feedback effects. In Chap. 5, substantial feedback effects and chaotic dynamics in semiconductor lasers are discussed and both numerical and experimental results are given under variations of the system parameters. Feedback induced chaos depending on the external cavity length and the feedback fraction is investigated. Chaos induced by external optical injection with frequency detuning is also an important issue in semiconductor laser systems. The theory of optical injection and their instabilities are discussed in Chap. 6. Unstable and chaotic oscillations are observed in the region outside the stable injection locking in the phase space of frequency detuning and the injection fraction. The coexistence state of chaotic attractors, which is known as one of the characteristics in nonlinear systems, is demonstrated in the injection locking systems. Enhancement of chaotic frequency is observed by strong optical injection locking in semiconductor laser systems. The effects for the modulation bandwidth in optically injection-locked semiconductor lasers are also discussed in this chapter. In Chap. 7, dynamic characteristics of optoelectronic feedback and injection current modulations are presented. Unstable chaotic pulsations induced by feedback and modulation to the injection current are investigated in relation to the characteristics of electronic feedback circuits.

The rate equations of semiconductor lasers with various laser structures are introduced in Chap. 8. We assume a single mode oscillation for a semiconductor laser in the preceding chapters, however the effects of the multimode oscillations in edge-emitting semiconductor lasers are considered and dynamic properties of multimode lasers are discussed in this chapter. Stable and unstable periodic oscillations, and chaotic pulsations of self-pulsating semiconductor lasers, which are developed as light sources for optical data storage systems, are studied. Vertical-cavity surface-emitting lasers (VCSELs) are promising light sources in optical communications and optical data storage systems. However, VCSELs have xspatial and polarization dynamics even in solitary oscillation. The dynamics of VCSELs are discussed in this chapter. Broad-area lasers and laser arrays are also interesting future devices in engineering applications. Their dynamics are also presented in this chapter. These new types of semiconductor lasers themselves contain instability arising from their structures and show chaotic behaviors even in the absence of external perturbations.

We cannot foresee the future of chaotic oscillations for time evolution, since chaos has a strong dependence on the initial condition of a system. However, chaos can be controlled. In Chap. 9, methods of chaos control are introduced. Control of chaotic oscillations in semiconductor lasers with optical feedback is discussed and the reduction of the feedback induced relative intensity noise (RIN) is demonstrated based on the method of chaos control. These methods can also be applied not only to ordinary edge-emitting semiconductor lasers but also to other semiconductor lasers with newly developed structures. Either stabilities or instabilities are enhanced by external perturbations in semiconductor lasers, and their dynamics is discussed in Chap. 10. In this chapter, methods of stabilization and control of semiconductor lasers are presented. Some of these are closely related to chaos control, as discussed in Chap. 9, and others involve forced control of stable oscillations. Stabilization for laser oscillations such as linewidth, frequency, spatial modes, polarization, and so on, are introduced in edge-emitting semiconductor lasers. Similar control techniques are also applied to newly developed semiconductor lasers. In semiconductor lasers with optical and optoelectronic feedback systems, periodic oscillations of the outputs preceding the onset of chaos are observed for the variations of chaotic parameters. These properties can be applied to various measurements, such as interferometric displacement and vibration measurements. In Chap. 11, applications of self-mixing interferometers and active interferometers are discussed, in which bistable states of the systems before the onset of chaotic bifurcations are used.

Synchronization of two chaotic nonlinear systems is interesting not only from the viewpoint of fundamental physics but also from applications. It is not self-evident that two chaotic nonlinear systems show synchronization and the theoretical background for chaos synchronization has not yet been fully established. However, synchronization of chaotic oscillations has been observed by experiments and numerical simulations. Chaos synchronization can be also observed in systems of chaotic semiconductor lasers. The systems and conditions for synchronization in chaotic semiconductor lasers are discussed in Chap. 12. Chapter 13 follows the applications of chaos synchronization. Since strict conditions must be satisfied for chaos synchronization in nonlinear systems, one can construct a secure communication channel in the sense of hardware levels. The possibility of chaos communications is presented based on chaos synchronization in semiconductor laser systems. Finally, chaotic communications through the existing public communication channel is demonstrated.

The origins of chaos are very unique for each nonlinear system, but there are common tools for the analyses of chaotic dynamics. For detailed descriptions of chaos and their analyses the reader is referred to appropriate books.

However, for those readers who are not familiar with chaos and its analyses, I finally attached an appendix on the origins of chaos in nonlinear systems and some of their common tools for chaotic data analyses. In this book, I treat main topics of chaos dynamics and applications in semiconductor lasers, however they are not the entirety of the research. Other related topics are dynamics in new types of semiconductor lasers, such as micro-cavity semiconductor lasers (Lee et al. 2002), random lasers (Cao 2003), multi-section lasers (Kawaguchi 1994), flared broad-area lasers (Levy and Hardy 1997), and others. These are also important issues related to stabilities, instabilities, and chaos in semiconductor lasers. Research into chaos in semiconductor lasers is still ongoing, and we can expect fruitful results both for basic physics and practical applications.

2 Chaos in Laser Systems

Starting from the Maxwell equation in a laser medium based on the model of two-level atoms, we derive the time dependent Maxwell-Bloch equations for field, polarization of matter, and population inversion. Then, we prove that the three differential equations are the same as those of Lorenz chaos. Well above the laser threshold, the laser reaches an unstable point at a certain pump level, which is called second laser threshold. However, only a few real lasers show chaotic dynamics with a second threshold and most other lasers do not have the second threshold, resulting in stable oscillations for the increase of the pump. Stable and unstable oscillations of lasers are related to the scales of the relaxation times for the laser variables. We discuss stability and instability of lasers based on the rate equations and present their classifications from the stability point of view.

2.1 Laser Model and Bloch Equations

2.1.1 Laser Model in a Ring Resonator

The theory of lasers should be treated by the interaction between matter and electro-magnetic field based on quantum mechanics. However, we employ here the semi-classical treatment followed by Haken (1985) and van Tartwijk and Agrawal (1998), which is very easy to understand. Figure 2.1 shows a ring resonator for a laser model with two-level atoms. The model treats only unidirectional wave propagation without considering the backward propagation of light, therefore the development of the equations for the model is very easy. Actual lasers are composed of a Fabry-Perot resonator and have forward and backward waves of light propagations in the laser medium. A few contain a unidirectional ring resonator. The semiconductor laser, which is the main issue of this book, is also basically a Fabry-Perot laser (Abraham et al. 1988). Although the model is not always applicable to real lasers, the description for a unidirectional traveling-wave ring resonator is very simple and the theory can be easily extended to ordinary Fabry-Perot lasers.

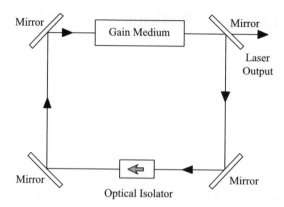

Fig. 2.1. Laser model with ring resonator

The light propagation equation in the laser medium is derived first. The electric field $\mathcal{E}$ (vector field) is written by a time dependent Maxwell equation as

$$\nabla^2 \mathcal{E} - \frac{1}{c^2}\frac{\partial^2 \varepsilon \mathcal{E}}{\partial t^2} = \mu_0 \frac{\partial^2 \mathcal{P}}{\partial t^2} \tag{2.1}$$

where $\mathcal{P}$ is the polarization vector of matter, ε is the electric permittivity tensor, c is the speed of light in vacuum, and μ_0 is the magnetic permeability in vacuum. Assuming a uniform refractive index of the laser medium and linearly polarized spatial modes for the x and y directions with the propagation for the z axis, the field and the polarization of matter reduce to scalar quantities propagating only to the z direction and (2.1) can be reduced to the following scalar equation:

$$\frac{\partial^2 \mathcal{E}}{\partial z^2} - \frac{\eta^2}{c^2}\frac{\partial^2 \mathcal{E}}{\partial t^2} = \mu_0 \frac{\partial^2 \mathcal{P}}{\partial t^2} \tag{2.2}$$

where η is the refractive index of the laser medium.

The field and the polarization propagate for the z direction with the wavenumber $k = \eta \omega_0/c$ and the angular oscillation frequency ω_0, are then written as

$$\mathcal{E}(z,t) = \frac{1}{2}E(z,t)\exp[\mathrm{i}(kz - \omega_0 t)] + c.c. \tag{2.3}$$

$$\mathcal{P}(z,t) = \frac{1}{2}P(z,t)\exp[\mathrm{i}(kz - \omega_0 t)] + c.c. \tag{2.4}$$

Here, $c.c.$ represents the complex conjugate of the preceding terms. $E(z,t)$ and $P(z,t)$ are the amplitudes of the respective variables and are assumed to vary slowly compared with the optical frequency (Slowly Varying Envelope Approximation: SVEA). Neglecting the second order small infinities and substituting (2.3) and (2.4) into (2.2), we obtain an equation for the amplitudes

$$\frac{\partial E}{\partial z} + \frac{\eta}{c}\frac{\partial E}{\partial t} = i\frac{k}{2\varepsilon_0\eta^2}P \tag{2.5}$$

2.1.2 Light Emission and Absorption in Two-Level Atoms

Before deriving the complete form of the propagation equation, we discuss absorption and emission of light from two-level atoms based on the semi-classical quantum theory and, then, derive the Bloch equation. The Hamiltonian $\mathcal{H}_0$ without perturbation for the electric field $\mathcal{E}$, the Hamiltonian $\mathcal{H}$ of the two-level atom is given by

$$\mathcal{H} = \mathcal{H}_0 - \boldsymbol{\mu} \cdot \boldsymbol{\mathcal{E}} \tag{2.6}$$

where $\boldsymbol{\mu} = e\mathbf{r}$ is the moment of the transition between the two levels ($\mathbf{r}$ and e are the position vector and the fundamental electric charge). For the eigenstates $\varphi_j(j = 1, 2)$ of the two levels and the energy of each level as $\hbar\omega_j$ ($\hbar$ being the Planck constant), the interaction between the two levels is written by

$$\langle\varphi_j|\mathcal{H}_0|\varphi_k\rangle = \hbar\omega_j\delta_{jk} \tag{2.7}$$

where δ_{ij} represents the Kronecker delta. The angular frequency of light emitted or absorbed in the two-level atoms is given by $\omega_A = \omega_2 - \omega_1$. In the presence of the optical field, the quantum state $|\psi>$ of the two-level atoms is written by the linear addition of the two states as

$$|\psi\rangle = c_1(t)\exp(-i\omega_1 t)|\varphi_1\rangle + c_2(t)\exp(-i\omega_2 t)|\varphi_2\rangle \tag{2.8}$$

Substituting the above equation into the Schrödinger equation, the coefficients c_1 and c_2 for the two states are calculated by solving the following coupled equations

$$\frac{dc_1}{dt} = \frac{ic_2}{\hbar}\exp(-i\omega_A t)\langle\varphi_1|\boldsymbol{\mu}\cdot\boldsymbol{\mathcal{E}}|\varphi_2\rangle \tag{2.9}$$

$$\frac{dc_2}{dt} = \frac{ic_1}{\hbar}\exp(i\omega_A\, t)\langle\varphi_2|\boldsymbol{\mu}\cdot\boldsymbol{\mathcal{E}}|\varphi_1\rangle \tag{2.10}$$

These are known as the Bloch equations (1946).

Using the number N_A of atoms in the unit volume, the macroscopic polarization of the medium is defined by

$$\boldsymbol{\mathcal{P}} = N_A\langle\psi|\boldsymbol{\mu}|\psi\rangle \tag{2.11}$$

From (2.8), the above equation reads as

$$\boldsymbol{\mathcal{P}} = N_A\{p(t)\mu_{12} + p^*(t)\mu_{21}\} \tag{2.12}$$

Then, the microscopic polarization $p(t)$ for each atom is given by

$$p(t) = c_1^*(t)c_2(t)\exp(-i\omega_A t) \tag{2.13}$$

$$\mu_{ij} = \langle \varphi_j | \boldsymbol{\mu} | \varphi_i \rangle \tag{2.14}$$

where $\mu_{ij}(i, j = 1, 2)$ is the moment of the transition from the lower to the upper state or vice versa. Finally, substituting the above equations into (2.9) and (2.10), we obtain the equation for the polarization of atoms

$$\frac{dp}{dt} = -i\omega_A p + \frac{i}{\hbar}E\mu_{21}w \tag{2.15}$$

and the distribution $w = |c_2(t)|^2 - |c_1(t)|^2$ for the population inversion of the two-level atoms

$$\frac{dw}{dt} = \frac{2}{i\hbar}E(p^*\mu_{21} - p\mu_{12}) \tag{2.16}$$

2.1.3 Maxwell-Bloch Equations

Rearranging the equations obtained for the field and the polarization and considering the time development of the population inversion in the laser medium, we derive the complete set of laser rate equations, which are the same expressions as those of Lorenz chaos.

Differentiating (2.4) with time and using the relations of (2.12) and (2.15), the macroscopic polarization equation is calculated as

$$\frac{dP}{dt} = -i(\omega_A - \omega_0)P + \frac{i\mu^2}{\hbar^2}W[E + E^*\exp\{-2i(kz - \omega_0 t)\}] \tag{2.17}$$

where $W = N_A w$ is the macroscopic population inversion and $\mu = |\mu_{12}|$. From (2.16), the equation for the population inversion is given by

$$\frac{dW}{dt} = \frac{1}{i\hbar}[EP^* - EP\exp\{2i(kz - \omega_0 t)\} - c.c.] \tag{2.18}$$

Since we are concerned with slowly varying variables compared with optical frequency (Rotating-Wave Approximation: RWA), we can omit the terms related to fast oscillation terms of the angular frequency $2\omega_0$ in (2.17) and (2.18) (Milloni and Eberly, 1988).

We need the external pump to lase, so that we add an extra term to (2.18) for lasing in the actual laser. Further, we add the phenomenological terms for the damping oscillations to (2.5), (2.17), and (2.18). The resulting equations, the Maxwell-Bloch equations, for field E, polarization P, and population inversion W are given by

$$\frac{\partial E}{\partial z} + \frac{\eta}{c}\frac{\partial E}{\partial t} = i\frac{k}{2\varepsilon_0\eta^2}P - \frac{n}{2T_{ph}c}E \tag{2.19}$$

$$\frac{\partial P}{\partial t} = -i(\omega_A - \omega_0)P + \frac{i\mu^2}{\hbar^2}EW - \frac{P}{T_2} \tag{2.20}$$

$$\frac{dW}{dt} = \frac{1}{i\hbar}(EP^* - E^*P) + \frac{W_0 - W}{T_1} \tag{2.21}$$

where W_0 is the population inversion induced by the pump at the laser threshold. T_{ph}, T_2, and T_1 are the relaxation times of the photons (photon lifetime), the polarization (transverse relaxation), and the population inversion (longitudinal relaxation), respectively. The actual laser exhibits spontaneous emission and, then, statistical Langevin noise terms are added to each equation to explain the noise effects (Pertermann 1988 and Risken 1996). However, statistical noises and irregular chaotic oscillations are of different origins and they can be discussed separately. Chaos is a phenomenon described by deterministic equations, so that such terms are excluded for investigating the pure laser dynamics. Noises are only introduced to account for the effects of laser oscillations when necessary. The Langevin noises will be briefly discussed in Chap. 3.

2.2 Lorenz-Haken Equations

2.2.1 Lorenz-Haken Equations

We have derived the laser equations for field amplitudes and polarization, and population inversion. In the following, we show that these equations are equivalent to Lorenz equations, which describe a model of the convective fluid flow for the atmosphere. Scaling the field E, the polarization P, and the population inversion W in (2.19), (2.20), and (2.21) as $\bar{E} = \sqrt{\varepsilon_0 c\eta/2}E$, $\bar{P} = k/\varepsilon_0\eta^2\sqrt{\varepsilon_0 c\eta/2}P$, and $w = \sigma_s W$ (with $\sigma_s = \mu^2\omega_0 T_2/2\varepsilon_0\hbar c\eta$), and neglecting the term $\partial E/\partial z$ as a small mean field that propagates in the z direction, the Maxwell-Bloch equations are written as follows (Haken 1975);

$$\frac{d\bar{E}}{dt} = i\frac{c}{2\eta}\bar{P} - \frac{1}{2T_{ph}}\bar{E} \tag{2.22}$$

$$T_2\frac{d\bar{P}}{dt} = -(1 - i\delta)\bar{P} - i\bar{E}w \tag{2.23}$$

$$T_1\frac{dw}{dt} = w_0 - w + \frac{\text{Im}[\bar{E}^*\bar{P}]}{I_{sat}} \tag{2.24}$$

where $\delta = (\omega_0 - \omega_A)T_2$ is the scaled atomic detuning and $I_{sat} = \hbar^2 c\eta\varepsilon_0/2\mu^2 T_1 T_2$ is the saturation intensity.

In the meantime, Lorenz proposed the differential equations for three variables X, Y, and Z as a model of atmospheric flow (Rayleigh-Bénard configuration) and proved the existence of chaos in the system (Lorenz 1963). Using chaotic parameters Σ, R, and β, the Lorenz equations are written as

$$\frac{dX}{dt} = -\Sigma(X - Y) \tag{2.25}$$

$$\frac{dY}{dt} = RX - Y - XZ \tag{2.26}$$

$$\frac{dZ}{dt} = -\beta Z + XY \tag{2.27}$$

Lorenz suggested that systems described by nonlinearly coupled differential equations with three variables are candidates for chaotic systems. By normalizing the variables as $x = \sqrt{b/I_{\text{sat}}}\bar{E}$, $y = (icT_{\text{ph}}/\eta)\sqrt{b/I_{\text{sat}}}\bar{P}$, and $z = (w_0 - w)cT_{\text{ph}}/\eta$, and replacing time by $t/T_2 \to t$, the Maxwell-Bloch equations in (2.22)–(2.24) are written as

$$\frac{dx}{dt} = -\sigma(x - y) \tag{2.28}$$

$$\frac{dy}{dt} = -(1 - i\delta)y + (r - z)x \tag{2.29}$$

$$\frac{dz}{dt} = -bz + \text{Re}[x^*y] \tag{2.30}$$

where $\sigma = T_2/2T_{\text{ph}}$, $b = T_2/T_1$, and $r = w_0cT_{\text{ph}}/\eta$. It is easily proved that the above three equations are the same as those of the Lorenz model and lasers described by two-level atoms are essentially the same chaotic system as the convective fluid in the atmospheric flow. Thus, (2.28)–(2.30) are called the Lorenz-Haken equations.

2.2.2 First Laser Threshold

A laser oscillation starts when the population inversion exceeds a certain level, namely the pumping threshold. The laser threshold can be calculated from (2.28)–(2.30) based on the linear stability analysis. The linear stability analysis, which applies small perturbations on the steady-states of the laser variables, is frequently used for obtaining the stability conditions. Assuming the stable solutions in (2.28)–(2.30) as x_s, y_s, and z_s, and applying small perturbations on the steady-state values, we write the time developments of the variables as $x(t) = x_s + \delta x(t)$, $y(t) = y_s + \delta y(t)$, and $z(t) = z_s + \delta z(t)$, where $\delta x(t)$, $\delta y(t)$, $\delta z(t)$ are small perturbations. Substituting these values into (2.28)–(2.30), we obtain the following differential equations for the perturbations

$$\frac{\mathrm{d}\delta x}{\mathrm{d}t} = -\sigma(\delta x - \delta y) \tag{2.31}$$

$$\frac{\mathrm{d}\delta y}{\mathrm{d}t} = -(1 - \mathrm{i}\delta)\delta y - (r - \delta z)\delta x \tag{2.32}$$

$$\frac{\mathrm{d}\delta z}{\mathrm{d}t} = -b\delta z + \mathrm{Re}[\delta x^* \delta y] \tag{2.33}$$

We can neglect the second small infinities such as $\delta z \delta x$ and $\delta x^* \delta y$, thus the equations are linearized.

When we put the time developments of the variables as $\delta h = \delta h_0 \exp(\gamma t)$ ($h = x, y, z$), the laser is stable for solutions of negative real parts of γ. On the other hand, it is unstable for solutions of positive real parts and the solutions diverge to infinities for the time development. Substituting the time developments $\delta h = \delta h_0 \exp(\gamma t)$ into (2.31)–(2.33), we obtain the following characteristic relation for the non-trivial solutions

$$\begin{vmatrix} \gamma + \sigma & -\delta & 0 \\ -r & \gamma + 1 - \mathrm{i}\delta & 0 \\ 0 & 0 & \gamma + b \end{vmatrix} = 0 \tag{2.34}$$

The real parts of the solutions in the above equations represent the measure for stability or instability of the solutions and the imaginary parts denote the oscillation frequencies of the corresponding solutions. Since $b = T_2/T_1$ is positive, one of the solutions $\gamma = -b$ is a stable solution with uniform convergence. The other solutions are calculated by solving the following equations

$$\gamma^2 + (\sigma + 1 - \mathrm{i}\delta)\gamma - \sigma(r - 1 + \mathrm{i}\delta) = 0 \tag{2.35}$$

When the pumping r reaches a certain value, the laser exceeds the threshold and laser oscillation starts. Above the threshold, the solutions of the imaginary parts are enough to take into account. Putting the form of the solutions as $\gamma = \mathrm{i}\Omega$ and substituting it into (2.35), we obtain the laser threshold from the conditions having zero values for the real and imaginary parts of (2.35) as

$$r_{\mathrm{th}}^{(1)} = 1 + \frac{\delta^2}{(\sigma + 1)^2} \tag{2.36}$$

For the laser oscillation, there is an accompanying frequency $\nu = \Omega/2\pi$ that corresponds to the solution of the imaginary part for the characteristic equation. Using the threshold, the frequency is given by

$$\nu_{\mathrm{R}} = \frac{\sigma}{2\pi}\sqrt{r_{\mathrm{th}}^{(1)} - 1} \tag{2.37}$$

The frequency ν_{R} is known as the relaxation oscillation frequency. When the detuning δ is zero, the threshold is $r_{\mathrm{th}}^{(1)} = 1$ or $w_0 = \eta/cT_{\mathrm{ph}}$, as expected. The extra term in the threshold in (2.36) is the increase of the threshold, which compensates the loss due to the detuning. As we discuss in the following section, there is another threshold that is called second laser threshold. Therefore, $r_{\mathrm{th}}^{(1)}$ is called first laser threshold.

2.2.3 Second Laser Threshold

Laser oscillation starts above the first threshold and shows a stable output power at a certain pump. Here, we again apply linear stability analysis for the laser operation. As we are considering the oscillation above the threshold, the field and the polarization vary with time at the same optical frequency for the steady-state values of x_s, y_s, and z_s. Assuming the difference of the angular detuning frequency $\Delta\omega$ between the laser oscillation and the internal cavity frequencies and the phase fluctuation ϕ_s of the complex field, we put the forms of the steady-state solutions as

$$x_s = x_0 \exp\{-\mathrm{i}(\Delta\omega_s t + \phi_s)\} \tag{2.38}$$

$$y_s = y_0 \exp(-\mathrm{i}\Delta\omega_s t) \tag{2.39}$$

$$z_s = z_0 \tag{2.40}$$

where $x_0 = \sqrt{bz_0}$, $y_0 = \sqrt{r_{\mathrm{th}}^{(1)} bz_0}$, $z_0 = r - r_{\mathrm{th}}^{(1)}$, $\Delta\omega_s = -\delta\sigma/(\sigma+1)$, and $\tan\phi_s = \delta/(\sigma+1)$. The laser output power is given by the square of x_0 and reads

$$x_0^2 = b\left(r - r_{\mathrm{th}}^{(1)}\right) \tag{2.41}$$

This is the well-known result that the laser output power linearly increases with the increase of the pump r well above the threshold $r_{\mathrm{th}}^{(1)}$.

For a pump below the laser threshold, the laser does not reach laser oscillation and it only exhibits a faint light output due to spontaneous emission, thus the laser is also under another stable condition. For the increase of the pump r over the threshold, whether the laser output power increases with the increase of the pump or not? In actual fact, there are nonlinear effects, such as saturation of gains of the laser material, to limit the optical output power. The effects also induce the change of laser parameter values describing the laser rate equations. Of course, what we are considering is not such effects, but the nonlinear effects intrinsically involved in the laser rate equations in (2.28)–(2.30). Here, consider the unstable phenomena induced by the increase of the pump r for these equations. For this purpose, we again employ the linear stability analysis for (2.38)–(2.40) near the steady-state values for the variables. The procedure is almost the same as the previous calculations. For simplicity, we calculate the stability solutions for the condition $\delta = 0$ (zero detuning condition). After some calculations, the same as the derivation for (2.35), the characteristic equation reads

$$\gamma^3 + a_2\gamma^2 + a_1\gamma + a_0 = 0 \tag{2.42}$$

where $a_2 = \sigma + b + 1$, $a_1 = b(\sigma + r)$, and $a_0 = 2b\sigma(r - 1)$. The stability solutions are calculated by solving the above equation.

At the threshold of the stable solution, the variable γ is purely imaginary, and it is assumed as $\gamma = i\Omega$. From the comparison between the real and imaginary parts for the solution, we obtain the threshold as

$$r_{\text{th}}^{(2)} = \frac{\sigma(\sigma + b + 3)}{\sigma - b - 1} \tag{2.43}$$

Over the pump r exceeding the threshold $r_{\text{th}}^{(2)}$, the laser gets unstable states and exhibits irregular oscillations of chaos via Hopf bifurcations (see Appendix A.1). In actual evolution processes for bifurcations, there are various routes to chaos, for example, chaos follows immediately after period-1 oscillation (quasi-period-doubling bifurcation). The other example is that instability to chaos follows after intermittent oscillations like spiky irregular oscillations. The details of routes to chaos in semiconductor lasers will be demonstrated in the following chapters. The threshold $r_{\text{th}}^{(2)}$ is called second threshold to distinguish it from the first laser threshold $r_{\text{th}}^{(1)}$. For example, for the conditions of $T_2 \gg T_1$, $b \approx 0$, and $\sigma = 2(T_2 = 4T_{\text{ph}})$, the threshold value is equal to $r_{\text{th}}^{(2)} = 10$ and it is much higher than the first threshold $r_{\text{th}}^{(1)} = 1$ without detuning. Actual unstable lasers have the second threshold values around tens to one hundred.

The typical frequency of the irregular pulsing can also be calculated from the characteristic equation for the pure imaginary part value of the variable Γ, and it is given by

$$\nu_{\text{R2}} = \frac{1}{2\pi}\sqrt{b\left(\sigma + r_{\text{th}}^{(2)}\right)} \tag{2.44}$$

For the existence of the second threshold, the condition of $\sigma > b + 1$ must be satisfied from (2.43). This is known as the bad-cavity condition of a laser that gives rise to unstable laser oscillations. The bad-cavity condition is rewritten by using the actual time constants as follows;

$$\frac{1}{2T_{\text{ph}}} > \frac{1}{T_2} + \frac{1}{T_1} \tag{2.45}$$

Namely, the bad-cavity of a laser oscillation is a lossy and dissipative system for photons having a low quality factor Q of the resonator. Further discussion of the bad-cavity conditions and instabilities above the second laser threshold can be found in the reference (van Tartwijk and Agrawal 1998). Equations (2.43) and (2.44) were derived for the condition of zero frequency detuning $\delta = 0$. For non-zero detuning $\delta \neq 0$, the analysis becomes much more complex, but the expression for this case has been given and almost the same order of the second laser threshold $r_{\text{th}}^{(2)}$ has been obtained (Mandel and Zeghlache 1983, Zeghlache and Mandel 1985, and Ning and Haken 1990).

2.3 Classifications of Lasers

2.3.1 Classes of Lasers

We have taken into consideration all of the time constants for the field, the polarization of matter, and the population inversion in the laser rate equations. The second laser threshold has been calculated for the inclusions of these parameters. However, lasers do not always show instabilities and chaotic behaviors with increased pumping, and most lasers are indeed stable. Only few lasers emitting infrared lines exhibit chaotic oscillations. For stability and instability of lasers, we have assumed the model of a ring laser with two-level atoms. On the other hand, most lasers in practical use are modeled by three- or four-level atoms. Therefore, lasers must be modeled by these in a strict sense and some modifications may be required for the above derivations. However, the results derived for the two-level atoms can here be extended to three- or four-level atoms and still be applicable for the discussion of the stability and instability for practical lasers.

Even for the same material, the laser may have several oscillation lines. In such a case, the laser has a different gain for each line and has different time constants for the relaxation oscillations depending on the oscillation frequency. Therefore, a laser with a certain material may be stable for a certain oscillation line and have no second threshold, while it may be unstable and have the second threshold for another line. The stability and instability of lasers intrinsically involved in laser rate equations are classified according to the scales of time constants for the relaxation oscillations T_{ph}, T_2, and T_1 introduced in Sect. 2.2.1. Namely, one or two of the time constants among the three in the differential equations may be adiabatically eliminated and one or two of the laser rate equations are enough to describe actual laser operations. Depending on the scales of the time constants, the stabilities of lasers are classified into the following three classes; class A, B, and C lasers (Arrechi et al. 1984a, b and Tredicce et al. 1985).

2.3.2 Class C Lasers

When the time constants of the relaxations are of the same order, we must consider all of the Lorenz-Haken differential equations. As already discussed, the laser oscillation starts at the first threshold with stable light output for a certain pump and it reaches the second laser threshold for the increase of the pump. Over pumping above the second threshold in the bad-cavity condition with low Q factor, the laser shows unstable oscillation like irregular pulsations and chaotic oscillations. According to the classifications of laser operations by Arecchi et al. (1984a), these lasers are called class C lasers. Class C lasers are generally infrared lasers and far-infrared lasers are almost classified into class C. This is originated from the fact that the three time constants of the relaxation oscillations for the field, the polarization of matter, and the population inversion tend to be the same order. Examples of class C lasers are NH_3 lasers (Weiss et al. 1985 and Hogenboom et al. 1985), Ne-Xe lasers (3.51 µm line) (Casperson 1978 and Special issue J. Opt. Soc. Am B 1985), and He-Ne lasers at 3.39 µm line (Weiss and King 1982, and Weiss et al. 1983). Though He-Ne lasers operating at infrared lines are class C lasers, He-Ne lasers at visible oscillations are categorized into a different class because the constants of the polarization and the population inversion have different time scales from those of the infrared operations. In general, these class C lasers do not have any commercial application.

Figure 2.2 is an example of experimentally observed chaotic waveforms in an infrared He-Ne laser at 3.39 µm oscillation (Weiss 1983). The bad-cavity condition was realized by tilting the angle of one of the mirrors in the laser resonator. A stable laser state **a** evolves into unstable oscillations **b–d** to chaotic state **e** with the increase of the mirror tilting angle. Figure 2.3 shows the oscillation spectra of the laser corresponding to period-doubling bifurcations to chaos for the increase of the mirror tilting angle. Figure 2.4

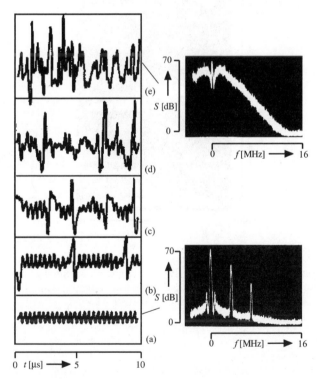

Fig. 2.2. Experimentally observed chaotic time series in an infrared He-Ne laser. Stable oscillation state (a) to chaotic state **e**. One of the mirrors in the laser cavity is tilted and the bad-cavity condition is realized (after Weiss CO, Godone A, Olafsson A (1983); © 1983 APS)

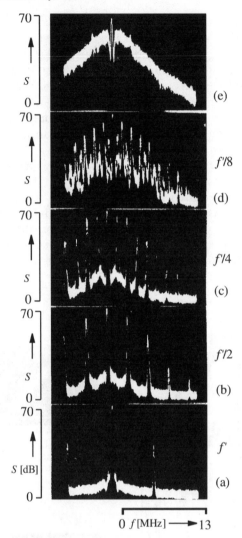

Fig. 2.3. Experimentally observed optical power spectra in an infrared He-Ne laser for period-doubling route to chaos. Tilting of one of the resonator mirrors leads to oscillations to **a** period-1, **b** period-2, **c** period-4, **d** period-8, and **e** chaos (after Weiss CO, Godone A, Olafsson A (1983); © 1983 APS)

is another experimental example of chaos showing pulsation instability in a Xe laser at 3.51 μm oscillation (Casperson 1978). With increasing pump, period-1 pulsation at first appears in Fig. 2.4a and the laser switches to period-2 pulsation in Fig. 2.4b. Thus, routes to chaos are not unique and depend on systems and parameters.

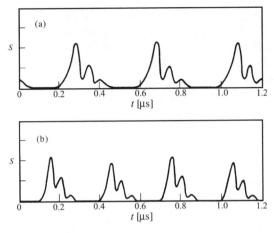

Fig. 2.4. Experimental plots of pulsation instabilities in an Xe laser at 3.51 μm oscillation for **a** period-1 pulsation at a discharge current of 40 mA and **b** period-2 pulsations at a discharge current of 50 mA (after Casperson LW (1978); © 1978 IEEE)

2.3.3 Class B Lasers

The time constant T_2 of the polarization of matter (transverse relaxation) is small enough compared with the other time constants, i.e., $T_{ph}, T_1 \gg T_2$, the differential equation for the polarization is adiabatically eliminated and we obtain for the representation of the polarization in (2.32) (Haken 1985)

$$y = \frac{r - z}{1 - i\delta}x \qquad (2.46)$$

Then, the laser rate equations can be described by the two differential equations for the field x and the population inversion z. These lasers are called class B lasers and they are stable in nature, since the lasers have the first threshold but do not have the second threshold. The electric field is complex and the complex field equation can be split into two differential equations, the amplitude and phase equations. However, the phase equation has no effect on other variables, so that these systems can still be characterized by two differential equations. Therefore, class B lasers are intrinsically stable.

However, they are easily destabilized by the introduction of external perturbations, resulting in the addition of extra degrees of freedom. If the equations for the field amplitude and the phase couple with each other through a perturbation, the laser must be described by the rate equations coupled with three variables. A laser coupled with three variables becomes a chaotic system and shows instabilities. Examples of external perturbations are modulation for the accessible laser parameters, external optical injection, and optical self-feedback from external optical components. Semiconductor laser, which is the main topic of this book, are classified into class B laser, as discussed later. Indeed, semiconductor lasers are easily destabilized and show chaotic behaviors by external perturbations, such as external optical feedback (van Tartwijk and Agrawal 1998 and Ohtsubo 2002a). One of the typical features of class B lasers is a relaxation oscillation of the laser output that is observed for a step time response when the population inversion does not follow the photon decay rate, i.e., $T_1 > T_{ph}$. Many lasers are classified into class B lasers and other examples are CO_2 lasers and solid state lasers including fiber lasers.

2.3.4 Class A Lasers

When the lifetime of photons in a laser medium is large enough compared with the other time constants of the relaxations, i.e., $T_{ph} \ll T_1, T_2$, the differential equations for the polarization of matter and the population inversion are adiabatically eliminated. In the same manner as class B lasers, the steady-state population inversion is given by (Haken 1985)

$$z = \frac{1}{b}\mathrm{Re}[x^*y] .$$ (2.47)

Then, the laser oscillation is only described by the differential equation for the field. Lasers satisfying the relations are called class A lasers and they are the most stable lasers with a high Q factor among the three classes. Even for class A lasers, they may be destabilized and show chaotic behaviors by external perturbations with two or more extra degrees of freedom as described in class B lasers. Visible He-Ne lasers, Ar-ion lasers, and dye lasers are examples of class A lasers.

3 Semiconductor Lasers and Theory

In this chapter, we discuss the oscillation conditions for semiconductor lasers and, then, derive the rate equations, which are the starting point of the study of chaotic dynamics in semiconductor lasers. The semiconductor laser described here is a Fabry-Perot type with a mono-layer of the active region, however other edge-emitting lasers such as multi-quantum well (MQW) lasers and distributed feedback (DFB) lasers can be treated in the same manner as Fabry-Perot lasers. Therefore, the macroscopic features of these lasers show the same behaviors from the viewpoint of chaotic dynamics, although the detailed characteristics strongly depend on the laser structure and the particular values of the device parameters. The relaxation oscillation frequency, which is calculated from the rate equations plays an important role for the dynamics of semiconductor lasers. The Langevin terms, which are stochastic noise effects, are introduced in the rate equations. Some other fundamental characteristics of semiconductor lasers are also discussed.

3.1 Semiconductor Lasers

We assume that the readers of this book are familiar with semiconductor lasers and how they operate. An extensive review for the characteristics of semiconductor lasers and the details of their operations can be found, for example, in the books of Agrawal and Dutta (1993), and Petermann (1988). Newly developed structures of semiconductor lasers, for example, vertical-cavity surface-emitting semiconductor lasers (VCSELs), show unique characteristics that are different from conventional edge-emitting semiconductor lasers. We need some modifications of the rate equations for such newly developed lasers. However, these new lasers exhibit the same or similar dynamics as those of edge-emitting lasers. We will derive the rate equations for these lasers and discuss their dynamics in Chap. 8.

The structure of semiconductor lasers is based on the p-n junction of the semiconductor materials and the laser oscillation is realized by the emission of light due to carrier recombination between the conduction and valence bands. The band structure of actual lasers is not modeled by a simple two-level system. Therefore, we cannot straightforwardly apply the results of Chap. 2 and we need some modification of the model and of the strict theoretical treatment. However, the intra-band relaxation within the medium of the semiconductor laser is fast enough of the order of 10^{-13} s compared with the carrier recombination rate of 10^{-9} s (Petermann 1988). This fact makes it possible to use approximately the model of two-level atoms for the theoretical investigation of the dynamics of semiconductor lasers (Agrawal and Dutta 1993). The rigorous treatment of semiconductor lasers can be found in the reference (Chow et al. 1993), but here we employ the approximate model of laser oscillations based on two-level atoms.

As discussed in the previous chapter, the relaxation time of the polarization in the semiconductor laser material is as small as $T_2 < 0.1$ ps, which is much smaller than the other time scales of the relaxations, $T_{ph} \sim$ several pico-seconds and $T_1 \sim$ several nano-seconds, and the laser is classified into class B. Therefore, the polarization term is adiabatically eliminated and the effect is simply replaced by the linear relation between the field and the polarization. The population inversion in general laser systems is simply replaced by the carrier density produced by electron-hole recombination. The photon number (which is equivalent to the absolute square of the field amplitude) and the carrier density are frequently used as the variables of the rate equations for semiconductor lasers. However, for the general descriptions of the dynamics in semiconductor lasers, we must employ the complex amplitude of the field (the amplitude and the phase of the field) instead of the photon number. There are several ways to derive the complex field equation, however they reach the same result. Here, we follow the derivation of the complex field equation by Petermann (1988).

3.2 Oscillation Conditions of Semiconductor Lasers

3.2.1 Laser Oscillation Conditions

In this chapter, we derive the rate equations for edge-emitting semiconductor lasers with a narrow stripe width. Even for edge-emitting lasers, there are several kinds of device structures, for example, the Fabry-Perot, multi-quantum well (MQW), and distributed feedback (DFB) structures. Also, there are two main structures for the guiding of light in the active layer, i.e., gain- and index-guiding structures. Even for these different device structures, the derivation of the rate equations for laser operations is similarly given and almost the same equations are obtained, although the parameters appearing in the rate equations may vary from one laser to the other. Before deriving the rate equations in semiconductor lasers, we consider several conditions and relations for laser oscillations. At first, we discuss the conditions for laser oscillations (Yariv 1997). The model of a Fabry-Perot resonator is shown in Fig. 3.1. Assuming a cavity length l, and reflectivities of the front and

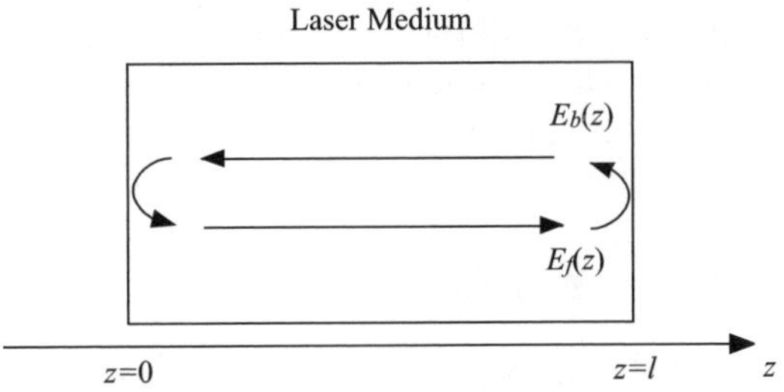

Laser Medium

Fig. 3.1. Model of Fabry-Perot laser

back facts r_1 and r_2, the fields E_f and E_b propagating forward and backward directions are written by

$$E_f(z) = E_{0f} \exp\left\{ ikz + \frac{1}{2}(g - a)z \right\} \tag{3.1}$$

$$E_b(z) = E_{0b} \exp\left\{ ik(l - z) + \frac{1}{2}(g - a)(l - z) \right\} \tag{3.2}$$

where g is the gain in the laser medium and a is the total loss due to absorption and scattering in the medium. All the parameters are defined for the laser intensity, so that a factor of $1/2$ is introduced in the above equations.

From the boundary conditions at the facets, $E_f(0) = r_1 E_b(0)$ and $E_b(l) = r_2 E_f(l)$, the steady-state condition for the laser oscillation is given by

$$r_1 r_2 \exp\{2ikl + (g - a)l\} = 1 \tag{3.3}$$

From the real part of the above equation, the amplitude condition of the laser oscillation for the threshold gain g_{th} is given by

$$g_{th} = a + \frac{1}{l} \ln \left(\frac{1}{r_1 r_2} \right) \tag{3.4}$$

Since there is spontaneous emission of light within the active region in the laser, the actual gain is slightly less than g_{th}. Also, the phase condition is calculated from the imaginary part and reads

$$kl = m\pi \tag{3.5}$$

where m is an integer. Equation (3.4) is the condition for the laser threshold and is interpreted as the balance of the gain with the losses of the internal absorption and reflection in the laser medium. For example, when we consider a InGaAsP semiconductor laser oscillating at a near-infrared line of $1.3 \sim 1.5\,\mu m$ with cleaved facets as a laser resonator, the threshold gain is calculated as $g_{th} \sim 75\,cm^{-1}$ for the facet intensity reflectivities of $R_1 = R_2 \sim 0.32$ (where $R_1 = r_1^2$ and $R_2 = r_2^2$, and the refractive index of the medium of 3.5) and the internal losses of $a \sim 30\,\mathrm{cm}^{-1}$ (Buss and Adams 1979). In actual fact, we roughly require the double of this amount for lasing (about $g'_{\mathrm{th}} \sim 150\,\mathrm{cm}^{-1}$), since the laser light is not only confined within the active region and the confinement factor is about 0.5 of the calculated gain.

3.2.2 Laser Oscillation Frequency

From (3.5), the possible frequency of laser oscillations is given by

$$\nu_m = m\frac{c}{2\eta l} \tag{3.6}$$

where m is again an integer. The laser will be oscillated at one of these possible frequencies or a few lines of them. ν_m is called the mode frequency and it corresponds to the m-th longitudinal mode of the laser oscillations. Usually, a laser oscillates at or near the maximum gain mode, which is very close to one of the resonator frequencies. Therefore, the actual frequency of laser oscillations deviates from the value calculated from (3.6), though the deviation is very small.

The refractive index η in the semiconductor laser is a function of the optical oscillation frequency ν due to the presence of the dispersion of laser materials. Using the notation of a small quantity such as Δ, we have the relation

$$\Delta(\eta\nu) = \eta\Delta\nu + \nu\Delta\eta \tag{3.7}$$

Here, we define the effective refractive index as

$$\eta_e = \eta + \nu\frac{\partial\eta}{\partial\nu} \tag{3.8}$$

With the above relation and the equality for the successive modes m and $m+1$, $\Delta(\eta\nu) = c/2l$, we obtain the mode separation between the m and $m+1$ modes

$$\Delta\nu = \frac{c}{2\eta_e l} \tag{3.9}$$

This corresponds to the round trip time of light within the internal cavity of lasers and the time is given by

$$\tau_{\mathrm{in}} = \frac{2\eta_e l}{c} \tag{3.10}$$

As an example, we consider the same InGaAsP lasers as in the previous section. We obtain a frequency separation of $\Delta\nu = 150\,\mathrm{GHz}$ for $l = 250\,\mu\mathrm{m}$, $\eta \sim 3.5$, and $\eta_e \sim 4$. The corresponding separation of the wavelengths in the laser oscillation is $\Delta\lambda = \lambda^2\Delta\nu/c = 1\,\mathrm{nm}$. Also the round trip time of light is calculated to be $\tau_{\mathrm{in}} = 6.7\,\mathrm{ps}$ (Petermann 1988).

3.2.3 Dependence of Oscillation Frequency on Carrier Density

In this section, we derive the dependence of the oscillation frequency on the carrier density, which plays an important role in laser operations. The refractive index of the active layer is a function of the carrier density n_{th} and the optical frequency ν_{th} at the laser threshold, and is written by

$$\eta = \eta_0 + \frac{\partial \eta}{\partial \nu}(\nu - \nu_{th}) + \frac{\partial \eta}{\partial n}(n - n_{th}) \qquad (3.11)$$

where n is the carrier density at the laser oscillation and η_0 is the refractive index at the threshold. Then, the threshold gain g_{th} and the oscillation frequency ν_{th} at the laser threshold is given by

$$g_{th} = g(n_{th}) \qquad (3.12)$$

$$\nu_{th} = m\frac{c}{2\eta_0 l} \qquad (3.13)$$

Substituting (3.6) and (3.13) into (3.11), and using the relation of (3.8), we obtain an important relation between the laser oscillation frequency and the carrier density as

$$(\nu - \nu_{th}) = -\frac{\nu_{th}}{\eta_e}\frac{\partial \eta}{\partial n}(n - n_{th}) \qquad (3.14)$$

Since the oscillation frequency has a linear relation to the carrier density as given by (3.14), the laser power is proportional to the frequency.

3.3 Derivation of Rate Equations

3.3.1 Gain at Laser Oscillation

To derive the rate equations of semiconductor lasers, we at first consider the gain at the lasing condition. The gain of light after the round trip within the cavity is given by the same equation for the steady-state laser oscillation condition as

$$G = r_1 r_2 \exp\{2ikl + (g - a)l\} \qquad (3.15)$$

The wavenumber k depends on the refractive index of the laser medium and is a function of the optical frequency ν (or the angular frequency $\omega = 2\pi\nu$) and also the carrier density n. The wavenumber can be expanded by the threshold values of those parameters as

$$k = \eta\frac{\omega}{c} = \frac{\omega_{th}}{c}\left\{\eta_0 + \frac{\partial \eta}{\partial n}(n - n_{th}) + \frac{\eta_e}{\omega_{th}}(\omega - \omega_{th})\right\} \qquad (3.16)$$

Here, we use the relation of (3.11) for the derivation of the above equation. Using (3.16), the gain G is written by the product of the frequency dependent and non-dependent terms, G_1 and G_2, as

$$G = G_1 G_2 \tag{3.17}$$

$$G_1 = r_1 r_2 \exp\{(g - a)l + i\phi_0\} \tag{3.18}$$

$$G_2 = \exp\left[i\frac{2\omega_{\text{th}}l}{c}\left\{\eta_0 + \frac{\eta_e}{\omega_{\text{th}}}(\omega - \omega_{\text{th}})\right\}\right] \tag{3.19}$$

The phase ϕ_0 of the above equation is given by

$$\phi_0 = \frac{2\omega_{\text{th}}l}{c}\frac{\partial\eta}{\partial n}(n - n_{\text{th}}) \tag{3.20}$$

In (3.19), we have used the condition for the laser oscillation that the phase $2\omega_{\text{th}}\eta_0 L/c$ must be equal to integer multiples of 2π. Then, replacing the quantity $-i\omega$ as the equivalent to the operator $\text{d}/\text{d}t$, (3.19) reads

$$G_2 = \exp\{i\tau_{\text{in}}(\omega - \omega_{\text{th}})\} = \exp(-i\omega_{\text{th}}\tau_{\text{in}})\exp\left(-\tau_{\text{in}}\frac{\text{d}}{\text{d}t}\right) \tag{3.21}$$

Since frequency and time are a Fourier transform pair, (3.22) is derived from the equivalence of the equation in the frequency domain with that in the time domain.

3.3.2 Rate Equation for the Field

To attain the laser oscillations, the complex field after the round trip within the laser cavity must coincide exactly with the previous field. Assuming the gain in (3.16) (3.17) and (3.20) as a kind of an operator, we can obtain the following relation:

$$E_{\text{f}}(t) = GE_{\text{f}}(t) \tag{3.22}$$

Then, using (3.20), the field after the round trip in the cavity is written as

$$E_{\text{f}}(t) = G_1 \exp(-i\omega_{\text{th}}\tau_{\text{in}})\exp\left(-\tau_{\text{in}}\frac{\text{d}}{\text{d}t}\right)E_{\text{f}}(t) \tag{3.23}$$

We divide the laser field $E_{\text{f}}(t)$ into two terms, the term changing with angular frequency ω_{th} and the term $\hat{E}_{\text{f}}(t)$ that varies slowly compared with the angular frequency. Then, the field is given by

$$E_{\text{f}}(t) = \hat{E}_{\text{f}}(t)\exp(-i\omega_{\text{th}}t) \tag{3.24}$$

With this expression and the fact that the operator $\exp(-\tau_{in}d/dt)$ is equivalent to the time delay effect of τ_{in}, (3.23) yields

$$
\hat{E}_f(t)\exp(-i\omega_{th}t) = G_1\exp(-i\omega_{th}\tau_{in}) \\
\hat{E}_f(t-\tau_{in})\exp\{-i\omega_{th}(t-\tau_{in})\}
\tag{3.25}
$$

Therefore, we can write the field $\hat{E}_f(t)$ as

$$
\hat{E}_f(t) = G_1\hat{E}_f(t-\tau_{in})
\tag{3.26}
$$

This equation means that the field $\hat{E}_f(t)$ after the round trip time of τ_{in} with SVEA approximation returns as the same form of the field with gain G_1. We have discussed the field propagation for the positive direction. Similarly, we obtain the same results for the field propagation for the negative direction and the same relation as that of (3.25) is derived for the field. Then, the total field $E(t)$ can be written as the same form as $\hat{E}_f(t)$ in (3.26).

When the round trip time τ_{in} is small enough, we can expand the total field around the delay time τ_{in} as

$$
E(t-\tau_{in}) = E(t) - \tau_{in}\frac{dE(t)}{dt}
\tag{3.27}
$$

Then, we obtain the differential form for the field as follows:

$$
\frac{dE(t)}{dt} = \frac{1}{\tau_{in}}\left(1-\frac{1}{G_1}\right)E(t)
\tag{3.28}
$$

Since the gain G_1 is very close to unity for laser oscillation, we approximate the gain as

$$
\frac{1}{G_1} = \exp\{-\ln(r_1r_2) - (g-a)l - i\phi_0\} \\
\approx 1 + \ln\frac{1}{r_1r_2} - gl + al - i\phi_0
\tag{3.29}
$$

Substituting the above equation into (3.28) and using the relations of (3.14) and (3.20), we finally obtain the rate equation for the field as

$$
\frac{dE(t)}{dt} = \left\{-i\left(\omega_0-\omega_{th}\right) + \frac{1}{2}\left(gv_g - \frac{1}{\tau_{ph}}\right)\right\}E(t)
\tag{3.30}
$$

where $v_g = c/\eta_e$ is the group velocity of light in vacuum and the laser is assumed to operate at the angular optical frequency of $\omega = \omega_0$. Here, τ_{ph} (which is the same as T_{ph} in the previous notation) is the photon lifetime describing the loss due to absorption and scattering of light in the cavity and it is written by

$$
\frac{1}{\tau_{ph}} = v_g\left\{a + \frac{1}{l}\ln\left(\frac{1}{r_1r_2}\right)\right\}
\tag{3.31}
$$

where the relation $c/\eta_e = v_g = 2l/\tau_{in}$ holds for the group velocity of light within the laser resonator.

3.3.3 Linewidth Enhancement Factor

For the derived rate equation of (3.30), we rewrite the field equation in a different form, which is frequently used in the following discussion. At the same time, we derive the equation for the carrier density. Before formulating them, we discuss the complex susceptibility of the medium at laser oscillation and derive the important parameter of semiconductor lasers known as the linewidth enhancement factor. We can define the susceptibility under laser oscillation in the same manner as that below the laser threshold. We consider the extra complex susceptibility $\chi_l = \chi_l' + i\chi_l''$ due to the laser oscillation and add it to that below threshold, $\chi_0 = \chi_0' + i\chi_0''$. The complex susceptibility is a function of the laser frequency, therefore we write the total susceptibility $\chi(\omega)$ as

$$\chi(\omega) = \chi_0(\omega) + \chi_l(\omega) = \chi_0'(\omega) + \chi_l'(\omega) + i\{\chi_0''(\omega) + \chi_l''(\omega)\} \tag{3.32}$$

In the following, we will explicitly write susceptibilities as a function of the laser frequency only when necessary. Otherwise, the equivalent quantity of the complex electric permittivity has the following form

$$\varepsilon = \varepsilon_b + \chi_l' + i(\chi_0'' + \chi_l'') \tag{3.33}$$

where $\sqrt{\varepsilon_b} = \eta_b$ is the refractive index below the laser threshold.

Assuming the propagation of light toward z direction along the laser resonator, the spatial field to that direction is given by

$$E(z) = |E(z)| \exp(ikz) \tag{3.34}$$

The propagation constant is written by

$$k = \eta_c k_0 = \eta k_0 + i\frac{a_{abs}}{2} \tag{3.35}$$

where k_0 is the propagation constant in vacuum, η_c is the complex refractivity, and a_{abs} is the intensity absorption coefficient in the medium. The refractive index η and the absorption coefficient a_{abs} are given by

$$\eta = \sqrt{\varepsilon_b + \chi_l'} \approx \eta_b + \frac{\chi_l'}{2\eta_b} \tag{3.36}$$

$$a_{abs} = \frac{k_0}{\eta}(\chi_0'' + \chi_l'') \tag{3.37}$$

Since the increment of the refractive index η for laser oscillation is equal to $n\partial\eta/\partial n$, χ_l' is expressed by $\chi_l' = 2\eta_b n\partial\eta/\partial n$. Also using the gain for starting the laser oscillation, $g = -a_{abs}$ (actually, this is the gain of the laser oscillation at transparency), χ_l'' is represented by $\chi_l'' = -(\eta_b n/k_0)(\partial g/\partial n)$.

The addition of the complex susceptibility due to laser oscillation is given by

$$\chi_l = 2\eta_b n \left(\frac{\partial \eta}{\partial n} - \frac{i}{2k_0} \frac{\partial g}{\partial n} \right) \tag{3.38}$$

The macroscopic complex refractive index at laser oscillation is written by

$$\eta_c = \eta - i\eta' \tag{3.39}$$

where η' is the imaginary part of the refractive index. The imaginary part of the refractive index has the relation with the gain g as

$$\eta' = -\frac{1}{2k_0} g \tag{3.40}$$

Then, with this relation together with (3.14), we have

$$(\omega - \omega_{\text{th}}) = -\frac{\omega_{\text{th}}}{\eta_e} \frac{\partial \eta}{\partial n} (n - n_{\text{th}}) = \frac{1}{2} \alpha v_{\text{g}} \frac{\partial g}{\partial n} (n - n_{\text{th}}) \tag{3.41}$$

The parameter α in (3.41) is an important parameter in semiconductor lasers, known as α parameter or linewidth enhancement factor, and plays a crucial role for laser oscillations. Easily understood from the above equation, the α parameter is defined by

$$\alpha = \frac{\text{Re}[\chi]}{\text{Im}[\chi]} = -2\frac{\omega}{c} \frac{\frac{\partial \eta}{\partial n}}{\frac{\partial g}{\partial n}} \tag{3.42}$$

where χ is the complex electric susceptibility defined by (3.32). The value of ordinary lasers, such as gas lasers, is almost equal to zero, while it has a non-zero value for semiconductor lasers and usual semiconductor lasers have positive values from $3 \sim 7$ (Cook 1975, and Osinski and Buss 1987). This non-zero value of the α parameter gives rise to complex dynamics of semiconductor lasers. The typical feature of semiconductor lasers is a broad linewidth of laser oscillations due to a non-zero α parameter. Therefore, the parameter is also called the linewidth enhancement factor. Indeed, the oscillation linewidth of semiconductor lasers is $1 + \alpha^2$ times larger than those of ordinary lasers (Henry 1982). We will return to this subject in Sect. 3.5.5.

3.3.4 Laser Rate Equations

Under operations close to laser threshold, the gain g is linearized for the carrier density as

$$g = g_{\text{th}} + \frac{\partial g}{\partial n} (n - n_{\text{th}}) \tag{3.43}$$

where g_{th} is the gain of the medium at the threshold and it has a relation through the carrier density n_0 at the transparency as

$$g_{th} = \frac{\partial g}{\partial n}(n_{th} - n_0) \tag{3.44}$$

Under this condition, the gain balances with the loss and $g_0 = a$, where g_0 is the gain at transparency. The gain must exceed this value for laser oscillations. However, the actual gain required for laser oscillations is slightly larger than this value, since photons dissipate from the laser facets. At a laser oscillation well above the threshold, the effect of gain saturation must be taken into account (Nakamura et al. 1978, Lang 1979, and Henry 1982). In that case, we use the coefficient of gain saturation ε_s and obtain the relation

$$g = \frac{g_{th}}{1 + \varepsilon_s|E|^2} \tag{3.45}$$

When the saturation effect is very small (as is often the case), we can approximate the gain as

$$g \approx g_{th}(1 - \varepsilon_s|E|^2) \tag{3.46}$$

This expression is frequently used in theoretical treatments.

In the following discussion, we use the linearized gain for the formulation of equations assuming that the laser operation is not so far from the laser threshold. If this is not the case, we take into account the gain saturation term of (3.45). From (3.41) and (3.43), the field equation is a function of the time dependent carrier density and it is rewritten as

$$\frac{dE(t)}{dt} = \frac{1}{2}[(1 - i\alpha)G_n\{n(t) - n_{th}\}]E(t) + E_{sp}(t) \tag{3.47}$$

where we define the linear gain $G_n = v_g\partial g/\partial n$. We must consider the effect of spontaneous emissions of light in laser oscillations. $E_{sp}(t)$ is the stochastic function corresponding to the zero-mean random field for spontaneous emissions. The field has the relation of $\langle E_{sp}(t)E^*(t)\rangle = R_{sp}/2$. The term R_{sp} is usually used for the effect of spontaneous emission in the photon number equation and is given by (Petermann 1988)

$$R_{sp} = \beta_{sp}\xi_{sp}\frac{n(t)}{\tau_s} \tag{3.48}$$

where β_{sp} is the coefficient of spontaneous emissions, ξ_{sp} is the internal quantum efficiency for spontaneous emissions, and τ_s is the carrier lifetime in the laser cavity. However, we frequently omitted the term unless necessary since it is usually as small as $\beta_{sp} \sim 10^{-5}$ (Thompson 1980). Furthermore, when we investigate the fundamental dynamics of instability and chaos in nonlinear systems, we can treat only the deterministic terms without considering statistical noises. Noise is essentially considered as a separate effect from chaotic oscillations in as far as it is small. Using the notation of the complex field $E(t) = A(t) \exp\{-i\phi(t)\}$, the amplitude $A(t)$ and the phase $\phi(t)$ of the field equation are separately given by

$$\frac{\mathrm{d}A(t)}{\mathrm{d}t} = \frac{1}{2}G_\mathrm{n}\{n(t) - n_\mathrm{th}\}A(t) \qquad (3.49)$$

$$\frac{\mathrm{d}\phi(t)}{\mathrm{d}t} = \frac{1}{2}\alpha G_\mathrm{n}\{n(t) - n_\mathrm{th}\} \qquad (3.50)$$

It should be noted that in this text, the time dependent term of the propagating electromagnetic field is defined as $\exp(-i\omega_0 t)$. However, some papers and texts use the notation of $\exp(i\omega_0 t)$ for the propagating term. In that case, the coefficient of the term related to the α-parameter in the right hand side of (3.47) is written by $(1 + i\alpha)$ instead of $(1 - i\alpha)$. Although starting from the different sign of the equation for the complex field, the amplitude and phase equations result in the same forms as (3.49) and (3.50).

From the physical model of two-level atoms in semiconductor lasers, the differential equation for the carrier density n, which is equivalent to the population inversion in common lasers is given by (Agrawal and Dutta 1993)

$$\frac{\mathrm{d}n(t)}{\mathrm{d}t} = \frac{J}{ed} - \frac{n(t)}{\tau_\mathrm{s}} - G_\mathrm{n}\{n(t) - n_0\}A^2(t) \qquad (3.51)$$

where J is the injection current density and d is the thickness of the active layer. The first term on the right side of the equation is the pumping by the injection current. The second term is the carrier recombination due to spontaneous emissions. In a strict sense, the carrier recombination includes various processes of carrier decays, for example, radiative and non-radiative carrier recombination, and Auger recombination (Agrawal and Dutta 1993). They may be a function of the carrier density, however we have treated the carrier lifetime as a constant coefficient with a good approximation. The third term represents the carrier recombination induced by the laser emission. The photon lifetime and the carrier densities at threshold and transparency have the relation

$$v_\mathrm{g}\frac{\partial g}{\partial n}(n_\mathrm{th} - n_0) = \frac{1}{\tau_\mathrm{ph}} \qquad (3.52)$$

The photon number inside the laser cavity is derived by the internal optical energy U and defined by the following relation:

$$S = \frac{U}{\hbar\omega} = \frac{\varepsilon_0\bar{\eta}\eta_\mathrm{e}}{2\hbar\omega}\int_\mathrm{cavity} d^3r\,|E_\mathrm{real}(r)|^2 \qquad (3.53)$$

where $\bar{\eta}$ is the refractive index for the mode, r is the three-dimensional coordinate, and E_real is the real optical field inside the cavity. In the derivations of (3.47) and (3.51), we assume that the field is normalized by the square of the photon numbers. Assuming that the field inside the laser cavity is constant over the coordinate at a fixed time, the relation of the photon number S and the real optical field E_{real} is approximated by

$$S = |E|^2 = \frac{\varepsilon_0 \bar{\eta} \eta_{\text{e}}}{2\hbar\omega} |E_{\text{real}}|^2 V \tag{3.54}$$

where V is the volume of the active layer (Agrawal and Dutta 1993). When we write the field amplitude A, it is given by $A = \sqrt{S} = \sqrt{\varepsilon_0 \bar{\eta} \eta_{\text{e}} V / 2\hbar\omega} |E_{\text{real}}|$. Using the internal photon number, the output power outside the laser cavity reads (Petermann 1988)

$$S_{\text{ext}} = \frac{\xi_{\text{ext}}}{\xi_{\text{int}}} \frac{1}{\tau_{\text{ph}}} S \tag{3.55}$$

where $\xi_{\text{ext}}/\xi_{\text{int}}$ is the ratio between the external and internal differential quantum efficiencies. Normally, a semiconductor laser is fabricated to take different facet reflectivities to obtain a maximum laser power. Assuming the different facet reflectivities R_1 and R_2, the photon number emitted from the facet with the reflectivity R_1 is calculated as follows (Petermann 1988):

$$S_{\text{R}_1}^{\text{out}} = \frac{(1 - R_1)\sqrt{R_2}}{(\sqrt{R_1} + \sqrt{R_2})(1 - \sqrt{R_1 R_2})} S_{\text{ext}} \tag{3.56}$$

If the two facets have the same reflectivity, i.e., $R_1 = R_2$, the photon numbers emitted from the facets are $S_{\text{R}_1}^{\text{out}} = S_{\text{R}_2}^{\text{out}} = S_{\text{ext}}/2$.

As described in Sect. 3.2.1, the gain in actual lasers must be multiplied by a factor less than unity (confinement factor Γ_T) and the effect must be taken into consideration for numerical simulations of semiconductor lasers (Botez 1981). The detailed discussion of the confinement factor can be found in the book by Agrawal and Dutta (1993). When we must consider the effects of statistical noises induced by spontaneous emission of light, the Langevin terms are added not only to (3.47) but also to (3.51). Since semiconductor laser is a class B laser, we need not consider the differential equation for the polarization of matter. The polarization is included through the relation $P = \varepsilon_0 \chi E$ in (3.47). However, the laser is described by the three equations for the field amplitude in (3.49), the phase in (3.50), and the population inversion (the carrier density) in (3.51). A nonlinear system described by three coupled differential equations is a candidate for a chaotic system. The semiconductor laser is described by three equations, but the phase does not affect the other rate equations. Therefore, the two equations, the field and carrier density equations, are enough to describe the operation of edge-emitting semiconductor lasers and they are stable lasers for solitary oscillation. We did not consider the carrier diffusion from the active layer in the carrier density equation. We need not take into account this effect for ordinary edge-emitting narrow-stripe lasers, because the stripe width of the active region is small enough ($\sim 3\,\mu$m). However, it plays an important role for the dynamics in VCSELs and broad-area lasers. For those lasers, the terms for carrier diffusion are included in the laser rate equations and the laser dynamics are greatly affected by the non-negligible finite stripe width. The effects will be discussed in Chap. 8.

3.4 Linear Stability Analysis and Relaxation Oscillation

3.4.1 Linear Stability Analysis

We have already introduced linear stability analysis for a steady-state laser operation in Sect. 2.2.2 and discussed the stability of laser for small perturbations. Here, we apply the method to semiconductor lasers (Ikegami and Suematsu 1967, Paoli and Ripper 1970, Arnold et al. 1982, and Tucker 1985). From (3.49)–(3.51), we summarize the rate equations for the field amplitude, the phase, and the carrier density as follows;

$$\frac{\mathrm{d}A(t)}{\mathrm{d}t} = \frac{1}{2}G_{\mathrm{n}}\{n(t) - n_{\mathrm{th}}\}A(t) \tag{3.57}$$

$$\frac{\mathrm{d}\phi(t)}{\mathrm{d}t} = \frac{1}{2}\alpha G_{\mathrm{n}}\{n(t) - n_{\mathrm{th}}\} \tag{3.58}$$

$$\frac{\mathrm{d}n(t)}{\mathrm{d}t} = \frac{J}{ed} - \frac{n(t)}{\tau_{\mathrm{s}}} - G_{\mathrm{n}}\{n(t) - n_0\}A^2(t) \tag{3.59}$$

Setting the differential terms of the above equations to zero, we obtain the steady-state solutions for the field amplitude A_{s}, the phase ϕ_{s}, and the carrier density n_{s}

$$A_{\mathrm{s}}^2 = \tau_{\mathrm{ph}}\left(\frac{J}{ed} - \frac{n_{\mathrm{th}}}{\tau_{\mathrm{s}}}\right) \tag{3.60}$$

$$\phi_{\mathrm{s}} = 0 \tag{3.61}$$

$$n_{\mathrm{s}} = n_{\mathrm{th}} \tag{3.62}$$

For a small perturbation δx, we write the variable as $x(t) = x_{\mathrm{s}} + \delta x(t)$ ($x = A$, ϕ, n), then we can calculate the differential equations for the perturbations as

$$\frac{\mathrm{d}\delta A(t)}{\mathrm{d}t} = \frac{1}{2}G_{\mathrm{n}}A_{\mathrm{s}}\delta n(t) \tag{3.63}$$

$$\frac{\mathrm{d}\delta\phi(t)}{\mathrm{d}t} = \frac{1}{2}\alpha G_{\mathrm{n}}\delta n(t) \tag{3.64}$$

$$\frac{\mathrm{d}\delta n(t)}{\mathrm{d}t} = -\frac{2A_{\mathrm{s}}}{\tau_{\mathrm{ph}}}\delta A - \left(\frac{1}{\tau_{\mathrm{s}}} + G_{\mathrm{n}}A_{\mathrm{s}}^2\right)\delta n(t) \tag{3.65}$$

where we neglect the second order small infinities in the same manner as in the previous chapter. As the condition for non-trivial solutions of the above equations, the determinant of the coefficient matrix in the differential equations must have the following equality

$$
\begin{vmatrix}
\gamma & 0 & -\frac{1}{2}G_{\mathrm{n}}A_{\mathrm{s}} \\[2ex]
0 & \gamma & -\frac{1}{2}\alpha G_{\mathrm{n}} \\[2ex]
\frac{2A_{\mathrm{s}}}{\tau_{\mathrm{ph}}} & 0 & \gamma + \frac{1}{\tau_{\mathrm{s}}} + G_{\mathrm{n}}A_{\mathrm{s}}^2
\end{vmatrix} = 0 \tag{3.66}
$$

γ is the solution of the following characteristic equation

$$
\gamma \left\{ \gamma^2 + \left(\frac{1}{\tau_{\mathrm{s}}} + G_{\mathrm{n}}A_{\mathrm{s}}^2 \right) \gamma + \frac{G_{\mathrm{n}}A_{\mathrm{s}}^2}{\tau_{\mathrm{ph}}} \right\} = 0 \tag{3.67}
$$

By solving the equation, we can discuss stability and instability of semiconductor lasers.

3.4.2 Relaxation Oscillation

We do not consider the trivial solution $\gamma = 0$ in (3.67). The other two solutions are easily calculated by putting the real and imaginary parts of the solutions as $\gamma = \Gamma_{\mathrm{R}} + i\omega_{\mathrm{R}}$. Then, we obtain a solution which satisfies the physical condition.

$$
\Gamma_{\mathrm{R}} = -\frac{1}{2} \left(G_{\mathrm{n}}A_{\mathrm{s}}^2 + \frac{1}{\tau_{\mathrm{s}}} \right) \tag{3.68}
$$

$$
\omega_{\mathrm{R}} = \sqrt{\frac{G_{\mathrm{n}}A_{\mathrm{s}}^2}{\tau_{\mathrm{ph}}} - \Gamma_{\mathrm{R}}^2} \tag{3.69}
$$

Since the value of the real part Γ_{R} is negative, the laser oscillation corresponding to this solution quickly decays out even once it is excited. Therefore, the laser does not become unstable even in the presence of relaxation oscillation. This oscillation is called relaxation oscillation and the frequency $\nu_{\mathrm{R}} = \omega_{\mathrm{R}}/2\pi$ is known as the relaxation oscillation frequency. For a low laser output power, the relation $\Gamma_{\mathrm{R}}^2 \ll G_{\mathrm{n}}A_{\mathrm{s}}^2/\tau_{\mathrm{ph}}$ holds, and then the relaxation oscillation frequency is approximated as $\nu_{\mathrm{R}} = \sqrt{G_{\mathrm{n}}A_{\mathrm{s}}^2/\tau_{\mathrm{ph}}}/2\pi$.

Relaxation oscillation in a semiconductor laser occurs because the carrier cannot follow the photon decay rate. A relaxation oscillation is easily excited by a step input to the injection current, a shot noise originating from the driving circuit, or perturbations such as small external optical feedback. These relaxation oscillations smoothly decay out as long as the disturbances are small enough. The relaxation oscillation frequency of ordinary semiconductor lasers is of the order of 1 GHz to 10 GHz. Only a few specific lasers have a relaxation oscillation frequency over 10 GHz. The relaxation oscillation frequency is a measure of the maximum modulation ability in semiconductor lasers through the injection current. Above the relaxation oscillation, the modulation efficiency is greatly degraded and intensity modulation through the injection current becomes difficult. Therefore, an external modulator, such as an electro-optic (EO) modulator is usually used for a high-speed signal modulation over 10 GHz.

When the laser oscillates not so far from the threshold, the photon number at the steady-state is calculated from (3.51) and given by

$$S = \frac{\tau_{\mathrm{ph}}}{ed}(J - J_{\mathrm{th}}) \tag{3.70}$$

where $J_{\mathrm{th}} = edn_{\mathrm{th}}/\tau_{\mathrm{s}}$ is the threshold current. The relaxation oscillation frequency is also written by

$$\nu_{\mathrm{R}} = \frac{1}{2\pi}\sqrt{\frac{G_{\mathrm{n}}}{ed}(J - J_{\mathrm{th}})} \tag{3.71}$$

The presence of the relaxation oscillation in semiconductor lasers gives rise to complex dynamics when the laser is constantly perturbed by external disturbances as discussed in the following chapters. For example, for a moderate amount of optical feedback from an external mirror to a semiconductor laser, the relaxation oscillation does not damp out and the laser shows various dynamic behaviors, not only simple sinusoidal oscillations but also chaotic oscillations. Figure 3.2 shows the example of calculated relaxation oscillations for a step input for the injection current. The actual laser oscillates at multimode lines close to the threshold but it recovers single mode oscillation well above the threshold (Marcuse and Lee 1983). We discuss single mode semiconductor lasers in this chapter, but the model used in Fig. 3.2 is a multi-mode operation, which will be discussed in Chap. 8. The numerical

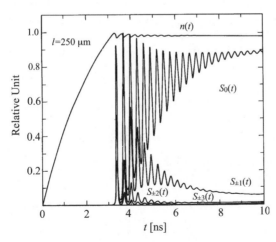

Fig. 3.2. Time evolution of carrier density n and photon population S under relaxation oscillations (after Marcuse D, Lee TP (1983); © 1983 IEEE)

simulation reflects the fact and the relaxation oscillations for the main and sub-modes starts when the carrier density exceeds the threshold. However, for a big enough pump, the relaxation oscillations rapidly decay out and the constant laser oscillation is achieved.

In real lasers, we could not ignore the effects of spontaneous emissions. Further, gain saturation effects are encountered in laser oscillations. In such a case, the relaxation oscillation angular frequency is given by almost the same form as (3.69), but we need some modifications for the damping factor (Yoon et al. 1989). In that case, instead of (3.68), the damping factor is given by

$$\Gamma_R' = -\frac{1}{2}\left\{ \frac{1}{\tau_s} + G_n A_s^2 + \frac{1}{\tau_{ph}}\left(\varepsilon_s G_n \tau_s A_s^2 + \beta_{sp}\frac{G_n \tau_{ph} n_0 + 1}{G_n \tau_s A_s^2} \right) \right\} \qquad (3.72)$$

The effect of spontaneous emissions is enhanced for a lower photon number, while the damping oscillation is much suppressed in the presence of gain saturation. Therefore, damping oscillation is remarkable for a lower injection current close to the laser threshold. At a higher pump, damping oscillation once kicks out relaxation oscillation by spontaneous emissions, but it is smeared out by the effect of gain saturation.

3.5 Langevin Noises

3.5.1 Rate Equations Including Langevin Noises

Langevin noises are defined for the photon number in laser oscillations. Since the differential equation for the field in (3.49) is obtained from an approximation of SVEA and the field amplitude is equivalent to the square root of the photon number $A(t) \propto \sqrt{S(t)}$, we can write the rate equations for the photon number S, the phase ϕ, and the carrier density n. The rate equations including the effects of spontaneous emission and statistical Langevin noises are given by

$$\frac{dS(t)}{dt} = [G_n\{n(t) - n_{th}\}]S(t) + R_{sp}(t) + F_S(t) \qquad (3.73)$$

$$\frac{d\phi(t)}{dt} = \frac{1}{2}\alpha[G_n\{n(t) - n_{th}\}] + F_\phi(t) \qquad (3.74)$$

$$\frac{dn(t)}{dt} = \frac{J(t)}{ed} - \frac{n(t)}{\tau_s} - G_n\{n(t) - n_0\}S(t) + F_n(t) \qquad (3.75)$$

where R_{sp} is the spontaneous emission term introduced in (3.48). The injection current may be a function of time through, e.g., a direct injection current modulation, therefore it is explicitly written by a time dependent function. The final term of each equation, $F_S(t)$, $F_\phi(t)$, and $F_n(t)$, is the effect of Langevin noises (Risken 1996). As already mentioned, the terms may be omitted to calculate pure dynamics of chaos in semiconductor lasers. Again note that the phase equation is not coupled with the other equations and the system still behaves stably even though it is described by three differential equations. Namely, the time development of phase does not affect the field amplitude and the carrier density. On the other hand, the field and the carrier density couples with each other. The effects of Langevin noises have been discussed in relation to AM and FM noises in laser modulations (McCumber 1966, Haug 1969, Henry 1982 and 1983, Yamamoto 1983, and Vahala and Yariv 1983a, b).

3.5.2 Langevin Noises

Real lasers include statistical Langevin noises and they are important to evaluate the performance of practical systems using semiconductor lasers as light sources. Langevin noises have common features observed not only in semiconductor lasers but also in all other lasers and they are formulated as the same equations (Risken 1996). Noises also sometimes trigger unstable oscillations in semiconductor lasers. In this section, we discuss the performance of semiconductor lasers induced by Langevin noises. The Langevin terms $F_S(t)$ and $F_\phi(t)$ are random noises induced by the quantum effects of spontaneous emission of light and the term $F_n(t)$ is induced by a noise originated by random carrier generation and recombination. They all originate from random shot noise effects. Since the effect of the carrier noise, $F_n(t)$, is negligible compared with that of $F_S(t)$, the term is sometimes omitted even when we consider the effects induced by Langevin noises in semiconductor lasers. Langevin noises have been extensively studied in detail (Lax 1960, McCumber 1966, Lax and Louisell 1969, Haug 1969, Saleh 1978, Yamamoto 1983, and Henry 1986). We need statistical methods to investigate the characteristics and effects of Langevin noises. We consider Langevin noises obeying the Markov process and assume that the correlation times of the noises are much less than the photon lifetime τ_{ph} and the carrier lifetime τ_s. We briefly show the derivations of the statistical noises and summarize them. In the rate equations from (3.73)–(3.75), denoting each Langevin noise term by $F_i(t)$ ($i = S$, ϕ, and n), the noise is a zero mean and the average is written by

$$\langle F_i(t) \rangle = 0 \tag{3.76}$$

We assume that the noise is like a shot noise and the correlation time is short enough, then we obtain the correlation between the two Langevin noises

$$\langle F_i(t) F_j(t') \rangle = 2D_{ij}\delta(t - t') \tag{3.77}$$

where D_{ij} is the diffusion coefficient for the differential diffusion equation in the presence of the Langevin force.

Since the explicit form of the diffusion coefficient D_{ij} is not easy to derive, we only show the procedure for the derivations. At first, we use the fact that the power spectrum is calculated by the Fourier transform relation with the correlation function and that it is equivalent to the correlation. For a differential equation of a variable $v(t)$ with a time dependent statistical force $F(t)$ (Langevin force)

$$\frac{\mathrm{d}v(t)}{\mathrm{d}t} + \gamma v(t) = F(t) \tag{3.78}$$

the average of the power spectrum $\Phi(\omega)$ is calculated from the Fourier transform of its correlation function of $F(t)$ as

$$\Phi(\omega) = \int_{-\infty}^{\infty} \langle F(t)F(t')\rangle \exp\{-\mathrm{i}\omega(t-t')\}\mathrm{d}(t-t') = 2D \tag{3.79}$$

where D is the diffusion coefficient of the correlation function. To derive the power spectra for the Langevin noises, we first calculate the Fourier transforms of the rate equations (3.73)–(3.75). Then, we take the ensemble averages for the equations to obtain the averaged power spectra. The detailed statistical processes of the photon number and the carrier are different to each other and they obey non-correlated quantum processes like shot noises. However, we assume Poisson random processes for them and calculate the averages, i.e., the statistical average is equal to the squared average. Calculating the correlation coefficient around the steady-state solution in each differential equation, we obtain the following relations for the coefficients:

$$D_{SS} = R_{sp}S \tag{3.80a}$$

$$D_{\varphi\varphi} = \frac{R_{sp}}{4S} \tag{3.80b}$$

$$D_{nn} = R_{sp}S + \frac{n}{\tau_s} \tag{3.80c}$$

$$D_{S\varphi} = 0 \tag{3.80d}$$

$$D_{Sn} = -R_{sp}S \tag{3.80e}$$

$$D_{n\varphi} = 0 \tag{3.80f}$$

The effects of Langevin noises are mainly generated by spontaneous emission and the effect of noise originating from carrier recombination is much smaller than that from spontaneous emission. Therefore, the noise due to carrier recombination is sometimes neglected in numerical simulations for the effects of noises in semiconductor lasers.

3.5.3 Noise Spectrum

Statistical noise characteristics of semiconductor lasers are investigated from the rate equations with Langevin noises. Here, we calculate the characteristics of intensity and phase noises from the rate equations by applying linear stability analysis. For the steady-state variables, we introduce small perturbations and linearize the rate equations. Then, we obtain the differential equations for the perturbations in the presence of Langevin noises

$$\frac{d\delta S(t)}{dt} = G_n A_s^2 \delta n(t) + F_S(t) \tag{3.81}$$

$$\frac{d\delta\phi(t)}{dt} = \frac{1}{2}\alpha G_n \delta n(t) + F_\varphi(t) \tag{3.82}$$

$$\frac{d\delta n(t)}{dt} = -\left(\frac{1}{\tau_s} + G_n A_s\right)\delta n(t) - \frac{1}{\tau_{ph}}\delta S(t) + F_n(t) \tag{3.83}$$

Due to the noises, the phase is not constant for the time development, which gives rise to unstable oscillations and sometimes induces instability and chaos in semiconductor lasers. We neglected the gain saturation and photons generated by spontaneous emissions such as small quantities. The noise characteristics are represented by using the Fourier components of the perturbations S, $\delta\phi$, and δn, then the corresponding Fourier components $\delta\tilde{S}$, $\delta\tilde{\phi}$, and $\delta\tilde{n}$ are calculated to be

$$\delta\tilde{S}(\omega) = \frac{(1/\tau_s + G_n S + i\omega)\tilde{F}_S(\omega) + G_n S\tilde{F}_n(\omega)}{(\omega_R + \omega - i\Gamma_R)(\omega_R - \omega + i\Gamma_R)} \tag{3.84}$$

$$\delta\tilde{\phi}(\omega) = \frac{1}{i\omega}\left\{\frac{1}{2}\alpha G_n \delta\tilde{n}(\omega) + \tilde{F}_\varphi(\omega)\right\} \tag{3.85}$$

$$\delta\tilde{n}(\omega) = \frac{i\omega\tilde{F}_n(\omega) - \tilde{F}_S(\omega)/\tau_{ph}}{(\omega_R + \omega - i\Gamma_R)(\omega_R - \omega + i\Gamma_R)} \tag{3.86}$$

where ω_R and Γ_R are the angular frequency of the relaxation oscillation and the damping coefficient as discussed before, respectively. From these equations, the noises are much enhanced at the angular frequency equal to the relaxation oscillation, $\omega = \omega_R$. In real lasers, it is sometimes difficult to apply the linear stability analysis for a direct injection current modulation when the modulation is not so small or the effect of spontaneous emissions is not negligible. In those cases, we must numerically solve the rate equations and such an analysis has been reported (Chinone et al. 1978)

3.5.4 Relative Intensity Noise (RIN)

The light output from a semiconductor laser is detected by a high-speed photo receiver. It is converted to an electric signal and analyzed by a spectrum analyzer. For the purpose of these analyses, we need a measure for the relative noise level to the average dc signal power, which is called relative intensity noise (RIN). For a certain angular frequency ω, RIN is defined by (Petermann 1988, and Agrawal and Dutta 1993)

$$\mathrm{RIN} = \frac{\varPhi_\mathrm{S}(\omega)}{\bar{S}^2} \tag{3.87}$$

where $\bar{S}$ is the average power of the laser output. The spectral density $\varPhi_\mathrm{S}$ for the noise component δS is given by (Papoulis 1984)

$$\varPhi_\mathrm{S}(\omega) = \int_{-\infty}^{\infty} \langle \delta S(t) \delta S(t') \rangle \exp\{-\mathrm{i}\omega(t - t')\} \mathrm{d}(t - t')$$
$$= \lim_{T \to \infty} \frac{1}{T} |\delta \tilde{S}(\omega)|^2 \tag{3.88}$$

From the spectral density calculated from (3.84) and (3.86) together with the relations in (3.78), the RIN is calculated to be

$$\mathrm{RIN} = \frac{2R_\mathrm{sp} \left[\omega^2 + (1/\tau_\mathrm{s} + G_\mathrm{n}\bar{S})^2 + G_\mathrm{n}^2 \bar{S}^2 (1 + n/\tau_\mathrm{s} R_\mathrm{sp}\bar{S}) - 2(1/\tau_\mathrm{s} + G_\mathrm{n}\bar{S}) G_\mathrm{n}\bar{S} \right]}{\bar{S} \left[(\omega_\mathrm{R} + \omega)^2 + \varGamma_\mathrm{R}^2 \right] \left[(\omega_\mathrm{R} - \omega)^2 + \varGamma_\mathrm{R}^2 \right]} \tag{3.89}$$

We have already mentioned that the effects of the noises are much enhanced at the relaxation oscillation frequency. We consider here the noise effects that are less than the relaxation oscillation frequency. As has already been derived, the relaxation oscillation frequency is proportional to the square root of the optical power from a laser and the damping factor is much less than the relaxation oscillation frequency, and then the denominator of (3.89) is proportional to S^3. Therefore, RIN is proportional to S^{-3} and the noise level rapidly decreases with the increase of the laser output power. Well above the laser threshold, the RIN is usually less than about 10^{-14} ($-140\,\mathrm{dB}$) for a 1 Hz bandwidth, which is low enough for the use of lasers as light sources to optical communications and optical data storage systems (Petermann and Arnold 1982, Elsäßer and Göbel 1985).

3.5.5 Phase Noise and Spectral Linewidth

Phase noise is calculated from the correlation of the complex field. For example, the power spectrum measured by a Fabry-Perot spectrometer in a real experiment is written by the Fourier transform of the field correlation as

$$\varPhi_\mathrm{E}(\omega) = \int_{-\infty}^{\infty} \langle E(t) E^*(t + \tau) \rangle \exp(-\mathrm{i}\omega\tau) \mathrm{d}\tau \tag{3.90}$$

Here, consider the field having a phase fluctuation of

$$E(t) = \sqrt{S} \exp[-\mathrm{i}\{\omega_0 t + \phi + \delta\phi(t)\}] \tag{3.91}$$

where $\delta\phi(t)$ is the time dependent random fluctuation and S and ϕ are set to be constants. ω_0 is the angular frequency of the laser oscillation. In actual fact, the amplitude (equivalently the photon number) fluctuates, accompanying the phase fluctuation. However, we neglect the effects as a small fluctuation. Substituting (3.91) into (3.90), the power spectrum is given by

$$\Phi_{\mathrm{E}}(\omega) = S \int_{-\infty}^{\infty} \langle \exp(\mathrm{i}\Delta\phi) \rangle \exp\{-\mathrm{i}(\omega - \omega_0)\tau\} \mathrm{d}\tau \qquad (3.92)$$

where $\Delta\phi = \delta\phi(t-\tau) - \delta\phi(t)$. Assuming Gaussian statistics for $\delta\phi$, we obtain the following relation (Papoulis 1984);

$$\langle \exp(\mathrm{i}\Delta\phi) \rangle = \exp\left\{ -\frac{1}{2}\langle (\Delta\phi)^2 \rangle \right\} \qquad (3.93)$$

Using (3.85), $\Delta\phi$ is written as

$$\Delta\phi = \frac{1}{2\pi} \int_{-\infty}^{\infty} \delta\tilde{\phi}(\omega)\{\exp(\mathrm{i}\omega\tau) - 1\}\exp(\mathrm{i}\omega t)\mathrm{d}\omega \qquad (3.94)$$

Then, the ensemble average of the square of $\Delta\phi$ is given by

$$\langle (\Delta\phi)^2 \rangle = \frac{1}{\pi} \int_{-\infty}^{\infty} \langle |\delta\tilde{\phi}(\omega)|^2 \rangle (1 - \cos\omega\tau)\mathrm{d}\omega \qquad (3.95)$$

The spectral power of the field due to the phase fluctuation is calculated by substituting (3.93) into (3.92). The calculation is not straightforward and, instead, it is usually obtained numerically (Henry 1986). Figure 3.3 shows an example of numerical calculations for the power spectrum. With the increase

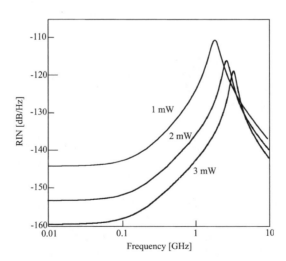

Fig. 3.3. Intensity noise spectrum for several power levels

of the output power, the RIN decreases as expected. The higher frequency peaks in the spectra correspond to the relaxation oscillations. Also the relaxation oscillation frequency increases with the increase of the laser output power. Similar spectral characteristics have also been observed by experiments (Vahala et al. 1983).

Semiconductor laser materials are usually homogeneous, so that the spectral shape of the laser oscillation is a Lorentzian, which is derived from the fact of an exponential time decay of the coherence of light. We calculate the spectral linewidth of the laser oscillation. Since the fluctuations of the angular frequency $\Delta\omega$ and the phase $\delta\phi$ have the relation $\Delta\omega = \mathrm{d}\delta\phi/\mathrm{d}t$, the power spectrum of the phase noise $\Phi_{\Delta\omega}(\omega)$ is written by

$$\Phi_{\Delta\omega}(\omega) = \omega^2 \langle |\delta\tilde{\phi}(\omega)|^2 \rangle \tag{3.96}$$

Substituting (3.85) into (3.96) and after lengthy calculations, we obtain the approximate form of the spectrum as

$$\Phi_{\Delta\omega}(\omega) = \frac{R_{\mathrm{sp}}}{2S} \left\{ 1 + \frac{\alpha^2 \omega_{\mathrm{R}}^4}{(\omega_{\mathrm{R}}^2 - \omega^2)^2 + (2\omega\Gamma_{\mathrm{R}})^2} \right\} \tag{3.97}$$

There is the following relation between $\Delta\phi$ and the phase noise spectrum $\Phi_{\Delta\omega}(\omega)$

$$\langle (\Delta\phi)^2 \rangle = \frac{1}{\pi} \int_{-\infty}^{\infty} \Phi_{\Delta\omega}(\omega) \frac{1 - \cos\omega\tau}{\omega^2} \mathrm{d}\omega \tag{3.98}$$

The above integral is not easily calculated in an explicit form, but it is approximated at an angular frequency close to $\omega = 0$ as

$$\langle (\Delta\phi)^2 \rangle = \tau \Phi_{\Delta\omega}(0) \tag{3.99}$$

From this result together with (3.92), the correlation function of the field in the presence of phase noise decays exponentially with a time constant of $2/\Phi_{\Delta\omega}(0)$. Also combining the result with (3.93), it is proved that the power spectral shape has a Lorentzian. The half-width of full maxima (HWFM) is calculated from the relation $\Delta\omega = 2\pi\Delta\nu = \Phi_{\Delta\omega}(0)$ as

$$\Delta\nu = \frac{1}{2\pi}\Phi_{\Delta\omega}(0) = \frac{R_{\mathrm{sp}}}{4\pi S}(1 + \alpha^2) \tag{3.100}$$

The spectral linewidth of laser oscillations is calculated from the well-known Schalow-Townes equation. The linewidth of gas lasers with an almost negligible value of the α parameter is given by $\Delta\nu_0 = R_{\mathrm{sp}}/4\pi S$. On the other hand, semiconductor lasers have a non-negligible value of the α parameter and the linewidth of semiconductor lasers is $(1+\alpha^2)$ times larger than that of ordinary lasers (Henry 1982, 1983, and 1986, Petermann 1988, and Agrawal and Dutta 1993). The spectral linewidth of semiconductor lasers is usually more than 10 times larger than of gas lasers and it ranges from several MHz to a hundred MHz without the control for stabilization (Fleming and Mooradian 1981, and Elsäßer and Göbel 1984). The spectral linewidth of laser oscillations decreases with an increase of the photon rate, so that the linewidth becomes narrow with an increase of the laser output power. Semiconductor lasers tend to become stable operations for higher output power. Therefore, they frequently show unstable oscillations at lower output power, although unstable behaviors are observed not only at low output power but also at high output power by the introduction of external perturbations.

3.6 Modulation Characteristics

3.6.1 Injection Current Modulation

The unique differentiating feature of semiconductor lasers is the direct modulation for the pump, namely the carrier that is the pump in semiconductor lasers is directly modulated through the injection current. Besides small and compact light sources, semiconductor lasers are suitable light sources of optical communications and information devices because of their direct modulation property. As will be discussed later, a modulation for the injection current itself is the introduction of an extra degree of freedom to semiconductor laser and it induces instability and chaos in the output. On the one hand, chaotic behaviors in semiconductor lasers can be also controlled to stable oscillations by direct injection current modulation. The relaxation oscillation plays a crucial role for the modulation properties in semiconductor lasers.

The efficiency of laser output by the injection current modulation is constant for a small to moderate modulation index when the modulation frequency is less than the relaxation oscillation frequency. On the other hand, the modulation efficiency is greatly degraded over the frequency of the relaxation oscillation and the modulation for the light output over that frequency is not possible (Boers and Vlaardinerbrek 1975 and Furuya et al. 1979). The effort to enhance relaxation oscillation frequency of semiconductor lasers has been made to device high efficiency lasers in practical use (Lau and Yariv 1985, and Uomi et al. 1985). The injection current modulation (amplitude modulation) results in the phase modulation (frequency modulation), since the laser oscillation frequency is a function of the injection current. The linear relation is established for the injection current modulation and the increment of the frequency is linearly proportional to the injection current without mode hop. The effect of coupling between the amplitude and the phase comes from the fact that the real and imaginary parts of the refractive index are not independent and they couple with each other. This unique characteristic originated from the non-zero value of the α parameter.

In the following, we discuss the modulation properties when a small signal is applied directly to the injection current. We again employ the method of a small signal modulation used in linear stability analysis. Since we are concerned with the modulation properties, we will not consider noise effects. As an injection current modulation, we assume a time dependent signal as

$$J(t) = J_b + J_m(t) = J_b + J_{0m} \sin \omega_m t \tag{3.101}$$

where J_b is the bias injection current, J_m is the small amplitude of the modulation ($J_{0m} \ll J_b$), ω_m is the modulation angular frequency. The small injection current modulation gives rise to a perturbation to the carrier density δn, and then fluctuations of the photon number δS and the phase $\delta \phi$ are induced through the change of the carrier density. Using the fact that the modulation is very small and from (3.73)–(3.75), the differential equations for the perturbations δS, $\delta \phi$, and δn are given by

$$\frac{d\delta S(t)}{dt} = G_n A_s^2 \delta n(t) \tag{3.102}$$

$$\frac{d\delta \phi(t)}{dt} = \frac{1}{2} \alpha G_n \delta n(t) \tag{3.103}$$

$$\frac{d\delta n(t)}{dt} = -\left(\frac{1}{\tau_s} + G_n A_s^2 \right) \delta n(t) - \frac{1}{\tau_{ph}} \delta S(t) + \frac{J_m(t)}{ed} \tag{3.104}$$

By Fourier transforming these equations, the corresponding components for these perturbations are written by

$$\delta \tilde{S}(\omega) = \frac{G_n A_s^2 \tilde{J}_m(\omega)/ed}{(\omega_R + \omega + i\Gamma_R)(\omega_R - \omega - i\Gamma_R)} \tag{3.105}$$

$$\delta \tilde{\phi}(\omega) = \frac{1}{2i\omega} \alpha G_n \delta \tilde{n}(\omega) \tag{3.106}$$

$$\delta \tilde{n}(\omega) = \frac{i\omega \tilde{J}_m(\omega)/ed}{(\omega_R + \omega + i\Gamma_R)(\omega_R - \omega - i\Gamma_R)} \tag{3.107}$$

where $\tilde{J}_m(\omega)$ is the Fourier transform of $J_m(t)$. As is easily understood from the above equations, the modulation for the injection current directly induces fluctuations to the carrier density and, then, it is coupled with the fluctuations of the photon number and the phase through the variation of the carrier density. Thus, instability is intrinsically included in semiconductor lasers as a coupled nonlinear system.

3.6.2 Intensity Modulation Characteristics

From the previous discussion, we calculate the modulation efficiency for the photon number. Using (3.101), the Fourier component of the injection current modulation $\tilde{J}_m(\omega)$ is written by

$$\tilde{J}_m(\omega) = -i\pi J_{0m} \{ \delta(\omega - \omega_m) + \delta(\omega + \omega_m) \} \tag{3.108}$$

Substituting the above equation into (3.105), the fluctuation of the photon number is calculated as

$$\delta S(t) = \delta S_0 \sin(\omega_m t + \theta_S) \qquad (3.109)$$

where the amplitude δS_0 and the phase θ_S are given by

$$\delta S_0 \frac{G_n S J_{0m}/ed}{\sqrt{(\omega_m^2 - \omega_R^2 - \Gamma_R^2)^2 + 4\omega_m^2 \Gamma_R^2}} \qquad (3.110)$$

$$\theta_S = \tan^{-1}\left(\frac{2\Gamma_R \omega_m}{\omega_m^2 - \omega_R^2 - \Gamma_R^2}\right) \qquad (3.111)$$

The intensity (photon number) of the output power from a semiconductor laser is modulated at the same frequency of the injection current modulation and it varies with time proportional to the modulation depth J_{0m}. However, the phase of the modulated intensity is not the same as the injection current modulation and, for example, the signal delay occurs for a modulation below the relaxation oscillation frequency ($\omega_m \ll \omega_R$). The modulation efficiency is greatly enhanced near the relaxation oscillation frequency and takes the maximum at the frequency (Ikegami and Suematsu 1968). Figure 3.4 shows examples of modulation response for different relaxation oscillation frequencies at a damping factor of $\Gamma_R = 3 \times 10^9 \, \mathrm{s}^{-1}$. The modulation power is given by $\delta S_0 = \tau_{ph} J_{0m}/ed$ for $\omega_m \ll \omega_R$. The modulation efficiency at the resonance frequency is about $\omega_R/2\Gamma_R$ times larger than that below the frequency.

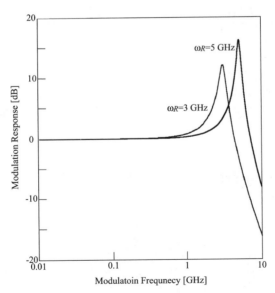

Fig. 3.4. Small signal modulation response at the damping factor Γ_R of $3 \times 10^9 \mathrm{s}^{-1}$

When the modulation exceeds the relaxation oscillation frequency, the carrier cannot follow the modulation speed of the injection current and the intensity of the modulation rapidly decreases with the increase of the modulation frequency. Thus, the relaxation oscillation frequency determines the capability of the maximum modulation in semiconductor lasers.

3.6.3 Phase Modulation Characteristics

The injection current modulation induces a change of the phase in laser oscillations known as frequency chirp (Dutta et al. 1984, Linke 1985, Agrawal 1985, and Kazarinov and Henry 1987). We write the fluctuation of the frequency as $\delta\nu(t) = d\delta\phi(t)/dt/2\pi$. The fluctuation is calculated from the Fourier transform of $\delta\tilde{\phi}(\omega)$ with the relation of (3.106) and is given by

$$
\delta\nu(t) = \frac{1}{4\pi^2} \int_{-\infty}^{\infty} i\omega\delta\tilde{\phi}(\omega)\exp(i\omega t)d\omega
$$

$$
= \frac{\alpha G_n}{8\pi^2 ed} \int_{-\infty}^{\infty} \frac{i\omega\tilde{J}_m(\omega)}{(\omega_R + \omega + i\Gamma_R)(\omega_R - \omega - i\Gamma_R)}\exp(i\omega t)d\omega \qquad (3.112)
$$

From this result, we easily understand that the α parameter, which has a non-zero value in semiconductor lasers plays an important role for the frequency chirp. For the sinusoidal modulation for the injection current, the frequency of the laser oscillation is modulated by the same frequency as

$$
\delta\nu(t) = \delta\nu_0 \sin(\omega_m + \theta_\nu) \qquad (3.113)
$$

where the amplitude $\delta\nu_0$ and the phase θ_ν are given by

$$
\delta\nu_0 = \frac{\alpha G_n J_{0m}}{4\pi ed} \frac{\omega_m}{\sqrt{(\omega_m^2 - \omega_R^2 - \Gamma_R^2)2 + (2\omega_m\Gamma_R)^2}} \qquad (3.114)
$$

$$
\theta_\nu = \frac{\pi}{2} + \tan^{-1}\left(\frac{2\Gamma_R\omega_m}{\omega_m^2 - \omega_R^2 - \Gamma_R^2}\right) \qquad (3.115)
$$

The modulation efficiency is proportional to the amplitude of the modulation J_{0m} and the amplitude is approximated for the modulation below the relaxation oscillation as

$$
\delta\nu_0 = \frac{\alpha\tau_{ph}\omega_m}{4\pi edS}J_{0m} \qquad (3.116)
$$

It is noted that the modulation amplitude is inversely proportional to the photon number and the effect of the modulation on the frequency becomes small for a higher injection current. The conversion efficiency from the injection current to the frequency is about several GHz/mA for ordinary edge-emitting semiconductor lasers. Figure 3.5 shows the dependence of frequency chirp on the modulation depth of the injection current observed experimentally in various laser structures. For a modulation frequency sufficiently below the relaxation oscillation, the phase has a negligible small value $\theta_\nu \sim 0$ and the modulated intensity is in-phase with the injection current modulation.

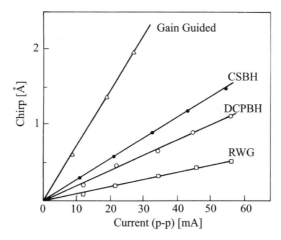

Fig. 3.5. Experimentally measured chirp as a function of the peak-to-peak value of the modulation current for various laser structures at a wavelength of 1.3 μm (after Dutta NK, Olsson NA, Koszi LA, Besomi P, Wilson RB (1984); © 1984 AIP)

3.7 Waveguide Models of Semiconductor Lasers

3.7.1 Index- and Gain-Guided Structures

Edge-emitting semiconductor lasers with narrow stripe width have common dynamics for external perturbations; however, the parameter range of each stable and unstable characteristic varies from one laser to the other, depending on the laser structures. The dynamics depend on the type of guided structures of light, i.e. gain- or index-guided structure. In addition, they are functions of types of cavity structures such as Fabry-Perot, multi-quantum well (MQW), or distributed feedback (DFB) cavity structures. In this section, we discuss some wave-guiding models and laser cavity types that affect the laser dynamics in the following chapters. Regarding wave-guiding structures in semiconductor lasers, there are two types of wave-guiding models; one is the index-guided structure, in which we can expect stable laser oscillation with a single longitudinal mode, and the other one is the gain-guided structure, in which the laser behaves rather unstably and it sometimes operates in multi-longitudinal modes (Petermann 1988 and Agrawal and Dutta 1993). Each device maker fabricates various types of gain- and index-guided structure lasers for the use of commercial applications.

Figure 3.6 shows typical examples of gain- and index-guided structures. Figure 3.6a is the schematic illustration of the structure of a gain-guide laser at its light exit facet. Due to the presence of insulator regions at p-type electrodes, injected carriers pass through a certain area in the active layer, so that the center of the active layer only has a high gain. Thus, the laser oscillates around the center of the active region. The laser is easy to fabricate, but it

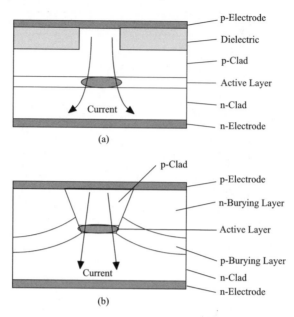

Fig. 3.6. Examples of gain- and index-guided laser structures. **a** Gain-guided laser and **b** index-guided laser with buried hetero structure

has a low efficiency of light confinement due to the decrease of the refractive index of the active layer by the carrier injection, resulting in leakage of light outside the lasing regions. For the same reason, the emitted light from the laser is usually astigmatic and, in the worst case, the laser has a twin-peaked far-field pattern. The laser easily shares gains for each lasing modes and it sometimes oscillates at multi-longitudinal modes. Another disadvantage of laser oscillations is a low noise performance induced by a high level of spontaneous emission of light. Nevertheless, gain-guided lasers are still used for commercial uses where there is no need to consider high laser performances.

Figure 3.6b is a schematic illustration of the structure of an index-guide laser at its light exit facet. This example is a type of buried hetero structure, which has a high refractive index and gain waveguide area at the center of the active layer, resulting in high efficiency of confinement of light, similar to an optical fiber structure. Therefore, an index-guided laser has a high quality of emitting beam better than that of a gain-guided laser. The advantages of index-guided lasers are as follows; high efficiency of stimulated emission with a small effect of spontaneous emission, low threshold, oscillation of single longitudinal mode, small astigma, single peak profile of far-filed pattern, and high modulation efficiency. Index-guided semiconductor lasers are widely used as light sources in optical communications and optical data storage systems. The dynamics of semiconductor lasers are strongly affected through the difference in gains in wave-guided structures. Though the universal dynamics and overall chaotic roots for the parameter variations look the same, regardless of the laser structures, each laser with specific structure has characteristic dynamics. The difference between gain- and index-guide structures plays an important role, especially in broad-area semiconductor lasers.

3.7.2 Waveguide Models

In this subsection, we discuss the theoretical treatments of the waveguide models with emphasis on the difference between gain- and index-guided laser structures. In the derivations of the rate equations, we assume that the laser propagates as a harmonic plane wave in the waveguide. A semiconductor laser generally oscillates at a fundamental transverse-electric (TE) mode in the laser cavity, except for the installation of special device structures to force the oscillation to the orthogonal mode, i.e. transverse-magnetic (TM) mode. Therefore, we can approximate plane wave propagation even in the study of laser dynamics. However, we must consider a finite extent of the wave amplitude with elliptic beam profile for propagation of the actual laser. Namely, we must take into account the spatial modes for the lateral and transverse oscillations in a strict sense. The gain for laser oscillation also has a spatial profile along the active layer and, as a result, the spatial mode profile is affected depending on the waveguide structures. In the following, we briefly develop a theoretical treatment for the lateral spatial modes in narrow-stripe edge-emitting semiconductor lasers by taking into consideration the laser waveguide structures. A narrow-stripe semiconductor laser defined here has a stripe width of $2 \sim 3\,\mu$m and the theoretical treatment is only valid for lasers with this range. For semiconductor lasers over the stripe width of several micron-meters, we cannot ignore the effects of the higher spatial mode oscillations and their competition in laser dynamics. Waveguide structures play crucial roles for the dynamic properties of vertical-cavity surface-emitting lasers (VCSELs) and broad-area semiconductor lasers, as will be discussed in Chap. 8. In particular, spatio-temporal instability is much enhanced for those lasers with gain-guided structures.

For the mathematical treatment of laser dynamics, we must consider both the temporal and spatial dependences of the Maxwell equation. But we are concerned here with the spatial effects of laser propagation, and the temporal effects may be ignored for a while. We also restrict the discussion of TE-mode propagation. To a good approximation, we can apply the separation of the variables for the electric field using the slab-waveguide model and assume scalar wave propagation (Agrawal and Dutta 1993). Then the solution of the electric field is approximated by

$$E(x, y, z) = E_0 \psi_y(y) \psi_x(x) \exp(\mathrm{i}\beta_z z) \tag{3.117}$$

where x is the lateral coordinate in the active layer, y is the orthogonal coordinate to x, β_z is the propagation constant along the laser cavity, and E_0

is the average amplitude. Substituting (3.117) into the Helmholtz equation, one obtains

$$\frac{1}{\psi_x(x)}\frac{\partial^2\psi_x(x)}{\partial x^2} + \frac{1}{\psi_y(y)}\frac{\partial^2\psi_y(y)}{\partial y^2} + \{\varepsilon(x,y)k_0^2 - \beta_z^2\} = 0 \qquad (3.118)$$

where $\varepsilon(x,y)$ is the complex dielectric constant generally having the form $\varepsilon(x,y) \approx \varepsilon(x)$ for a slab-waveguide. In the effective index approximation, the transverse field distribution $\psi_y(y)$ is obtained first by solving

$$\frac{\partial^2\psi_y(y)}{\partial y^2} + \{\varepsilon(x,y)k_0^2 - \beta_{eff}^2(y)\}\psi_y(y) = 0 \qquad (3.119)$$

where $\beta_{eff}(x)$ is the effective propagation constant at the coordinate x. Substituting (3.119) into (3.118), the lateral field distribution is then obtained by solving

$$\frac{\partial^2\psi_x(x)}{\partial x^2} + \{\beta_{eff}^2(x) - \beta_z^2\}\psi_x(x) = 0 \qquad (3.120)$$

From the assumption of a thin active layer, i.e. the constant carrier density for the y direction at a fixed x coordinate, the non-time dependent carrier density equation along the lateral coordinate x is given by the following equation:

$$D_e\frac{\partial^2 n(x)}{\partial x^2} + \frac{J(x)}{ed} - \frac{n(x)}{\tau_s} - G_n\{n(x) - n_0\}E_0^2 |\psi_x(x)|^2 = 0 \qquad (3.121)$$

where the first term of the above equation is the carrier diffusion along the active layer and D_e is the diffusion constant of the carrier density. Usually, the diffusion constant D_e and the gain G_n are also functions of the lateral coordinate x. Non-time-dependent spatial mode analyses can be conducted using (3.119–3.121).

3.7.3 Spatial Modes of Gain- and Index-Guided Lasers

The mode analysis is simplified for the assumption of the separation of the variables. For the fundamental TE-mode, the solution for (3.120) is given by

$$\psi_x(x) = \psi_0 \cos(\kappa_x x) \qquad\qquad\qquad (|x| \leq w/2)$$
$$= \psi_0 \cos\left(\frac{\kappa_x w}{2}\right) \exp\left\{-\gamma_x\left(|x| - \frac{w}{2}\right)\right\} \qquad (|x| > w/2) \qquad (3.122)$$

where ψ_0 is the normalization constant, w is the width of the active layer, and κ_x and γ_x are defined by

$$\kappa_x = \sqrt{\beta_{core}^2 - \beta_{eff}^2} \qquad (3.123a)$$

$$\gamma_x = \sqrt{\beta_{eff}^2 - \beta_{clad}^2} \qquad (3.123b)$$

In the above equations, β_{core} and β_{clad} are the propagation constants both in the active layer and outside the active layer. The solution can be considered as a Gaussian beam profile with a good approximation. The beam profile for the transverse direction can also be considered as a fundamental Gaussian mode. However, assumption of the fundamental TE-mode is only valid for a narrow stripe semiconductor laser. For broad stripe width or even for a gain-guided laser with narrow stripe width, we must consider higher spatial modes derived form (3.120) due to effects such as hole-burning of the carrier. The excitations of higher spatial modes significantly affect laser instabilities.

For an index-guided laser, the wave fronts for the lateral and transverse directions at the exit face are approximated as plane waves. On the other hand, the approximation of plane wave is only valid for the transverse direction in a gain-guided laser. The wave front for the lateral direction must be treated as a divergent spherical wave (Agrawal and Dutta 1993). In a gain-guided laser, the effective propagation constant becomes a function of the lateral coordinate. We assume the parabolic profile for the propagation constant written by

$$\beta_{eff} = \beta_{0eff} - k_0^2 a^2 x^2 \tag{3.124}$$

where β_{0eff} is the propagation constant at $x = 0$. The complex parameter a is defined by $a = a_r - i a_i$ and has a relation with the complex refractive index in (3.39) as follows (Cook and Nash 1975):

$$\mathrm{Re}[\eta_c] = \eta - \frac{a_r^2 - a_i^2}{2\eta} x^2 \tag{3.125a}$$

$$\mathrm{Im}[\eta_c] = -\frac{\eta' + 2 a_r a_i x^2}{2\eta} \tag{3.125b}$$

Solving the differential equation for the fundamental spatial mode, one obtains

$$\psi_{x0}(x) = B_0 \exp\left(\frac{1}{2} i k_0 a_i x^2\right) \exp\left(-\frac{1}{2} k_0 a_r x^2\right) \tag{3.126}$$

where B_0 is the normalization constant. The solutions for the higher spatial modes are given by the following Gauss-Hermite function:

$$\psi_m(x) = H_m\left(\sqrt{k_0 a} x\right) \psi_0(x) \tag{3.127}$$

Here, H_m is the Hermite polynomial of order m. As already discussed, the wave front of the transverse direction in a gain-guided laser is assumed to be plane wave, therefore we can only consider the spatial dependence for the lateral direction. From (3.117), the surfaces of constant phase of the complex field in gain-guided lasers are written as

$$\beta_z z + \frac{1}{2} k_0 a_i x^2 = \text{const.} \tag{3.128}$$

The surfaces are cylindrical with a radius of a certain curvature near the coordinate at $x = 0$. Therefore, the phase front of the gain-guided laser at the exit face must be approximately treated as a spherical wave along the lateral direction. Assuming that the difference of the refractive indices between the center and edge of the active layer is small, the curvature is approximated as

$$R_m = \frac{\beta_z}{k_0 a_i} \approx \frac{\eta}{a_i} \tag{3.129}$$

For an index-guided laser, $a_i = 0$ resulting a plane wave front for the lateral direction. A gain-guided laser has a divergent wave front and it is unstable laser compared with an index-guided laser. The difference of the instabilities can be implemented by the variations of the gains between the two laser models in the rate equations.

3.7.4 Effects of Spontaneous Emission in Gain- and Index-Guided Lasers

The effects of spontaneous emission play crucial roles for laser oscillations. The increase in spontaneous emission in the cavity is counted as a loss of the laser, and results in the increase of the laser threshold. The effects of spontaneous emission are quite different depending on whether the laser structure is a gain-guided type or an index-guided one. The stationary solution for the photon number S_s is related to the spontaneous emission, and is easily derived from (3.75) as

$$S_s = \frac{1}{G_n(n_s - n_{th})} R_{sp} \tag{3.130}$$

In actual lasers, the effect of spontaneous emission on the lasing oscillation (photon number) is enhanced due to non-negligible losses from the laser cavity. The enhancement factor K_c is calculated from the traveling wave amplifier model and the corrected stationary solution for the photon number is given by Petermann (1988)

$$S = \frac{1}{G_n(n_s - n_{th})} K_c R_{sp} \tag{3.131}$$

with

$$K_c = \left\{ \frac{(r_1 + r_2)(1 - r_1 r_2)}{2 r_1 r_2 \ln(1/r_1 r_2)} \right\}^2 \tag{3.132}$$

where r_1 and r_2 are the facet reflectivities of the laser cavity. For the reflectivities $r_1 = r_2 \approx 1$, the correction factor is $K_c = 1$ and no correction for spontaneous emission is required. However, in the case of semiconductor lasers, K_c is usually larger than unity. For example, for the cleaved facets of AlGaAs lasers of $r_1^2 = r_2^2 = 0.32$, one obtains $K_c = 1.11$, which is still not significant. However, the situation is valid for index-guided semiconductor lasers.

For gain-guided semiconductor lasers, we must consider an additional enhancement of spontaneous emission. The phase fronts in gain-guided lasers are curved, so that energy is carried not only in the axial direction but also in the lateral direction. As a result, the lasing mode in the gain-guided laser captures a larger fraction of spontaneous emission than the lasing mode in the index-guided laser. This is somewhat similar to the effect of aberrations in lenses. The additional enhancement factor K_{gain} for the spontaneous emission noise of a gain-guided laser is given by Petermann (1979)

$$K_{gain} = \left| \frac{\int |\psi_x(x)|^2 \, dx}{\int \psi_x^2(x) dx} \right|^2 \tag{3.133}$$

For index-guided lasers, one has the plane phase fronts and $\psi_x^2 = |\psi_x|^2$ yielding $K_{gain} = 1$. On the other hand, ψ_x is a complex for gain-guided lasers and the value of K_{gain} is always larger than unity. For the fundamental spatial mode oscillation, the factor K_{gain} is calculated as (Petermann 1979)

$$K_{gain} = \sqrt{1 + \left(\frac{a_i}{a_r} \right)^2} \tag{3.134}$$

Here, a_i/a_r is defined as the strength of astigmatism. The strength of astigmatism has the relation with the linewidth enhancement factor (Kirkby et al. 1977)

$$\frac{a_i}{a_r} = \sqrt{1 + \alpha^2} + \alpha \tag{3.135}$$

Using this relation, the additional enhancement factor for spontaneous emission is written as

$$K_{gain} = \sqrt{2(1 + \alpha^2) + 2\alpha\sqrt{1 + \alpha^2}} \tag{3.136}$$

For a typical value of the linewidth enhancement factor of $\alpha = 4$, the additional enhancement factor is calculated to be $K_{gain} = 8.2$. Normally, the typical value of the additional enhancement factor for spontaneous emission has a value of the order of 10. As a whole, gain-guided lasers have a strong dependence on spontaneous emission and have higher threshold for laser oscillations as shown in Fig. 3.7 (Petermann and Arnold 1982). The gain-guided lasers are unstable lasers having nuclear laser threshold and even larger radiation less than the threshold. These characteristics strongly affect the laser dynamics for time development through the gain factor.

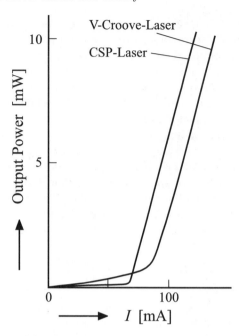

Fig. 3.7. Experimental L-I characteristics of gain- and index-guided semiconductor lasers (after Petermann K, Arnold G; © 1982 IEEE)

3.7.5 Laser Types

Fabry-Perot lasers, MQW (multi-quantum well) lasers and DFB (distributed feedback) lasers are typical models of edge-emitting lasers currently commercially available. These lasers are widely used for light sources of optical communications, optical data storage systems and optical information systems. Although the theoretical treatments discussed here have been developed for a type of Fabry-Perot semiconductor lasers, the laser rate equations are commonly applicable to other types of narrow-stripe semiconductor lasers including MQW and DFB lasers. However, each type of laser has different device characteristics and the stable and unstable features of laser oscillations for external perturbations are strongly dependent on each device parameters. For example, a semiconductor laser is easily destabilized by self-optical feedback as will be discussed in Chap. 4, but the external feedback level for unstable chaotic oscillations is different for each laser depending on the values of the device parameters.

Fabry-Perot lasers are easily fabricated and were used in the early days. The theory for the dynamics discussed in the preceding sections is based on Fabry-Perot structures. The gain for semiconductor laser materials is broad, up to several tens of nanometers and the gain difference for the laser modes in Fabry-Perot lasers is small, so that they tend to oscillate at multi-longitudinal modes. In particular, gain-guided lasers with Fabry-Perot cavity structure show multimode oscillation without exception. Even a single mode Fabry-Perot laser exhibits multimode oscillations at a fast modulation for the bias injection current. From the viewpoint of chaotic dynamics, the Fabry-Perot laser is an unstable laser, showing a wide range of unstable oscillations for the parameter variations from external perturbations.

MQW lasers have many thin quantum-well layers parallel to the active layer. Each quantum-well layer has a thickness ranging from ten to several tens of nanometers, where the confined electrons play like waves within the quantum potentials. Figure 3.8 shows a schematic diagram of AlGaAs multi-quantum well lasers. Figure 3.8a shows the band model for multi-quantum well structure and Fig. 3.8b is the schematic representation of the confined particle energy levels of electrons (c), heavy holes (hh) and light holes (lh) in a single quantum-well. Due to the narrow potential well of a GaAs layer sandwiched by AlGaAs layers, the energy levels are quantized and have discrete distributions. The use of multi-quantum wells has several advantages over Fabry-Perot lasers with a single thick ($\sim$100 nm) active layer. The gain for a particular oscillation wavelength can be much enhanced due to the quantum-well structure, and the gain coefficient is usually greater than

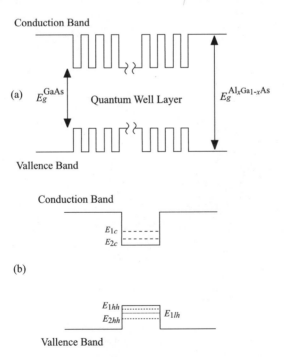

Fig. 3.8. Schematic diagram of AlGaAs multi-quantum well lasers. **a** Band model for multi-quantum well structure and **b** energy band of a single quantum well with confined particle energy levels of electrons (c), heavy holes (hh) and light holes (lh)

3 times or more compared with a single layer Fabry-Perot laser. The shorter oscillation wavelength is attained compared with a bulk semiconductor laser by the quantization of the energy levels. Figure 3.9 shows modal gain as a function of the injection current with the number of quantum wells in the active layer. As is easily recognized, the differential gain $\partial g/\partial n$ increases with the increase of the number of quantum wells. From (3.71), the relaxation oscillation frequency, which limits the modulation performance, is proportional to the square root of the differential gain, so that we can expect enhancement of the modulation property for multi-quantum well lasers. A multi-quantum well laser has a small linewidth enhancement factor, and unstable parameter regions are much reduced compared with bulk semiconductor lasers. As a result of efficient optical confinement, we can achieve lower threshold lasers with small optical losses. With such excellent performances and reasonable prices for fabrication cost, multi-quantum well lasers are widely used in optical data storage systems and optical information processing systems.

In optical communication systems, a semiconductor laser with single mode and narrow linewidth is desired for good quality of data transmissions. For such purpose, distributed feedback (DFB) lasers have been developed. Figure 3.10 shows a schematic diagram of the typical $\pi/2$ shifted DFB laser structure along the laser cavity. A DFB laser has a grating structure close

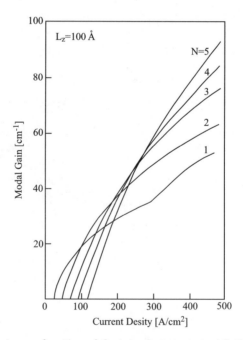

Fig. 3.9. Modal gain as a function of the injection current with the number of quantum wells N. The quantum well thickness is assumed to be 10 nm (after Arakawa Y, Yariv A; © 1985 IEEE)

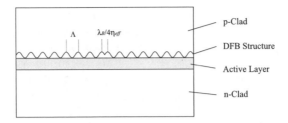

Fig. 3.10. Schematic diagram of the typical $\pi/2$ shifted DFB laser structure. $\Lambda = m\lambda_B/2\eta_{\text{eff}}$ (m is the integer number)

to the active layer in the axial direction. Therefore, a preference is obtained for an oscillation wavelength, which fits the grating's period. If the grating has a continuous periodic structure without non-phase shift, the light corresponding to the Bragg wavelength λ_B is not an oscillation mode due to the stop-band. However, light with wavelength λ_B becomes the transmission mode due to the opening of the transmission window in the presence of $\pi/2$ phase shift at the center of the grating structure. This results in stable laser oscillations at the Bragg wavelength. As a disadvantage of the DFB structure, the amplitude of the standing wave is fixed by the grating, so that the effect of carrier hole-burning is much enhanced at the regions of large field amplitudes. Then the gain at those points decreases and the effect is to lower the laser threshold. This is not too serious, as will be discussed in the following. As a whole, DFB semiconductor lasers are excellent light sources as a dynamic single mode laser with low chirp suitable for optical communications.

The oscillation frequency reflects the phase condition for the modes of a DFB structure. The required threshold gain g_{th} for the DFB laser modes strongly depends on the respective modes, yielding oscillation angular frequency ω_0. This behavior differs from a Fabry-Perot type laser, in which equal threshold gain is required for all the resonant modes. The threshold gain for a DFB laser is written by

$$g_{\text{th}} = a + a_{\text{m}} \tag{3.137}$$

where a is the loss in the cavity due to optical scattering and is already defined in Eq. (3.4). a_{m} is the loss due to the reflection in the DFB structure. In the case of a Fabry-Perot laser, the reflection loss is explicitly given as $a_{\text{m}} = \ln(1/r_1 r_2)/L$. The detailed analysis for a DFB laser oscillation can be treated by using coupled-wave equations. However, the reflection loss in the DFB laser is not given by an analytical form, and it is calculated numerically (McCall and Platzman 1985).

The threshold gain of a DFB laser depends on the coupling strength of the grating structure, namely $\kappa_{\text{DFB}}l$ (κ_{DFB} and l being the coupling coefficient and the cavity length). The factor $\kappa_{\text{DFB}}L$ is a function of the cavity loss $a_{\text{m}}l$. For a typical coupling strength around $\kappa_{\text{DFB}}l = 2$, one obtains the value of the cavity loss $a_m l = 1.4$ for a $\pi/2$ phase shifted DFB laser, which corresponds to the case of the equal facet reflectivity of $R = r^2 = 0.25$ in a Fabry-Perot laser. Therefore, the threshold gain of DFB is higher than that of a Fabry-Perot laser, but it is very close to the threshold of Fabry-Perot laser (Petermann 1988). Due to strong non-uniformity of the optical intensity for DFB lasers along the laser length, the spontaneous emission enhancement factor equivalent to the coefficient K_c defined in (3.132) is usually larger then that for Fabry-Perot type lasers with equal cavity loss $a_m l$. If the coupling strength is taken as $\kappa_{\mathrm{DFB}} l = 2$ (for a non-phase shifted DFB laser, the spontaneous emission factor $K_c = 1.94$ is obtained (Petermann 1988). It is noted that the internal reflectivities at the laser facets have complex values in a strict sense, so that the reflectivity from the external mirror must be also treated as a complex value.

4 Theory of Optical Feedback
in Semiconductor Lasers

A semiconductor laser with optical feedback is an excellent model for generating chaos in its output power and the system has proven to be very useful in practical applications. This chapter concerns the theoretical background for instability and chaos induced by optical feedback in edge-emitting narrow-stripe semiconductor lasers, such as Pabry-Perot lasers, multi-quantum well (MQW) lasers, and distributed feedback (DFB) lasers. Particular dynamics of feedback induced instability and chaos in semiconductor lasers are separately discussed in the following chapter. In this chapter, we focus on the theoretical treatment of optical feedback effects in semiconductor lasers. Lasers show the same or similar dynamics as far as rate equations are described by the same equations. We here assume single mode operations for semiconductor lasers. The dynamics for multimode cases will be discussed in Chap. 8.

4.1 Theory of Optical Feedback

4.1.1 Optical Feedback Effects and Classifications
of Optical Feedback Phenomena

The effects of optical feedback in semiconductor lasers have been studied from the beginning of their development (Risch and Voumard 1977, Voumard 1977, and Gavrielides et al. 1977). In early 1980, Lang and Kobayashi published a milestone paper on the effects of optical feedback in semiconductor lasers, which initiated an enormous research effort devoted to the study of the dynamics induced by optical feedback. Since then, bistability, instability, self-pulsations, and coherence collapse states have been observed in feedback induced irregular oscillations in semiconductor lasers (Miles et al. 1980, Glas et al. 1983, Lenstra et al. 1985, and Cho and Umeda 1986). In semiconductor lasers, self-optical-feedback effects are frequently used for the control of oscillation frequency, selection of mode, and suppression of side modes. Indeed, the linewidth of laser oscillations can be stabilized by a strong optical feedback and chirping of oscillation frequency can be compensated by optical feedback (Goldberg et al. 1982, Tamburrini et al. 1983, Agrawal 1984, and Lin et al. 1984). On the other hand, the semiconductor laser shows unstable oscillations for a certain range of optical feedback levels. The dynamics of semiconductor lasers induced by optical feedback in this range are very interesting not only from the viewpoint of fundamental physics but also for practical applications, since optical feedback effects appear everywhere in optical systems including optical communication systems, optical data storages, and optical measurements. These irregular oscillations are induced by the dynamics involved in laser systems known as chaos described by nonlinear delay differential equations.

Cleaved facets were frequently used as a laser resonator in semiconductor lasers in the early days. Therefore, the reflectivity of laser facets of semiconductor lasers is much lower than that of other lasers such as gas lasers. Since light in a cavity of a semiconductor laser is reflected perpendicularly to the laser facet, the internal amplitude reflectivity r_0 is given by

$$r_0 = \frac{\eta - 1}{\eta + 1} \tag{4.1}$$

where η is the refractive index of the laser material. For example, the refractive index η of the AlGaAs semiconductor laser without any optical coating is about 3.6 and the amplitude reflectivity of the facet is calculated to be $r_0 = 0.57$. The corresponding intensity reflectivity is $R_0 = r_0^2 = 0.32$. Only 32% of the light generated by the stimulated emission is fed back into the laser cavity and the other photons dissipate from the laser cavity (Zah et al. 1987). To make a high power laser, the laser facets are coated appropriately by dielectric films. Then, the rear facet of the cavity usually has a high reflectivity of more than 90% and the front fact has a low reflectivity of less than 10%. This is quite different from other lasers where both facets have high reflectivities close to 100%.

In spite of such a dissipative laser structure, laser oscillations are still possible in semiconductor lasers due to the high efficiency of the conversion from pump to light. This makes semiconductor lasers different from other lasers. Thus, light goes away from the cavity after a few reflections within the resonator. In other words, semiconductor lasers are easily affected by external light due to optical feedback or optical injection from a different laser. Indeed, the use of optical isolators is essential in optical communication systems to prevent unstable laser operations generated by feedback light from optical components and optical fiber facets. Optical feedback induces various instabilities in semiconductor lasers, for example, noises (actually they are chaotic fluctuations as discussed later) are much enhanced by optical feedback. In optical communications, the quality of signal transmissions has priority, so that optical isolators are used at the expensive of system sizes and costs to reduce feedback noises. On the other hand, optical information equipment, for example, optical data storages, in which serious problems by optical feedback are encountered for the performance of operations, the cost of the system is most important. In those systems, the reduction and control of noises (actually chaotic oscillations) are essential issues for good systems.

There are many parameters to characterize instabilities and chaos in semi-conductor lasers. Every parameter is important for describing the character-istics, however one important and most useful parameter to figure out the characteristics is the reflectivity of the external mirror. Tkach and Chraplyvy (1986) investigated the instabilities of semiconductor lasers with optical feed-back and categorized them into the following five regimes, depending on the feedback fraction.

Regime I.   Very small feedback (the feedback fraction of the amplitude is less than 0.01%) and small effects. The linewidth of the laser oscillation becomes broad or narrow, depending on the feedback fraction (Kikuchi and Okoshi 1982).

Regime II. Small, but not negligible effects (less than ~0.1% and the case for $C > 1$, where the C parameter is a measure of instability, dis-cussed in Sect. 4.2). Generation of the external modes gives rise to mode hopping among internal and external modes (Takach and Chraplyvy 1985).

Regime III. This is a narrow region around ~ 0.1% feedback. The mode hop-ping noise is suppressed and the laser may oscillate with a narrow linewidth (Takach and Chraplyvy 1986).

Regime IV. Moderate feedback (around 1%). The relaxation oscillation be-comes undamped and the laser linewidth is broadened greatly. The laser shows chaotic behavior and sometimes evolves into un-stable oscillations in a coherence collapse state. The noise level is enhanced greatly under this condition (Lenstra et al. 1985).

Regime V.   Strong feedback regime (higher than 10% feedback). The internal and external cavities behave like a single cavity and the laser oscillates in a single mode. The linewidth of the laser is narrowed greatly (Fleming and Mooradian 1981).

In the above regimes, the quoted fraction is that of the actual optical feedback level into the active layer and it does not mean the reflectivity of the exter-nal mirror, since there are scattering and absorption losses of light through optical components. Furthermore, a diffraction loss of light due to a collima-tor lens usually put in front of the laser facet is not negligible, because the thickness of the active layer is as small as 0.1 μm in ordinary edge-emitting lasers. Therefore, the fraction of optical feedback actually fed back into the active layer becomes one tenth or less than the intensity reflectivity of the external mirror. However, semiconductor lasers are sensitive enough to desta-bilize their output power by a small amount of optical feedback of less than 1% of the amplitude. Usually, an isolation of 40 dB is required in optical communication systems to avoid optical feedback effects.

The investigated dynamics of the above regimes were for a DFB laser with a wavelength of 1.55 μm, so that the feedback fraction corresponding to each dynamics scenario described above is not always true for other lasers. However, the dynamics for other lasers show similar trends for the variations of feedback fraction. The lasers show the same or similar dynamics as far as the rate equations are written in the same forms. As has already been discussed, the rate equations for edge-emitting semiconductor lasers, such as Fabry-Perot, MQW, and DFB lasers, are described by the same forms. Therefore, these lasers exhibit similar chaotic dynamics, though the parameters may have different values. We are very interested in regime IV that shows chaotic dynamics (Sacher et al. 1989, and Mørk et al. 1990 and 1992), though it is a small level of optical feedback (the intensity fraction of the feedback is only 0.01%). In actual applications of semiconductor lasers, this regime is important because, for example, the feedback fraction of laser amplitude in Compact Disk systems corresponds to regime IV (Gray et al. 1993). Thus, regime IV is important both for the studies of nonlinear dynamics and applications.

4.1.2 Theoretical Model

The static characteristics of semiconductor lasers with optical feedback can be theoretically investigated with the relations among the reflectivities of internal cavity and external reflector, the gain in a medium, and other static laser parameters. However, the dynamic characteristics must be described by time dependent equations of the systems. The equations for semiconductor lasers in the presence of optical feedback are easily obtained by modifying the rate equations for the solitary laser discussed in Chap. 3. The schematic model of a semiconductor laser with optical feedback is shown in Fig. 4.1. For a while, we consider that the external reflector is a conventional plain reflection mirror. The effects of other reflectors such as grating and phase-conjugate mirrors will be discussed later. Light from a laser is reflected from an external mirror and fed back into the laser cavity with time delay. We assume that the mirror is positioned within the coherence length of the laser. Also, the laser is assumed to be operated at a single mode, although this is not always true in actual situations. The laser sometimes oscillates at multimode under certain parameter conditions of optical feedback even when the laser oscillates at a single mode in the solitary condition. The external feedback effect is added to the equation for the complex field of (3.47) and the field

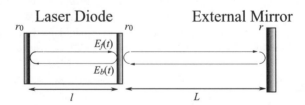

Fig. 4.1. Model of semiconductor laser with optical feedback

equation is written in the following form (Lang and Kobayashi 1980):

$$\frac{\mathrm{d}E(t)}{\mathrm{d}t} = \frac{1}{2}(1 - i\alpha)G_{\mathrm{n}}\{n(t) - n_{\mathrm{th}}\}E(t) + \frac{\kappa}{\tau_{\mathrm{in}}}E(t - \tau)\exp(i\omega_0\tau) \qquad (4.2)$$

where κ is the feedback coefficient due to the external optical feedback, $\tau = 2L/c$ is the round trip time of light within the external cavity, ω_0 is the angular oscillation frequency of the laser. The extra term has a delay time τ and the complex field is described by a delay differential equation and this is the origin of instability and chaotic dynamics in semiconductor lasers. The equation is known as the Lang-Kobayashi equation after their derivation.

The feedback coefficient κ can be calculated from considering the multiple-reflection effects of light in the external cavity. In Fig. 4.1, we consider the fields propagating forward and backward within the cavity and the extra term added to the laser field from the optical feedback in front of the facet of the resonator. For the steady-state oscillation in the presence of external feedback, the relation between the forward and backward traveling fields at the laser facet, $E_{\mathrm{f}}(t)\exp(-i\omega_0t)$ and $E_{\mathrm{b}}(t)\exp(-i\omega_0t)$, is given by (Lang and Kobayashi 1980)

$$E_{\mathrm{b}}(t) = r_0\left\{E_{\mathrm{f}}(t) + \frac{1 - r_0^2}{r_0}r\sum_{m=1}^{\infty}(-r_0r)^{m-1}E_{\mathrm{f}}(t - m\tau)\exp(im\omega_0\tau)\right\} \qquad (4.3)$$

where r is the amplitude reflectivity of the external mirror. In the parenthesis of the above equation, the first term is the ordinary field of reflection in the internal cavity and the second is the effect of the external optical feedback. The semiconductor laser is easily destabilized and shows chaotic dynamics even for a small level of feedback less than a few percent of the amplitude reflectivity. We here consider a steady-state solution as $E_{\mathrm{f}}(t - m\tau) \sim E_{\mathrm{f}}(t)$ and only assume a single reflection for a small external reflection r. Then, the feedback coefficient κ is written by (van Tartwijk and Lenstra 1995)

$$\kappa = (1 - r_0^2)\frac{r}{r_0} \qquad (4.4)$$

We assume that the reflectivities for the front and back facets of the laser cavity are the same at r_0. It is not always true for actual lasers, but the feedback rate for different reflectivities can be calculated straightforwardly. Recent semiconductor lasers have a low intensity reflectivity of the front facet as small as 10% or less by optical coating and, therefore, the lasers are much affected by optical feedback.

The time dependent phase in the presence of optical feedback plays an important role, since the phase couples with the other variables. For the carrier density, we need not consider the modification of the equation. Similar to the derivations for the rate equations in (3.57)–(3.59), we obtain the rate equations in the presence of optical feedback as follows (Ohtsubo 2002a):

$$\frac{dA(t)}{dt} = \frac{1}{2}G_n\{n(t) - n_{th}\}A(t) + \frac{\kappa}{\tau_{in}}A(t - \tau)\cos\theta(t) \qquad (4.5)$$

$$\frac{d\phi(t)}{dt} = \frac{1}{2}\alpha G_n\{n(t) - n_{th}\} - \frac{\kappa}{\tau_{in}}\frac{A(t - \tau)}{A(t)}\sin\theta(t) \qquad (4.6)$$

$$\frac{dn(t)}{dt} = \frac{J}{ed} - \frac{n(t)}{\tau_s} - G_n\{n(t) - n_0\}A^2(t) \qquad (4.7)$$

$$\theta(t) = \omega_0\tau + \phi(t) - \phi(t - \tau) \qquad (4.8)$$

We can investigate the dynamics of semiconductor lasers with optical feedback by numerically solving the above equations. In the rate equations for a solitary laser derived from (3.57)–(3.59), the phase does not affect the other variables and, therefore, a semiconductor laser is only described by the field amplitude and carrier density equations. However, we must consider the phase for a time development in the presence of optical feedback, since the phase is related to the other variables through the optical feedback term as shown in the above equations. Then, three coupled equations are essential for semiconductor lasers with optical feedback and they show unstable oscillations and chaotic dynamics in their output powers like three coupled equations in Lorenz systems. In the numerical simulations, the fourth-order Runge-Kutta algorithm is frequently used for the sake of the accuracy of the calculations (Press et al. 1986).

4.2 Linear Stability Analysis for Optical Feedback Systems

4.2.1 Linear Stability Analysis

When the fluctuation of the output power is small even in the presence of optical feedback in a semiconductor laser, we assume a steady-state solution for the average field. In this case, we obtain the steady-state solutions for $A(t) = A_s$, $\phi(t) = (\omega_s - \omega_{th})t$, and $n(t) = n_s$ from (4.5)–(4.7) as follows (Tromborg et al. 1984 and 1987, and Agrawal and Dutta 1993):

$$A_s^2 = \frac{J/ed - n_s/\tau_s}{G_n(n_s - n_0)} \qquad (4.9)$$

$$\omega_s - \omega_{th} = -\frac{\kappa}{\tau_{in}}\{\alpha\cos(\omega_s\tau) + \sin(\omega_s\tau)\} \qquad (4.10)$$

$$n_s = n_{th} - \frac{2\kappa}{\tau_{in}G_n}\cos(\omega_s\tau) \qquad (4.11)$$

For zero feedback coefficient $\kappa = 0$, the above equations reduce to the solutions for the solitary laser already given by (3.60)–(3.62). We rewrite (4.10) as

$$\omega_{\text{th}}\tau = \omega_{\text{s}}\tau + C\sin(\omega_{\text{s}}\tau + \tan^{-1}\alpha) \tag{4.12}$$

where the C parameter already mentioned in the regimes of the dynamics for the optical feedback level is defined by (van Tartwijk and Lenstra 1995)

$$C = \frac{\kappa\tau}{\tau_{\text{in}}}\sqrt{1+\alpha^2} \tag{4.13}$$

From (4.12), we can calculate modes for laser oscillations in the presence of optical feedback.

The relation in (4.12) can be written by

$$\Delta\omega_{\text{s}}\tau = -C\sin(\varphi_0 + \Delta\omega_{\text{s}}\tau) \tag{4.14}$$

where $\Delta\omega_{\text{s}} = \omega_{\text{s}} - \omega_{\text{th}}$ corresponds to the steady-state value of the phase difference $\phi(t) - \phi(t - \tau)$ and $\varphi_0 = \omega_{\text{th}}\tau + \tan^{-1}\alpha$. When $C < 1$, there is only one solution, as already discussed, which is dynamically stable and can be identified as a slightly changed solitary laser state. By increasing the C parameter value, the number of the mode solutions increases but is always an odd number. The curves in (C, φ_0) space in Fig. 4.2 that separate the regions of equal number of solutions are given by

$$\varphi_0 = (2m+1)\pi \pm \cos^{-1}\left(\frac{1}{C}\right) \mp C\sin\left\{\cos^{-1}\left(\frac{1}{C}\right)\right\} \tag{4.15}$$

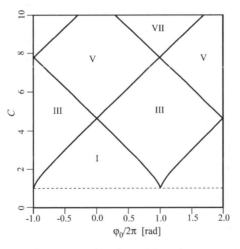

Fig. 4.2. Number of solutions for (4.12) in (C, φ_0) space. The roman numbers represent the number of solutions

where $C \geq 1$ and m is an integer number. This causes a pattern to arise in (C, φ_0) space, as shown in Fig. 4.2 where the roman numbers represent the number of solutions. For $C > 1$, multiple steady-state solutions appear.

The solutions in (4.12) are also graphically calculated as intersections of the curves $y = \omega_{th}\tau$ and $y = \omega_s\tau + C\sin(\omega_s\tau + \tan^{-1}\alpha)$ as shown in Fig. 4.3 (Favre 1987 and Murakami et al. 1997). When $C < 1$ (for a small optical feedback and a short external cavity), (4.12) has only a single solution and the laser exhibits stable oscillation. If $C > 1$, many possible modes for the laser oscillations (external modes and anti-modes) are generated with the relation among the internal laser modes and the excited external modes, and then the laser shows unstable operations. By adjusting the position of the external mirror (which is equivalent to appropriate selection of the round trip time τ) and setting $\omega_0\tau = \tan^{-1}\alpha$, the higher bound of the coefficient C for a single mode oscillation of the laser is easily obtained from Fig. 4.3 as $C \sim 3\pi/2$ (Pertermann 1988). Above this value $C > 3\pi/2$, many modes are excited and the laser becomes unstable. Complicated dynamics are observable in the output power, and the laser does not always exhibit unstable oscillations. Even for such unstable regimes, the laser may show stable oscillations. The details of the dynamics will be discussed in Sect. 5.2.

When the C parameter well exceeds the value of unity, many modes are excited in the laser output and the laser becomes unstable. Another representation for possible oscillation modes is frequently used in the phase space of the oscillation frequency and the carrier density. Figure 4.4 is such a representation for the parameter space in the $\Delta\omega_s\tau$ versus Δn_s plane. The relation is calculated from (4.10) and (4.11) by eliminating the sine and cosine functions and is given by (Henry and Kazarinov 1986)

$$\left(\Delta\omega_s\tau - \frac{\alpha\tau}{2}G_n\Delta n\right)^2 + \left(\frac{\tau}{2}G_n\Delta n\right)^2 = \left(\frac{\kappa\tau}{\tau_{in}}\right)^2 \tag{4.16}$$

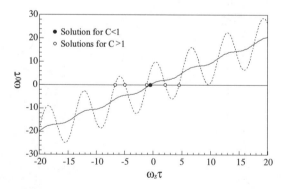

Fig. 4.3. Dependence of steady-state solutions for the phase on the parameter value C. *Solid* and *dashed lines* correspond to $C = 0.76$ and $C = 9.50$, respectively. The *black circle* denotes only one solution for $C < 1$ and *white circles* represent multiple solutions for $C > 1$

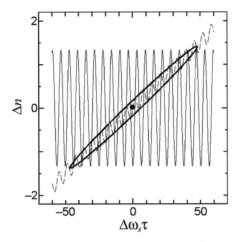

Fig. 4.4. Carrier density Δn change versus frequency change $\Delta \omega$ for the possible steady-states under external feedback. Crossing points of the solid and broken sinusoidal waves are the locations of the modes. Modes are on an ellipsoid. The *solid dot* at center is the solitary oscillation mode

where $\Delta \omega_s = \omega_s - \omega_{th}$ and $\Delta n_s = n_s - n_{th}$. The broken sinusoidal curve in the figure denotes the deviation from the steady-state of the oscillation angular frequency $\Delta \omega_s$ and the other sinusoidal curve represents that of the carrier density Δn_s. Crossing points of these two curves are the locations of possible oscillations and they are on the ellipsoid given by (4.16) (thick solid curve in the figure). Those in the lower half are the solutions for stable oscillations (external modes) and those in the upper half are unstable oscillations. Solutions for unstable oscillations are sometimes called anti-mode. The laser oscillates at one of the external modes and the maximum gain mode is the most probable mode for laser oscillation. However, when the laser oscillation is unstable due to external feedback, the mode hops around among the external modes and the anti-modes, thus the laser exhibits chaotic oscillations. One typical instability is the phenomenon known as low-frequency fluctuations (LFFs), in which the laser output power shows frequent irregular dropouts having the frequency from MHz to hundred MHz (Mørk et al. 1988 and Fischer et al. 1996a). The solid dot at the center of the ellipsoid in the figure is the solution for the laser oscillation in the solitary laser (solitary mode). The laser without optical feedback, of course, has no fluctuation in the sense of chaotic dynamics and oscillates only at this mode.

The stability and instability of laser oscillations in the presence of optical feedback are theoretically studied by the linear stability analysis for the steady-state solutions of the laser variables. In the same manner as a solitary laser, using the rate equations and taking the first order small infinities for the perturbations, the equations for the field δE, the phase $\delta \varphi$, and the carrier density δn are calculated as (Tromborg et al. 1984)

$$\frac{d\delta A(t)}{dt} = \frac{1}{2}G_n A_s \delta n(t) - \frac{\kappa}{\tau_{in}}\cos(\omega_s\tau)\{\delta A(t) - \delta A(t-\tau)\}$$
$$- \frac{\kappa}{\tau_{in}}A_s\sin(\omega_s\tau)\{\delta\phi(t) - \delta\phi(t-\tau)\} \qquad (4.17)$$

$$\frac{d\delta\phi(t)}{dt} = \frac{\alpha}{2}G_n A_s \delta n(t) + \frac{\kappa}{\tau_{in}}\frac{\sin(\omega_s\tau)}{A_s}\{\delta A(t) - \delta A(t-\tau)\}$$
$$- \frac{\kappa}{\tau_{in}}\cos(\omega_s\tau)\{\delta\phi(t) - \delta\phi(t-\tau)\} \qquad (4.18)$$

$$\frac{d\delta n(t)}{dt} = -2G_n A_s(n_s - n_0)\delta A(t) - \left(G_n A_s^2 + \frac{1}{\tau_s}\right)\delta n(t) \qquad (4.19)$$

Assuming that the perturbations take the forms of $\delta x(t) = \delta x\exp(\gamma t)$ ($x = A$, ϕ, and n), the characteristic equations for the condition having non-trivial solutions for the variables δA, $\delta\phi$, and δn are calculated from the following matrix:

$$\begin{pmatrix} \gamma + \frac{\kappa}{\tau_{in}}K\cos(\omega_s\tau) & \frac{\kappa}{\tau_{in}}K A_s\sin(\omega_s\tau) & -\frac{1}{2}G_n A_s \\ -\frac{\kappa}{\tau_{in}}\frac{K}{A_s}\sin(\omega_s\tau) & \gamma + \frac{\kappa}{\tau_{in}}K\cos(\omega_s\tau) & -\frac{1}{2}G_n \\ 2A_s G_n(n_s - n_0) & 0 & \gamma + G_n A_s^2 + \frac{1}{\tau_s} \end{pmatrix} = 0 \quad (4.20)$$

where $K = 1 - \exp(-\gamma\tau)$. The oscillation modes for the perturbations are calculated by solving the characteristic equation

$$D(\gamma) = \gamma^3 + 2\{-\Gamma_R + \frac{\kappa}{\tau_{in}}K\cos(\omega_s\tau)\}\gamma^2$$
$$+ \left\{\omega_R^2 - \frac{4\kappa K\Gamma_R}{\tau_{in}}\cos(\omega_s\tau) + (\frac{\kappa}{\tau_{in}}K)^2\right\}\gamma$$
$$- \frac{2\kappa K^2\Gamma_R}{\tau_{in}} + \frac{\kappa K\omega_R^2}{\tau_{in}}\{\cos(\omega_s\tau) - \alpha\sin(\omega_s\tau)\} = 0 \qquad (4.21)$$

In the above equation, Γ_R and ω_R are the previously defined parameters of the damping factor and angular frequency of the relaxation oscillation at the solitary mode.

We cannot calculate explicit forms of the solutions for (4.21), since the equation includes the exponential form for the variable γ and, then, the solutions are numerically calculated. The real part of the solution is related to the stability of the mode and the imaginary part of it represents the oscillation frequency of the mode as has already been discussed. When the real part (damping factor) takes a negative value, the mode is stable and the excited oscillation damps out for the time development with a frequency calculated from the imaginary part. On the other hand, the mode is unstable for a positive value of the real part and the laser shows either regular or irregular

oscillations with a typical frequency corresponding to the imaginary part. If
the level of optical feedback is low or the condition $\kappa\tau/\tau_{\text{in}} \ll 1$ is satisfied,
we can assume $\gamma\tau \ll 1$ and obtain the analytical form of the solution for γ.
Then, the real and imaginary parts, Γ'_{R} and ω'_{R}, of the solution are given by
(Agrawal and Dutta 1993)

$$\Gamma'_{\text{R}} = \Gamma_{\text{R}} ,\tag{4.22}$$

$$\omega'_{\text{R}} = \omega_{\text{R}}\sqrt{\frac{1 + (\kappa_c - \alpha\kappa_s)\tau/\tau_{\text{in}}}{(1 + \kappa_c\tau/\tau_{\text{in}})^2 + (\kappa_s\tau/\tau_{\text{in}})^2}}\tag{4.23}$$

where $\kappa_c = \kappa\cos(\omega_s\tau)$ and $\kappa_s = \kappa\sin(\omega_\sigma\tau)$.

The relaxation frequency in the presence of optical feedback shifts from
that of the solitary oscillation. Increase or decrease of the frequency shift depends on the signs of κ_c and κ_s, however it is usually enhanced at moderate
optical feedback and takes a larger value than that of the solitary oscillation. For a laser oscillation, the sign of the expression inside the square root
in (4.23) must be positive and we obtain the stability condition (Acket et al.
1984 and Lenstra et al. 1984)

$$1 + C\cos(\omega_s\tau + \tan^{-1}\alpha) > 0\tag{4.24}$$

Equation (4.24) denotes that the laser becomes unstable for $C > 1$ as expected, while it is stable for $C < 1$ even if optical feedback is present in
semiconductor lasers. We calculated oscillation modes for perturbations of
the steady-state values for the variables. The solutions obtained from such
characteristic equations are called linear modes, the name comes from the
linear stability analysis.

4.2.2 Linear Mode, and Stability and Instability in Semiconductor Lasers

For certain ranges of optical feedback level, the output of a semiconductor laser evolves from stable states to chaotic states via unstable periodic
oscillations. One or a few frequencies for the solutions derived from the characteristic equation in (4.21) are equal to or close to the typical frequency
corresponding to the response of the system. Periodic oscillations in chaotic
states are generally not harmonic oscillations, but they include an obscure
fundamental frequency and its higher harmonics. In quasi-periodic oscillations, frequency peaks become obscured due to irregular oscillations and
no clear spectral peak is observable in complete chaotic states, like white
noises. In semiconductor lasers with optical feedback, modes generated by
the internal and external cavities are mixed and the laser oscillates at one
or several modes. The other important frequency of laser oscillations besides these modes is the frequency of the relaxation oscillation. Since chaos
is a nonlinear phenomenon, many modes not only related to the internal and

external modes and the relaxation oscillation modes but also their sums and differences, and higher harmonics are excited (Cohen et al. 1988, Helms and Petermann 1990, and Levine et al. 1995). For a chaotic bifurcation, the laser first becomes unstable with a frequency close to the relaxation oscillation, which is called period-1 oscillation. Next, the external mode is also excited. After that, many modes are excited and the laser oscillates at quasi-periodic oscillation. Then, the laser evolves into chaotic oscillations with complicated and broadened frequency components.

Figure 4.5 is an example of numerically calculated linear modes from (4.21) (Murakami and Ohtsubo 1998). In the figure, the change of modes is shown for the increase of the amplitude reflectivity from the external mirror. The vertical axis is the damping factor (the real part of the solution of the characteristic equation) and the horizontal axis is the frequency of the oscillation (the imaginary part of the solution). For a negative value of the damping factor, the mode damps out for a time evolution even if it is once excited. The value of the real part is negative for the highest mode in the absence of optical feedback (around the frequency of 2.5 GHz in this case) and the laser never gets into unstable oscillations. The frequency corresponds to the relaxation oscillation at the solitary mode. With the increase of the external feedback, the real part of the highest mode at first exceeds zero and the laser becomes unstable with a frequency of the relaxation oscillation (period-1 oscillation). Under the condition in this figure, the C parameter at which the laser at first exhibits unstable oscillation has a value of $C = 2.8$ (calculated from the external reflection of 0.4%). The value is slightly less than

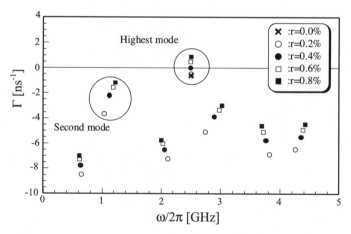

Fig. 4.5. Linear mode distributions at the external cavity length of $L = 10$ cm and the bias injection current of $J = 1.3J_{th}$. The highest mode corresponds to the relaxation oscillation and the second mode to the external cavity mode. With increasing external feedback, the real part of each mode increases and the laser becomes less stable

$C = 3\pi/2$, which was estimated in the previous section, but the assumption in the previous section is proved to be reasonable. With further increase of the external reflectivity, the damping rates for all the modes increase and the laser becomes less stable. The second highest mode in the presence of optical feedback corresponds to the external cavity mode and the frequency is about 1.5 GHz (approximately equal to the frequency calculated from the external cavity length of 10 cm). With this mode, the laser shows higher periodic oscillations and it evolves into chaotic oscillations through a bifurcation for the increase of optical feedback (Ye and Ohtsubo 1998). As we recognize from the figure, the external frequency does not have a fixed value, but shifts with the increase of the reflectivity, except for the relaxation oscillation mode, which always almost has a fixed value.

4.2.3 Gain Reduction Due to Optical Feedback

Though direct analyses for the rate equations are essential for investigating the dynamics of semiconductor lasers with optical feedback, the steady-state analysis is still useful and important to obtain parameter conditions for stable and unstable laser operations. Here, we calculate the gain in the presence of optical feedback under a steady-state condition. We assume the same reflectivities calculated in (4.4) (the internal reflectivity r_0 and the external reflectivity r), the effective reflectivity at the front facet taking into account the external mirror at steady-state is given by (Koelink et al. 1992, Osmundsen and Gade 1983, Kakiuchida and Ohsubo 1994, and Katagiri and Hara 1994)

$$r_{\text{eff}} = \frac{r_0 + r \exp(i\omega_0\tau)}{1 + r_0 r \exp(i\omega_0\tau)} \tag{4.25}$$

We investigate the gain of laser oscillation in the presence of optical feedback under the condition of a small external reflectivity $r \ll 1$. From the above equation, the effective reflectivity is written by

$$r_{\text{eff}} = |r_{\text{eff}}| \exp(i\phi_r) = r_0 + (1 - r_0^2) r \exp(i\omega_0\tau) \tag{4.26}$$

where ϕ_r is the phase of the effective reflectivity. Also, the effective reflectivity $\kappa = (1 - r_0^2)r/r_0$ defined in (4.4) is small enough. Then, the absolute value and phase of the effective reflectivity are approximated as

$$|r_{\text{eff}}| = r_0\{1 + \kappa \cos(\omega_0\tau)\} \tag{4.27}$$

$$\phi_r = \kappa \sin(\omega_0\tau) \tag{4.28}$$

The condition of laser oscillation under optical feedback is also given by the same equation as (3.3) and reads

$$r_0 r_{\text{eff}} \exp\{2ikL + (g - a)L\} = 1 \tag{4.29}$$

Therefore, the condition of the gain is

$$g_c = a + \frac{1}{L} \ln \left(\frac{1}{r_0 |r_{\text{eff}}|} \right) \tag{4.30}$$

The difference between the gains with and without optical feedback for a small value of κ is given by

$$g_c - g_{\text{th}} = -\frac{\kappa}{L} \cos(\omega_0 \tau) \tag{4.31}$$

The gain in the presence of optical feedback depends on the round trip time τ and it changes periodically for the variation of the external cavity length. The mode for the maximum gain is attained at $\omega_0 \tau = 2m\pi$ (m being an integer). As the gain varies depending on the optical feedback level, we can control or suppress the adjacent modes from the main oscillation mode by using the gain difference in accordance with (4.31) when the external mirror is positioned close to the laser facet. The difference of gains between successive modes in edge-emitting semiconductor lasers is as small as $0.1\,\text{cm}^{-1}$ and the condition $\kappa/L < 0.1$ is required for stable laser oscillations (Petermann 1988). For example, with an internal reflectivity of the laser facet of $r_0 = 0.32$ and the external cavity length of $L = 30\,\text{cm}$, we obtain the condition of the stable laser oscillation for the external amplitude reflectivity as $r < 2 \times 10^{-3}$. This value corresponds to that in regimes III to IV already discussed in Sect. 4.1.1 and is equal to the boundary of the regimes between the stable and unstable oscillations.

4.2.4 Linewidth in the Presence of Optical Feedback

The linewidth of laser oscillations in the presence of optical feedback is also calculated in the same manner as in Sect. 3.5.5. We consider small perturbations for the steady-state values of the variables in the presence of optical feedback and derive the linewidth from the power spectrum for the time derivative equations for the perturbations. The calculation is rather lengthy but straightforward, so that only the result is given here (Tormborg et al. 1987). Using the linewidth $\Delta\nu$ without optical feedback, the linewidth $\Delta\nu_{\text{ex}}$ in the presence of optical feedback is calculated as

$$\Delta\nu_{\text{ex}} = \frac{\Delta\nu}{F^2} \tag{4.32}$$

The coefficient $F = d\omega_{\text{th}}/d\omega_{\text{s}}$ for the reduction (or the broadening) of the spectral line width is calculated from (4.12) and given by

$$F = \frac{d\omega_{\text{th}}}{d\omega_{\text{s}}} = 1 + C \cos(\omega_{\text{s}}\tau + \tan^{-1}\alpha) \tag{4.33}$$

The minimum spectral linewidth is attained when the phase adjustment condition $\omega_s \tau = -\tan^{-1} \alpha$ is satisfied. Then, the spectral linewidth at the minimum condition is given by

$$\Delta \nu_{\text{ex}} = \frac{\Delta \nu}{(1 + C)^2} \tag{4.34}$$

On the other hand, the linewidth for the maximum gain condition at $\omega_s \tau = 2m\pi$ is calculated to be

$$\Delta \nu_{\text{ex}} = \frac{\Delta \nu}{\left(1 + \kappa \frac{\tau}{\tau_{\text{in}}}\right)^2} \tag{4.35}$$

The linewidth with optical feedback at the maximum gain condition is always less than the value of the solitary oscillation. These results hold for stable laser operations even when the laser is subjected to optical feedback. However, for optical feedback above a certain level, the laser does not oscillate at one of the modes but many modes are simultaneously excited or even drifting or wandering among the modes (external modes and anti-modes) occur. Such oscillations give rise to much noise (actually chaotic fluctuations) and even result in the collapse of coherence. These are the typical features in regimes III and IV in the preceding discussion. At this state, the linewidth of the laser is much broadened to as large as over GHz or more. However, the coherence of the laser recovers and the linewidth becomes narrow for a sufficiently strong optical feedback at regime V.

4.3 Feedback from a Grating Mirror

Other than conventional optical feedback reflectors, a grating mirror is frequently used to select the oscillation line in a semiconductor laser or stabilize the oscillation frequency. Grating optical feedback is originally applied for the stabilization of laser oscillations, however it sometimes induces instabilities in lasers. Before discussing instabilities, we present the theoretical background of grating feedback and stabilization of optical frequency. For a small feedback coefficient and also small detuning between the laser and grating frequencies, the complex field equation can be approximately written by a similar equation of conventional optical feedback as

$$\frac{dE(t)}{dt} = \frac{1}{2}(1 - i\alpha)G_n\{n(t) - n_{\text{th}}\}E(t) + \frac{\kappa_g}{\tau_{\text{in}}}E(t - \tau)\exp(-i\Delta\omega t + i\omega_g\tau) \tag{4.36}$$

where κ_g is the feedback coefficient from the grating mirror and $\Delta\omega$ is the angular frequency detuning given by $\Delta\omega = \omega_g - \omega_0$ (ω_g being the angular frequency of the grating feedback). However, in general, the optical feedback from the grating mirror is strong and the frequency detuning between the laser oscillation and the grating is as large as up to several nano-meters in wavelength. Therefore, the approximation in (4.36) is only valid within a small range of grating feedback. To treat the dynamics of grating feedback in a strict sense, the relation of the phase between the complex fields for the forward and backward propagations as a multiple-reflection model must be taken into account (Pittioni et al. 2001). Instead, we here consider the static model of the grating feedback and some stable and unstable features of the dynamics are presented.

The effective reflectivity of the static model in grating feedback including the laser facet and the grating mirror with multiple reflections is calculated as the same manner as the conventional mirror in (4.25) and is given by (Binder et al. 1990 and Genty et al. 2000)

$$r_{\text{eff}} = |r_{\text{eff}}| \exp(i\phi_r) = \frac{r_0 + r(\omega)\exp(i\omega\tau)}{1 + r_0 r(\omega)\exp(i\omega\tau)} \tag{4.37}$$

The above equation has the same form as (4.25), but the external reflectivity by the grating mirror is a function of the frequency $\nu = \omega/2\pi$. The condition of the laser oscillation can be written in the same form as (4.29) and the gain is also given by (4.30). We here apply the steady-state analysis and calculate the conditions for the phase and the gain. From the relation $\Delta(\eta\omega_g) = \omega_{\text{th}}\Delta\eta + (\omega_g - \omega_{\text{th}})\eta$, the change of the phase $\Delta\phi_d$ is written by

$$\Delta\phi_d = \frac{2L}{c}\{\omega_{\text{th}}\Delta\eta + (\omega_g - \omega_{\text{th}})\eta\} + \phi_r \tag{4.38}$$

where we put the angular frequency at the laser oscillation $\omega = \omega_g$. $\Delta\eta$ is expanded by the carrier density and the angular frequency as

$$\Delta\eta = \frac{\partial\eta}{\partial n}(n - n_{\text{th}}) + \frac{\partial\eta}{\partial\omega}(\omega_g - \omega_{\text{th}}) \tag{4.39}$$

Using the definition of the refractive index in (3.39), i.e., $\eta_c = \eta - i\eta'$, together with the equalities

$$\frac{\partial\eta}{\partial n} = \alpha\frac{\partial\eta'}{\partial n} = -\frac{\alpha c}{2\omega_{\text{th}}}\frac{\partial g}{\partial n} \tag{4.40}$$

the relation between the carrier density and the gain is written as

$$\frac{\partial\eta}{\partial n}(n - n_{\text{th}}) = -\frac{\alpha c}{2\omega_{\text{th}}}\frac{\partial g}{\partial n}(g_g - g_{\text{th}}) \tag{4.41}$$

where g_g is the gain in the presence of grating feedback. Substituting (4.39)–(4.41) into (4.38), the phase change reads

$$\Delta\phi_d = -\alpha(g - g_{\text{th}})L + \frac{2\eta_e L}{c}(\omega_g - \omega_{\text{th}}) + \phi_r \tag{4.42}$$

Putting $\Delta\phi_d = 0$ for a possible solution for the laser oscillation and using $\tau_{in} = 2\eta_e l/c$, we obtain

$$\omega_g - \omega_{th} = \frac{1}{\tau_{in}}\{\alpha(g_g - g_{th})L - \phi_r\} \tag{4.43}$$

Then, the reduction of gain in the presence of grating feedback is given by

$$g_g - g_{th} = \frac{1}{L}\ln\frac{1}{r_0|r_{eff}(\omega_g)|} \tag{4.44}$$

and the linewidth reduction factor is calculated as

$$F_g = \frac{d\omega_{th}}{d\omega_g} = 1 + \frac{1}{\tau_{in}}\frac{d\phi_r}{d\omega_g} - \frac{\alpha}{\tau_{in}}\frac{d}{d\omega_g}\{\ln\frac{1}{|r_{eff}(\omega_g)|}\} \tag{4.45}$$

The linewidth of a semiconductor laser with grating optical feedback is finally written by

$$\Delta\nu_g = \frac{\Delta\nu}{F_g^2} \tag{4.46}$$

where $\Delta\nu$ is again the linewidth of the solitary laser. When the laser beam has a Gaussian profile and a certain diffraction order is selected by the grating as a feedback light, the reflectivity is explicitly given by

$$r(\omega_g) = r_g \exp\left\{-\frac{(\omega_g - \omega_G)^2}{\Delta\omega_G^2}\right\} \tag{4.47}$$

where ω_G is the selected angular frequency of the grating, r_g is its reflectivity, and $\Delta\omega_G$ is the width of the grating resolution at that angular frequency defined by $\Delta\omega_G = c\tan\theta/w_0$ (θ is the incidence angle of light onto the grating and $2w_0$ is the diameter of the Gaussian beam). The linewidth of a semiconductor laser is narrowed by a grating feedback under stable oscillation. However, it is again noted that the laser becomes unstable even by a grating feedback for a certain range of the feedback strength, either for small or strong grating feedback.

4.4 Phase-Conjugate Feedback

A semiconductor laser is frequently used as a light source of phase-conjugate optics (Pochi 1993). Or a phase-conjugate mirror is positively used to return light exactly into the active region in a semiconductor laser, since the light reflected from the phase-conjugate mirror is automatically fed back into the laser cavity due to the generation of the conjugate wave without any additional optical components in the external optical path. The phase-conjugate feedback induces instabilities in the laser oscillation and the dynamics of the laser are not always the same as those from the ordinary feedback reflector. The typical time scale in semiconductor lasers with optical feedback is of the order of a nano-second, defined by the laser relaxation oscillation frequency. Therefore, typical effects of phase-conjugate feedback occur when the phase-conjugate mirrors respond as fast as this time scale. Such phase-conjugate mirrors are realized in quick-response Kerr media with large third-order susceptibility and also quick-response photorefractive mirrors of semiconductor materials (Agrawal and Klaus 1991, Agrawa and Gray 1992, van Tartwijk et al. 1992, Langley and Shore 1994, Gray et al. 1994, and Bochove 1997). On the other hand, the dynamics for slow-response photorefractive mirrors, where the response is much slower than the time variations of the laser dynamics, are the same as those for ordinary plain reflection mirrors. For a slow-response photorefractive crystal, for example a TiBaO$_3$ crystal, the laser light automatically returns into the laser cavity, however the mirror produces the same dynamics of optical feedback as an ordinary reflection mirror (Miltyeni et al. 1995, Liby and Statman 1996, and Murakami and Ohtsubo 1999). Only the spatial phase-conjugate characteristic is effective in such optical feedback. In either case of fast or slow response phase-conjugate mirrors, phase-conjugate feedback can be also applied to control the quality of oscillations for semiconductor lasers (Gray et al. 1995, Kurz and Mukai 1996, and Anderson et al. 1999).

Figure 4.6 shows an optical setup for generating a phase-conjugate wave by four-wave mixing from a phase-conjugate mirror. We assume that the phase-conjugate mirror responds much faster than the typical chaotic fluctuations of semiconductor lasers. The angular frequencies of the signal and pump beams at the phase-conjugate mirror are set to be ω_0 and ω_p, respectively, and the generated phase-conjugate wave has a frequency $\omega_c = 2\omega_p - \omega_0$. Therefore, we consider the angular frequency detuning $2\delta = 2(\omega_p - \omega_0)$ between the laser angular frequency and that of the feedback light. Thus, the equation of the complex field $E(t)$ for the semiconductor laser with phase-conjugate feedback is given by

$$
\begin{aligned}
\frac{\mathrm{d}E(t)}{\mathrm{d}t} = {} & \frac{1}{2}(1 - \mathrm{i}\alpha)G_\mathrm{n}\{n(t) - n_\mathrm{th}\}E(t) \\
& + \frac{\kappa}{\tau_\mathrm{in}}E^*(t - \tau)\exp\left\{-\mathrm{i}2\delta\left(t - \frac{\tau}{2}\right) + \mathrm{i}\phi_\mathrm{PCM}\right\}
\end{aligned}
\tag{4.48}
$$

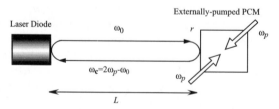

Fig. 4.6. Optical setup in a semiconductor laser with phase-conjugate optical feedback

where ϕ_{PCM} is the phase shift induced by the reflection at the phase-conjugate mirror. The final term in the above equation is the effect of phase-conjugate feedback. The rate equations for the field amplitude, the phase, and the carrier density are written by

$$\frac{\mathrm{d}A(t)}{\mathrm{d}t} = \frac{1}{2}G_{\mathrm{n}}\{n(t) - n_{\mathrm{th}}\}A(t) + \frac{\kappa}{\tau_{\mathrm{in}}}A(t - \tau)\cos\theta(t) \tag{4.49}$$

$$\frac{\mathrm{d}\phi(t)}{\mathrm{d}t} = \frac{1}{2}\alpha G_{\mathrm{n}}\{n(t) - n_{\mathrm{th}}\} - \frac{\kappa}{\tau_{\mathrm{in}}}\frac{A(t - \tau)}{A(t)}\sin\theta(t) \tag{4.50}$$

$$\frac{\mathrm{d}n(t)}{\mathrm{d}t} = \frac{J}{ed} - \frac{n(t)}{\tau_{\mathrm{s}}} - G_{\mathrm{n}}\{n(t) - n_0\}A^2(t) \tag{4.51}$$

$$\theta(t) = 2\delta\left(t - \frac{\tau}{2}\right) + \phi(t) + \phi(t - \tau) + \phi_{\mathrm{PCM}} \tag{4.52}$$

Equations (4.49)–(4.51) are in the same form as (4.5)–(4.7), however, (4.52) is different from (4.8) even for zero detuning ($\delta = 0$). This makes the laser dynamics of phase-conjugate feedback different from those of an ordinary optical feedback reflector.

A typical feature of the dynamics in phase-conjugate feedback is the phase locking phenomenon. The steady-state solutions for the field, the phase, and the carrier density at zero detuning $\delta = 0$ are given by

$$A_{\mathrm{s}}^2 = \frac{J/ed - n_{\mathrm{s}}/\tau_{\mathrm{s}}}{1/\tau_{\mathrm{p}} + G_{\mathrm{n}}(n_{\mathrm{s}} - n_{\mathrm{th}})} \tag{4.53}$$

$$\phi_{\mathrm{s}} = \frac{1}{2}\tan^{-1}(-\alpha) \tag{4.54}$$

$$n_{\mathrm{s}} = n_{\mathrm{th}} - \frac{2\kappa\cos(2\phi_{\mathrm{s}})}{G_{\mathrm{n}}} \tag{4.55}$$

Namely, the phase is locked to a certain value given by (4.54), while it changes depending on the time of the feedback loop in the conventional external reflector and it has multiple solutions for the laser oscillations. The laser for ordinary optical feedback is very sensitive to short variations of the external mirror compatible with optical wavelength. However, the phase of the laser with phase-conjugate feedback does not show any change for such a small variation of the external mirror. Here, we discussed the case when the phase-conjugate mirror responds immediately after the arrival of the signal beam. The laser dynamics of semiconductor lasers with a finite response time in a phase-conjugate mirror have also been discussed (DeTienne et al. 1997, and van der Graaf et al. 1998). For a finite response of a phase-conjugate mirror, the equation for the complex field is given by

$$\frac{\mathrm{d}E(t)}{\mathrm{d}t} = \frac{1}{2}(1 - i\alpha)G_{\mathrm{n}}\{n(t) - n_{\mathrm{th}}\}E(t) + \frac{\kappa}{\tau_{\mathrm{in}}}\exp\left\{-i2\delta\left(t - \frac{\tau}{2}\right)\right\}$$

$$\times \int_{-\infty}^{t} E^*(t' - \tau)\exp\left\{-(1 - i\delta t_{\mathrm{m}})\frac{(t - t')}{t_{\mathrm{m}}}\right\}\mathrm{d}t' \tag{4.56}$$

where t_m is the time it takes the light to penetrate the phase-conjugate mirror. Above, we assume a fast-response phase-conjugate mirror, but similar dynamics are obtained for a finite-response phase-conjugate mirror.

4.5 Incoherent Feedback and Polarization-Rotated Optical Feedback

4.5.1 Incoherent Feedback

Coherent optical feedback effects are important in applications of semiconductor lasers. For a long external cavity when the feedback light has an incoherent coupling with the original light in the laser cavity, the rate equations in (4.5)–(4.7) are still applicable for investigating the laser dynamics. Even in incoherent optical feedback, a laser becomes unstable and shows instability and chaos in its output. Besides incoherent feedback from a distant reflector, another incoherent system can be considered as shown in Fig. 4.7. In the figure, a laser operating at a single longitudinal mode is subjected to optical feedback from an external reflector. However, in this case, the returned light has crossed polarization of the original oscillation due to the polarization optics, such as a quarter waveplate placed into the path of the external feedback. The returned laser field does not interfere with the inner oscillation field, but it acts as the perturbation for carriers and has the coupling with them. Through this interaction, the laser shows instabilities.

The model is described by the following rate equations (Otsuka and Chen 1991):

$$\frac{\mathrm{d}S(t)}{\mathrm{d}t} = G_\mathrm{n}\{n(t) - n_\mathrm{th}\}S(t) \tag{4.57}$$

$$\frac{\mathrm{d}n(t)}{\mathrm{d}t} = \frac{J}{ed} - \frac{n(t)}{\tau_\mathrm{s}} - G_\mathrm{n}\{n(t) - n_0\}\{S(t) + \kappa' S(t - \tau)\} \tag{4.58}$$

where κ' is the feedback coefficient coupled with the carrier density and τ has the same definition as before (the round trip time of light in the external cavity). We do not have to consider the phase, since the phenomena come from

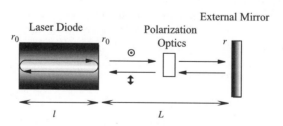

Fig. 4.7. Optical setup in a semiconductor laser with incoherent optical feedback

the incoherent origin. The rate equations are only written by two differential equations, however they are coupled with each other by the delay differential term. Thus, we can expect instabilities and chaos in semiconductor lasers. One of the typical features in incoherent optical feedback is sustained pulsations in the laser output. The gain saturation term discussed in Sect. 3.3.4 must be taken into account for such pulsations. In incoherent optical feedback in semiconductor lasers, we obtain not only irregular or chaotic pulsations in the laser output but also regular pulsings (such as period-1 oscillations) with high-speed oscillations as fast as pico-seconds (Otsuka and Chen 1991). Those regular high-speed pulsing oscillations are important for the application of light sources in high-speed optical communications.

4.5.2 Polarization-Rotated Optical Feedback

The effect is incoherent when the orthogonal-polarization feedback is small, as discussed above. We consider here the case of a large optical feedback of the crossed-polarization component, where the crossed-polarization component becomes the lasing mode. Figure 4.8 shows examples of single path systems with orthogonal-polarization optical feedback. Figure 4.8a is a ring-loop model for orthogonal-polarization optical feedback, by which we can avoid multiple-reflection scheme within optical feedback loop. The main oscillated TE mode(transverse electric mode) from a narrow-stripe edge-emitting laser

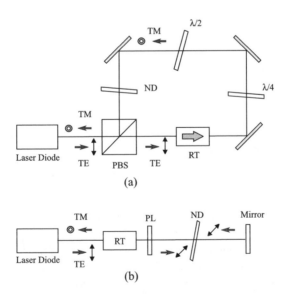

Fig. 4.8. Optical setups of orthogonal polarization feedback in semiconductor laser. **a** Ring-loop feedback system. PBS: polarization beam splitter, RT: Faraday rotator, $\lambda/4$: $\lambda/4$ waveplate, $\lambda/2$: $\lambda/2$ waveplate, ND: neutral density filter. **b** Single-pass feedback system. PL: polarizer

goes through a polarization beam splitter and is converted to a TM mode (transverse magnetic) by $\lambda/4$ and $\lambda/2$ waveplates. Figure 4.8b is another example of orthogonal-polarization feedback systems. The TE polarized beam enters a Faraday rotator (RT), whose input polarizer is removed, and the beam's polarization rotates 45°. The beam reflected by the feedback mirror is reinjected to the rotator, and this creates an orthogonal polarized beam to the laser oscillation mode (i.e. TM mode). In this configuration, the reflected vertical beam from the laser facet is once passed through the rotator, but it is blocked by the polarizer (PL). Thus, a single feedback loop is guaranteed in this setup. For both systems, the effect of orthogonal-polarization feedback can be described by the same rate equations. For a strong crossed-polarizing optical feedback (say, for example, 10 times larger than ordinary parallel-polarization optical feedback to induce chaotic oscillations), the TM oscillation merges in the laser output power besides the TE oscillation mode. In this situation, we can observe quite different dynamics compared with ordinary parallel-polarizing optical feedback and the detail of the dynamics will be discussed in Chap. 5 (Heil et al. 2003c).

For the crossed-polarization scheme with strong optical feedback, we must use a coherent model for the laser oscillations, since the both amplitudes of TE- and TM-modes are time dependent functions. Then the rate equations of crossed-polarization feedback system are written as

$$\frac{dA_{\text{TE}}(t)}{dt} = \frac{1}{2}G_{\text{n,TE}}\left\{n(t) - n_{\text{th,TE}}\right\}A_{\text{TE}}(t) \tag{4.59}$$

$$\frac{d\phi_{\text{TE}}(t)}{dt} = \frac{1}{2}\alpha G_{\text{n,TE}}\left\{n(t) - n_{\text{th,TE}}\right\} \tag{4.60}$$

$$\frac{dA_{\text{TM}}(t)}{dt} = \frac{1}{2}G_{\text{n,TM}}\left\{n(t) - n_{\text{th,TM}}\right\}A_{\text{TM}}(t) + \frac{\kappa}{\tau_{\text{in}}}A_{\text{TE}}(t-\tau)\cos\theta(t) \tag{4.61}$$

$$\frac{d\phi_{\text{TM}}(t)}{dt} = \frac{1}{2}\alpha G_{\text{n,TM}}\{n(t) - n_{\text{th,TM}}\} - \frac{\kappa}{\tau_{\text{in}}}\frac{A_{\text{TE}}(t-\tau)}{A_{\text{TM}}(t)}\sin\theta(t) \tag{4.62}$$

$$\frac{dn(t)}{dt} = \frac{J}{ed} - \frac{n(t)}{\tau_{\text{s}}} - \{n(t) - n_0\}\{G_{\text{n,TE}}A_{\text{TE}}^2(t) + G_{\text{n,TM}}A_{\text{TM}}^2(t)\} \tag{4.63}$$

$$\theta(t) = \omega_0\tau + \phi_{\text{TM}}(t) - \phi_{\text{TE}}(t-\tau) \tag{4.64}$$

where the subscripts TE and TM represent the variables and parameters for TE- and TM-modes. The gain G_{n} and the carrier density at threshold n_{th} have different values for the TE- and TM-modes in a strict sense. When an optical feedback is small, the terms for the TM-mode in (4.61) and (4.62) is adiabatically eliminated and we can put $A_{\text{TM}}^2(t) \propto A_{\text{TE}}^2(t-\tau)$. Then replacing equation (4.59) for the photon number, the relations of (4.55) and (4.56) hold. Crossed-polarization optical feedback plays an important role in VCSELs as will be discussed in Chap. 8. In VCSELs, typical polarization dynamics are observable even for a small amount of optical feedback with crossed-polarization.

4.6 Filtered Feedback

We have discussed several optical feedback schemes and formulated the equations for the models. We can consider systematic treatments for these models (Yousefi and Lenstra 1999, Lenstra et al. 2005 and Green and Krauskopf 2006). We here formulate the preceding optical feedback models. Also the formulation can be extended other feedback models such as opto-electronic feedback models, which will be discussed in Chap. 7. Through the introduction of systematic descriptions, we can give rise to good perspective for universal understanding of the dynamics in feedback phenomena in semiconductor lasers, i.e. coherent and incoherent optical feedback, phase-conjugate feedback, grating feedback, etc. Figure 4.9 shows the notation of the system for filtered feedback. Assuming that the laser field E and the feedback function given by an external device F are slowly time-dependent amplitudes, the filtered feedback system is written by

$$\frac{\mathrm{d}E(t)}{\mathrm{d}t} = \frac{1}{2}(1 - i\alpha)G_n \left\{n(t) - n_{\mathrm{th}}\right\} E(t) + \frac{\kappa_{\mathrm{feedback}}}{\tau_{\mathrm{in}}} F(t) \qquad (4.65)$$

We assume that the emitted laser field is $E(t)\mathrm{e}^{-i\omega_0 t} + c.c.$ and the feedback field $F(t)\mathrm{e}^{-i\omega_0 t} + c.c.$. For a linearly responding device, the function is given by

$$F(t) = \int_{-\infty}^{t} r(t' - t)E(t')\mathrm{d}t \qquad (4.66)$$

where $r(t)$ represents the response function of the external devices. It is noted that, in a case of phase-conjugate optical feedback, E in (4.66) must be replaced by E^*. The carrier density equation remains the same and given by

$$\frac{\mathrm{d}n(t)}{\mathrm{d}t} = \frac{J}{ed} - \frac{n(t)}{\tau_s} - G_n\{n(t) - n_0\}|E(t)|^2 \qquad (4.67)$$

For simplicity, the response function assumed to be given by a simple Lorentzian frequency filter. Indeed, the spectral form of the transfer function induced by optical feedback from a grating or Fabry-Perot filter can be

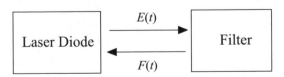

Fig. 4.9. Notation of filtered feedback

given by a Lorentzian shape as will be discussed in Chap. 5. From the Fourier transform relation, the time dependent response function is given by

$$r(t) = \Lambda \exp\left\{-\Lambda|t| - i(\omega_c - \omega_0)t\right\} \tag{4.68}$$

where ω_c is the central frequency of the Lorentz spectrum and Λ is the half-width at half-maximum (HWHM) of the spectrum. Under this assumption, one obtains the differential equation for the feedback as

$$\frac{dF(t)}{dt} = \Lambda E(t - \tau)\exp(i\omega_0\tau) - \{\Lambda + i(\omega_c - \omega_0)\}F(t) \tag{4.69}$$

In general, the response does not always have a Lorentzian spectral function in coherent optical feedback. However, a general response function can be expanded by a linear superposition of Lorentz functions and one can generally decompose the response function as a sum of exponential functions of the same type of the equation

$$r(t) = \Lambda \exp\left\{-\Lambda|t| - i(\omega_c - \omega_0)t\right\} \ .$$

From the above discussions, we can figure out general descriptions for the systems with filtered optical feedback. In the following, we will study the explicit forms of the feedback function for some limiting cases. In a conventional optical feedback without frequency filter (usual plane mirror feedback), Λ is assumed to be infinity. In this limit, the differential equation is simply reduced as

$$F(t) = E(t - \tau)\exp(i\omega_0\tau) \tag{4.70}$$

The expression, of course, is the same as the extra term added to the field equation of a semiconductor laser with optical feedback in (4.2). For a very narrow filter case, i.e. $\Lambda \to 0$, (one of such examples is optical injection from a different laser), the feedback function is easily calculated as

$$F(t) = E_{\mathrm{inj}}(t)\exp\{-i(\omega_m - \omega_0)t\} \tag{4.71}$$

Injection locking instability will be discussed in Chap. 6. The third example is optical feedback from a four-wave mixing phase-conjugate mirror with finite time response time and where the feedback field is detuned from the solitary laser, which was discussed in Sect. 4.4. In a four-wave mixing phase-conjugate optical feedback, the differential equation of the response function is modified as

$$\frac{dF(t)}{dt} = \Lambda E^*(t - \tau)\exp\left\{-2i\delta\left(t - \frac{\tau}{2}\right)\right\} - (\Lambda + i\delta)F(t) \tag{4.72}$$

where δ is the detuning of the angular frequency between the four-wave mixing pump beam ω_p and the reference frequency ω_0, i.e. $\delta = \omega_p - \omega_0$. A system with opto-electronic feedback is also written by the same feedback function as discussed here, and the dynamics of such systems will be discussed in Chap. 7.

5 Dynamics of Semiconductor Lasers with Optical Feedback

Optical feedback in semiconductor lasers gives rise to rich varieties of dynamics and the effects have been extensively studied for the past two decades. The theoretical background was discussed in the preceding chapter. In this chapter, substantial feedback effects and chaotic dynamics in semiconductor lasers are presented and theoretical and experimental results are given. As fundamental characteristics, feedback induced chaos is investigated for variations of feedback strength and position of the external cavity. Coherent and incoherent feedback effects are also taken into account in the dynamics. The external feedback mirror to a semiconductor laser may not be always a simple reflector (conventional plain mirror) but may be a grating or phase-conjugate mirror. Instabilities are also induced by such reflectors and the dynamics induced by grating and phase-conjugate mirrors are presented.

5.1 Optical Feedback from a Conventional Reflector

5.1.1 Optical Feedback Effects

In this section, we discuss the dynamics of chaos in semiconductor lasers with optical feedback and show various routes to chaos for parameter variations. Figure 5.1 shows experimental examples of chaotic oscillations in a MQW laser with optical feedback. Figure 5.1a shows the laser output for a negligible small optical feedback. The excited relaxation oscillation smoothly decays out after the laser is switched on and we can see only statistical noise induced by spontaneous emissions (in actual fact, the detector noise is included in the waveform). With the increase of optical feedback, the laser shows periodic oscillation in Fig. 5.1b. The scheme corresponds to regimes III to IV discussed in Sect. 4.1.1. The relaxation oscillation frequency of the laser is about 3 GHz, but higher oscillations above GHz are not observable due to the slow response of the oscilloscope in this experiment. Therefore, the observed periodic oscillation of the peak frequency of 451 MHz corresponds to the excited external mode. The external feedback fraction is about 1% in this case. With a further increase of optical feedback, an irregular oscillation of the laser output power is shown in Fig. 5.1c. The oscillation corresponds to

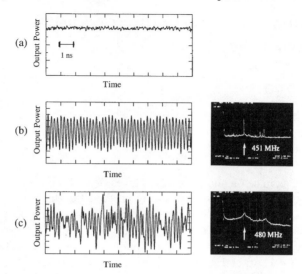

Fig. 5.1. Experimental time series and rf spectra at an injection current of $J = 1.5J_{th}$ and an external cavity length of $L = 30\,\text{cm}$. **a** Negligible small feedback, **b** period-1 oscillation with small feedback, and **c** chaotic oscillations with strong feedback

regime IV and the coherence of the laser is almost destroyed. Though the fundamental spectral peak of 480 MHz is still visible, the spectrum spreads out, which is a typical feature of quasi-chaotic or weak chaotic oscillations. For optical feedback above this level, we could see no clear peaks in the spectrum and the laser exhibits fully chaotic oscillations.

5.1.2 Potential Model in Feedback Induced Instability

In Subsect. 4.2.4, we derived the linewidth of a semiconductor laser with weak optical feedback (regimes I $\sim$ II). The linewidth is reduced according to (4.32) as far as the optical feedback is small enough. When the feedback is in-phase in regime II, the laser is well stabilized and the linewidth is much reduced. But, for an out-of-phase feedback, the linewidth is not reduced any more and the laser shows hopping between the two oscillation modes arising from the external feedback, according to the relation $\omega_s - \omega_0 = -\kappa \left\{ \sin(\omega_s \tau) + \alpha \cos(\omega_s \tau) \right\}$. Figure 5.2 shows an example of optical spectra in regime II (Tkach and Chraplyvy 1986). By a small in-phase optical feedback, the linewidth of the laser is greatly reduced as seen from Fig. 5.2a and b. On the other hand, the laser shows two oscillation modes for a small out-of-phase optical feedback. The separation of the two peaks corresponds to the external cavity frequency.

Mode hopping between two external modes can be explained by a potential model for the optical feedback system (Lenstra 1991 and van Tartwijk

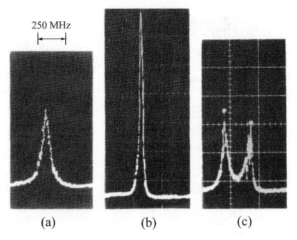

250 MHz

(a)   (b)   (c)

Fig. 5.2. Optical spectra of a laser with week external feedback in regime II. **a** No feedback, **b** −62 dB in-phase feedback, and **c** −62 dB out-of-phase feedback (after Tkach RW, Chraplyvy AR (1986); © 1986 IEEE)

and Lenstra 1994). In their model, the hopping is induced by noises and the oscillation is kicked out from one of the wells among the possible external modes and falls in another state. To derive the potential model to explain the mode hopping, let us consider the difference between the present and the delayed phases $\Delta\phi(t) = \phi(t) - \phi(t-\tau)$. Assuming that the delay is very small and expanding the delayed phase by the delay time τ, the difference can be expressed as

$$\Delta\phi(t) = \phi(t) - \phi(t-\tau)$$
$$\approx \phi(t) - \left\{\phi(t) - \tau\frac{d\phi(t)}{dt} + \frac{1}{2}\tau^2\frac{d^2\phi(t)}{dt^2}\right\} \approx \tau\frac{d\phi(t)}{dt} - \frac{1}{2}\tau\frac{d\Delta\phi(t)}{dt}$$
$$(5.1)$$

Here we used the relation $d^2\phi(t)/dt^2 = \{dt\phi(t)/dt - d\phi(t-\tau)/dt\}/\tau = d\Delta\phi(t)/dt/\tau$. Assuming the steady-state solutions except for the phase and taking into consideration (4.6) and (5.1), one obtains the relation

$$\frac{d\phi(t)}{dt} = \frac{1}{\tau}\Delta\phi(t) + \frac{1}{2}\frac{d\Delta\phi(t)}{dt} = -\frac{\kappa}{\tau_{\text{in}}}\sqrt{1+\alpha^2}\sin\{\theta_0 + \Delta\phi(t)\} + F_\phi(t) \quad (5.2)$$

where F_ϕ is the Langevin noise function for the phase and $\theta_0 = \omega_s\tau + \tan^{-1}\alpha$. Then the equation of the potential function of the phase difference can be written by

$$\frac{d\Delta\phi(t)}{dt} = -\frac{2}{\tau}\Delta\phi(t) - 2\frac{\kappa}{\tau_{\text{in}}}\sqrt{1+\alpha^2}\sin\{\theta_0 + \Delta\phi(t)\} + 2F_\phi(t)$$
$$= -\frac{1}{\tau}\frac{d}{d\Delta\phi}\left[\Delta\phi^2(t) - 2C\cos\{\theta_0 + \Delta\phi(t)\}\right] + 2F_\phi(t)$$
$$(5.3)$$

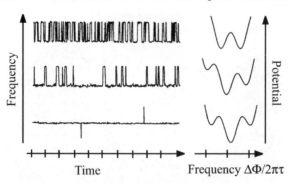

Fig. 5.3. Variations of frequency calculated from the potential model under the condition of $J = 1.84J_{\text{th}}$ and $L = 60$ cm. *Left*: time series of frequency, *right*: potential. The optical feedback is -45.7 dB. *Lower trace* is for a relative phase of $\theta_0 = 0$ (in-phase), *middle trace* $\theta_0 = 1.4$, and *upper trace* $\theta_0 = \pi$ (out-of-phase) (after Mørk J, Semkow M, Tromborg B (1990); © 1990 IEE)

Considering the term $U(\Delta\phi(t)) = \Delta\phi^2(t) - 2C \cos\{\theta_0 + \Delta\phi(t)\}$ as a potential function and interpreting that a particle moves with the coordinate $\Delta\phi$ in the potential, one obtains the equation for the time development of the phase difference in the potential as

$$\frac{\mathrm{d}\Delta\phi(t)}{\mathrm{d}t} = -\frac{1}{\tau}\frac{\mathrm{d}}{\mathrm{d}\Delta\phi}U(\Delta\phi(t)) + 2F_\phi(t) \tag{5.4}$$

From this equation, the statistical behaviors of mode hopping can be simulated. Figure 5.3 is an example of numerical simulations based on the potential model for a small optical feedback of -45.7 dB (Mørk et al. 1990b). The right figure shows the plots of the potential for different phase values of θ_0. For the in-phase case, the state is trapped to a deep potential well, so that the laser stays stable and oscillates almost at a single frequency. Thus, the laser has a narrow oscillation linewidth. On the other hand, two states compete with each other for the out-of-phase case and the probability for the dwelling time in each state becomes the same. Thus the laser frequency frequently switches between the two states due to noise. A mode hop to a neighboring mode occurs when the noise takes the phase delay beyond one of the potential barriers confining the mode. The barriers for the central mode in the lower potential are higher than the barrier separating the two modes in the upper potential. Since higher barriers are harder to pass, this explains the large difference in mode hopping rates.

5.1.3 Optical Spectrum in Stable and Unstable Feedback Regimes

For a feedback parameter C less than unity at regime I, only one solution $\omega\tau$ exists for the phase condition, yielding a linewidth narrowing or broadening depending on the phase $\omega_s \tau$. For the condition $C > 1$ at regime II, multiple solutions exist for the phase condition and it turns out that the emission frequency locks to the solution with the lowest phase noise. At this feedback, mode hopping may occur between modes with a similar amount of phase noise. Especially, for a phase change $\theta_0 = \pi$, two solutions exist with just the same linewidth, yielding strong mode hopping between these two modes as already discussed in the previous subsection. With increasing feedback level at regime III, the frequency splitting of the hopping modes converges to the frequency separation between the external cavity modes and the feedback phase $\omega_s \tau$ for these modes converges the phase $\theta_0 = 2m\pi$. Therefore, at this range of feedback, the laser will lock more and more to the feedback phase adjusted for minimum linewidth and mode hopping disappears. A certain saturation in linewidth reduction is obtained for a larger feedback level above $-50\,\mathrm{dB}$.

With further increase of the feedback in regime IV, the frequency fluctuations dramatically increase, yielding a tremendous linewidth broadening, which is characteristic for the coherence collapse regime. Figure 5.4 shows experimentally obtained optical spectra at feedback regimes III $\sim$ IV (Tkach and Chraplyvy 1986). The laser first destabilized with the relaxation oscillation for the increase in the feedback fraction, as shown in Fig. 5.4a. Then the laser evolves into quasi-periodic oscillations with several spectral peaks in Fig. 5.4b. Finally, the laser shows chaotic oscillation and the linewidth of the spectrum is greatly broadened by the optical feedback. In a sense of laser oscillations, the laser is coherence collapsed and the linewidth is as much as $100\,\mathrm{GHz}$ in Fig. 5.4c. However, the laser may be still single in a sense of longitudinal laser oscillation observed such as an optical spectrum analyzer with

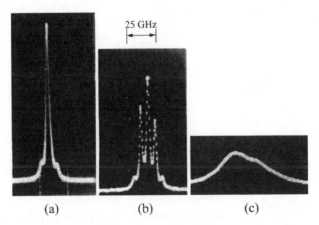

Fig. 5.4. Unstable optical spectra in regimes III $\sim$ IV. **a** Periodic state with relaxation oscillation at optical feedback of approximately $-40\,\mathrm{dB}$, **b** quasi-periodic oscillations at approximately $-30\,\mathrm{dB}$, and chaotic state at $-20\,\mathrm{dB}$ (after Tkach RW, Chraplyvy AR (1986); © 1986 IEEE)

a resolution of nanometers. The range of each regime strongly depends on the condition of the laser operations, such as laser device structure, bias injection current, and feedback length. However, each regime is clearly identified under the different conditions of the laser operations.

In a summary, Fig. 5.5 shows the experimental and theoretical results for spectral narrowing and broadening in the presence of optical feedback in semiconductor lasers (Schunk and Petermann 1988). Spectral linewidth is plotted as a function of feedback fraction. The experiments were conducted for various phase values of $\omega_0\tau$. The roman numbers denote the feedback regimes discussed in Sect. 4.1. The dashed line is the linewidth of the solitary

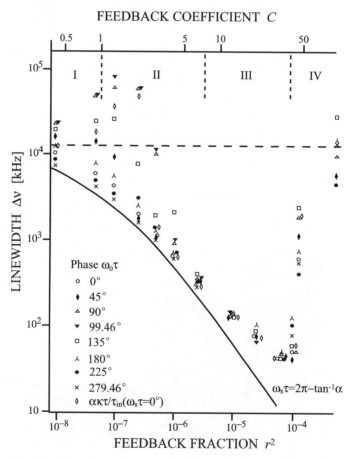

Fig. 5.5. Spectral linewidth versus feedback fraction at an output power of 5 mW and external cavity length of 50 cm for various $\omega_0\tau$. The roman numbers denote the feedback regimes. *Dashed line*: linewidth of the solitary laser, *solid curve*: linewidth of the external cavity laser according to (4.32) at an in-phase optical feedback (after Schunk N, Petermann K (1988); © 1988 IEEE)

laser. The solid curve is the expected minimum linewidth calculated from the theoretical equation in (4.32) under the assumption of an in-phase condition. The theory well agrees with the experiments at regimes I and II. However, it deviates from the experiments in regime III and complete spectral broadening is observed in regime IV, where the theoretical equation (4.32) cannot be applied. For moderate optical feedback in regimes III and IV, it is difficult to obtain analytical forms to describe the dynamics of semiconductor lasers with optical feedback. In usual, the laser has a rather broad spectral linewidth of several tens of MHz at solitary oscillation. By optical feedback, the linewidth is greatly reduced to 100 kHz at regime III (narrowing of 1/1000 of the original spectral linewidth). This situation is easily understood from analysis for the relative intensity noise (RIN) of the laser output. The RIN abruptly increases at the external feedback level of 10^{-4}. The RIN here is obtained as an average over the frequency range of $5 \sim 500\,\mathrm{MHz}$ (Schunk and Petermann 1988). In a strong optical feedback regime V, the linewidth is again strongly narrowed and the spectral narrowing by strong optical feedback will be discussed in Chap. 10.

5.1.4 Chaos in Semiconductor Lasers with Optical Feedback

Next, we numerically investigate routes to chaos for the optical feedback level. Here, we used parameter values of AlGaAs MQW semiconductor lasers of a Channeled Substrate Planer (CSP) type as shown in Table 5.1 (Liu et al. 1995). Figure 5.6 shows the result of dynamic behaviors of the laser output power at a fixed bias injection current and an external mirror position for the variations of the feedback level. The Langevin noises are excluded in the calculations in order to see the pure dynamics produced by the nonlinear equations. The left row in the figure is time series, the middle row is its attractor, and the right row is the power spectrum. For a small optical feedback, the laser output power is constant. For the external feedback level of 0.5%

Table 5.1. Some parameter values for an edge-emitting narrow-stripe semiconductor laser (GaAs-GaAlAs780nm wavelength)

Symbol	Parameter	Value
G_n	gain coefficient	$7.00 \times 10^{-13}\,\mathrm{m^3\,s^{-1}}$
α	linewidth enhancement factor	3.00
r_0	facet reflectivity	0.556
n_{th}	carrier density at threshold	$2.02 \times 10^{24}\,\mathrm{m^{-3}}$
n_0	carrier density at transparency	$1.40 \times 10^{24}\,\mathrm{m^{-3}}$
τ_s	lifetime of carrier	$2.04\,\mathrm{ns}$
τ_{ph}	lifetime of photon	$1.93\,\mathrm{ps}$
τ_{in}	round trip time in laser cavity	$8.00\,\mathrm{ps}$
V	volume of active region	$1.2 \times 10^{-16}\,\mathrm{m^3}$
ε	gain saturation coefficient	$8.4 \times 10^3\,\mathrm{s^{-1}}$

$(2.5 \times 10^{-5}$ in intensity) shown in Fig. 5.6a, the laser becomes unstable and exhibits a period-1 oscillation. The main frequency of the oscillation is 2.53 GHz and it is very close to the relaxation oscillation frequency of 2.50 GHz at the solitary mode. When the feedback level is raised at 1.0%, a period-2 oscillation appears as shown in Fig. 5.6b. Figure 5.6c shows a chaotic oscillation at the feedback level of 2.0%. When the laser output power shows periodic oscillations, we can see clear spectral peaks, however, at the chaotic state, clear spectral peaks are not observable but the optical spectrum is broadened around the relaxation oscillation frequency.

A chaotic attractor is a trajectory in the phase space of chaotic variables and is frequently used of the analysis of chaotic oscillations (see Appendix A.2). Since the laser output power at a stable oscillation is constant, the attractor (not shown here) is a fixed point in the phase space of the output power and the carrier density. A period-1 signal, as is the case in Fig. 5.6a, is a closed loop. The attractor of a period-2 oscillation is a double-loop as shown in Fig. 5.6b. However, the chaotic attractor behaves in a rather different way from fixed state or periodic oscillations. At chaotic oscillations, the state goes around points within the closed compact space in the attractor, however, it never visits the same point in the space. The trajectory crosses in the attractor in Fig. 5.6c, since it is a projection onto only two-dimensional space. In actual fact, the chaotic trajectory goes around in a multi-dimensional space and never crosses in such a space (Mørk et al. 1990, Ye et al. 1993, and Li et al.

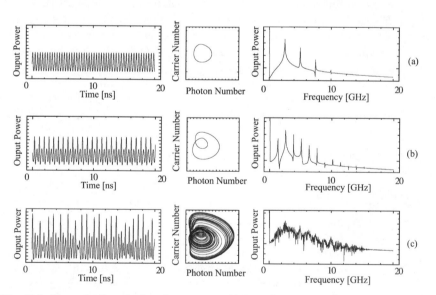

Fig. 5.6. Numerically calculated time series (*left row*), attractors (*middle row*), and power spectra (*right row*) for different feedback ratios at a bias injection current of $J = 1.3 J_{\text{th}}$ and an external cavity length of $L = 3$ cm. Feedback fractions of the amplitude are **a** $r = 0.005$, **b** $r = 0.01$, and **c** $r = 0.02$

1993). A chaotic attractor is quite different from other periodic oscillations and looks very strange. Therefore, it is sometimes called a strange attractor (Appendix A.2).

5.1.5 Chaotic Bifurcations

The plot of a bifurcation diagram is used to investigate chaotic evolutions for the change of a certain parameter (see Appendix A.1). A bifurcation diagram is obtained from a time series by sampling and plotting local peaks and valleys of the waveform for a parameter change. Figure 5.7 shows such an instance. The vertical axis is the local peaks and valleys of the waveform. The horizontal axis is the parameter of the optical feedback level. In Fig. 5.7a the laser is stable for an external feedback of less than 0.35%. The state is called a fixed point. Above this value, a relaxation oscillation appears in the laser output and the diagram has two points corresponding to the peaks and valleys of the period-1 oscillation. When the feedback level exceeds the value of 0.94%, a period-2 oscillation starts and the output has four states of peaks and valleys. For a further increase of the feedback, the laser evolves into quasi-periodic states and finally chaotic oscillations over the feedback level of 1.36%. The chaotic laser oscillates at the mixed frequencies

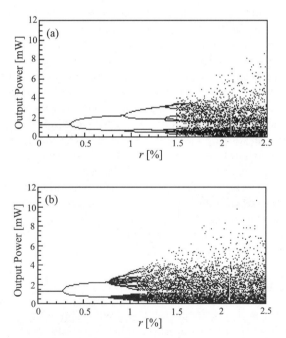

Fig. 5.7. Calculated chaotic bifurcation diagrams at a bias injection current of $J = 1.3J_{\text{th}}$. The external cavity length of **a** $L = 9\,\text{cm}$ (period-doubling bifurcation) and **b** $L = 15\,\text{cm}$ (quasi-periodic bifurcation)

of the internal and external modes and the relaxation oscillation mode. The evolution such as shown in Fig. 5.7a is called a period-doubling bifurcation or a Hopf bifurcation (the Hopf bifurcation actually has a rigid definition for chaotic evolutions and the example shown here may not be an exact Hopf bifurcation). A test for Hopf bifurcation is frequently used to check chaotic routes and evolutions of the output in nonlinear systems for variations of chaotic parameters (Appendix A.1).

Hopf bifurcation is not the only chaotic evolution but various routes to chaos exist in nonlinear systems. Figure 5.7b is another example. The external cavity length is different from that in Fig. 5.7a, but the other parameters are the same. In this figure, the fixed point evolves into period-1 oscillations, however the laser becomes quasi-periodic and chaotic states occur immediately after period-1 oscillations for the increase of the feedback parameter. The oscillation is called a quasi-periodic bifurcation to distinguish it from a Hopf bifurcation. Another route is an intermittent route to chaos, which is known as low-frequency fluctuations in semiconductor lasers (Mørk et al. 1988 and Fischer et al. 1996a). Chaotic bifurcations highly depend on chaos parameters in nonlinear systems (Helms and Petermann 1990, Ritter and Haug 1993a, b, and Levine et al. 1995).

We ignored Langevin noises in the numerical calculations to investigate the pure dynamics involved in the nonlinear system. In the presence of noises, chaotic dynamics are barely affected as long as the effect of spontaneous emission is small. In such a case, the maxima and minima of the output in the fixed and periodic states in the bifurcation have finite widths. Thus, the overall features of chaotic dynamics are unchanged by noises. However, spontaneous emission of light plays a crucial role for the dynamics when the photon number in the cavity is small (Yu 1999). In actual experimental situations, it is not easy to obtain a bifurcation diagram for a waveform with high frequency fluctuations such as in a semiconductor laser, since we require a very high-speed digitizing oscilloscope to fully reconstruct the diagram from the time series. One of the characteristic frequencies of chaotic oscillations in semiconductor lasers with optical feedback is the relaxation oscillation frequency. It usually ranges form several GHz to ten GHz depending on the device parameters and the driving condition. Instead, to analyze chaotic oscillations (oscillations for the relaxation and external-cavity modes) in semiconductor lasers with optical feedback, a rf spectrum analyzer or a Faby-Perot spectrum analyzer is frequently used in experiments.

5.1.6 Dynamics for Injection Current Variations

The laser output power of semiconductor lasers at solitary oscillations is linearly proportional to the bias injection current. A typical feature of optical feedback in semiconductor lasers is the threshold reduction, which is related to the gain reduction as discussed in Sect. 4.2.3 (Hegarty et al. 1998). Figure 5.8 shows an example of L-I characteristics experimentally obtained in

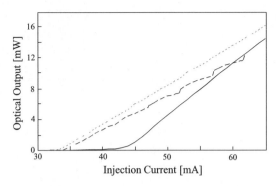

Fig. 5.8. Experimentally obtained light-injection current (L-I) characteristics of semiconductor lasers with optical feedback. *Solid line:* solitary oscillation, *dotted line:* external cavity length of $L = 15\,\mathrm{cm}$, *broken line:* external cavity length of $L = 150\,\mathrm{cm}$

a semiconductor laser with and without optical feedback. The solid curve is the L-I characteristics for the solitary oscillation and the other curves are with optical feedback. The reductions of the threshold current are about 30% in this case. The angular frequency of the laser oscillation at solitary oscillation is a function of the bias injection current and is written as (Petermann 1988)

$$\omega_0 = \omega_{\mathrm{c}} - \frac{\partial \omega_0}{\partial J} J \qquad (5.5)$$

where ω_{c} is the offset angular frequency and $\partial \omega_0/\partial J$ is the conversion coefficient of the injection current to the angular frequency. Due to the change of the frequency for the increase or decrease of the injection current, the successive external modes are sequentially selected. Then, mode hops occur at certain bias injection currents. This induces instabilities in semiconductor lasers and chaotic oscillations in the laser output power are observed in-between the mode jumps.

Figure 5.9a shows an example of output power jumps numerically calculated from the rate equations for the increase of the injection current. The conditions of the calculations correspond to regime III discussed in Sect. 4.1.1. Figure 5.9b is the bifurcation diagram for Fig. 5.9a. A chaotic scenario between successive jumps is observed. At the position after a mode jump, the laser oscillates with period-1, the laser becomes unstable with the increase of the injection current, and, finally, it evolves into chaotic states. The frequency of the period-1 oscillation is almost equal to the relaxation oscillation frequency. The averaged output power is proportional to the injection current, however periodic power jumps are observed in the output power with a period equal to the increment of the injection current corresponding to the frequency of the external mode (Fukuchi et al. 1999). The conversion coefficient $\partial \nu_0/\partial J$ is of the order of GHz/mA in ordinary edge-emitting semiconductor lasers (Petermann 1988). The jump of the output

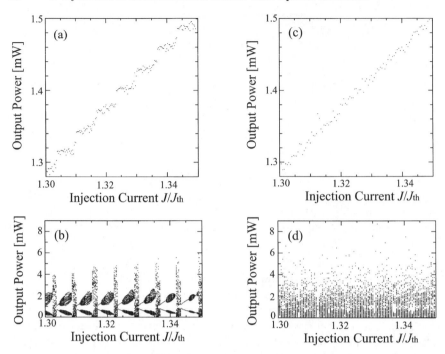

Fig. 5.9. Numerically calculated laser output powers and chaotic bifurcations for injection current at the external cavity length of $L = 15$ cm. **a** L-I characteristic at an external reflectivity of $r = 0.01$. **b** Chaotic bifurcation diagram corresponding to **a**. **c** L-I characteristic at an external reflectivity of $r = 0.02$. **d** Chaotic bifurcation diagram corresponding to **c**. **c** and **d** are coherence collapse states

power originates from the external mode alternation, namely, one external mode switches to the next at the jump position. Therefore, for example, a mode hop occurs about every 1 mA for the increment of the bias injection current at an external cavity length of 15 cm. With a further increase of the external optical feedback in regime IV, the laser output power becomes completely unstable and no jump is observable in the L-I characteristic as shown in Fig. 5.9c. In the corresponding bifurcation diagram in Fig. 5.9d, the laser output shows only chaotic oscillations throughout the range in the bias injection current.

Figure 5.10 shows the example of mode jumps obtained from experiments (Fukuchi et al. 1999). In Fig. 5.10a, clear mode jumps occur in the L-I characteristic. In actual fact, the laser output power has a hysteresis either for the increase or decrease of the injection current. Figure 5.10b shows optical spectra observed by a Fabry-Perot spectrometer at the injection currents marked in Fig. 5.10a. At point 'a' followed by a mode jump, the laser oscillation stays stable. With an increase of the injection current at 'd', the relaxation

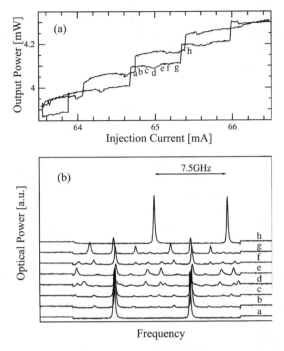

Fig. 5.10. Experimentally obtained L-I characteristic and optical spectra at an external cavity length of $L = 3\,\text{cm}$ and an external reflectivity of $r^2 = 0.03$. **a** Output power versus injection current. **b** Optical spectra corresponding to the marked position in **a**

oscillation increases. After that, the laser shows quasi-periodic oscillations with mixed frequencies of the relaxation and external modes, and then unstable oscillations before returning to a stable oscillation at 'h.' A similar process is repeated as the injection current increases. The measured relaxation oscillation frequency (for example in spectrum 'd') is 4.5 GHz. Since the external mode frequency (5.0 GHz) is close to the relaxation oscillation frequency in this case, it is not easy to distinguish the external modes with the relaxation oscillation in the spectra. However, the difference between periodic and quasi-periodic oscillations is clear in the figure. The result well coincides with the numerical simulations in Fig. 5.9. For a further increase of the optical feedback, no jump is observed in the L-I characteristic. No distinct peak is observable by a Fabry-Perot spectrometer and the coherence of the laser is completely destroyed.

Finally, we briefly refer to the effect of coherence induced by optical feedback in semiconductor lasers. Normally, the spectral linewidth of a semiconductor laser without any stabilization is much broadened due to a finite value of the α parameter compared to other lasers, such as gas lasers. The linewidth of a semiconductor laser with solitary oscillation ranges from several MHz to several tens of MHz. In this case, the averaged coherence length of the laser is over several meters. On the other hand, the linewidth of a semiconductor laser in the presence optical feedback is further broadened over several to several tens of GHz, even if the laser is assumed to be oscillated at single longitudinal mode. When a laser is oscillated at quasi-periodic or weak chaotic states by optical feedback, the averaged coherence is maintained around several tens of centimeters. However, the averaged coherence is much reduced to less than several centimeters for strong chaotic oscillations or low-frequency fluctuation (LFF) states which are discussed in the following section. The degree of the averaged coherence is easily obtained from observation of the visibility for the laser output by employing a Michelson interferometer. Figure 5.11 is an example of experimental observations of visibility curves in Michelson interferometer (Lenstra et al. 1985). For the increase of optical feedback, the coherence length becomes small. The spectral linewidths calculated from the observation are 11 GHz (corresponding coherence length of 2.0 cm) and 15 GHz (coherence length of 2.7 cm) for Fig. 5.11a and b, respectively. In Fig. 5.11b, the second spectral peak (4.4 GHz) corresponds to the laser relaxation oscillation of the laser. Thus, the large degradation of the averaged coherence due to chaotic oscillations induced by optical feedback must be taken into account.

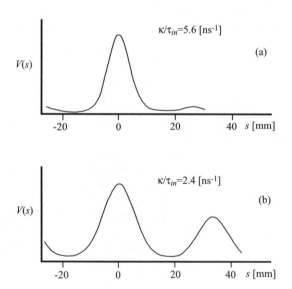

Fig. 5.11. Experimentally observed visibility curve for the arm length difference of a Michelson interferometer. Feedback strengths are **a** $\kappa/\tau_{\mathrm{in}} = 5.6$ [ns^{-1}] and **b** $\kappa/\tau_{\mathrm{in}} = 2.4$ [ns^{-1}]. The external cavity length is $L = 80$ cm and the injection current is $J = 1.51 J_{\mathrm{th}}$ (after Lenstra D, Verbeck BH, den Boff AJ (1985); © 1988 IEEE)

5.2 Dependence of Chaotic Dynamics on the External Mirror Position

5.2.1 Periodic Stability Enhancement for Variations of the External Cavity Length

We discuss here the dependence of chaotic dynamics on the external cavity length in semiconductor lasers. We consider the case where the change of the external cavity length is much larger than the internal laser cavity length ($\sim$cm) and investigate the stability and instability for the change. Assuming a small external feedback and satisfying the conditions $\omega_R^2 \gg (\kappa/\tau_{in})^2$, $-\kappa\Gamma_R/\tau_{in}$, the characteristic equation (4.19) is approximated by

$$D(\gamma) = \gamma^3 + 2\left\{-\Gamma_R + \frac{\kappa K}{\tau_{in}}\cos(\omega_s\tau)\right\}\gamma^2 + \omega_R^2\gamma$$
$$+ \frac{\kappa K\omega_R^2}{\tau_{in}}\{\cos(\omega_s\tau) - \alpha\sin(\omega_s\tau)\} \tag{5.6}$$

Assuming $\gamma = i\omega$ and substituting it into (5.6), the boundary condition for stable and unstable oscillations is calculated. From the real part of the equation, we obtain

$$-\alpha\sin(\omega_s\tau) + \left\{1 - 2\left(\frac{\omega}{\omega_R}\right)^2\right\}\cos(\omega_s\tau) = -\frac{\left(\frac{\omega}{\omega_R}\right)^2}{\Gamma_R\frac{\kappa}{\tau_{in}}\sin^2(\frac{\omega\tau}{2})} \tag{5.7}$$

From the condition for the imaginary part together with the above equation, the boundary condition is given by (Tromborg et al. 1984)

$$\omega^2 - \omega_R^2 = -2\Gamma_R\omega\cot\left(\frac{\omega\tau}{2}\right) \tag{5.8}$$

To investigate the stability and instability dependence, we must numerically solve the condition for τ from (5.8), but we can conjecture from the above equation that the critical point changes periodically for the external cavity length. The round trip time τ corresponding to this period is written by

$$\tau = \frac{2}{\omega}\left[\cot^{-1}\left\{-\frac{1}{2\Gamma_R\omega}(\omega^2 - \omega_R^2)\right\} + m\pi\right] \tag{5.9}$$

From (5.8), $\omega = \omega_R$ is the solution for the periodic boundary and the condition is also given by $\omega_R\tau = 2m\pi$ (m is an integer). Namely, the boundary changes periodically for the external cavity length satisfied as $L = mc/2\nu_R$ (ν_R is the relaxation oscillation frequency). Again, the relaxation oscillation frequency plays an important role for the stability and instability of the laser systems.

From the graphical relation between ω and τ in (5.9), we can numerically obtain the solution for the resonance of the laser oscillations (Tromborg et al. 1984, Mørk et al. 1992, and Murakami et al. 1997). The intersections of the graph $\omega = \omega_R$ along the round trip time τ are the points where constructive interference is achieved and the stability of the laser for the external reflectivity is much enhanced at these points. In the graph, the intersections crossing at $\omega/\omega_R = 1$ are the values of τ where the boundary of the stability enhancement occurs. We obtain periodic solutions for the intersections and the separation between the successive solutions is exactly equal to the time corresponding to the frequency of the relaxation oscillation. This periodicity is also calculated from the direct numerical simulation for the rate equations. A periodic enhancement of stability for the laser oscillation can be observed for the change of the external mirror position of the order of centimeters as shown in the previous section. Figure 5.12a shows the numerical result of the phase diagram of stable and unstable oscillations for the external mirror position and the optical feedback level (Murakami et al. 1997 and Murakami 1999). The diagram is numerically obtained for $\omega_0\tau = 0$ in the rate equations in (4.8) and it does not include the effects of a small mirror variation compatible with the optical wavelength. In the figure, "periodic state" within chaotic oscillations denotes the periodic windows frequently observed in chaotic bifurcations.

With the increase of optical feedback at a fixed external cavity length, the laser also becomes unstable at a rather larger external feedback level and fixed point of the laser oscillation evolves into the period-1 state after crossing the boundary. At the external cavity lengths satisfying the condition $L = mc/2\nu_R$, the laser constructively couples with the external cavity and a larger fraction of feedback light is required to destabilize it. The stable area greatly expands at this location, but the laser rapidly turns out to be unstable after the feedback exceeds the critical point. This corresponds to the quasi-periodic bifurcations of chaos that are shown in Fig. 5.7b. At other mirror locations, the laser easily becomes unstable and rapidly evolves into chaotic states with a small amount of external feedback. The laser evolves rather slowly into unstable and chaotic states for the increase of the external reflectivity at these locations. We can see clear period-doubling bifurcations of the laser oscillations in the mirror position at the bottom of the periodic enhancement. This corresponds to the Hopf bifurcations shown in Fig. 5.7a. The excited relaxation oscillation at the critical point also changes periodically for the external cavity length as shown in Fig. 5.12b. The relaxation oscillation decreases with the increase of the external cavity length. Then, it jumps up at the stability peak and repeats the same process for the increase of the external cavity length. The periodic feature of the stability enhancement was also confirmed by experiments (Ye and Ohtusbo 1998). The origin of the phenomena is also presented in the next section.

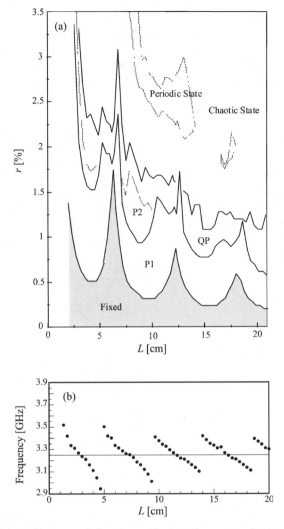

Fig. 5.12. a Phase diagram of chaotic oscillations for external mirror position and optical feedback level at a bias injection current of $J = 1.3J_{\mathrm{th}}$. Fixed: fixed state, P1: period-1 oscillation, P2: period-2 oscillation, QP: quasi-periodic oscillation. **b** Excited relaxation oscillation at boundary between fixed and period-1 states. The relaxation oscillation frequency at free running state which is shown as a straight line is 3.25 GHz

5.2.2 Origin of Periodic Stability Enhancement

The periodic stability enhancement for the change of the external cavity length observed in the previous section is explained by competitions of linear modes derived from the linear stability analysis for the rate equations

(Murakami and Ohtsubo 1998). Figure 5.13 shows mode transition around the third stability peak at the external cavity length of $L = 14.0$ cm with the external reflectivity of 0.7% corresponding to Fig. 5.12a. In Fig. 5.13a at $L = 13.75$ cm, mode A is the highest mode and the real part of the mode has a positive value. Then, the oscillation becomes unstable at mode A and this mode plays role in the relaxation oscillation mode. Figure 5.13b is the mode distribution at $L = 14.0$ cm where the stability of the laser oscillation is locally much enhanced. The reflectivity of 0.7% is selected as slightly less than the feedback level for unstable oscillations. Two modes A and B are competing and the real part of mode A becomes negative. Then, the laser output stays at a stable fixed state and the relaxation oscillation mode is not excited. Figure 5.13c is the mode distribution at the position of the external mirror of $L = 14.25$ cm. The highest mode switched from A to B and the laser again shows unstable oscillations at mode B. From these results, the stability enhancement along the external cavity length is explained by the mode

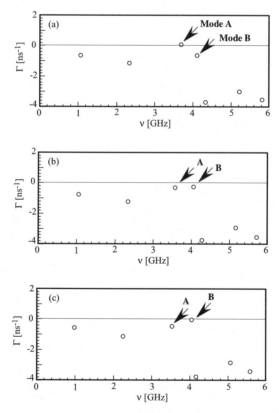

Fig. 5.13. Competition of two modes related to resonance frequency at $r = 0.05$ and $J = 1.5J_{\text{th}}$ for variations of the external cavity length. **a** $L = 13.75$ cm, **b** $L = 14.0$ cm at the stability peak, and **c** $L = 14.25$ cm

competition between linear modes near the relaxation oscillation frequency. Once two modes have positive values of their real parts for a larger external optical feedback, the two modes become unstable simultaneously. Thus, the laser is rapidly destabilized and shows chaotic oscillations above the fraction of external feedback. On the other hand, at and around the stability bottoms in Fig. 5.12a, there is only one mode near the relaxation oscillation frequency and the laser shows typical period-doubling bifurcation to chaos.

For a longer external cavity length, the laser is less stabilized with a small fraction of optical feedback as seen from Fig. 5.12a. Usually, a semiconductor laser without any special control for stabilization has a spectral bandwith of several to several tens of MHz and the coherence is in the order of meters. The optical feedback phenomena from a distant reflector longer than the laser coherence are attributed to the incoherent effect. The rate equations for the two variables of the photon number and the carrier density are enough to describe such phenomena. However, the effects of coherence both for coherent and incoherent feedback are included in the rate equations in (4.5)–(4.7), and we can still use them for numerical calculation of the dynamics in incoherent feedback.

5.2.3 Effects of Linewidth Enhancement Factor

The linewidth enhancement factor α of semiconductor lasers plays an important role in the laser dynamics and the laser instability is greatly enhanced for larger value of the linewidth enhancement factor. In this section, we investigate the dependence of onset of chaos on the linewidth enhancement factor in the presence of optical feedback to semiconductor lasers. For that purpose, we first consider a transfer function from the modulation current to the modulated optical power in a semiconductor laser. Assuming a long external cavity limit (the length of the external mirror is much larger than $c/2f_R$), the small-signal transfer function like general control systems can be defined by

$$H_K(i\omega) = \{1 - K(i\omega)\} \frac{H(i\omega)}{1 - K(i\omega)H(i\omega)} \qquad (5.10)$$

where $H(i\omega)$ corresponds to the normalized transfer function of the semiconductor laser without optical feedback ($K(i\omega) = 0$) and is given by (Helms and Petermann 1990)

$$H(i\omega) = \frac{1}{\left(i\frac{\omega}{\omega_R}\right)^2 + i\frac{\omega}{\omega_d} + 1} \qquad (5.11)$$

Here, $\omega_R = 2\pi f_R$ is the resonance angular frequency and $\omega_d = -\omega_R^2/\Gamma_R$ [Γ_R being the damping coefficient defined in (3.68)]. We are considering the case of a long external cavity regime, so that $\omega_d > \omega_R$. The function $H_K(i\omega)$ is derived for the minimum linewidth mode, since this is the most stable external cavity mode. Also weak feedback and a linewidth enhancement factor $\alpha > 1$ are assumed to derive this equation. Then the feedback term $K(i\omega)$ is given by

$$K(i\omega) = i\frac{k_c\sqrt{1+\alpha^2}}{\omega}\{1 - \exp(-i\omega\tau)\} \tag{5.12}$$

with the round trip delay τ of the external cavity and the feedback rate $k_c = \kappa/\tau_{in}$.

At a certain feedback level, the transfer function (5.11) exhibits an unstable pole. The existence of such a pole does not necessarily mean that a coherence collapse occurs, since the coherence collapse is described by a very complicated dynamic process. However, the minimum feedback level at which an unstable pole occurs, actually corresponds to the onset of the coherence collapse. Since $\omega_R < \omega_d(\omega_R > \Gamma_R)$, $H(i\omega)$ takes its maximum at $\omega \approx \omega_R$, where the unstable pole will also occur. A pole occurs for

$$K(i\omega)H(i\omega) = 1 \tag{5.13}$$

A long external cavity is assumed here, so that $K(i\omega)$ gets maximum for $\exp(i\omega\tau) = 1$ in (5.12), where $\omega\tau$ is an odd multiple of π. Using $\omega \approx \omega_R$ and the condition in (5.13), the critical feedback coefficient $k_{c,critical}$ reads

$$k_{c,critical} = \frac{\omega_R^2}{2\omega_d}\frac{1}{\sqrt{1+\alpha^2}} = \frac{1}{2}\frac{\Gamma_R}{\sqrt{1+\alpha^2}} \tag{5.14}$$

Noted that the $k_{c,\,critical}$ is derived from the assumption for a large number of the linewidth enhancement factor. Therefore (5.14) is only valid for $\alpha > 1$. To deduce the expression applicable to a small value of the linewidth enhancement factor, we employ a following empirical approximation applicable for a small value of the linewidth enhancement factor (Helms and Petermann 1990):

$$\frac{1}{\sqrt{1+\alpha^2}} \approx \frac{\sqrt{1+\alpha^2}}{\alpha^2} \tag{5.15}$$

Using this relation with (5.14), the critical feedback coefficient is expressed as

$$k_{c,critical} = \frac{\Gamma_R}{2}\frac{\sqrt{1+\alpha^2}}{\alpha^2} \tag{5.16}$$

For a large value of the linewidth enhancement factor, (5.16) reduces to the same expression as (5.14). For the feedback parameter of a Fabry-Perot laser in (4.4), the critical feedback level where the semiconductor laser starts to show unstable oscillations is given by

$$r_{critical} = \frac{\Gamma_R^2\tau_{in}^2\sqrt{1+\alpha^2}}{4\alpha^2\frac{1-r_0^2}{r_0}} \tag{5.17}$$

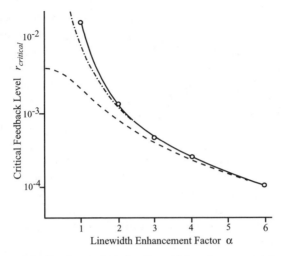

Fig. 5.14. Critical feedback strength for linewidth enhancement factor. *Solid line*: numerically determined critical feedback strength, *dashed line*: theoretical prediction from (5.14), *dash-dotted line*: theoretical prediction from (5.16) (after Helms J, Petermann K (1990); © 1990 IEEE)

Figure 5.14 is the plot of the critical feedback strength for the linewidth enhancement factor. The solid line shows the numerically calculated critical feedback strength. While the dashed line is the theoretical prediction obtained from (5.14) and the dash-dotted line is the one from (5.16). The equation (5.16) well represents the critical feedback strength for the linewidth enhancement factor where the unstable laser oscillation starts. This tolerable feedback level increases with the increasing damping of the relaxation oscillations.

5.2.4 Sensitivity of the Optical Phase

We showed the dependence of the dynamics in semiconductor lasers with optical feedback on the external mirror position in the previous section. The dynamics is strongly dependent on the position of the external mirror not only for the long range of the mirror movement in the order of centimeter to meter, but also for short variation of the external mirror comparable with optical wavelength. When the length of the external mirror is short enough and it is less than the internal cavity length (effective cavity length), the mode separation of the internal cavity becomes shorter than that of the external cavity. Then, one of the successive internal modes is sequentially selected by the continuous small change of the external mirror position. When the mode hops, a noise known as mode-hop noise is induced and the laser performance is much deteriorated. In this regime, the coupling between the internal and external cavity is strong, though the laser output power shows a periodic variation with a period of the half wavelength of $\lambda/2$. The laser under this condition is usually stable and rarely shows chaotic oscillations. When the external mirror is located at a length longer than the internal cavity length but it is within the length corresponding to the frequency of the relaxation oscillation, i.e., $l < L < c/2\nu_R$, the laser is still stable as far as the external feedback stays small. However, for a larger feedback level, the laser shows instabilities and it evolves into chaotic states for a further increase of optical feedback. For $L > c/2\nu_R$ but within the coherence length of the laser oscillation, the laser becomes less stable at a smaller feedback level and it is easily destabilized by optical feedback. As will be discussed in the following, the laser output shows chaotic dynamics for a small variation of the external mirror position compatible with optical wavelength λ. Namely, the laser has a phase sensitivity to the change of the external cavity length (Lang and Kobayashi 1980, Osumndsen and Gade 1983, Arimoto and Ojima 1984, and Kaikuchida and Ohtsubo 1994).

Figure 5.15 schematically shows the dependence of the optical output for the increase of external optical feedback when the external mirror is positioned at several centimeters to meters in ordinary semiconductor lasers. As far as the optical feedback is very small but not negligible, very smoothly varying undulation is observed as shown in Fig. 5.15a. With the increase of optical feedback, the amplitude of the modulation grows but shows asymmetric features in the waveform (see Fig. 5.15b). Then, a further increase of optical feedback but still at a modest feedback level, sudden jumps of the optical output appear for the variation of the external mirror position. Also, the

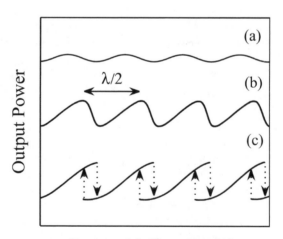

Fig. 5.15. Schematic plot of the dependence of output power for a change of external mirror position compatible with optical wavelength. a Small feedback with sinusoidal oscillation, b moderate feedback with asymmetric oscillation, and c strong feedback with hysteresis

waveform shows hysteresis depending on the upward or downward variation of the mirror position as shown in Fig. 5.15c (Lang and Kobayashi 1980). The duration of the periodic cycle of the output power is not always equal to $\lambda/2$ but also $\lambda/4$, $\lambda/6$, and so on (Kakiuchida and Ohtsubo 1994). When the optical feedback fraction becomes very large, we can observe such higher order periodic undulations in the optical output.

5.2.5 Chaotic Dynamics for a Small Change of the External Cavity Length

In Sect. 5.2.1, we assumed a fixed phase of $\omega_0\tau = 0$ and obtained the dependence of the dynamics on the external mirror length for a range of the order of centimeters to meters. We here consider the effects for a small change of the external mirror position, i.e., $\omega_0\tau$ is not zero and the change of the external mirror position is compatible with the optical wavelength. The semiconductor laser is very sensitive to a small change of the external mirror position and the output power varies periodically with length corresponding to half of the optical wavelength. In actual fact, not only periodic oscillations of the laser output power but also chaotic instabilities are observed within such a short variation. When the fraction of optical feedback is small enough, the laser output shows smooth periodic undulations for a change of the external mirror position as discussed in the previous section. Various methods for metrological applications (for example, micro-vibration measurement of surfaces compatible with optical wavelength) have been proposed based on the periodic change of the laser output power (Donati et al. 1995). These techniques are only applicable for the external feedback fraction in the regimes from II to III discussed in Sect. 4.1.1. On the other hand, smooth undulations are no longer observed in regimes III and IV and, instead, chaotic oscillation appears for the change of the external mirror length (Ikuma and Ohtsubo 1998). We will return to the subject and demonstrate some applications based on the phenomena in Chap. 11. We show various routes to chaos for such small changes of the external mirror position in the following and also compare the theory with experiments.

Figure 5.16 shows the dynamics of the laser output power numerically calculated for the small change of the external mirror position (Ikuma and Ohtsubo 1998). The wavelength of the laser used is $\lambda = 780$ nm, so that the period of undulations is half of the wavelength, $\lambda/2$. The left column is the variation of the averaged laser output power. The right column is the bifurcation diagram corresponding to the left. When the external feedback is small (Fig. 5.16a), the laser output power smoothly changes for the increase of the external mirror length with a period of $\lambda/2$. The laser stably oscillates at a fixed point even though it shows periodic undulation for the change of the external mirror position (Fig. 5.16a right). For larger optical feedback, the laser output power periodically changes with a period of $\lambda/2$, however it shows sudden drops and irregular fluctuations (Fig. 5.16b–d). Chaotic bifurcations

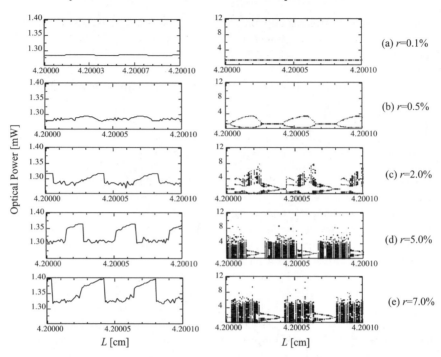

Fig. 5.16. Numerical results of averaged laser output power and bifurcation diagram for a small change of external mirror position at $L = 4.2000$ cm and the bias injection current at $J = 1.3J_{\text{th}}$

are clearly visible at larger optical feedback. For example, at the external mirror length around 4.2000 cm in Fig. 5.16d, the laser output power is stable and it oscillates at a single mode. As the increase of the external cavity length, it suddenly becomes unstable and shows chaotic oscillations. With the further increase of the external cavity length, it reduces from the quasi-periodic state to periodic oscillation and finally reaches the stable state. This process repeats with the period equal to $\lambda/2$. In actual fact, there exists a hysteresis of the laser output power for upward and downward changes of the external mirror position. The power drops observed in Fig. 5.16d and e originate from this instability. The waveforms are the time-averaged signals. Hence, the time-resolved signals at a certain external mirror position show fast irregular oscillations with a typical frequency of the relaxation oscillation as is the case for chaos in semiconductor lasers with optical feedback, as has already been discussed. The maximum Lyapunov exponent is frequently used to help the understanding of the dynamic behaviors of nonlinear systems (Appendix A.2). The maximum Lyapunov exponent has a positive value in chaotic regions, while it has zero or negative value in periodic and stable oscillations. At the point where the Lyapunov exponent has a value of less than zero, the laser output power becomes fixed states (single straight line) in

the bifurcation diagram. Indeed, the maximum Lyapunov exponent is positive
for the chaotic region in Fig. 5.16, while it is negative for stable states (Ikuma
and Ohtsubo 1998).

The optical spectrum at stable oscillation in Fig. 5.16a is completely a single mode, though the optical frequency changes with the variation of the optical output power. On the other hand, the coherence of the laser is destroyed
due to chaotic oscillations and the laser has several broadened spectral peaks
mixed with the relaxation oscillation and external cavity modes (coherence
collapse state). We can experimentally observe the evidence of chaotic bifurcations from the laser output power by using a Fabry-Perot spectrometer.
Figure 5.17 shows an example of optical spectra obtained by a Fabry-Perot
spectrometer for the small change of the mirror position. The external cavity length is increased from bottom to top in this figure. The unstable state
(weak chaotic or quasi-periodic oscillation) reduces to periodic oscillation and
once stable state, then it again becomes unstable oscillations with a period
equal to half of the wavelength.

We assume a single internal mode oscillation for a semiconductor laser
in the above discussion. However, for large external feedback reflectivity,

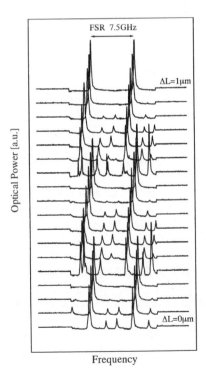

Fig. 5.17. Experimentally observed optical spectra obtained by a Fabry-Perot
spectrometer for small change of external mirror position

a semiconductor laser sometimes oscillates with multimode. In such a case, we observe periodic undulations with periods not only of $\lambda/2$ but also $\lambda/4$, $\lambda/6$, and so on, for a small change of the external cavity length. Which period emerges in the laser output power variations depends on the absolute external mirror position from the laser facet. For a long range variation of the external cavity length equal to the order of ~mm or more, there also exists a periodic change of the laser output power with a period equal to the effective internal cavity length when the laser oscillates with multimode (Murakami 1999). At moderate optical feedback, when only a single external mode contributes to the laser oscillation (which occurs at integer multiples of the internal cavity length $m\eta L$ with integer number m), an undulation with period $\lambda/2$ is observed. When two external modes play an important role for the laser oscillation at the external mirror position of $(m + 1/2)\eta L$, an undulation with a period of $\lambda/4$ is observed. Further, at the external mirror position of $(m+1/3)\eta L$ and $(m+2/3)\eta L$, an undulation with a period of $\lambda/6$ appears. Similarly, higher order periodic undulations are observed depending on the external mirror position and the excited internal modes. Even in a higher order periodic undulation, there exist chaotic bifurcations within that period. Those effects are well demonstrated by the numerical simulation of the multimode laser rate equations and can be compared with experimental results (Kakiuchida and Ohtsubo 1994).

5.3 Low-Frequency Fluctuations (LFFs)

5.3.1 Low-Frequency Fluctuation Phenomena

Periodic and quasi-periodic bifurcations are typical routes to chaos in nonlinear systems. However, these are not only the chaotic routes. Folding and stretching variables in the projection process in nonlinear systems, and various possibilities for chaotic routes are generated. Other than periodic and quasi-periodic bifurcations, one of the well-known routes is the intermittent route to chaos (Risch and Voumard 1977, Fujiwara et al. 1981, Temkin et al. 1986, and Sano 1994). Low-frequency fluctuation (LFF) is one of chaotic oscillation known as the intermittent chaos of saddle node instability also observed in semiconductor lasers with optical feedback. To study LFFs is very important from the viewpoint of practical applications, since LFFs induce much noise in the laser output power. A typical feature of LFFs is a sudden power dropout with a following gradual power recovery. LFFs occur irregularly in time depending on the system parameters and the frequency of LFFs is the order of MHz to a hundred MHz. Since the frequency of LFFs is much lower than ordinary chaotic fluctuations related to the relaxation oscillations, the phenomena are called low-frequency fluctuations. When waveforms of LFFs are observed by a fast digital oscilloscope, they seem to be continuous signals. However, it is proved that LFF has very fast time structures within the waveform and it consists of a series of fast pulses on the order of pico-second. Indeed, this fast pulsation has been experimentally observed by using a streak camera (Fischer et al. 1996a). Though LFFs were first recognized as sudden power dropouts with low frequency in the early days and this was the origin of the name of the phenomena, LFFs have quite different features from the ordinary chaotic behaviors and show a rich variety of dynamics. In the following, we show various dynamics of the phenomena and discuss the origin of LFFs.

LFFs are first observed in the output power above but close to the laser threshold. However, LFFs are phenomena induced by saddle node instability in nonlinear systems, therefore they are also observable in laser oscillations well above the threshold under appropriate conditions of the parameters when the system has saddle node unstable periodic orbits (Pan et al. 1977). LFFs are also observed by injection current modulation and optical injection from a different laser when the laser is subjected to optical feedback (Takiguchi et al. 1998). Various models for the origin of LFFs have been proposed. One of them is the model that LFF is the instability driven by noise in the nonlinear laser system (Tromborg et al. 1997, Eguia et al. 1998, and Mørk et al. 1999). Other ones are explained by competitions between modes in a multimode model of the laser (Hegarty et al. 1998), and the crisis between two attractors

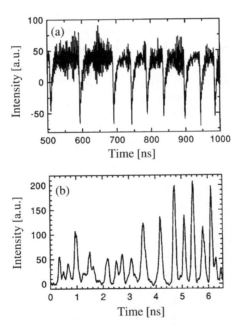

Fig. 5.18. Low-frequency fluctuations in semiconductor lasers with optical feed-back at $J = 1.03J_{th}$. **a** Detected by a slow response detector. **b** Observed by a fast streak camera. Delay time is $\tau = 3.6$ ns (after Fischer I, van Tartwijk GHM, Levine AM, Elsäßer W, Gobel EO, Lenstra D (1996a); © 1996 APS)

(van Tartwijk et al. 1995). However, LFFs have been experimentally observed in a single mode laser with optical feedback and they have also been proved to appear in a single mode model by numerical simulations without noise effects. Therefore, LFFs are the deterministic chaos induced by saddle node instability with time-inverted type-II intermittency in semiconductor lasers with optical feedback (Sacher et al. 1989 and 1992).

Figure 5.18a is an example of typical LFF waveforms obtained by experiment (Fischer et al. 1996a). This is a low-pass filtered waveform and a higher component over nano-second oscillations is not observable. The typical features of LFFs are frequent power dropouts after stationary output and subsequent gradual power recovery processes as shown in the figure. When the laser is operated close to the threshold, the output power breaks down even below the threshold at the solitary operation. The frequency strongly depends on the conditions of the laser parameters and the external mirror. The power dropouts occur irregularly and their average frequency is about 5 MHz in this case, hence low-frequency fluctuations. Figure 5.18b shows a time-resolved waveform of LFFs observed by a high-speed streak camera. The waveform consists of a pulse train with an average period of 300 ps. Figure 5.19a is a one-shot of LFFs. To show the power recovery process clearly, one-shot of LFFs is averaged and plotted in Fig. 5.19b. Stepwise power recovery is clearly visible in the figure. In actual fact, relaxation oscillations are included at the onset of each jump in the power recovery, although it is averaged out in this figure. The time duration of each step in the power recovery is equal to the round trip time of light in the external loop. LFFs tend to appear in the laser output when the external cavity length is sufficiently long

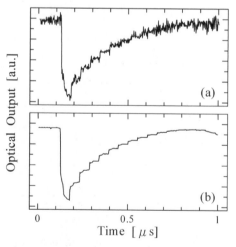

Fig. 5.19. a Single shot of LFFs and **b** one-shot of LFFs averaged over 3000 events. The external cavity length and the bias injection current are $L = 8.10$ m and $J = 1.07 J_{\mathrm{th}}$. The external reflectivity is $r^2 = 0.12$

and the external feedback is large enough (Takiguchi et al. 1999a). Under such conditions, saddle node instabilities are easily induced in the system and many external and anti-modes are generated. As will be discussed later, LFFs are explained as chaotic itinerary among these modes. On the other hand, LFFs are rarely observed for short external cavity length with small optical feedback, since the oscillation modes are sparsely distributed in the phase space. From these results, LFF, which is first recognized as fluctuations of the laser output with a low frequency, is composed of three components of time scales. One is a component of a low-frequency fluctuation with a period around MHz; the second is a component related to the external cavity length; the third is a high-frequency component with a period of pico-seconds. In the following, several characteristics of LFFs and their origin are described.

5.3.2 LFF Characteristics

One typical chaotic evolution is an intermittent route to chaos and low-frequency fluctuations (LFFs) are frequently observed as such chaotic oscillations in semiconductor lasers. Feedback induced LFFs in semiconductor lasers depend on the bias injection current, the external cavity length and the feedback level. As a general trend, the frequency of LFFs linearly increases with the increase of the bias injection current. The frequency of LFFs is about several to several tens of MHz when the external mirror is positioned around 10 to 100 cm. On the other hand, the frequency is less than MHz for a longer external cavity length over 1 m (but the mirror is still positioned within coherence length of the laser). The linear relation between the frequency and the bias injection current is also held in this case. The frequency of LFFs linearly decreases for the increase of the external reflectivity at the bias injection current well above the threshold, however it stays almost constant for the bias injection current close to the threshold (Takiguchi 2002).

LFFs were initially observed near the solitary laser threshold. However, it was later proved that LFFs occur everywhere along the boundary between the optical feedback regimes of IV and V discussed in Sect. 4.1.1. Figure 5.20a shows the diagram of the possible area for the occurrence of LFFs obtained by experiment (Pan et al. 1997). Due to the low internal reflectivity of the front facet of semiconductor lasers, LFFs are observed not only for low injection current close to the laser threshold, but also for a higher injection current. The laser threshold is reduced by the external feedback as discussed, but, under certain conditions of the laser parameters, the slope efficiency decreases and the laser power becomes lower than that of the solitary laser at a higher injection current (see Fig. 5.7). In such cases, LFFs occur at a high injection current and the laser power shows not dropouts but jumpups in LFFs as shown in Fig. 5.20b. Thus, LFFs are universal phenomena observed in semiconductor lasers with optical feedback and they are a typical feature of saddle node instabilities involved in nonlinear systems. When LFFs occur, the laser usually oscillates at multimode and a single mode operation of the

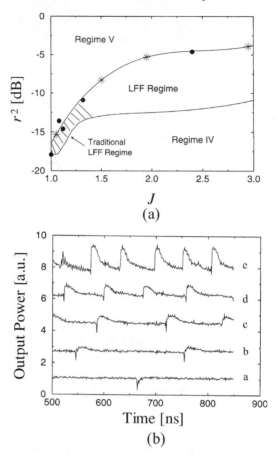

Fig. 5.20. a LFF regions between regimes IV and V at $L = 60$ cm. The horizontal axis is the injection current and the vertical axis is the ratio of the feedback light. Symbols denote experimental results and *solid lines* theoretical results. **b** Time series of LFFs for bias injection currents of $J = 1.20J_{th}$, $1.34J_{th}$, $1.48J_{th}$, $1.66J_{th}$, and $2.08J_{th}$ from *bottom* to *top* at $\kappa\tau/\tau_{in} = 240$ (after Pan MW, Shi BP, Gray GR (1997); © 1997 OSA)

laser is not always an appropriate model to describe it. However, we can observe LFFs at a single mode operation of the laser as mentioned before (Heil et al. 1998). For example, when a semiconductor laser with optical feedback is operated by selecting a single mode from a grating mirror feedback, we can still observe LFFs under this condition. Furthermore, LFFs are reproduced by the numerical simulation in the laser output power with a single mode model for the rate equations (Sano 1994). LFFs are also simulated at a multimode operation of the laser. In the multimode case, switching of the output power among the modes occurs and the total power shows LFF waveforms, which is the same as those of ordinary chaotic oscillations.

5.3.3 Origin of LFFs

LFFs are explained by saddle node and intermittent instability involved in the nonlinear system of semiconductor lasers with optical feedback. LFFs occur not only by self-optical feedback but also by optical injection from a different laser. Another example of the occurrence of LFFs is an injection current modulation in the presence of optical feedback and, indeed, the laser also shows LFFs by the injection current modulation. However, we focus on the discussion for the origin of LFFs induced by optical feedback. The same or a similar explanation is also applicable for other cases. Phenomena of LFFs are generally explained by employing the model that the laser output power hops around external- and anti-modes of the laser oscillations due to unstable saddle node instability generated by the optical feedback (Sano 1994).

We start the discussion for the external- and anti-modes in the phase space of the oscillation frequency and the carrier density. In semiconductor lasers with optical feedback, we obtain the steady-state solutions from (4.10) and (4.11) as

$$\Delta\omega\tau = -\frac{\kappa\tau}{\tau_{in}}\sqrt{1+\alpha^2}\sin(\omega_s\tau + \tan^{-1}\alpha) , \qquad (5.18)$$

$$\Delta n = -\frac{2\kappa}{\tau_{in}G_n}\cos(\omega_s\tau) , \qquad (5.19)$$

where $\Delta\omega = \omega_s - \omega_{th}$ and $\Delta n = n_s - n_{th}$. Figure 5.21a shows about half the distribution of the modes in the phase space of $\Delta\omega - \Delta n$ calculated

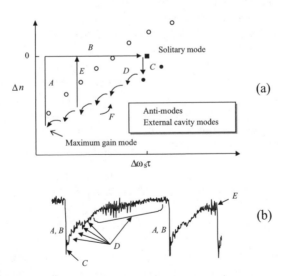

Fig. 5.21. **a** External- and anti-modes in the phase space of frequency and carrier density. Only half of the ellipsoid of the mode distribution is shown in the graph. **b** Corresponding LFF waveform to **a**

from (5.18) and (5.19). This corresponds to the case for a long external cavity length with a high level of external feedback. Therefore, the modes are densely distributed. For the case of short external cavity length with a low external feedback, the mode distribution will be sparse. The laser oscillates at one of the stable modes (lower half modes on the ellipsoid). The modes of the upper half on the ellipsoid are unstable and they are not generally stable lasing modes. The black dot at the center of the ellipsoid is the mode of the solitary oscillation. The laser without optical feedback oscillates at this single mode. Among the modes in Fig. 5.21a, the most possible mode of the laser oscillation in the presence of optical feedback is the maximum gain mode. However, the laser does not always oscillate at this mode. Folding and stretching the variables induced by nonlinear characteristics in the system due to small fluctuations, the laser may be suddenly trapped to a spiky orbit with a large amplitude oscillation.

At first, we think that the laser initially oscillates around the maximum gain mode and the state of the laser oscillation fluctuates near this mode with small amplitude. In actual fact, this is not the fixed stable mode. Once, the state reaches a point very close to the counterpart anti-mode and, then, the laser may be trapped in the anti-mode. The phase remains unchanged, but the carrier density abruptly jumps up to the value of the solitary oscillation. The sudden jump of the carrier density induces the increase of the phase and the laser shows a sudden power dropout. At this state, the laser output power is almost equal to the free running oscillation. This corresponds to a sudden power dropout of LFF. After that, the laser is trapped by one of the external modes close to the solitary oscillation mode. Then, the state goes around the successive external modes toward the maximum gain mode. This corresponds to the power recovery process of LFF. When the laser reaches the maximum gain mode, the above process is repeated. The occurrence of LFFs is not periodic but irregular, since the fluctuations exist in the chaotic itinerary due to the nonlinear effects. For example, when the chaotic oscillation has a large amplitude in the power recovery process, the laser may be trapped in the associate anti-mode and a power dropout may occur even before reaching the maximum gain mode. Another case is a reversion of the power recovery and the state goes up against the direction for the maximum gain mode. The corresponding LFF waveform to Fig. 5.21a is also shown in Fig. 5.21b. Figure 5.22 shows a numerically calculated chaotic itinerary of the laser output power in the phase space of $\Delta\omega - \Delta n$ under a LFF regime. The figure corresponds to almost two cycles of LFFs.

As has already been mentioned, LFFs have three time scales of fluctuations. The first one is a fast pulse-like oscillation of the order of several tens of pico-seconds. When the laser is oscillated close to one of the external modes, the carrier density changes with large amplitude and it reaches the associated anti-mode, while the phase stays almost constant. This corresponds to a series of fast pulsations. The time scale depends on the fluctuations of the carrier number in the active layer. The second is the transition time between

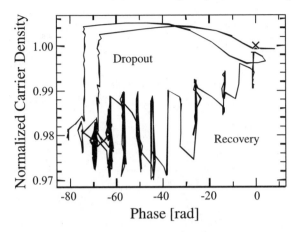

Fig. 5.22. Calculated chaotic itinerary in the phase space of phase and carrier density at $J = 1.01J_{\text{th}}$, $L = 30$ cm, and $r = 0.1$. Symbol $\times$ is the solitary mode

the successive external modes in the stepwise time recovery process. The time is related to the external cavity length and, for example, it is on the order of nano-seconds for the external cavity length of several tens of centimeters. The third time scale is the duration time of power dropout events. This time scale is usually of the order of micro-seconds, but the period is not fixed, as has already been discussed. We have assumed the steady-state condition to calculate the external- and anti-modes in Fig. 5.21. However, the occurrence of LFFs is of dynamic nature, so that the discussion developed here may not be applicable for such dynamic states in the strict sense. Nonetheless, the approach can explain the overall feature of the LFF dynamics, because chaos itself is generated from small perturbations to the initial state of the laser oscillations.

Finally, we will show a return map of the LFF characteristics to the demonstrate typical folding and stretching process in the nonlinear system (Mørk et al. 1999). The relation between the present field and the previous field before time τ can be calculated from (4.2). Reducing the delay differential equations into difference equations, we obtain the relation by Fourier transforming the difference equation. The relation between the Fourier components $\hat{E}_n(\omega)$ and $\hat{E}_{n-1}(\omega)$ is given by

$$\hat{E}_n(\omega) = \frac{\kappa \exp(-\mathrm{i}\omega\tau)\hat{E}_{n-1}(\omega) + \hat{F}_{\mathrm{E}}(\omega)}{\tau_{\mathrm{in}}\left\{\mathrm{i}\omega - \frac{1}{2}(1 - \mathrm{i}\alpha)G_{\mathrm{n}}(n - n_{\mathrm{th}})\right\}} \tag{5.20}$$

The equation includes the Langevin noise component $\hat{F}_{\mathrm{E}}(\omega)$, although it is not essential. Calculating the square modulus of (5.20) and converting to the form for the intensity, we obtain a return map for delay time τ. The result is shown in Fig. 5.23. Dots in the figure are the modes of laser oscillations. The modes along the line are stable external modes and the modes below the

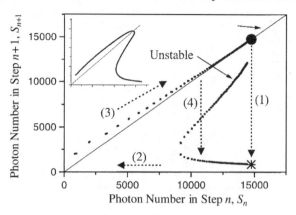

Fig. 5.23. Calculated return-map for LFFs at the feedback level of $\kappa = 0.15$ and injection current of $J = 1.0J_{\mathrm{th}}$. The arrows with numbering illustrate how the map is traversed starting from the steady-state point. Notice that the switching at 4 can take place for any value of the photon number in the range where three solutions exist. The inset shows a schematic of the full map, which is not accessible in the calculations or in experiments (after Mørk J, Sabbatier H, Sørensen MP, Tromborg B (1999); © 1999 Elsevier)

line are anti-modes. The state of the laser oscillations goes up from small to large intensity and the intensity breaks down to a level close to the threshold. The chain of dots shows the nonlinear nature of the system. The qualitative explanation can be given by this model. To reproduce the irregular occurrence of LFFs, the Langevin term is introduced in the model. When the noise term is not included, LFFs occur periodically with an exact period. Therefore, it may not reflect real oscillations of LFFs, since the equation is derived from the simplification of the discrete sampling for the rate equations. On the other hand, periodic LFFs are experimentally observed for a strong coupling of light with a short external cavity length (Heil et al. 2001a).

5.4 Chaotic Dynamics in Short External Cavity Limit

5.4.1 Stable and Unstable Conditions in Short External Cavity

Most investigations of the dynamics of semiconductor lasers subject to delayed optical feedback have been focused on the so-called long external cavity regime, in which the external cavity length is larger than the corresponding frequency of the relaxation oscillation frequency ν_{R}. However, in many practical applications in engineering, for example, fiber communications or optical storage systems, the typical external cavity is only a few centimeters long. In Sect. 5.2.1, we observed the periodicity of the stability enhancement for the variations of external cavity length. The first period of the stability is almost equal to the length calculated from the relaxation oscillation frequency.

Namely, within this optical feedback length, the semiconductor laser becomes stable and shows robustness for the feedback. In the following section, we discuss the dynamics and stability of semiconductor lasers from short external cavity. As a measure of stability and instability in semiconductor lasers, we have already introduced the C parameter. The C parameter includes the round trip time τ and the time is proportional to the external cavity length as $\tau = 2L/c$. The laser is stable for external optical feedback as far as $C < 1$, namely the short cavity limit. The C parameter is also proportional to the feedback strength κ as (4.13), so that even for a feedback from a short cavity, the laser is not always stable and the stability strongly depends on the external feedback strength. To begin the discussion of the short cavity limit, we first consider the boundary for the stability and instability in the phase space of the feedback time τ and the feedback fraction r.

Here, the condition for the short external cavity in a semiconductor laser is $\tau \nu_R < 1$. Even under a small cavity condition, the laser shows unstable oscillations when the external optical feedback is strong enough. In other word, the laser always has unstable solutions of the oscillation for $\tau \nu_R > 1$. Schunk and Petermann (1989) theoretically predicted that there is an external cavity length below which coherence collapse should not be observed. Figure 5.24 shows the boundary condition below which coherence collapse should not be observed. The boundary is obtained from numerically solving the rate equations in the presence of optical feedback in a semiconductor laser. The relaxation oscillation and α parameter of the laser are assumed

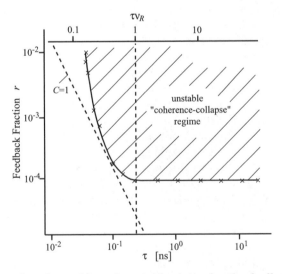

Fig. 5.24. Boundary for stable and unstable states for the feedback time τ and the external reflectivity r. *Solid line*: unstable boundary calculated form the rate equations, *dashed line*: line for the feedback coefficient of $C = 1$ (after Schunk N, Petermann K (1989); © 1989 IEEE)

to be $\nu_R = 4.4\,\text{GHz}$ and $\alpha = 6$, respectively. The shaded part is the area where unstable laser oscillations are observable. The laser does not always show unstable oscillations in all of this area, but mixed states with stable and unstable oscillations are observable depending on the parameter conditions. For comparison, the boundary of the stable condition for the C parameter ($C \propto r\tau$) is shown as a broken line. A semiconductor laser should be always stable for $C < 1$. As is seen from the figure, the line $C = 1$ is not at all suitable for the criterion of the boundary between the stable and unstable regimes in the strict sense.

5.4.2 Regular Pulse Package Oscillations in Short External Cavity

The stability of a semiconductor laser is greatly enhanced for a short external cavity limit, as has been already pointed out. In this limit, the laser tends to oscillate at a LFF regime rather than irregular fast chaotic oscillations. Furthermore, the laser becomes very sensitive to the optical phase due to the strong coupling between the internal and external cavities. Heil et al. (2001a) experimentally and theoretically investigated the dynamics of a semiconductor laser with short external cavity. As a typical feature of the dynamics, the laser showed a regular pulse package with a frequency of several hundreds of MHz like a LFF oscillation and each package contained about 10 pulses with a period corresponding to the external cavity length. The frequency of regular pulse package is considerably higher than a frequency with long external cavity ($\sim$10 MHz or less).

Figure 5.25 shows the results of the dynamic characteristics in a semiconductor laser with optical feedback from a short external cavity. Figure 5.25a and b is the experimental results of observed spectra. In Fig. 5.25a, the laser shows a regular pulse package of an LFF frequency of 390 MHz. The pulse package is very regular and the trajectory variations from pulse package to pulse packages are very small. Also the spectral peak for the external cavity of 4.5 GHz, which corresponds to the short external cavity length of 3.3 cm, is seen. The inset is the direct waveform observed by a digital oscilloscope, but the waveform is smeared out due to the resolution of a digitizer (the bandwidth of 4.5 GHz). The regular pulsating oscillations with fully modulated waveform were confirmed by the observation from a streak camera and each pulse width was observed to be around 10 ps. In Fig. 5.25b, the LFF frequency is 1.195 GHz, which is quite fast, compared with the case of a long external cavity feedback. Only the LFF frequency is shown in the figure, since the external cavity frequency is 14 GHz and the spectral peak is outside the scope of this plot. Under regular pulse package emission, the lasers operate on several longitudinal oscillation modes, which is similar to ordinary LFFs with long external cavity. In all cases of external cavity length in semiconductor lasers, the dynamics is sensitive to optical phase compatible to the optical wavelength. However, strong phase sensitivity is observed in the case of short external cavity optical feedback. In this experiment, the

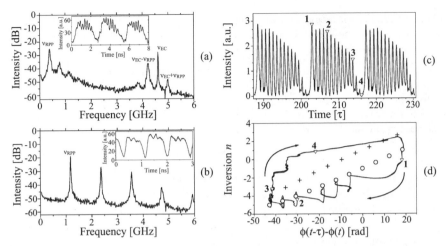

Fig. 5.25. Experimentally observed rf spectra and intensity time series for **a** $L = 3.3$ cm and $J = 1.08 J_{\text{th}}$ corresponding to $\nu_R = 1.1$ GHz, and **b** $L = 1.1$ cm and $J = 1.80 J_{\text{th}}$ corresponding to $\nu_R = 3.8$ GHz. **c** Numerical calculation of regular pulse package state and **d** trajectory plotted on the normalized inversion n and phase difference $\phi(t - \tau) - \phi(t)$ plane. The parameters used in the simulations are $J = 1.15 J_{\text{th}}$ (corresponding relaxation oscillation frequency of $\nu_R = 3.42$ GHz), $L = 1.79$ cm, $1/\tau_{\text{ph}} = 5.86 \times 10^{11}$ s^{-1}, $\kappa/\tau_{\text{in}} = 7.91 \times 10^{10}$ s^{-1}, $\tau/\tau_{\text{ph}} = 70$, and $\tau_s/\tau_{\text{ph}} \sim 1710$ (after Heil T, Fischer I, Elsäßer W, Gavrielides A (2001); © 2001 APS)

regular pulse package occurs only over a certain phase interval for the ranges of $1.3 J_{\text{th}} < J < 1.5 J_{\text{th}}$. For $J > 1.5 J_{\text{th}}$, the regular pulse package is present for all feedback phase.

Figure 5.25c is the simulation results for regular pulse package oscillations based on the laser rate equations. The threshold reduction due to the optical feedback is 11%. It is noted that the laser intensity is plotted against time normalized to the photon life time τ_{ph}. Figure 5.25d shows the trajectory of chaotic oscillations. In the phase space, the location of external modes is indicated by circles and the anti-modes by crosses. The temporal evolution occurs clockwise and the numbers in the figure provide a one-to-one correspondence of time series and phase space portrait. The LFF itinerary for one cycle is similar to that of a long external cavity. One of the remarkable characteristics is that the trajectory always visits the same external cavity modes and the laser shows regular pulses.

Increasing the length of the external cavity, the transition from regular pulse package dynamics to LFF dynamics occurs roughly at $\nu_R \geq \nu_{\text{ex}}$ (ν_{ex} being external cavity frequency); the short time scale dynamics become irregular and are dominated by relaxation oscillations. We find remarkable common features in the dynamics of regular pulse package and LFF. In both cases, low-frequency phenomena are present, which correspond to a directed global

Table 5.2. Classification of semiconductor laser with moderate optical feedback: short cavity regime versus long cavity regime.

	Short cavity	Long cavity
Waveform	Regular pulse package	Low-frequency fluctuation
Frequency relation	$\nu_{ex} \gg \nu_R$	$\nu_{ex} \ll \nu_R$
Mode number	Small ($\sim$10)	Large ($\sim$100)
Phase sensitivity	Qualitative changes	Little qualitative changes
Dynamics on short time scales	Regular fast pulses ($\nu = \nu_{ex}$)	Irregular fast pulses ($\nu_{ave} \approx \nu_R$)

trajectory along several attractor ruins, resulting in fast intensity pulsations underlying the low-frequency envelope. In the LFF, the trajectory usually backtracks irregularly, and visits different attractor ruins, whereas the trajectory in regular pulse package evolves in a regular way within a well-defined looped channel always along the same series of attractor ruins. Considering the complicated phase space structure of a delay system, this is remarkable and indicates a global orbit underlying to the regular pulse package dynamics. For certain parameter sets, regular states have also been observed in the long cavity regime. However, in the long cavity regime, the fast time scales are dominated by relaxation oscillations, whereas the fast time scales in the short cavity regime are dominated by the external cavity roundtrip frequency. The merging of attractor ruins to a global orbit appears to be a basic phenomenon series expansion of the delay terms in the optical feedback model.

The LFF envelope frequency in the short cavity limit has the dependence of the bias injection current (Heil et al. 2003b). The LFF frequency has a linear relation to the bias injection current and the frequency increases with the increase of the bias injection current. In addition, the linear dependence is independent of the cavity length. The trends are similar to the the the case of LFFs in long external cavity regimes. The linear relation is remarkable; in spite of the solitary relaxation oscillation frequency, it shows a square-root scaling with $\nu_R \propto \sqrt{J - J_{th}}$. The relation is also well reproduced by numerical simulations using the rate equations with short cavity external optical feedback. Thus, the regular pulse package envelope frequency cannot be simply associated with the relaxation oscillation frequency. In addition, the frequency is significantly smaller than the relaxation oscillation frequency. The linear scaling and the slow time scales of the regular pulse package envelope frequency already indicate that not the solitary laser characteristics, but rather the structure of the phase space and the corresponding unstable manifolds govern the dynamics of the pulse package. The main differences between long and short cavity limits are summarized in Table 5.2.

5.4.3 Bifurcations of Regular Pulse Package

Regular pulse package in a short cavity is induced by the wandering among the external modes and anti-modes, which is quite similar to the occurrence of LFFs in a long external cavity. In both cases, low frequency phenomena are present connected with a global trajectory along several attractor ruins, and producing pulses when visiting each of them. The transition from regular pulse package to LFF occurs when the delay becomes larger than the relaxation oscillation period. What makes regular pulse package distinct from LFF is the sensitivity to phase of the back-reflected light. Tabaka et al. (2004) investigated the dynamics and bifurcation scenarios of regular pulse packages in the short external regimes. With increased feedback strength in the short optical feedback regime, regular pulse package may undergo a period-doubling bifurcation cascade. On the bifurcation cascade, they found chaotic coexistent states with a time-periodic solution that originates from a newly born external cavity mode. The external cavity modes exhibit supercritical Hopf bifurcations. From these supercritical Hopf bifurcations emerge branches of time-periodic solutions. They found that the largest region of regular pulse package occurs for delays around half of the relaxation oscillation period of the solitary laser.

Figure 5.26 shows the plot of the point representing the detected regular pulse package dynamics, including regular pulse package of higher order periods (up to period of 10), to obtain representative picture of the regions of the regular pulse package dynamics. The map reveals that regular pulse package dynamics appears for intermediate levels of feedback strength hand at delays smaller than the relaxation oscillation period $\tau_R/\tau_{ph} = 170$. This is in agreement with previous observations by Heil et al. (2001a), in which they suggest that regular pulse package disappears when the delay approaches

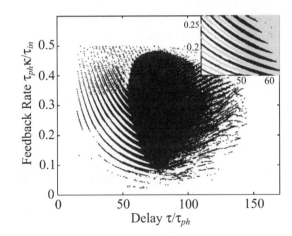

Fig. 5.26. Map of the regular pulse package dynamics in the space of the feedback rate and the delay. The parameters are $J = 1.155 J_{th}$, $\alpha = 5$, and $\omega_0\tau = \arctan\alpha + \pi$. The inset illustrates the phase dependency of regular pulse package: in black the map is plotted for $\omega_0\tau = \arctan\alpha$ and in gray $\omega_0\tau = \arctan\alpha + \pi$ (after Tabaka A, Panajotov K, Veretennicoff I, Sciamanna M (2004); © 2004 APS)

the relaxation oscillation period and then the transition to LFF takes place. Moreover, in Fig. 5.26, one finds that the largest region of regular pulse package is for delays around half of the relaxation oscillation period. For smaller delays, small windows of regular pulse package are very well identified in Fig. 5.26. For changing phase of the back reflected light, the structure of this part of the map may be significantly altered. When the phase changes by $\pm\pi$ the windows of regular pulse package shift in such a way that they fit the space in the map where regular pulse package was not previously present. This phase sensitivity is illustrated in the inset of Fig. 5.26, where the regions of regular pulse package are plotted for two different phase conditions that differ by π. We observe that changing the phase shifts the regions of regular pulse package up and down such that regular pulse package may be found in that interval in the space of the feedback rate and the delay, only if the proper phase condition is chosen. However, from the map we see that it is only true for small delays. For delay times larger than half the relaxation oscillation period the occurrence of regular pulse package dynamics becomes phase insensitive.

5.5 Dynamics in Semiconductor Lasers with Grating Mirror Feedback

The theoretical background of grating feedback in semiconductor lasers has already been discussed in Sect. 4.3. Grating feedback is frequently used to select a certain mode among the possible oscillation lines and to stabilize the oscillation of a semiconductor laser. However, even in a semiconductor laser with grating feedback, instabilities are also observed for a certain range of the feedback level and experimental configurations, since the feedback is the introduction of an additional degree of freedom. The general purpose of grating mirror feedback is the stabilization of semiconductor lasers and a few studies have reported on instability and chaos in semiconductor lasers with grating feedback (Zorabedian et al. 1987 and Binder et al. 1990; Detoma et al. 2005). These are the results for conventional feedback from plain grating mirrors. While the effects of grating optical feedback from fiber Bragg grating and volume holographic grating were reported (Naumenko et al. 2003a; Ewald et al. 2005), in which similar dynamic behaviors as those of plain grating mirror were observed.). Here, we only focus on the origin of instability in semiconductor lasers induced by grating optical feedback. In ordinary grating feedback to stabilize laser oscillation, the feedback fraction is not so small and (4.34) is not appropriate to describe such a system, since we must take into account multiple reflections between the external grating mirror and the front laser facet. For such a model, the complex fields for the forward and backward propagations as a multiple reflection must be used and the matrix transformation technique taking into account the grating mirror reflection is applied (Pittioni et al. 2001). The method is also effective for the treatment of optical feedback from distributed fiber grating into a semiconductor laser. However, we currently have few tools for analyzing exactly the dynamic properties of grating feedback and we require further formulation for grating feedback in semiconductor lasers. Here, we discuss instabilities in semiconductor lasers with grating feedback by introducing a steady-state analysis.

The relation of the output power versus the detuning between laser and grating mirror frequencies is calculated from (4.35). Figure 5.27 shows the results (Binder et al. 1990). When the reflectivity of a grating mirror is small, the laser output power shows a continuous variation with small amplitude, since the gain for the laser oscillation smoothly changes. In the figure of the curve P_L, ω_G is the angular frequency of the maximum reflection of the grating mirror given by (4.45). The period of oscillatory power change along the detuning is related to the wavelength of the laser, the maximum grating frequency, and the external cavity length. In the calculation, the feedback to the active layer is very small, since the internal reflectivity of the laser facet is set to be 2% and the fraction of grating feedback is as small as 3%. However, the laser output power shows hysteresis, as shown in the curve P_H when the grating feedback increases (the internal reflectivity of the laser facet and the reflectivity of the grating mirror are assumed to be 0.05% and 13%, respectively, in this calculation). The bistability of the laser output power may predict instability of the laser oscillations in grating optical feedback for the change of the parameters.

The lasing condition for the compound cavity of a semiconductor laser with a grating mirror is also obtained from (4.35). From the phase condition in (4.41) and the amplitude condition in (4.42), the gain and the phase of

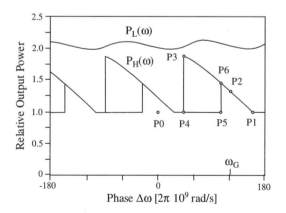

Fig. 5.27. Calculated output power versus tuning characteristics showing continuous tunability for low reflectivity laser (P_L) and hysteresis for high reflectivity laser (P_H) at an external cavity length of $L = 19\,\mathrm{cm}$ and an optical wavelength of $\lambda = 1.3\,\mu\mathrm{m}$ (after Binder JO, Cormack D, and Somani A (1990); © 1990 IEEE)

the laser oscillations in the presence of grating optical feedback are plotted
as a function of the frequency detuning. The results are shown in Fig. 5.28
(Genty et al. 2000). The detuning of the grating center frequency is $\nu_G = \omega_G/2\pi$ and the solitary laser frequency $\nu_0 = \omega_0/2\pi$ is set to be $\nu_G - \nu_0 = 20$ GHz. Due to the presence of grating optical feedback, the gain (solid line)
and the phase (dotted line) show undulations and the cross points between
them are the conditions for possible laser oscillations. The oscillation modes
are discretely distributed and each mode has a different gain. The symbol of
an open triangle shows the grating mode, but the maximum gain mode, which
is denoted by an open rectangle, is different from this mode. Such periodic
and skewed gain modes will become the origin of instability in semiconductor
lasers with grating optical feedback.

The linewidth of the laser oscillation in the presence of grating feedback
is also calculated from (4.44). Figure 5.29 shows the linewidth for the frequency detuning at a steady-state oscillation under the same condition as
in Fig. 5.28 (Genty et al. 2000). The original linewidth of the laser oscillation at the solitary mode is 20 MHz. Therefore, the linewidth of the laser
oscillation is much narrowed and the laser is stabilized for negative detuning
with a larger absolute value. Possible stable oscillation modes are discretely
distributed along the detuning. Also, oscillation modes denoted by the vertical lines (crossed marks) are not continuous lines but discretely distributed.
Further, the crossed marks are not on the vertical lines, but have a negative
slope as plotted by the inset in Fig. 5.29. The continuous tuning range for
each group is about 20 MHz for the external cavity length of $L = 12$ cm. If
the external cavity length is reduced to 1 cm, we can expand the fine tuning range and it becomes 2.5 GHz. When the feedback fraction increases, the
effective detuning increases so as to compensate the increase of the photon

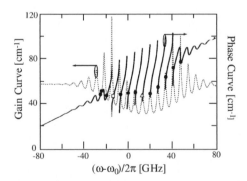

Fig. 5.28. Gain and phase curves as a function of frequency detuning for $\nu_G - \nu_0 = 20$ GHz. The external cavity length is $L = 12$ cm. The reflectivities of the grating
mirror and the internal cavity are $r = 0.25$ and $r_0 = 0.05$, respectively. *Solid line:*
gain, *dotted line:* phase. The symbols are the possible oscillation modes (after Genty
G, Gröhn A, Talvitie H, Kaivola M, Ludvigsen H (2000); © 2000 IEEE)

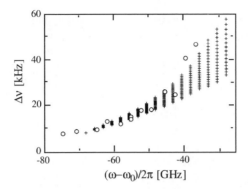

Fig. 5.29. Calculated linewidth as a function of frequency detuning for variations of grating orientation. The *dotted line* represents the linewidth computed in the case of a conventional reflector. The condition is the same as in Fig. 5.28 (after Genty G, Gröhn A, Talvitie H, Kaivola M, Ludvigsen H (2000); © 2000 IEEE)

number due to the optical feedback. As a result, the linewidth of the laser oscillation becomes narrow. For a wide tuning range of the spectral response, we were able attain a wide tuning range by the feedback for the wide spectral response of the grating, although the laser hops around among these oscillation modes and instability is enhanced by such hoppings. As a result, chaotic oscillation may be encountered in the dynamics.

5.6 Dynamics in Semiconductor Lasers with Phase-Conjugate Mirror Feedback

5.6.1 Linear Stability Analysis

A linear stability analysis is applied for semiconductor lasers with optical feedback from a phase-conjugate mirror. The procedure is the same as that in Sect. 4.2.1. For simplicity, we first assume a case of zero detuning $\delta = 0$. For small perturbations of δA, $\delta \phi$, and δn in (4.51)–(4.53), we obtain differential equations for the perturbations. Assuming the form of the solutions as $\delta x(t) = \delta x_0 \exp(\gamma t)(x = A, \phi, n)$, the condition for non-trivial solutions of the equations is given by (Murakami et al. 1997)

$$D(\gamma) = \gamma^3 + \gamma^2 \left(\frac{1}{\tau_R} + \frac{2\kappa}{\tau_{in}} \cos 2\phi_s \right) + \gamma \left\{ \omega_R^2 + \frac{2\kappa}{\tau_R \tau_{in}} + K_1 K_2 \left(\frac{\kappa}{\tau_{in}} \right)^2 \right\}$$

$$+ \left\{ \frac{K_1 K_2}{\tau_R} \left(\frac{\kappa}{\tau_{in}} \right)^2 + \omega_R^2 \frac{K_2 \kappa}{\tau_{in}} (\cos 2\phi_s - \alpha \sin 2\phi_s) \right\} = 0 \qquad (5.21)$$

where $K_1 = 1 - \exp(-\gamma\tau)$, $K_2 = 1 + \exp(-\gamma\tau)$, $1/\tau_R = -2\Gamma_R$, and ϕ_s is the phase defined by (4.52). The other parameters are the same as those in (4.19). Setting $\gamma = i\omega$ the stability condition for phase-conjugate feedback is calculated from the above equation and given by

$$\omega^2 - \omega_R^2 = -\left(\frac{\omega}{\tau_R} + 2\omega\frac{\kappa}{\tau_{in}}\cos 2\phi_s\right)\tan\left(\frac{\omega\tau}{2}\right) \qquad (5.22)$$

Here, we assume that the feedback is small and the relations of $\omega_R^2 \gg (\kappa/\tau_{in})^2$, $\kappa/\tau_{in}\tau_R$ hold.

For conventional optical feedback, $\omega^2 - \omega_R^2$ is proportional to $\cot(\omega\tau/2)$ (see (5.4)), while it is proportional to $\tan(\omega\tau/2)$ in phase-conjugate feedback. Periodic stability conditions with period $\tau = 2\pi/\omega_R$ are obtained both for conventional and phase-conjugate feedback. However, the positions of the external mirror at which the stability condition is satisfied are located alternately along the time axis of τ. Figure 5.30a is the calculated result from the rate equations for the boundary between stable fixed state and period-1 oscillation. The solid line shows the result for the phase-conjugate feedback,

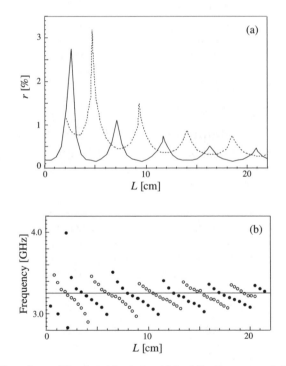

Fig. 5.30. a Boundary of fixed stable state at bias injection current $J = 1.3J_{th}$. The *solid line* corresponds to the case of phase-conjugate feedback, while the *dotted line* to optical feedback with a conventional plain mirror (corresponding to the stable fixed boundary in Fig. 5.12a). **b** Variations of excited resonance mode frequency. *Black circles* are for phase-conjugate feedback and white ones for conventional mirror feedback

while the broken line is for the conventional feedback as already shown in Fig. 5.12a. Periodic stability enhancement is observed with a period equal to the length corresponding to the relaxation oscillation frequency of the solitary laser, but stability peaks between the cases for phase-conjugate and conventional mirrors are located alternately with each other. Figure 5.30b shows the accompanying frequency at which the laser is destabilized, i.e., the frequency excited at the stability boundary. Closed circles are for phase-conjugate feedback and the open circles are for conventional optical feedback. The positions of jumps are periodic and the frequency jumps at the stability peaks. Also the jump positions between the cases for phase-conjugate and conventional mirrors are located alternately with each other.

5.6.2 Dynamics Induced by Phase-Conjugate Feedback

In this section, we briefly remark on the dynamics induced by phase-conjugate feedback in the special case of zero detuning $\delta = 0$. Bifurcation diagrams were obtained for optical feedback from a phase-conjugate mirror (Murakami et al. 1997). When the reflectivity is small, the laser stays stable and it evolves into periodic oscillations for the increase of the reflectivity. However, clear period-doubling bifurcation is not visible and the laser soon evolves into quasi-periodic and chaotic oscillations after stable fixed state. Furthermore, the laser shows less stable oscillations in the phase-conjugate feedback compared with conventional optical feedback. Namely, the laser is easily destabilized by phase-conjugate feedback.

The phase diagram of the chaotic evolutions was calculated and a similar diagram to Fig. 5.12a was obtained, though the positions of periodic stability peaks are different from those of conventional optical feedback. The differences of dynamics between the systems of conventional and phase-conjugate feedback are recognized from their mode distributions. As discussed in Fig. 4.5, the laser was first destabilized by the excitation of the relaxation oscillation modes for the increase of the external reflectivity. The remarkable difference is the distribution of the second modes, which are excited after the relaxation oscillation mode. For conventional optical feedback, the laser evolves into unstable oscillations by the excitation of this mode after period-1 oscillation. The mode corresponds to the fundamental oscillation of the external cavity mode and the frequency is approximately calculated as $\nu = c/2L$ (L being the external cavity length). The frequency of the second mode in the case of phase-conjugate feedback clearly differs from that of conventional optical feedback and the value of the real part of the solution is much higher than that of conventional optical feedback. Therefore, the laser is quickly destabilized by the increase of the phase-conjugate reflectivity (Murakami and Ohtsubo 1998).

The excited second mode frequency in conventional optical feedback is linearly proportional to the external mode frequency with a slope of unity. However, in the phase-conjugate case, the excited mode frequency is less than that for the conventional feedback and it shows a periodic undulation for the change of the external cavity length. The period is also equal to the relaxation oscillation frequency of the solitary laser (Murakami et al. 1997). Chaotic bifurcations also existed for the change of the bias injection current in the presence of phase-conjugate feedback (like in Figs. 5.9 and 5.10). The laser frequency is a function of the bias injection current and modes generated by phase-conjugate feedback are successively selected for the increase of the bias injection current. Then, the output power jumps at the positions of the mode jumps. Also, the positions of jumps in the L-I characteristic in phase-conjugate feedback occur alternately to those of conventional mirror feedback.

5.6.3 Dynamics in the Presence of Frequency Detuning

In the preceding section, we ignored the frequency detuning between the pump frequency of the four-wave mixing and the laser oscillation to calculate the dynamics of phase-conjugate feedback. For four-wave mixing phase-conjugate feedback having frequency detuning, different dynamics from a phase-conjugate mirror with degenerate four-wave mixing are observed. In the case of phase-conjugate optical feedback with zero detuning, the laser evolves from stable state to chaotic oscillations through a period-doubling like bifurcation. However, different behaviors are observed for the dynamics of non-degenerate four-wave mixing (Murakami et al. 1997). Though it is not clear, we can observe period-1 oscillations with very small amplitude in a region of small phase-conjugate reflectivity. The frequency of the oscillations depends on the external reflectivity, but it is close to the frequency detuning of 2δ. In the presence of feedback from a non-degenerate four-wave mixing mirror, the laser first oscillates at the frequency of the four-wave mixing of $\omega_0 + 2\delta$. For the increase of the feedback reflectivity, the laser oscillation is locked to the pump frequency of $\omega_0 + \delta$ and, then, the laser evolves into chaotic oscillations with the same frequency as the pump. Figure 5.31 shows the phase diagram of the oscillation states in the phase space of the frequency detuning and the phase-conjugate reflectivity. The gray area in the figure corresponds to stable oscillations with locking state to the laser frequency. Above the area, the region of chaotic bifurcations locked to the pump frequency exists. We did not consider the gain saturation effects in the calculations. When we take into consideration the gain saturation in the rate equations, the detailed dynamics are slightly different from those without the effects, but the fundamental dynamics are almost the same.

5.6.4 Finite and Slow Response Phase-Conjugate Feedback

We here consider a finite or slow response phase-conjugate feedback in semiconductor lasers. For a finite response of a phase-conjugate mirror, the equation for the complex field has already been given by (4.56). The important

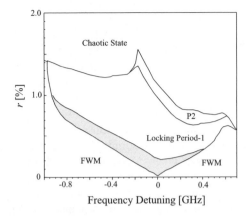

Fig. 5.31. Phase diagram for frequency locking with variation of detuning in the phase-conjugate mirror. The injection current and the external cavity length are $J = 1.4J_{th}$ and $L = 5$ cm. P2: period-2 oscillation, FWM: four wave mixing

parameter of finite response phase-conjugate feedback is the penetration time t_m of light into the phase-conjugate medium. Depending on this parameter, we observe remarkable shifts of peaks in the stability enhancement curve such as shown in Fig. 5.30. Figure 5.32 shows the stability diagram for the feedback rate $\gamma = \kappa/\tau_{in}$ and the delay time τ (unit of $2\pi/\omega_R$) at three different penetration times (van der Graff et al. 2001). The parameter Λ in the figure is the same definition as the spectral linewidth defined in the filtered optical feedback in (4.69). For the instantaneous response ($t_m = 0$), the stability peak enhancements are compatible with those of Fig. 5.30, while the peaks shift for the increase of the penetration time. Comparing PCF from an in-

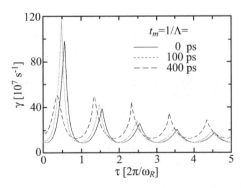

Fig. 5.32. Feedback rate at which the laser changes stability as a function of the cavity round-trip time τ, normalized to the relaxation oscillation period. The laser is stable for low feedback and becomes unstable upon increasing γ. In the simulations, the parameters are $\delta = 0$ and $J = 1.05J_{th}$ (after van der Graaf WA, Pesquera L, Lenstra D (2001); © 2001 IEEE)

stantaneous response mirror with that from a slow mirror, we see that the stability enhances slightly with increasing $t_{\rm m}$. This is caused by the mirror-induced spectral filtering of the reflected field, suppressing frequencies larger than $1/t_{\rm m}$. Similar behavior was found in the case of conventional optical feedback (Yousefi and Lenstra 1999). More striking is the shifted location of the stability peaks, which, in view of the time delay in the mirror, resembles a situation of an external round-trip length larger than in an instantaneous mirror. The effective round-trip length enhancement is not sharply defined, which reduces the quality of the resonance for large $t_{\rm m}$. Since the relative importance of this increases when τ gets smaller, this may explain why the peak at $\omega_{\rm R}\tau/2\pi = 1/2$ is lower for 400 ps than for shorter response times.

When the response time of a phase-conjugate mirror is very large compared with chaotic variations of semiconductor lasers, the laser behaves completely different from a laser with fast response phase-conjugate optical feedback. For slower time response of phase-conjugate mirrors, the time dependent features in the dynamics like phase locking phenomenon are lost. For example, the time response of a photorefractive phase conjugate medium, such as a photorefractive crystal of $BaTiO_3$ and $LiNbO_3$, and other photorefractive polymer materials, is very slow compared with time fluctuations of chaotic semiconductor lasers. Therefore, the grating formed in a photorefractive crystal can be considered as a static grating. Once the grating is formed in a photorefractive mirror, the dynamics are only governed by the total feedback loop of the pump beam. This loop plays the role of an external cavity length. As a result, the dynamics are completely the same as those for a conventional mirror feedback except for the generation of a spatial phase-conjugate wave. The dynamics in such photorefractive phase-conjugate feedback are quite different from those for a fast response phase-conjugate mirror. Indeed, the same results as those for conventional optical feedback have been experimentally obtained for a slow response phase-conjugate mirror using a $BaTiO_3$ crystal (Murakami and Ohtsubo 1999). However, a photorefractive mirror with slow time response may be a good medium as a feedback reflector for positive use of optical feedback effects, since it functions as an automatical feedback reflector without any adjustment for optical components in the systems. The reflection fraction of the feedback beam can be easily controlled by changing the pump ratio for phase-conjugation.

5.7 Dynamics of Semiconductor Lasers with Incoherent Optical Feedback

5.7.1 Dynamics of Incoherent Optical Feedback

For incoherent optical feedback in semiconductor lasers, the dynamics can be described by only two rate equations for the photon number and the carrier density. The feedback effect is included in the field equation for coherent feedback, while it appears as a modulation to carrier density for the case of incoherent optical feedback. The nonlinearity of the system is much enhanced by the delay differential term in the carrier density equation induced by incoherent optical feedback. Thus, instability and chaos can be observed in such a system. The dynamics of semiconductor lasers with incoherent optical feedback is investigated by using the rate equations in (4.55) and (4.56). Instabilities of semiconductor lasers with incoherent feedback have been studied by Otuska et al. (1991) and Ishiyama (1999). They carried out numerical simulations for incoherent optical feedback in semiconductor lasers and found sustained pulsations in the laser output. Even without optical feedback in a semiconductor laser, damping oscillation due to relaxation oscillation arises in the laser output. When there is incoherent feedback in a semiconductor laser, the damping oscillation is memorized and then fed back to the laser. The carrier density is intensely modulated by the feedback light, since the feedback is to the carrier density and the laser produces modulated output. The returned light further modulates the carrier density. When the growth rate of pulsations by the positive feedback balances with the damping forces induced by the relaxation oscillation, the laser shows sustained pulsating oscillations.

Figure 5.33 is a numerical example of pulsating oscillations of the output power in a semiconductor laser with incoherent optical feedback. Changing the bias injection current, the relaxation oscillation frequency increases and, as a result, both the pulsating frequency and the amplitude of pulses increase. In this simulation, the feedback strength and the feedback time are assumed to be $\kappa' = 0.4$ and $\tau = 0.1\tau_s$, respectively. Regular pulsing is not always observed for all the parameter ranges of the bias injection current and the coupling strength. The region of unstable and pulsating oscillations is calculated in the phase space of the bias injection current and the feedback strength. The phase diagram in Fig. 5.33b is obtained from the linear stability analysis for the rate equations. At a certain feedback strength, the laser shows stable oscillation close to the laser threshold. It evolves into period-1 oscillation (regular pulsing) at the threshold $w_{\text{th},2}$. With the increase of the bias injection current, the pulsing frequency increases and the pulse amplitude grows. However, with further increase of the bias injection current, the damping rate of the relaxation oscillation decreases and the decay rate of the carrier density also decreases. This suppresses the sustained pulsation in the laser output. Above a certain bias injection current ($w_{\text{th},3}$), the feedback-induced modulation is damped out within the delay time and finally the instability vanishes. Since the occurrence of sustained pulsating oscillations is related to the relaxation oscillation, the bias injection current for the threshold $w_{\text{th},3}$ at which the regular pulsing diminishes decreases with the increases of the delay time and the nonlinear gain.

Sustained pulsating oscillations in semiconductor lasers induced by incoherent optical feedback are used to generate a series of fast pulsing in optical communications. We can easily generate fast pulses shorter than nano-second (as fast as pico-second) under the conditions for small nonlinear gain and

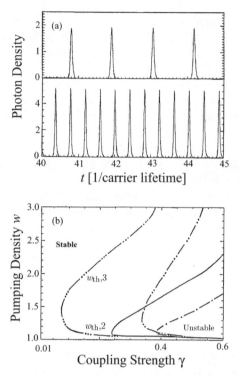

Fig. 5.33. Numerical example for sustained pulsating oscillations in a semiconductor laser with incoherent optical feedback. **a** $J = 1.1 J_{th}$ and **b** $J = 1.5 J_{th}$. The time axis is normalized by the carrier lifetime τ_s. **c** Stability boundaries for various conditions. *Solid curve:* $g = \varepsilon_s \Gamma V = 10$, $T = \tau / \tau_s = 0.1$, and $\beta = \tau / \tau_s = 0.375$. *Dashed-dotted curve:* $g = 20$, $T = 0.1$, and $\beta = 0.375$. *Dashed-double-dotted curve:* $g = 10$, $T = 0.05$, and $\beta = 0.375$. *Dashed-triple-dotted curve:* $g = 0$, $T = 0.1$, and $\beta = 0$. The normalized pump is defined by $w = J/J_{th}$ and the feedback parameter is $\gamma = \kappa' V$ (V being the volume of the active layer) (after Otsuka K, Chern JL (1991); © 1991 OSA)

also small confinement factor. In incoherent optical feedback, we have an advantage of designing the pulse sequence with an arbitrary pulsing frequency. Also, the pulse width can be designed by appropriately setting the parameter values. Chern et al. (1993) investigated the dynamics of class B lasers with incoherent optical feedback including semiconductor lasers. They showed the coexistence of two attractors under the same parameter conditions for pulsating oscillations induced by the Q switching effect and sustained relaxation oscillations. They demonstrated that either oscillation occurs depending on the initial condition of the laser oscillation.

Some research have been conducted into the experimental studies of incoherent optical feedback in semiconductor lasers (Cohen et al. 1990 and Yen 1998). An experimental study on the effects of semiconductor lasers with incoherent optical feedback has been recently conducted and pulsating oscillations have been demonstrated (Cheng et al. 2003). Figure 5.34a shows examples of experimentally obtained pulsating oscillations at the bias injection current of $J = 1.3J_{th}$. With the increase of the external cavity length, the frequency of pulses decreases. However, the feedback strength is weak and the pulsing frequency is only about 250 MHz. To obtain pulses with a much higher repetition rate as fast as nano-second, a stronger feedback ratio is required. The laser used is the single mode near-infrared semiconductor laser of the Fary-Perot type. The laser oscillates at a single mode as shown in the inset of the figure even when it shows pulsing oscillations. Figure 5.34b also presents experimentally obtained relations between the pulse width and the pulse intensity for the optical feedback ratio. As a general trend, for the increase of the feedback level, the pulse width decreases and the pulse height increases.

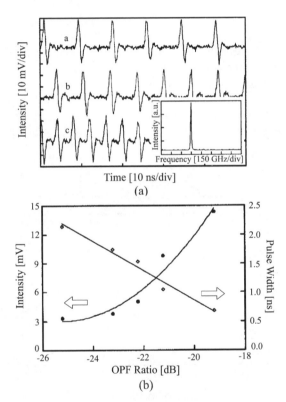

Fig. 5.34. a Experimentally obtained sustained pulsating oscillations in semiconductor lasers with incoherent feedback for different external cavity length. The external cavity lengths are $L = 133.3$, 100, and 60 cm from *top* to *bottom*. The optical feedback ratio is -19.2 dB. Inset is the optical spectrum obtained by a Fabry-Perot spectrometer. **b** Dependence of pulse width and pulse intensity on optical feedback ratio at the bias injection current of $J = 1.25J_{th}$ and external cavity length of $L = 75$ cm (after Cheng DL, Yen TC, Chang JW, Tsai JK (2003); © 2003 Elsevier)

5.7.2 Dynamics of Polarization-Rotated Optical Feedback

In Sect. 4.5.2, we formulated the scheme for polarization-rotated optical feed-back as one of example of incoherent optical feedback. Here, we discuss some dynamic properties of polarization-rotated optical feedback in semiconductor lasers. The dynamics of polarization-rotated optical feedback are quite different from those of ordinary optical feedback discussed in the previous section. Figure 5.35 shows an example of light-injection current (L-I) characteristics calculated from the rate equations of polarization-rotated feedback in (4.59–4.64) (Heil et al. 2003c). The intensities with optical feedback contain instabilities for the time development, but the plotted L-I characteristics are averaged intensities. As was already discussed, one of the typical features of parallel-polarization optical feedback is the threshold reduction of the injection current for the laser oscillation. However, there is no threshold reduction in the case of polarization-rotated feedback, and the laser oscillation starts at the same bias injection current as the solitary laser. Furthermore, the slope efficiency remains the same as that of the solitary laser. The level of polarization-rotated feedback required to induce instability in the laser output is much higher than that for the case of ordinary optical feedback, as shown in the following discussion, and the level is usually 10 times higher than that in ordinary optical feedback (in amplitude). Another typical fea-

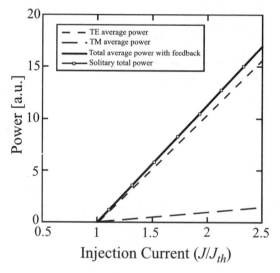

Fig. 5.35. Numerical result of light-injection current (L-I) characteristics. *Black thick line*: total output of the laser with TM-mode feedback, *solid line with squares*: solitary laser output, *dotted line*: TE-component output of the laser with TM-mode feedback, *dashed line*: TM-component output of the laser with TM-mode feedback. The optical feedback ratio is $\kappa/\tau_{in} = 7 \times 10^{10}\,\mathrm{s}^{-1}$ (after Heil T, Uchida A, Davis P, Aida T; © 2003 APS)

ture in ordinary optical feedback is the increase of the laser output power due to the coherent optical injection. On the other hand, the laser output power of polarization-rotated feedback shows little change compared with the solitary case in spite of such large optical feedback. The laser oscillation is divided into the TE- and TM-modes and the total power remains unchanged. It is noted that the temporal waveform of the TM-mode is delayed with respect to that of the TE-mode by the propagation time τ of the external feedback loop, showing that the TM-mode is following the delayed feedback signal. Namely, the TM-mode is just a copy of the TE-mode with the delay time τ. These results mean that the interaction between the TE- and TM-mode intensities through the carrier density generates chaotic intensities and the dynamics of both the amplitude and phase are governed by the dynamics of the carrier density. Therefore, the dynamics induced by polarization-rotated feedback come from incoherent origin and the scheme is categorized as incoherent feedback.

To induce instability in polarization-rotated optical feedback, one requires a strong optical feedback; however, the fully developed chaotic state, as is observed in ordinary optical feedback, is scarcely seen. In most cases, the instability observed in polarization-rotated feedback is quasi-periodic oscillation or week chaotic state. Figure 5.36 shows bifurcation diagrams of chaotic evolutions for the parameters (Heil et al. 2003c). Figure 5.36a is the bifurcation diagram of the TE laser output for the bias injection current at a fixed feedback ratio of $\kappa/\tau_{\mathrm{in}} = 7 \times 10^{10}\,\mathrm{s}^{-1}$. From the figure, when the injection current is increased, one recognizes that the value of the steady-state solution of the carrier density is increased and the laser tends to be less sensitive to the feedback light. Finally, the laser reached stable oscillation state even in the presence of orthogonal-polarization feedback. In ordinary optical feedback, the laser tends to less unstable for the increase of the bias injection current, but instabilities of the laser persist and never disappear. This is the big difference between polarization-parallel and polarization-rotated optical feedback. Another difference of polarization-rotated feedback is that the po-

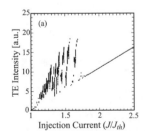 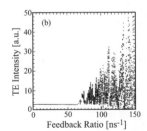

Fig. 5.36. a Bifurcation diagram of the intensity dynamics as a function of the injection current. The optical feedback ratio is $\kappa/\tau_{\mathrm{in}} = 7 \times 10^{10}\,\mathrm{s}^{-1}$. **b** Bifurcation diagram of the intensity dynamics as function of the feedback power ratio. The bias injection current is $J = 1.4 J_{\mathrm{th}}$ (after Heil T, Uchida A, Davis P, Aida T; © 2003 APS)

sitions of the spectral peaks associated the round-trip frequency do not shift with the increase of the bias injection current. This is in contrast to the dynamics of coherent parallel-polarization feedback, where a significant shift and broadening of these peaks occurs for increasing bias injection current. Figure 5.36b shows the bifurcation diagram of TE laser output for the feedback ratio (κ/τ_{in}) at a fixed bias injection current of $J = 1.4J_{th}$. The feedback ratio at which onset of chaos starts is much higher than one order or more for the case of parallel polarization feedback. The theoretical results discussed here are supported by the experimental results (Heil et al. 2003c).

To induce instability in polarization-rotated optical feedback, one requires a strong optical feedback; however, the fully developed chaotic state, as is observed in ordinary optical feedback, is scarcely seen. In most cases, the instability observed in polarization-rotated feedback is quasi-periodic oscillation or week chaotic state. Figure 5.36 shows bifurcation diagrams of chaotic evolutions for the parameters (Heil et al. 2003c). Figure 5.36a is the bifurcation diagram of the TE laser output for the bias injection current at a fixed feedback ratio of $\kappa/\tau_{in} = 7 \times 10^{10}\,\mathrm{s}^{-1}$. From the figure, when the injection current is increased, one recognizes that the value of the steady-state solution of the carrier density is increased and the laser tends to be less sensitive to the feedback light. Finally, the laser reached stable oscillation state even in the presence of orthogonal-polarization feedback. In ordinary optical feedback, the laser tends to less unstable for the increase of the bias injection current, but instabilities of the laser persist and never disappear. This is the big difference between polarization-parallel and polarization-rotated optical feedback. Another difference of polarization-rotated feedback is that the positions of the spectral peaks associated the round-trip frequency do not shift with the increase of the bias injection current. This is in contrast to the dynamics of coherent parallel-polarization feedback, where a significant shift and broadening of these peaks occurs for increasing bias injection current. Figure 5.36b shows the bifurcation diagram of TE laser output for the feedback ratio (κ/τ_{in}) at a fixed bias injection current of $J = 1.4J_{th}$. The feedback ratio at which onset of chaos starts is much higher than one order or more for the case of parallel polarization feedback. The theoretical results discussed here are supported by the experimental results (Heil et al. 2003c).

5.8 Dynamics of Filtered Optical Feedback

5.8.1 Filtered Optical Feedback

In this section, as an example of filtered optical feedback, we describe the dynamics of coherent optical feedback from a Fabry-Perot resonator. Filtered optical feedback has become a topic of interest, since it offers the potential to control the laser dynamics through the spectral width of the filter and the detuning from the original laser oscillation frequency. The dynamics of filtered optical feedback from a Fabry-Perot resonator have been extensively investigated in relation to the filter width compared with the relaxation oscillation frequency and external cavity frequency (Fischer et al. 2000b, 2004a,b, Yousefi and Lenstra 2003, and Erzgräber et al. 2006). The treatments of the filtered feedback from a Fabry-Perot resonator can be applied to other case of coherent optical feedback filter, such as grating optical feedback (Yousefi and Lenstra 1999). For example, using feedback from a diffraction grating is a common method for obtaining single-mode operation of a semiconductor laser. In such cases, a specific spectral component of the diffracted light is fed back to the laser, and this forms a good example of filtered optical feedback. The use of a Michelson interferometer is another example of a filtered optical feedback to control the behaviors of low frequency fluctuations (LFFs) in a semiconductor laser subject to optical feedback (Rogister et al. 1999). In other applications, an alkali vapor or a filter is placed within the external cavity laser system to frequency stabilize the laser. In such configurations, the laser frequency is locked to one of the transition lines of for example, rubidium or barium, or to a resonance of a Fabry-Perot interferometer (Andersen et al. 1999). All of these systems are equivalent to a semiconductor laser with an external cavity containing a frequency-selective filter.

The typical feature of filtered coherent optical feedback is frequency oscillations in which the laser intensity is almost constant in spite of the periodic laser frequency oscillations. The complex dynamics behavior in the lasers is rooted in the undamping of the intrinsic relaxations, the relatively large self-phase modulation property of semiconductor lasers, and the feedback delay. The self-modulation property manifests itself as a coupling between amplitude and phase of the laser field and originates from the existence of the non-zero α parameter. The filter can be viewed as a mechanism for restricting the phase space. By the introduction of the filter, the number of external cavity modes decreases and also the modes are also moved around in the phase space. Here, we focus on the effects of coherent optical feedback from frequency filter, and those for filtered optoelectronic feedback will be presented in Chap. 7. Figure 5.37 shows a schematic setup of filtered optical feedback in semiconductor lasers. In wide variety of applications for laser control, an optical feedback from a Fabry-Perot resonator to a semiconductor laser is used instead of a conventional reflector. Usually, we must consider multiple reflections from the filter and equation (4.68) is replaced by the reflectance for multiple loops for the filter feedback (Yousefi and Lenstra 1999). In the figure, we consider only a system of a single unidirectional ring loop for simplicity and the optical feedback through a single-path Fabry-Perot resonator. With a good approximation, the optical feedback from the Fabry-Perot filter is given by the following Lorentzian shape:

$$R(\omega) = \frac{\Lambda}{\Lambda + \mathrm{i}(\omega - \omega_c)} \qquad (5.23)$$

where Λ is the hull-width at half-maximum (HWHM) of the spectral transmission through the Pabry-Perot filter and ω_c is its center frequency relative

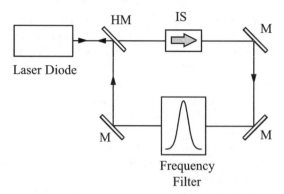

Fig. 5.37. Schematic diagram of unidirectional ring configuration for the filtered optical feedback setup. HM: half mirror, IS: isolator, M: mirror

to the solitary laser frequency. Consequently, we can assume the relation of (4.68) as the time response function. In the following, we present several dynamics for filtered optical feedback from Fabry-Perot resonator in semiconductor lasers.

5.8.2 External Cavity Modes

In this section, we find the relation of the actual laser oscillations with the solitary oscillation frequency. The first step is to investigate the steady-state regime of the filter external optical feedback system by means of the fixed point analysis for the rate equations including the filtered optical feedback effect described by (4.65), (4.57), and (4.69), i.e. the linear stability analysis, which is the same procedure in Sect. 4.2. We assume the following forms for the steady-state solutions for the optical field, the filter, and the carrier density:

$$E(t) = \sqrt{P_s} \exp(-i\Delta\omega_s t) \tag{5.24}$$

$$F(t) = \sqrt{P_{Fs}} \exp\{-i(\Delta\omega_s t + \theta_s)\} \tag{5.25}$$

$$n(t) = n_s \tag{5.26}$$

where $\Delta\omega_s = \omega_s - \omega_0$ (ω_s being the laser oscillation frequency) and P_s, P_{Fs}, and n_s are the steady-state values for the optical power, the filter transmission optical power, and the carrier density, respectively. Substituting (5.24)–(5.26) into the rate equations, the transcendental equation in $\Delta\omega_s$ can be derived as

$$\Delta\omega_s\tau = -h(\Delta\omega_s)C\sin\left(\Delta\omega_s\tau + \omega_0\tau + \tan^{-1}\alpha - \tan^{-1}\frac{\Delta\omega_s - \omega_c}{\Lambda}\right) \tag{5.27}$$

with

$$h(\Delta\omega_s) = \frac{\Lambda}{\sqrt{\Lambda^2 + (\Delta\omega_s^2 - \omega_c)^2}} \tag{5.28}$$

The C parameter in (5.27) is the measure of the fraction of feedback and has already been defined by (4.13). The constant phase θ_s is also obtained from the linear stability analysis and given by

$$\theta_s = -\omega_0\tau - \tan^{-1}\frac{\Delta\omega_s - \omega_c}{\Lambda} \tag{5.29}$$

For the infinite filter bandwith ($\Lambda \to \infty$), equation (5.27) reduces to (4.14) of the case of conventional optical feedback. Since $h(\Delta\omega_s)$ in (5.27) is smaller than unity, the relation $h(\Delta\omega_s)C < C = \kappa\tau/\tau_{in}\sqrt{1+\alpha^2}$ always holds. Therefore, a comparison with conventional optical feedback shows that the number of fixed points for filtered optical feedback is always smaller than for conventional optical feedback. Figure 5.38 is an example of the fixed point frequencies calculated from (5.27) (Yousefi and Lenstra 1999). In the case of frequency optical feedback, the snake-like contour in Fig. 5.38 is modulated by the filter profile. This is due to the envelope $h(\Delta\omega_s)$ in (5.27) modulating the sine function. This means that, for a given ω_0, the number of external

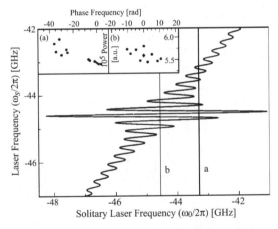

Fig. 5.38. Fixed-point solutions to the phase relation (5.27) in the (ω_s, ω_0) plane at $C = 51$ and $\Lambda = 2\,\text{GHz}$. The insets show the fixed points in the ($\phi(t) - \phi(t-\tau)$, P) plane for the two cases **a** and **b** as indicated in the main frame. In **a**, the detuning is large enough to split the fixed-point body into two separate islands, creating a potentially globally bistable situation. In **b**, the filter profile is seen to be superimposed on the standard fixed-point ellipse of conventional optical feedback. The diamond in **b** is the solitary mode (after Yousefi M, Lenstra D (2003); © 2003 AIP)

cavity modes is strongly dependent on the detuning of ω_0 with respect to ω_c. For the frequency ω_0 far away from the central frequency ω_c of the filter, $h(\Delta\omega_s)$ will be almost zero, and there will only be a few solutions, while several modes are available for $\omega_0 \approx \omega_c$. Another difference from conventional optical feedback is the phase shift introduced by $\tan^{-1}\{(\Delta\omega_s - \omega_c)/\Lambda\}$. This term does not exist in the case of conventional optical feedback and it is a consequence of the causality principle (Kramers–Kronig relations) applied to the Lorentzian filter. Figure 5.38 is the case of moderate feedback of intermediate filter where the relation $\nu_{ex} < \Delta\nu_f < \nu_R$ holds (ν_{ex}, $\Delta\nu_f$, and ν_R being the frequencies of the external cavity, the filter bandwidth, and the relaxation oscillation).

5.8.3 Frequency Oscillations and Chaotic Dynamics

A typical feature of optical feedback from a Fabry-Perot filter is frequency oscillations for which the laser intensity is almost constant, and only the laser frequency oscillates. The origin of this phenomenon has been unclear at present and the further study for understanding the mechanism is still on going. Figure 5.39 shows an experimental example of frequency oscillations in filtered optical feedback in semiconductor lasers (Fischer et al. 2004a). The example corresponds to an intermediate filter case of the frequencies of $\nu_{ex} = 50\,\text{MHz}$, $\Delta\nu_f{=}700\,\text{MHz}$, and $\nu_R = 4 \sim 5\,\text{GHz}$. The relative center frequency ω_f of the Fabry-Perot filter is 200 MHz. The observed periodic frequency oscillation is 19.6 ns, which quite agrees with the round trip time of the external delay. From the detailed analysis, the frequency is not simply

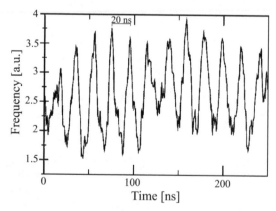

Fig. 5.39. Experimentally observed time series showing oscillations in the frequency of the laser when subject to filtered optical feedback. The period of the oscillations corresponds to an external delay of 6 m ($\sim$20 ns). The HWHM bandwidth of the filter is 700 MHz (after Fischer APA, Yousefi M, Lenstra D, Carter MW, Vemuri G (2004); © 2004 AIP)

equal to the round trip time but the round trip time pulse the time corresponding to the frequency shift of the filter, $\sim 1/\Lambda$. Even in the frequency oscillation state, the measured laser power shows no evidence of oscillatory behaviors, thus the phenomenon is called "frequency oscillation".

The dynamics of filtered optical feedback in semiconductor lasers are strongly affected both by the frequency relations among the bandwidth of the filter, the external cavity, and the relaxation oscillation of the laser, and the feedback strength from the filter. Erzgräber et al. (2006) investigated the dynamics of filtered optical feedback in semiconductor lasers based on the numerical simulations from the rate equations of (4.65), (4.57), and (4.69). Figure 5.40 shows their results for an intermediate filter case of the frequencies of $\nu_{ex} = 200\,\mathrm{MHz}$, $\Delta\nu_f = 700\,\mathrm{MHz}$, and $\nu_R = 4.2\,\mathrm{GHz}$. For a weak but not such small optical feedback case in Fig. 5.40a, the laser shows relaxation oscillations at a frequency of 4.2 GHz. Since the HWHM of the filter is narrow enough compared to the relaxation oscillation frequency, the intensity I_F transmitted through the filter (and also fed back in the laser cavity) is almost constant. Figure 5.40b is an example of frequency oscillations for the dynamics of filtered optical feedback. In contrast to the relaxation oscillations in Fig. 5.40a, the laser intensity is almost constant, while its frequency $\dot{\phi}_E$ oscillated at the typical time period of the addition of the round trip time and the time shift in the frequency filter. Note that the laser frequency $\dot{\phi}_E$ and the feedback intensity I_F are approximately in anti-phase for this frequency oscillation. Fischer et al. (2004a) postulated that the frequency oscillations are interpreted as an interplay between the filter and the laser that compensates for the effects of the amplitude-phase coupling, leading to an effectively zero α parameter. Also, Rogister et al. (1999) argued that filtered optical feedback, in this case from a double mirror, effectively reduces the value of α. Finally, Fig. 5.40c shows that frequency oscillations can undergo further bifurcations, i.e. a torus bifurcation. This dynamical regime differs from the

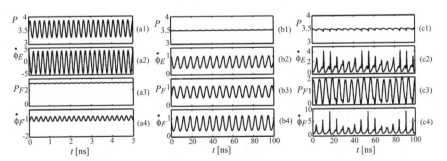

Fig. 5.40. Time series of P, P_F, $\dot{\phi}_E$, and $\dot{\phi}_F$ of **a** relaxation oscillations, **b** frequency oscillations, and **c** quasi-periodic frequency oscillations. From **a** to **c**, $(\tau_{ph}\kappa/\tau_{in}, \omega_0\tau)$ takes the values $(0.02, 4\pi/3)$, $(0.007, -2\pi)$, and $(0.014, -2\pi)$, respectively (after Erzgräber H, Krauskoph B, Lenstra D, Fischer PA, Vemuri G (2006); © 2006 AIP)

quasi-periodic dynamics associated with relaxation oscillations, in that there is again only very small intensity dynamics. Notice that the frequency oscillations exhibit a slow modulation with a period about 6 times larger than the basic oscillation. This ratio may crucially depend on other parameters.

In the following, we summarize the dynamics of filtered optical feedback in semiconductor lasers. Fischer et al. (2000b) systematically investigated the dynamics for the strength of filtered optical feedback. Similar trends to conventional optical feedback have been observed for the increase in feedback strength, although the dynamics are more or less affected by the filter profile and several frequencies related to the filter, the laser device, and the feedback length. Namely, for the increase of the feedback strength (the increase of the C parameter), we can observe discrete external mode selections and hopping, hysteresis for increase or decrease of the bias injection current versus the laser power, and coherence collapse states. Also, following Fischer et al. (2004b), the dynamics of filtered optical feedback in semiconductor lasers are summarized for the frequency relations among the bandwidth of the filter, the external cavity, and the relaxation oscillation of the laser, and the feedback strength from the filter.

1) Wide filter case: $\Delta\nu_f > \nu_R$

Since the filter bandwidth is larger than the relaxation oscillation frequency, the relaxation-oscillation-side-peak falls within the filter profile when the filter is centered at the solitary laser frequency. This case is very close to conventional optical feedback, wherein, one can have several hundred external cavity modes under the filter profile, and so the dynamics can be quite complicated.

2) Intermediate filter case: $\nu_{ex} < \Delta\nu_f < \nu_R$

One can have a few tens of external cavity modes under the filter profile, such that the dynamics are more complicated than the narrow filter case, but in this regime one has the possibility to control dynamics.

3) Narrow filter case: $\Delta\nu_f \ll \nu_R$ and $\Delta\nu_f < \nu_{ex}$

In this case, the filter is so narrow that at most one external cavity mode lies under the filter profile and so the laser prefers to operate on that single external cavity mode. Yet even in this case, dynamics are still possible. One mechanism for dynamics is the destabilization of the external cavity mode through a Hopf bifurcation, leading to undamped relaxation oscillations. Another mechanism is to detune the solitary laser by one relaxation oscillation frequency with respect to the filter center. The feedback channel via the relaxation oscillation side peak will likely give rise to undamped relaxation oscillations. In fact, these two mechanisms are also active in the intermediate case.

6 Dynamics in Semiconductor Lasers with Optical Injection

Since the semiconductor laser has unique features of high gain, low facet reflectivity, and amplitude-phase coupling through the α parameter, it is also sensitive to optical injection from a different laser. Locking and unlocking phenomena in optically injected semiconductor lasers have been extensively studied. Especially optical injection locking has been appreciated as a useful tool for controlling and stabilizing laser oscillations. The general application of optical injection is to control the laser and the locking condition is extensively investigated to distinguish the unlocking phenomena. Little attention has been paid to the unlocking dynamics. However, recent studies proved that rich varieties of dynamics such as the four-wave mixing, period-doubling route to chaos, and non-locking beating, are involved in the unlocking region. In this chapter, we focus on the dynamic characteristics of locking and unlocking regimes in optically injected semiconductor lasers.

6.1 Optical Injection

6.1.1 Optical Injection Locking

The technique of optical injection locking is frequently used to lock the frequency and stabilize the oscillation of a slave laser. The injection locking system is very simple, as shown in Fig. 6.1. Injection-locked semiconductor lasers are very useful for stabilizing the laser, however they sometimes shows a rich variety of dynamics. For optical injection locking, we prepare two lasers with almost the same oscillation frequencies and the frequency detuning between them must usually be within several GHz. A light from a laser under a single mode oscillation (master laser) is fed into the active layer of the other laser (slave laser). Then, the two lasers synchronize with

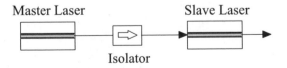

Master Laser   Slave Laser

Isolator

Fig. 6.1. Optical injection system in semiconductor lasers

each other in the same optical frequency under the appropriate conditions of the frequency detuning and the injection strength. The remarkable characteristics of optical injection locking in semiconductor lasers originated from the fact that the α parameter (linewidth enhancement factor) has a non-zero definite value, which makes semiconductor lasers very different from other lasers. As a viewpoint of laser dynamics, an optical injection from a different laser means the introduction of an extra degree of freedom to the semiconductor laser. Therefore, various dynamics are observed by optical injection, including stable and unstable injection locking, instabilities and chaos, and four-wave mixing depending on the locking conditions (Mogensen et al. 1985, Sacher et al. 1992, Lee et al. 1993, Annovazzi-Lodi et al. 1994, Liu and Simpson 1994, Simpson et al. 1995b, Kovanis et al. 1995, Erneux et al. 1996, De Jagher et al. 1996, Simpson et al. 1997, Gavirielides et al. 1997, and Eriksson and Lindberg 2001).

Optical injection technique is originally developed for the stabilization of the injected slave laser, so that, at first glance, it may be surprising that the laser is destabilized by the optical injection. However, optical injection to a laser is an introduction of the extra degree of freedom from a viewpoint of nonlinear dynamics and the perturbed laser is a candidate of a chaotic system. Figure 6.2a shows an example of bifurcation diagram of the slave laser for the change of the frequency detuning between the master and slave lasers at a fixed optical injection rate. In this map, the number of sampling points is not large enough to show the states and each bifurcation may not be clear. However, we can see stable and unlocking oscillations, and various unstable oscillation states for the change of the frequency detuning. Figure 6.2b and c shows the time series and rf spectrum at the frequency detuning of $\Delta f = 2.56\,\mathrm{GHz}$. Similarly to chaotic oscillations for the case of optical feedback in the previous section, we can observe chaotic oscillations. As we will see in the following, periodic and unstable oscillations are observed in adjacent to the stable injection locking state. Also, unlocking oscillations are distributed for large values of the large frequency detuning. We will later present the instabilities and chaotic dynamics by optical injection and, here, we investigate the principle of optical injection locking in semiconductor lasers.

Again the tool for investigating the characteristics of optical injection locking is the linear stability analysis. We assume that the frequency detuning $\Delta\nu = \Delta\tilde{\omega}/2\pi$ is small and the fraction of the photon number S_m (optical injection power $S_\mathrm{m} = |E_\mathrm{m}|^2 = A_\mathrm{m}^2$) from a master laser is also small compared with the photon number S_s of a slave laser. As usual, the injection strength to the slave laser may or may not be small, but for the moment, we consider the case of a rather small injection. Here, we discuss the effects to understand the principle of optical injection locking. It is also noted that a laser may show unstable and chaotic oscillations for a small injection fraction under certain injection conditions, as we will see in the following sections. We use a steady-state complex field $E_\mathrm{m}(t) = \sqrt{S_\mathrm{m}}\exp\{-\mathrm{i}\phi_\mathrm{m}(t)\}$ for a master laser and assume

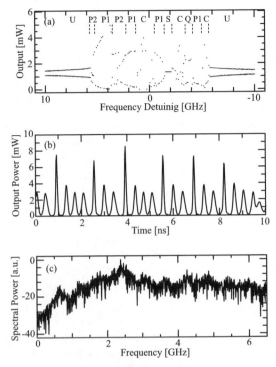

Fig. 6.2. a Bifurcation diagram for the frequency detuning at a fixed optical injection ration of $r_{inj} = 3\%$ and a bias injection current of $J = 1.3J_{th}$. **b** Example of the time series of chaotic states at the frequency detuning of $\Delta f = 2.56$ GHz. **c** The rf spectrum corresponding to **b**. S: stable state, U: unlocking state, P1: period-1 oscillation, P2: period-2 oscillation, Q: quasi-periodic oscillation, C: chaotic oscillation

the complex field $E_s(t) = \sqrt{S_s}\exp\{-i\phi_s(t)\}$ for a slave laser. The phases ϕ_m and ϕ_s are generally time dependent functions, but the master laser is under steady-state operation and its phase is assumed to be $\phi_m = 0$. Though the phase of the slave laser fluctuates with time, it is approximated as a small fluctuation and assumed to be a constant value in the following. Taking these assumptions into consideration, the rate equation for the slave field is written by

$$\frac{dE_s(t)}{dt} = \frac{1}{2}(1 - i\alpha)G_n\{n(t) - n_{th}\}E_s(t) + \frac{\kappa_{inj}}{\tau_{in}}E_m(t)\exp(-i\Delta\omega t) \quad (6.1)$$

where $\Delta\omega = 2\pi\Delta\nu = \omega_m - \omega_s$ is the detuning between the angular frequencies, ω_m and ω_s, for the master and slave lasers, respectively, κ_{inj} is the injection coefficient, and τ_{in} is the round trip time of light in the laser cavity as introduced before.

6.1.2 Injection Locking Condition

Optical injection locking is a coherent phenomenon, so that the discussion must be based on the complex field instead of the photon number. As the carrier density of the slave laser is affected by optical injection, we put the fluctuation of it as δn. We introduce a phase $\psi(t) = \phi_{\rm s}(t) - \phi_{\rm m}(t) - \Delta\omega$ and a small deviation between the photon numbers with and without the optical injection as $S_{\rm s} - S_{\rm 0s}$ ($S_{\rm 0s}$ is the photon number of the slave laser in the absence of the optical injection). By the use of the representation of $\psi(t)$ instead of $\phi_{\rm s}(t) - \phi_{\rm m}(t)$, we can define the rate equations as autonomous equations. Then, we obtain the solutions $S_{\rm s} - S_{\rm 0s}$, $\psi_{\rm s}$, and δn for the steady-state values

$$\frac{1}{2}G_n\delta n + \frac{1}{\tau_{\rm in}}\sqrt{\frac{S_{\rm inj}}{S_{\rm s}}}\cos\psi_{\rm s} = 0 \tag{6.2}$$

$$\Delta\omega = \frac{1}{2}\alpha G_n\delta n - \frac{1}{\tau_{\rm in}}\sqrt{\frac{S_{\rm inj}}{S_{\rm s}}}\sin\psi_{\rm s} \tag{6.3}$$

$$\delta n = -\left(\frac{1}{\tau_{\rm s}} + G_n S_{\rm s}\right)\delta n - \left(\frac{1}{\tau_{\rm ph}} + G_n\delta n\right)(S_{\rm s} - S_{\rm 0s}) \tag{6.4}$$

where $S_{\rm inj} = \kappa_{\rm inj}^2 S_{\rm m}$. In the above equation, replacing the fluctuation of the carrier density with $x = G_n\delta n$ and eliminating the variables $\phi_{\rm s}$ and $S_{\rm s}$, we obtain the characteristic equation as follows;

$$-\frac{1}{4\tau_{\rm s}}\left(1+\alpha^2\right)x^3 + \left\{\frac{1}{4}\left(1+\alpha^2\right)\omega_{\rm R}^2 + \frac{1}{\tau_{\rm s}}\right\}x^2$$
$$-\left(\alpha\Delta\omega\omega_{\rm R}^2 + \frac{\Delta\omega^2}{\tau_{\rm s}} + \frac{1}{\tau_{\rm in}^2}G_n S_{\rm inj}\right)x + \omega_{\rm R}^2\left(\Delta\omega^2 - \frac{1}{\tau_{\rm in}^2}\frac{S_{\rm inj}}{S_{\rm 0s}}\right) = 0 \tag{6.5}$$

From the above equation, we obtain the solutions for the fluctuation of the carrier density. Eliminating δn in (6.2) and (6.3), we also obtain the relation between the phase $\psi_{\rm s}$ and the laser powers as

$$\sin(\psi_{\rm s} + \tan^{-1}\alpha) = -\frac{\tau_{\rm in}\Delta\omega}{\sqrt{1+\alpha^2}}\sqrt{\frac{S_{\rm s}}{S_{\rm inj}}} \tag{6.6}$$

For the condition of optical injection locking in real laser systems, the absolute value of the right hand side of (6.6) must be less than unity. The condition reads

$$|\Delta\omega| \le \Delta\omega_{\rm L} = \frac{\sqrt{1+\alpha^2}}{\tau_{\rm in}}\sqrt{\frac{S_{\rm inj}}{S_{\rm s}}} \tag{6.7}$$

Successful optical injection locking occurs at a frequency satisfying the above equation for the injection fraction $S_{\rm inj}/S_{\rm s}$. The α parameter encountered in the above equation plays an important role in optical injection locking.

Using the relation of (6.7), the fluctuation of the carrier density δn is given by

$$\delta n = \frac{2\alpha\Delta\omega \pm 2\sqrt{\Delta\omega_{\mathrm{L}}^2 - \Delta\omega^2}}{G_{\mathrm{n}}(1 + \alpha^2)} \tag{6.8}$$

In the above equation, the plus and minus signs denote that the corresponding solution for (6.6) has a phase value of zero or π radian. There exist two solutions for the same photon number S_{s}. One is a stable solution and the other is unstable. In general, optical injection locking occurs at or close to the stable solution. Figure 6.3 shows the areas of optical injection locking in the phase space for the frequency detuning between the master and slave lasers and the injection ratio. The solid curves show the boundaries between optical injection locking and non-locking regions. In the non-locking region, we can expect various dynamics such as chaotic oscillations and four-wave mixing when the detuning is not so far from zero. Indeed, we can observe various dynamics when the frequency detuning and the injection ratio are small in these regions. Within the region of the optical injection locking, there are stable and unstable locking areas. The boundary of the unstable and stable injection locking areas is denoted by a dotted curve. In the unstable injection locking area, we can observe chaotic bifurcations for certain parameter ranges. The asymmetric feature of stable injection locking again originated from the fact that the α parameter has a non-zero value in semiconductor lasers.

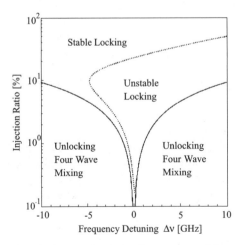

Fig. 6.3. Locking and unlocking regions in phase space of frequency detuning and injection field

6.2 Stability and Instability in Optical Injection Systems

6.2.1 Rate Equations

Side modes are sometimes excited in the oscillation of a semiconductor laser subjected to optical injection even if it operates at a single mode under a solitary condition. Therefore, we must take into account the effect of side modes into the rate equations. However, we first rewrite the rate equations of an optically injected semiconductor laser for a single mode operation and, after that, we introduce the side mode effect. To investigate the dynamics of optically injected semiconductor lasers, we again introduce the equations of the field amplitudes for the master and slave lasers, $A_m(t)$ and $A_s(t)$, the phase $\psi(t)$, and the carrier density $n(t)$ as

$$\frac{dA_s(t)}{dt} = \frac{1}{2}G_n\{n(t) - n_{th}\}A_s(t) + \frac{\kappa_{inj}}{\tau_{in}}A_m(t)\cos\psi(t) \tag{6.9}$$

$$\frac{d\psi(t)}{dt} = \frac{1}{2}\alpha G_n\{n(t) - n_{th}\} - \frac{\kappa_{inj}}{\tau_{in}}\frac{A_m(t)}{A_s(t)}\sin\psi(t) - \Delta\omega \tag{6.10}$$

$$\frac{dn(t)}{dt} = \frac{J}{ed} - \frac{n(t)}{\tau_s} - G_n\{n(t) - n_0\}A_s^2(t) \tag{6.11}$$

$$\psi(t) = \phi(t) - \Delta\omega t \tag{6.12}$$

In the above equations, we do not consider the gain saturation terms, however the gain saturation also plays an important role in multimode oscillations in semiconductor lasers with optical injection. For such a case, we can use the relation of (3.45) or (3.46) for the gain saturation effect. The assumption of a single mode operation can be well applied to a DFB semiconductor laser. However, a single mode Fabry-Perot semiconductor laser, it is easily destabilized and oscillated at a multimode by the introduction of optical injection.

When the side mode effects are important, the complex field for the main mode is rewritten as (Ryan et al. 1994)

$$\frac{dE_s(t)}{dt} = \frac{1}{2}(1 - i\alpha)G_n\{n(t) - n_{th}\}E_s(t) - \frac{1}{2}\{\varepsilon'|E_s(t)|^2 + \theta_c|E_{s'}(t)|^2\}E_s(t)$$
$$+ \frac{\kappa_{inj}}{\tau_{in}}E_m(t)\exp(-i\Delta\omega t) \tag{6.13}$$

where $E_{s'}$ is the complex field of the side mode and θ_c is the cross-saturation coefficient for the gain. Here, we consider the excitation of one side mode and also take into account the self-saturation effect $\varepsilon' = \varepsilon_s(1 - i\alpha)G_n$. The rate equation for the complex field of the side mode is given by

$$\frac{dE_{s'}(t)}{dt} = \frac{1}{2}(1 - i\alpha)G_n\{n(t) - n_{th}\}E_{s'}(t)$$
$$- \frac{1}{2}\{\varepsilon'|E_{s'}(t)|^2 + \theta|E_s(t)|^2\}E_{s'}(t) - \mu_d E_{s'}(t) \tag{6.14}$$

where the final term is the gain defect of the secondary mode and μ_{d} is the coupling coefficient called gain defect.

Due to this gain defect, mode switching will be suppressed in this model and the laser is assumed to always be oscillated at the main mode as far as the coefficient has a significant value. Using these two modes, the carrier density equation is written by

$$\frac{dn(t)}{dt} = \frac{J}{ed} - \frac{n(t)}{\tau_{\mathrm{s}}} - G_{\mathrm{n}}\{n(t) - n_0\}\{|E_{\mathrm{s}}(t)|^2 + |E_{\mathrm{s}'}(t)|^2\} \qquad (6.15)$$

In actual fact, many side modes may be excited in the laser oscillations due to optical injection. However, it is proved that the model introduced here well explains mode excitations for real oscillations in a Fabry-Perot semiconductor laser subjected to optical injection.

The laser gain is usually linearized for the carrier density. However, in a strict sense, it is also a function of the photon number and the gain term $g' = (1 - i\alpha)G_{\mathrm{n}}\{n(t) - n_{\mathrm{th}}\}$ is replaced by

$$g' = (1 - i\alpha)G_{\mathrm{n}}\{n(t) - n_{\mathrm{th}}\} + (1 - i\alpha')G_{\mathrm{P}}\{|E(t)|^2 - |E_0|^2\} \qquad (6.16)$$

where G_{P} is the expansion coefficient for the photon number $S = |E|^2$, α' is the coefficient for the saturation of the output power, and E_0 is the steady-state field amplitude. For the model of a two-level atom in laser oscillations, we can approximate the coefficient α' equal to α, while it reduces to zero under the resonance condition (Simpson et al. 2001). Stability and instability of semiconductor lasers for optical injection are strongly dependent on the linewidth enhancement factor α and also on the coefficient α' of the saturation. It is proved in the following that this nonlinear coefficient α' is related to the suppression of the laser instabilities. Namely, the laser is stabilized for a larger value of this factor, while it shows instabilities for a small value of it. The damping term μ_{a} introduced in the side mode equation in (6.14) also plays an important role in the dynamics as discussed in the following.

In semiconductor laser systems of optical injection, optical feedback, and optoelectronic feedback, we can observe multistability and coexistence states of chaotic oscillations in the dynamics. At coexistence states, the respective chaotic attractor is completely different from others even for a particular set of the parameter values. Which state we can observe is strongly dependent on the initial conditions of the systems. Coexistence states of attractors are not only simulated by numerical calculations, but also experimentally observed. Some such examples in optical injection systems will be discussed in Sect. 6.2.4 (Wieczorek et al. 1999 and 2000).

6.2.2 Chaotic Bifurcations by Optical Injection

The important parameters in the dynamics of optically injected semiconductor lasers are the frequency detuning between the master and slave lasers and the injection strength from the master to the slave. Figure 6.4 shows the experimental results of the dynamic characteristics in a semiconductor laser subjected to optical injection (Simpson 2003). The figure shows the plots of optical frequencies observed by a Fabry-Perot spectrometer (left column) and rf power spectra obtained by a spectrum analyzer (right column). Chaotic bifurcations are well demonstrated by the plots. We can assume a single mode operation for the semiconductor laser even in the presence of optical injection, since the laser used is a DFB laser. As is easily recognized from the stability map in Fig. 6.3, the slave laser operates outside of the stable locking region for a small optical injection. When the injection fraction exceeds a certain threshold, the laser is injection-locked by the master laser and operates stably. It is noted that the injection strength defined in the figure is not exactly equal to the intensity fed back into the active layer.

In Fig. 6.4a, for a small level of the injection of 0.14, the slave laser shows four-wave and multiwave mixing associated with the unlocked slave laser frequency and has a side peak in the spectrum due to regenerative amplification. The effect of multiwave mixing becomes distinct in the laser output power at the injection of 0.23 in Fig. 6.4b. At the same time, the component corresponding to the relaxation oscillation becomes non-vanishing and the oscillation close to the relaxation oscillation frequency of 4.7 GHz is excited. Also, the spectrum is much broadened. The multiwave mixing effect is recognized as the phase-modulation like Adler-type frequency pulling towards locking (Simpson 2003). However, the frequency pulling here is somewhat different from the ordinary effect and it is an unstable phenomenon accompanying the relaxation resonance. Frequency-pulled multiwave mixing components disappear at the injection of 0.41 and the multiwave mixing features are pulled to the injection frequency as shown in Fig. 6.4c. As a result, a sharp and enhanced component of the relaxation oscillation is observed. Therefore, the laser shows a stable oscillation under the condition. Incommensurate frequency is encountered in the dynamics at the injection of 0.52 in Fig. 6.4d and the floor of the spectrum becomes broadened. This is a typical feature of the onset of quasi-periodic bifurcation and chaos. The floor of the spectrum further becomes broadened at the injection of 0.77 in Fig. 6.4e and several spectral peaks appear except for the relaxation oscillation component. Within the main peak, we can see two visible peaks. This indicates that the laser corresponds to period-3 oscillation. The oscillation mode within the relaxation oscillation frequency reduces as a single peak at the injection of 1.02 and the laser shows period-2 oscillation as shown in Fig. 6.4f. When the injection fraction is large enough at the injection of 1.30 in Fig. 6.4g, the laser oscillates at period-1 oscillation with the main frequency corresponding to the relaxation oscillation. The higher harmonics of the period-1 oscillation is also visible. Finally in Fig. 6.4h, at a strong injection field of 3.01, the laser is completely locked to a certain frequency and shows period-1 oscillation. The locked frequency is different from the relaxation oscillation frequency at the solitary oscillation. This phenomenon is related to the enhancement of the

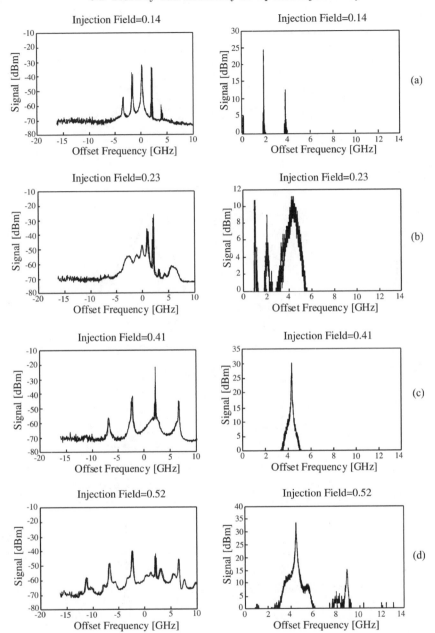

Fig. 6.4. Experimentally observed optical frequencies and rf power spectra corresponding to chaotic bifurcation in semiconductor lasers under optical injection. On the left are optical spectra and right are rf spectra. The laser is a single mode DFB laser at a wavelength of $1.557\,\mu\text{m}$ and a bias injection current of $J = 2.0J_{\text{th}}$. The relaxation oscillation frequency at solitary oscillation is $4.7\,\text{GHz}$. The injection rate (intensity) is changed as **a** 0.14, **b** 0.23, **c** 0.41, **d** 0.52

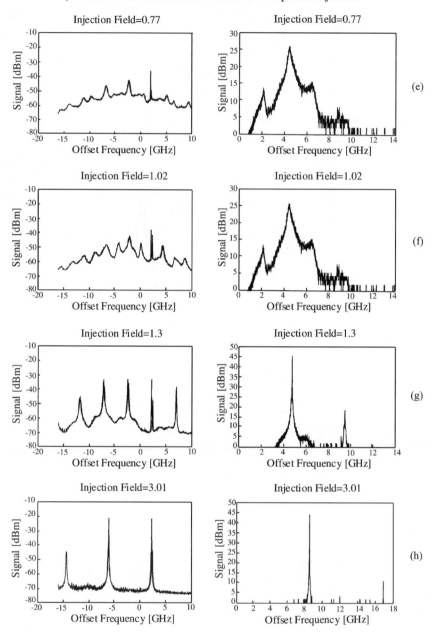

Fig. 6.4. (continued) **e** 0.77, **f** 1.02, **g** 1.30, and **h** 3.01, respectively, at the fixed frequency detuning of +2 GHz (after Simpson TB (2003); © 2003 Elsevier)

cutoff frequency for the modulation bandwidth of the laser as discussed in the following. In this example, the chaotic evolution is observed for a fixed frequency detuning (+2 GHz). The dynamics are not always the same as those for other conditions of the detuning, but they exhibit typical chaotic routes when the absolute value of the frequency detuning is within several GHz.

In accordance with stable and unstable oscillations in optically injected semiconductor lasers, chaotic bifurcations are numerically calculated taking into consideration the effects of side mode excitation. Figure 6.5 shows chaotic bifurcations for a change of the injection ratio at a fixed frequency detuning of +2 GHz (Simpson 2003). In Fig. 6.5a, the laser is assumed to be oscillated at a single mode, since it includes the effect of a larger defect with $\mu_d = 0.1$. The laser once evolves from periodic to chaotic oscillations and, then, takes an inverse route of chaotic bifurcations for the increase of the injection ratio. Finally, it reduces to the period-1 state. The behaviors are quite similar to the chaotic route for a single mode laser discussed in Fig. 6.4. On the other hand, the instability of the laser is suppressed because of the leakage of the power from the main mode to the side mode when the effect of the defect is as small as $\mu_d = 0.001$ in Fig. 6.5b. Under this condition, the laser shows no typical chaotic bifurcations. Figure 6.5c shows the plot of the relative circulating power level in the main mode for single and multimode operations. The power of the main mode is transferred to the side mode and the instability of the laser oscillation is greatly suppressed, when there is a side mode and the injection strength is small. However, the side mode is never excited for a larger value of the defect and the assumption of a single mode oscillation is well established.

6.2.3 Chaos Map in the Phase Space of Frequency Detuning and Injection

We discuss chaos maps in the phase diagram of the frequency detuning and the injection ratio. Stable injection locking is achieved in a region for a certain combination of the frequency detuning and the injection ratio, however various unstable and chaotic dynamics are observed in unstable locking and unlocking regions. Figure 6.6 shows the chaotic map obtained experimentally from the behaviors of the optical spectra in Fig. 6.5 (Simpson 2003). The laser is operated at a single mode even in the presence of optical injection and the side mode is suppressed in this laser. It is noted that the vertical axis and the horizontal axis are replaced compared with the plot in Fig. 6.3. The diamond-filled symbol shows in the negative frequency detuning is the boundary between unstable and stable operations. This corresponds to the saddle node boundary between stable locked and unlocked operations. Open diamonds show the unlocking-locking transition in a region of bistability and torus bifurcation. The square mark close to zero detuning is the Hopf bifurcation boundary between stable locked and limit cycle dynamics. The triangle is the boundaries for regions of period-2 dynamics. These period-2 regions in-

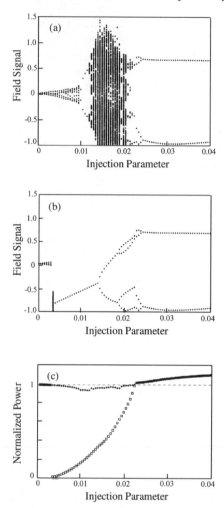

Fig. 6.5. Bifurcation diagram as a function of the injection ratio at a frequency detuning of $+2\,\mathrm{GHz}$. **a** Bifurcation diagram for large gain defect of $\mu_{\mathrm{d}} = 0.1$. The laser oscillates at a single mode. **b** Bifurcation diagram for small gain defect of $\mu_{\mathrm{d}} = 0.001$. Significant power leaks into the side mode. **c** Relative circulating power. The symbol of diamonds is for single mode oscillation, while open squares denote the case of multimode oscillations (after Simpson TB (2003); © 2003 Elsevier)

clude complex dynamics and they are shown by the shaded lines and crosses in the figure. Bounded by the circles is a region of period-4 operation. At injection levels below the saddle node bifurcation line and at low offset frequencies, multiwave mixing and Adler-type frequency pulling to locking are observed in the lightly shaded regions.

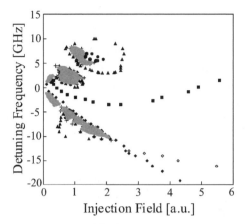

Fig. 6.6. Experimentally obtained chaotic map from measured optical spectra of a single mode DFB laser under optical injection. The meaning of each symbol is referred to the text (after Simpson TB (2003); © 2003 Elsevier)

Figure 6.7 shows the experimental result for the map in a semiconductor laser with side mode excitation by optical injection (Hwang and Liu 2000). The laser used is a conventional Fabry-Perot type edge-emitting laser with a quantum well structure. The back facet of the laser is coated for high reflection and the front output facet is coated for a reflection of a few percent. At the free running state of the laser oscillation, the side mode is suppressed as low as less than 0.5%. The symbols in the figure are 4: a perturbation spectrum with weak regenerative amplification and four-wave mixing sidebands, S: stable injection locking, P1: limit-cycle oscillation, P2: period-doubling, P4: period-quadrupling, chaos: deterministic chaos, M: multiwave mixing with most output on another longitudinal mode, SR: subharmonic resonance, hatched regions: principal output on another longitudinal mode, thin curves: smooth transition between dynamic regions, thick dotted curves: abrupt mode-hop transitions with minor hysteresis, thick broken curves with an arrow: one-way mode hops out of mode, and thick full curves: abrupt transition to/from a region of chaos or multiwave mixing where there is significant power in another longitudinal mode, from/to a region with power primarily in the principal mode.

For a small injection, the optical injection acts as a perturbation generating weak sidebands at the offset frequency, regenerative amplification, and equally and oppositely shifted four-wave mixing. With increasing both of the frequency detuning and the injection ratio, various instabilities appear in the laser output power. The tendency of periodic bifurcations and chaotic islands in the unstable region is the same as that for a DFB laser. However, distinct chaotic bifurcation is not observable for a Fabry-Perot laser in the region of negative frequency detuning along the stable boundary, while it was observed for a DFB laser (see Fig. 6.6). There is an abrupt mode hop near the locking–unlocking boundary at negative detuning which has a small hysteresis. Analytical studies of the locking–unlocking boundary at negative detuning have shown that there is a region of bistability associated with the locking–unlocking transition (Li 1994a, b). The bistability results from competing attractors representing locked and unlocked solutions for the coupled equations (Lenstra et al. 1993). The carrier density is larger than that for the steady-state n_s under the unlocked solution, while it stays a smaller value for the locked solution. The gain of the side mode increases with the increase of the carrier density and the refractive index of the active layer accordingly changes. The change induces the transfer of the optical energy from the main mode to side modes. Then, the gain of the main mode is reduced and this sometimes results in frequent mode hop. However, the chaotic dynamics disappears in the output power. On the contrary, instabilities still remain in the dynamics of a single mode laser without the excitation of the side mode as shown in Fig. 6.6.

Considering the gain defect and using the equations (6.13)–(6.15), stable and unstable maps in the phase space of the frequency detuning and the injection ratio like in Figs. 6.6 and 6.7 can be calculated (Simpson 2003). The results are quite consistent with the experimental results. Namely, the suppression of chaotic dynamics for the excitation of the side mode is well reproduced. Hwang and Liu (2000) numerically calculated maps of stable and unstable regions in the phase space of the frequency detuning and the injection ratio by changing the parameters in the rate equations. They studied the dependence of the parameters for the cavity decay rate $\gamma_c = 1/\tau_{ph}$, the carrier relaxation rate $\gamma_s = 1/\tau_s$, the differential relaxation rate γ_n, and the nonlinear carrier relaxation rate γ_p. For the change of those parameters,

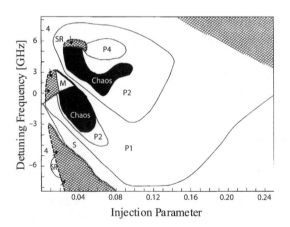

Fig. 6.7. Experimentally obtained chaotic map from measured optical spectra in a Fabry-Perot laser operating at 827.6 nm. A side mode is excited by optical injection. The bias injection current is $J = 1.67 J_{th}$. The detail of the map is discussed in the text (after Hwang SK, Liu JM (2000); © 2000 Elsevier)

they obtained the following results. (1) The carrier decay rate γ_s affects little change, since the carrier relaxation is usually induced by spontaneous emission of light and it is a small perturbation for the field strength. (2) The laser is stabilized for a small value of the differential relaxation rate γ_n, since the fast carrier diffusion reduces unstable regions. (3) When the nonlinear carrier relaxation rate γ_p increases, the unstable region shrinks and the laser is stabilized. The increase of the nonlinear carrier relaxation rate γ_p results in the change of the carrier density in the active layer and the fluctuation of the optical phase is suppressed. This results in the suppression of frequency fluctuations. Therefore, the laser is stabilized. (4) For a larger value of the α parameter, instability of laser oscillation is enhanced. (5) Stability and instability of semiconductor lasers subjected to optical injection are also dependent on the bias injection current. With increasing the bias injection current, the stable region in the map expands. When a laser is operated at a higher injection current level, the coherent optical power stored in the cavity is higher, thus allowing the laser to be more resistant to the perturbation of the externally injected optical field. This is why stronger externally optical perturbation is required to observe instabilities and chaos in the system at a higher injection current level.

6.2.4 Coexistence of Chaotic Attractors
in Optically Injected Semiconductor Lasers

A nonlinear system has the nature of multistability under a certain condition of the parameters. Namely, the system may have coexistent states of different chaotic attractors for the same parameter set. Which attractor the system converges to strongly depends on the initial conditions. Indeed, coexistence of chaotic orbits has been observed in various systems of semiconductor lasers (Masoller and Abraham 1998, Heil et al. 1998 and 1999, Sukow et al. 1999, and Viktorov and Mandel 2000). In optically injected semiconductor lasers, multistability and coexistence of chaotic attractors have also been studied (Wieczorek et al. 2000, 2001a, b, c, and 2002). Here, we present such examples. Under a certain experimental configuration, we always observe a particular chaotic attractor, since the process of obtaining chaotic oscillation is generally the same. Therefore, we usually observe a chaotic oscillation for one of the chaotic attractors in a fixed experimental condition even if multistabilities are involved in the system. However, if the separation between the two coexisting attractors in the high dimensional phase space is not so far away, switching from one attractor to the other may occur due to, for example, noises involved in the system. Indeed, transition of the state from one chaotic oscillation to another has been experimentally observed (Heil et al. 1998 and 1999).

Using the bifurcation theory, it is easy to know whether a nonlinear system has coexisting attractors under the operating condition when the system exhibits bistability or multistability. However, the characteristics of the coexisting attractors cannot be obtained using the bifurcation analysis. With the simulation method, the existence of the bistability or multistability is found by numerically simulating the system with different initial conditions under the same operating conditions. Figure 6.8 shows the numerical result for the map of coexistence states in semiconductor lasers subjected to optical injection (Wieczorek et al. 2000). The plot is a similar one in the phase space as shown in Figs. 6.6 and 6.7, but normalized axes are used. Each point in this plane corresponds to a particular phase portrait, which contains more than one attractor. Black parts of bifurcation curves correspond to supercritical bifurcations in which attractors bifurcate and grey parts correspond to subcritical bifurcations of repelling objects. Subcritical bifurcations are less important from an experimental point of view, but we trace them out as they may produce stable objects for instance in a subcritical torus bifurcation or change to supercritical.

In Fig. 6.8, the region inside the straight line from zero detuning to negative detuning is the stable injection locking area. A detailed explanation of the map is found in the reference (Wieczorek et al. 2000). We here focus on the points that weigh with actual observations. In the stable region, there exist areas for the saddle-node bifurcation (SN) and the Hopf bifurcation (H). When the black part of SN is crossed, one of the bifurcating stationary points is an attractor. It physically corresponds to the laser operating at constant power and at the frequency of the injected light, meaning that the laser

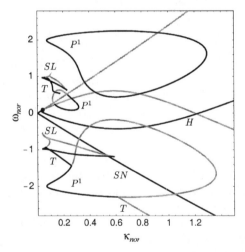

Fig. 6.8. Bifurcation diagram showing coexistence states in a phase space of normalized frequency detuning and injection ratio. The vertical axis is the normalized frequency of $\omega_{\mathrm{nor}} = \Delta\omega/\omega_{\mathrm{R}}$, and the horizontal axis of the injection ratio is also normalized of $\kappa_{\mathrm{nor}} = \kappa_{\mathrm{inj}}A_{\mathrm{m}}/\omega_{\mathrm{R}}\tau_{\mathrm{in}}A_{\mathrm{0s}}$. P^1: period-doubling bifurcations, SL: saddle-nodes of limit cycles, T: torus, H: Hopf bifurcation, and SN: saddle-node bifurcation (after Wieczorek S, Krauskopf B, Lenstra D (2000); ©️ 2000 Elsevier)

locks to the input signal. On the other hand, along the gray part of the curve *SN*, a repellor and a saddle point bifurcate. Along the black part of *H*, an attracting periodic orbit is born from the attracting stationary point and this corresponds physically to the undamping of the relaxation oscillation. Physically, the appearance of a new orbit means that some resonance in the laser gets excited, often because the operational parameters κ_{nor} and ω_{nor} drive the laser close to the relaxation frequency or its multiples. In Fig. 6.8, the two saddle nodes of the limit cycle bifurcation curves starting with a cusp at $\omega_{\mathrm{nor}} \approx \pm 1$ represent a resonance between the relaxation oscillation frequency of the laser and the detuning of the injected light from the free running laser frequency.

Starting from different initial conditions, the nonlinear system may have different attractors in the phase space, even if the parameters have the same values as shown in the previous figure. Figure 6.9 shows examples of attractors in multistability states in the phase space of the imaginary part of the field, E_{y}, and the carrier density n (Weiczorek et al. 2000). The plots are the same conditions as those in Fig. 6.8. Figure 6.9a shows the plot of three attractors. Figures 6.9b and c are periodic states of period-1 with small amplitude and large periodic orbit, respectively. Fig. 6.9d corresponds to a quasi-periodic oscillation on a torus. As has already been discussed, which of the attractors the system settles down to depends on the initial conditions. Furthermore, when a parameter is swept gradually through a region of multistability, then one will find hysteresis loops with sudden jumps from one attractor to another at different values of the parameter, depending on the direction of the sweep.

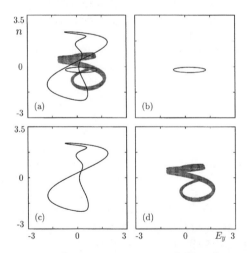

Fig. 6.9. Coexistence state of attractors at $\kappa_{\mathrm{nor}} = 0.29$ and $\omega_{\mathrm{nor}} = -1.37$. **a** Simultaneous plot of three attractors, **b** running phase solution, **c** large periodic orbit, and **d** quasi-periodic motion on a torus (after Wieczorek S, Krauskopf B, Lenstra D (2000); © 2000 Elsevier)

6.3 Enhancement of Modulation Bandwidth and Generation of High Frequency Chaotic Oscillation by Strong Optical Injection

6.3.1 Enhancement of Modulation Bandwidth by Strong Optical Injection

The modulation bandwidth of a semiconductor laser at free running state is limited by the relaxation oscillation frequency. However, when a semiconductor laser is strongly injected under stable injection locking conditions, the modulation bandwidth of the slave laser is greatly enhanced. At the same time, the suppression of laser noises is achieved, but the strong modulation gives rise to frequency chirping in the laser oscillation. The effects of noises and frequency chirping under optical injection are critical for the laser operation (Piazzolia et al. 1986 and Yabre 1996). In a locking–unlocking bistable state, a large modulation current can unlock the laser. In a state near or beyond the Hopf bifurcation boundary, the dynamic instability of the laser can lead to high broadband noise and large frequency chirping. Also, the enhancement of the modulation bandwidth of semiconductor lasers subjected to strong injection has been demonstrated (Simpson et al. 1995b and 1996, Simpson and Liu 1997, Wang et al. 1996, and Chen et al. 2000, Wang et al. 2004). For weak optical injection and optical feedback, the modulation bandwidth is increased due to the increase of the photon number within the internal cavity, since the relaxation oscillation frequency is proportional to the square root of the photon number (see (3.69) and (4.21)). The amount of the shift of the cutoff frequency is up to ten percent at most. However, the cutoff frequency of the laser under strong optical injection is greatly enhanced up to several times the relaxation oscillation frequency of the free running laser. Therefore, a different explanation for the origin of the enhanced modulation bandwidth may be required to understand the phenomenon. The bandwidth-enhanced semiconductor laser is very useful as a broadband light source for optical communications.

Figure 6.10 is an example of experimental results of the enhancement of the modulation bandwidth. For a modulation of a small sinusoidal wave of 12 GHz to the bias injection current, the modulated laser output attenuated and is only −27.49 dBm without optical injection as shown in Fig. 6.10a, since the modulation is far away from the relaxation oscillation frequency (about 3 GHz). On the other hand, the modulation efficiency is increased up to 10 dBm by a strong optical injection (Fig. 6.10b). As will be discussed in Chap. 13, chaotic carrier frequency is the measure of the maximum data transmission rate in secure optical communications based on chaos synchronization in semiconductor laser systems. The chaotic carrier frequency is also increased by a strong optical injection and a large capacity of the channels for the communication is expected.

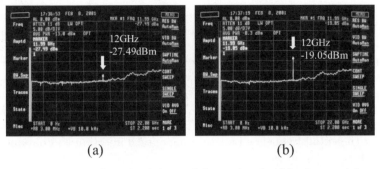

(a)   (b)

Fig. 6.10. Experimental result of the modulation bandwidth of a strongly optical-injection-locked semiconductor laser. The laser is modulated by a small sinusoidal signal at 12 GHz. The modulation efficiency without optical injection is −27.49 dBm. The efficiency with optical injection is −19.05 dBm. The relaxation oscillation of the DFB laser used is 3 GHz at free running state

The enhancement of the modulation bandwidth in a semiconductor laser under strong optical injection is numerically studied based on the rate equations. Wang et al. (1996) investigated the modulation response for a small signal to the bias injection current using a linear stability analysis. Figure 6.11 is the result. The frequency detuning between the master and slave lasers is assumed to be zero in this case. The cutoff frequency read from the resonance frequency is 12.6 GHz for the injection ratio of $S_{\mathrm{inj}}/S_{\mathrm{s}} = 0.44$ (curve d), while the relaxation oscillation frequency is 3.4 GHz at the free running state (curve a). In addition, the response is almost flat well below the cutoff frequency and the modulation bandwidth is enhanced up to four times compared with that of the free running state. As has already been discussed, the relaxation

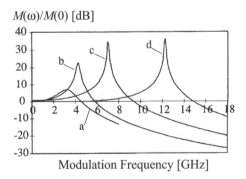

Modulation Frequency [GHz]

Fig. 6.11. Normalized modulation response of a semiconductor laser at $J = 2.4J_{\mathrm{th}}$. a: free running laser, b: $S_{\mathrm{inj}}/S_{\mathrm{s}} = 0.011$, c: $S_{\mathrm{inj}}/S_{\mathrm{s}} = 0.092$, d: $S_{\mathrm{inj}}/S_{\mathrm{s}} = 0.44$. S_{inj} and S_{s} are the photon numbers injected from the master laser and the steady-state value of the free running slave laser (after Wang J, Haldar MK, Li L, Mendis VC (1996); © 1996 IEEE)

oscillation frequency is proportional to the square root of the photon number and the photon number is a function of the bias injection current. To obtain the equivalent modulation bandwidth of 12 GHz for the free running laser, we would require the bias injection current to be seven times larger than that of the free running state, which corresponds to almost 13 times the threshold injection current and might damage the laser. Thus, the method of strong optical injection is effective for greatly enhancing the modulation bandwidth in semiconductor lasers. Wang et al. (1996) conducted a linear stability analysis for the cutoff frequency under strong optical injection and obtained the approximate solution as

$$
\nu_{\text{enhanced}} = \frac{1}{2\sqrt{3}\pi} \left[K_e - \left(\frac{K_a}{S_s} \right)^2 + \left\{ \left(\frac{K_a}{S_s} \right)^2 - 4K_e \left(\frac{K_a}{S_s} \right)^2 \right. \right.
$$
$$
\left. \left. + K_e^2 - 6K_a \, K_b \alpha G_n^2 (n_s - n_0) \right\}^{1/2} \right]^{1/2} \tag{6.17}
$$

where the parameters in the above equation are given by

$$
K_e = \left(\frac{\kappa_{\text{inj}}}{\tau_{\text{in}}} \right)^2 \frac{S_m}{S_s} \tag{6.18}
$$

$$
K_a = 2k_c \sqrt{S_m S_s} \cos(\phi_s - \phi_m) = - \left\{ G_n(n_s - n_0)(1 - \varepsilon_s S_s) - \frac{1}{\tau_{\text{ph}}} \right\} S_s - R_{\text{sp}} \tag{6.19}
$$

$$
K_b = k_c \sqrt{\frac{S_m}{S_s}} \sin(\phi_s - \phi_m) = \frac{1}{2} \alpha G_n(n_s - n_0) - \Delta\omega \tag{6.20}
$$

As will be shown later, the cutoff frequency is linearly proportional to the injection power. Therefore, the origin of the enhancement of the modulation bandwidth does not simply come from the increase of the photon number in the active layer. It is explained by the interference between the optical frequency of the original laser oscillation and the shifted frequency due to the strong optical injection.

The enhancement of the modulation bandwith is also strongly dependent on the frequency detuning between the master and slave lasers. Figure 6.12 shows the dependence of the modulation response in the presence of frequency detuning between the master and slave lasers (Chen et al. 2000). The conditions of the numerical simulations are as follows: the laser is assumed to be an index-guided GaAs/AlGaAs quantum-well laser and is biased at $\hat{J} = 0.67$ (corresponding to $J = 1.67 J_{\text{th}}$), where $\hat{J}$ is the scaled injection current defined by $\hat{J} = (J/ed - n_s/\tau_s)/(n_s/\tau_s)$. The injection parameter defined by $\kappa_{\text{nor}} = (\kappa_{\text{inj}} \tau_{\text{ph}} A_m)/(\tau_{\text{in}} A_s)$ is fixed at a moderate level of $\kappa_{\text{nor}} = 0.2$, while different values of frequency detuning representing different locking conditions are chosen. At $\kappa'_{\text{nor}} = 0.2$, the stable locking region is bounded by

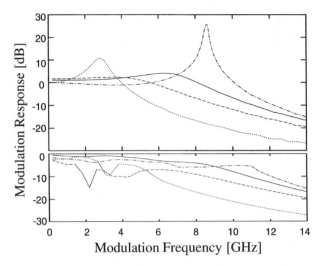

Fig. 6.12. Normalized current modulation response in the presence of frequency detuning $\Delta\nu$ between master and slave lasers. *Dash-dotted curve*: Injection locking at $\Delta\nu = 1$ GHz. *Solid curve*: injection locking at $\Delta\nu = -10$ GHz. *Dashed curve*: injection locking at $\Delta\nu = -18$ GHz. *Dotted curve*: free running. The curves in the *upper plot* are the response for $m = 1\%$, and those in the *lower plot* are the responses for $m = 100\%$. The 0 dB in the lower plot corresponds to the 0 dB in the upper plot in order to make all the response curves comparable (after Chen HF, Liu JM, Simpson TB (2000); © 2000 Elsevier)

$\Delta\nu = 1$ GHz and $\Delta\nu = -13$ GHz, where $\Delta\nu = 1$ GHz is the Hopf bifurcation boundary. Between $\Delta\nu = -13$ GHz and $\Delta\nu = -22$ GHz is a region of locking–unlocking bistability, where the laser can be either locked or unlocked depending on the initial condition. Under that condition, the laser cannot be locked when the frequency detuning is more negative than -22 GHz. Relative to the free running laser, a broadband noise reduction occurs in the locked region when the injection field is negatively detuned beyond $\Delta\nu = -3$ GHz. The three representative values of frequency detuning chosen in this case are $\Delta\nu = 1$ GHz on the Hopf bifurcation boundary (dash-dotted curves in the figures), $\Delta\nu = -10$ GHz in the stable locking region (solid curves), and $\Delta\nu = -18$ GHz in the locking-unlocking bistability region (dashed curves).

In the figure, the upper plot shows the response for a small modulation index of $m = 1\%$ and the graphs are normalized to the low frequency response of the laser in its free running condition, under the four different operating conditions. The lower plot in Fig. 6.12 shows the distorted current modulation response when the modulation index reaches $m = 100\%$. At a given modulation strength, negatively shifting the frequency detuning of the injected optical field generally reduces the distortion in the current modulation response if the laser remains stably locked. However, when the laser is injection-locked in the bistability region, a high modulation index can cause instability by unlocking the laser. As a result, the modulation response in such an operating condition becomes very irregular, as can be seen from the dashed curve (-18 GHz) in the lower plot of Fig. 6.12. For weak current modulation with small values of the modulation index, the modulation response will be obscured by the intrinsic laser noise. For the change of the modulation index, the laser noise induces insignificant differences between the overall response due to the combined modulation current and intrinsic noise and the modulation response alone when the modulation index m is larger than 1%. Below $m = 1\%$, the relative importance of the laser noise gradually increases and the laser noise induces fluctuations in the response that obscure the modulation response.

6.3.2 Origin of Modulation Bandwidth Enhancement

The origin of the enhancement of modulation bandwidth by strong optical injection is explained by Murakami et al. (2003). They consider the frequency shift of the slave laser induced by strong optical injection. The expansion of the modulation bandwidth is realized by the interference between the original optical frequency at the free running state and the shifted frequency after the injection. According to their explanation, the difference between the two frequencies corresponds to the expanded modulation bandwidth. Figure 6.13 schematically shows the model of the frequency shift. Let the angular frequency of the slave laser at the free running state be given by ω_0 and that of the master laser be ω_{inj}. In the figure, the frequency detuning is assumed to be positive, but the other case will be reduced to the same result. By a strong optical injection, the carrier density in the slave laser increases. This induces the change of the optical frequency of the laser oscillation and results in redshift of the oscillation frequency. Using the change of the carrier density δn, the laser once oscillates at an optical angular frequency ω_{shift} and the shift of the laser angular frequency after the injection is given by

$$\Delta\omega_{shift} = \frac{1}{2}\alpha G_n \delta n \tag{6.21}$$

The change of the carrier density δn is proportional to the strength of optical injection. The frequency shift given by (6.21) has the same form as the first term in (6.3). In actual fact, the frequency of the slave laser is locked to the frequency of the injection laser (angular frequency of ω_{inj}). Accordingly, the injection-locked laser may operate at a frequency different from its cavity resonance condition, namely operating at ω_{inj}, not at ω_{shift}. Such frequency detuning between ω_{inj} and ω_{shift} influences the modulation bandwidth, as predicted by Simpson et al. (1996).

Here, we consider the transient situation. The field corresponding to the shifted cavity resonance ω_{shift} is once excited and interference between the two components of the angular frequencies ω_{shift} and ω_{inj} occurs. Then, the beat between the two frequencies is induced in the output of the slave laser.

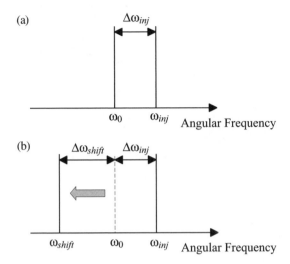

Fig. 6.13. Resonant condition of a semiconductor laser in the presence of optical injection. **a** Spectrum before optical injection. ω_0 is the angular frequency of the solitary laser, ω_{inj} is the frequency of the injected light, and $\Delta\omega_{\text{inj}}$ is the frequency detuning between them. **b** Cavity resonant condition under injection locking. ω_{shift} is the cavity resonance frequency shifted from ω_0 by $\Delta\omega_{\text{shift}}(n)$ due to optical injection

However, sufficient gain is not allocated to this mode and the oscillation of the mode rapidly decays out, since this is a transient field. The oscillation angular frequency of the slave laser is restored to ω_{inj}. The laser output may exhibit a damping oscillation at the beat frequency due to such transient interference. Note that this damped oscillation differs from the relaxation oscillation in the physical mechanism, because the relaxation oscillation results from an interaction or coupling between photon and carrier through the stimulated emission. Therefore, from (6.3), the resonance angular frequency produced by the interference $\omega_{\text{res}} = \omega_{\text{inj}} - \omega_{\text{shift}}$ is given by

$$\omega_{\text{res}} = \Delta\omega_{\text{inj}} - \Delta\omega_{\text{shift}} = -\frac{1}{\tau_{\text{in}}}\sqrt{\frac{S_{\text{inj}}}{S_{\text{s}}}}\sin\psi_{\text{s}} \tag{6.22}$$

Following the above explanation, the dependence of the resonance frequency in the presence of strong optical injection is calculated (Murakami et al. 2003). Figure 6.14 shows the plots of dependence of the injection ratio (amplitude) and the frequency detuning on the cutoff frequency. Fig. 6.14 a is the dependence of the cutoff frequency on the injection ratio at the frequency detuning of +0.5 GHz. Under a strong optical injection condition, the cutoff frequency is linearly proportional to the injection ratio in accordance with the prediction in (6.21). In the figure, three data are plotted: the solid line is the prediction calculated from (6.21), circles are the direct numerical calculation

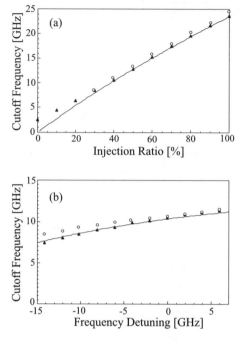

Fig. 6.14. Dependence of resonance frequency on **a** injection rate and **b** frequency detuning. Circles represent the numerical results, triangles are the results obtained from the stability analysis, and the solid line is the theoretical curve of (6.22)

from the rate equations, and triangles are obtained from the linear stability analysis. In strong optical injection of over 30%, the three plots coincide well with each other. Figure 6.14b is the plot of the cutoff frequency for the frequency detuning at a high optical injection ratio of 40%. The cutoff frequency tends to increase with increased detuning from negative to positive values. Thus, the enhancement of the cutoff frequency under strong optical injection is explained by the interference between the injection laser frequency and the implicit frequency shift of the slave laser induced by the strong optical injection.

6.3.3 Modulation Response by Strong Optical Injection

The effects of the injection current modulation on the response distortion and the noise compression can be evaluated by eye patterns with digital signals. The capability of data transmission in optical communications is calculated in Fig. 6.15 (Chen et al. 2000). The eye patterns are numerically calculated from the rate equations with strong optical injection. In the numerical simulations, eye patterns are generated by modulating the injection current of the semiconductor laser under the various operating conditions with a train of random raised-cosine functions:

$$J_\mathrm{m} = \sum_{K=0}^{n} C_\mathrm{K} h(t - KT) \qquad (6.23)$$

$$h(t) = \frac{\sin(\frac{\pi t}{T})\cos(\frac{\pi \beta_\mathrm{t} t}{T})}{\frac{\pi t}{T}[1 - (\frac{2\pi \beta_\mathrm{t} t}{T})^2]} \qquad (6.24)$$

where C_K is a series of random numbers with the value 0 or 1, which represents digitized information, and $T = 1/f_\mathrm{m}$, where the modulation frequency f_m represents the bit rate. The value of the parameter β_t is chosen to be 0.3 in this simulation.

Figure 6.15 shows the eye patterns for different modulation indexes at the modulation frequency of $f_\mathrm{m} = 2.9\,\mathrm{GHz}$. The modulation frequency is equal to the relaxation oscillation frequency of the laser the free running state. Three different modulation indexes $m = 10\%$, 50%, and 100% are chosen to present the advantages of the injection-locked laser in the stable locking region. The eye patterns obtained in the condition when the laser is injection-locked at $\Delta\nu = -10\,\mathrm{GHz}$ have clearer eye opening with less distortion or less noise than those obtained in other operating conditions. The relative eye opening shows

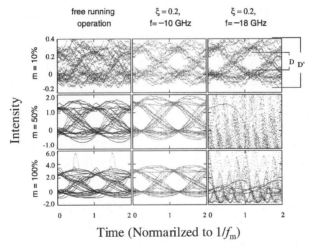

Fig. 6.15. Eye pattern for square wave modulations. The bit rate is chosen as the corresponding resonance frequency for each operating condition: the bit rate $f_\mathrm{m} = 2.9\,\mathrm{GHz}$ is chosen for the free running operation, $f_\mathrm{m} = 4.7\,\mathrm{GHz}$ for $\Delta\nu = -18\,\mathrm{GHz}$, and $f_\mathrm{m} = 6.8\,\mathrm{GHz}$ for $\Delta\nu = -10\,\mathrm{GHz}$. The eye opening obtained from the operating condition $\Delta\nu = 1\,\mathrm{GHz}$ is zero for the range of the bit rate we are concerned with, so the eye patterns are not shown. The intensity of the eye patterns is the differential intensity above or below the corresponding field intensity of the injection-locked laser without current modulation for each operating condition (after Chen HF, Liu JM, Simpson TB (2000); © 2000 Elsevier)

the same tendency as the noise compression. For a small modulation index of $m = 10\%$, the modulation efficiency is much improved compared with that at the free running state. However, when the modulation index increases, the modulation signal with a modulation index of $m = 50\%$ unlocks the laser for a large negative frequency detuning ($\Delta\nu = -18\,\mathrm{GHz}$), resulting in zero eye opening. When the frequency detuning is positively shifted beyond the noise compression region, the eye opening also rapidly decreases to zero. Therefore, the eye opening obtained in the operating condition with $\Delta\nu = 1\,\mathrm{GHz}$ is zero for a modulation index of any value.

6.3.4 Suppression of Frequency Chirping by Strong Optical Injection

It has already been noted that the chirping of frequency due to injection current modulation in a semiconductor laser is much suppressed by a strong optical injection. We here demonstrate an example. Usually, a change in the carrier density causes a change in the refractive index of the laser medium. This change in the index generates frequency chirping, which can place a considerable limitation on the modulation bit rate. The frequency chirping is measured by the normalized chirp to the power ratio CPR, which is defined as follows (Piazzolia et al. 1986):

$$\mathrm{CPR} = \frac{1}{2\pi R_\mathrm{P}} \left| \frac{\mathrm{d}\phi}{\mathrm{d}t} \right| \qquad (6.25)$$

where R_P is the modulation response. The frequency chirp originates from the linewidth enhancement factor α, which has a non-zero value in a semiconductor laser. Therefore, it is proved that the CRP is related to the linewidth enhancement factor. Neglecting noise effects in a semiconductor laser and applying a small signal analysis, the relation between the linewidth enhancement factor α and CPR can be obtained by linearizing the rate equation around the locking operating point (Simpson et al. 1996). This relationship between α and the CPR can be expressed as follows:

$$\frac{1}{2\pi R_\mathrm{P}} \left| \frac{\mathrm{d}\phi}{\mathrm{d}t} \right| \approx f_\mathrm{m} \alpha \sqrt{\frac{f_\mathrm{m}^2 + (u - v/\alpha)^2}{f_\mathrm{m}^2 + (u + v/\alpha)^2}} \qquad (6.26)$$

where u and v are given by

$$u = \frac{\kappa_\mathrm{inj}}{2\pi\tau_\mathrm{in}} \left| \frac{A_\mathrm{m}}{A_\mathrm{s}} \right| \cos\phi_\mathrm{L} \qquad (6.27)$$

$$v = \frac{\kappa_\mathrm{inj}}{2\pi\tau_\mathrm{in}} \left| \frac{A_\mathrm{m}}{A_\mathrm{s}} \right| \sin\phi_\mathrm{L} \qquad (6.28)$$

Here ϕ_L is the phase of the intracavity laser field relative to the injection field. An effective linewidth enhancement factor α_eff, which is the modified chirping parameter under injection locking, can be defined as follows:

$$\alpha_{\text{eff}} \approx \alpha \sqrt{\frac{f_{\text{m}}^2 + (u - v/\alpha)^2}{f_{\text{m}}^2 + (u + v/\alpha)^2}} \qquad (6.29)$$

Of course, the effective chirping parameter α_{eff} is equal to α when the laser is at the free running state ($u = v = 0$). The dependence of the effective chirping parameter on the modulation frequency f_{m} is shown in Fig. 6.16 (Chen et al. 2000). The injection locking of the laser at $\Delta\nu = -10\,\text{GHz}$ reduces the effective chirping parameter more than injection locking the laser at $\Delta\nu = -18\,\text{GHz}$ does. Positive shifting of the frequency detuning reduces the effective chirping parameter further until the boundary of the Hopf bifurcation is reached. For a large modulation index, the effective chirping parameter finally reaches that of the free running state. Therefore, the effect of the suppression for the chirping is remarkable for lower modulation frequency. When the effect of the intrinsic noise or that of the nonlinearity of the laser on the frequency chirping are significant, the simple relationship in (6.26) between the CPR and α is no longer valid. In this situation, it is not possible to simply represent the frequency chirping with an effective chirping parameter. Then, the measurement of the frequency chirping including the effects of the intrinsic noise and the nonlinearity of the laser dynamics under a large modulation current is better quantified directly with the CPR.

From the detailed analysis for CPR, if the laser noise were not present, a significant reduction of the frequency chirping could be achieved by optical injection, and positively shifting the frequency detuning could further reduce the frequency chirping. In reality, however, when the modulation index is small, the chirp is dominated by the laser noise. As a result, the chirp

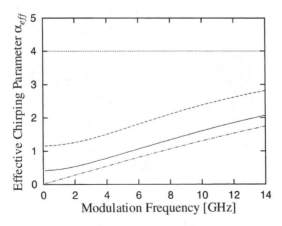

Fig. 6.16. Effective chirping parameter. Each curve corresponds directly to the curves in Fig. 6.12 that have the same style (after Chen HF, Liu JM, Simpson TB (2000); © 2000 Elsevier)

follows the same tendency as the power noise. Therefore, reduction of the frequency chirping in a semiconductor laser is not always guaranteed by injection locking (Chen et al. 2000). A semiconductor laser injection-locked in a locking-unlocking bistable state cannot fully take such benefits because a large modulation current can unlock the laser. Further, one cannot operate in a state near or beyond the Hopf bifurcation boundary because of the high broadband noise and the large frequency chirping associated with the instability of the laser. A semiconductor laser operated in a stable injection locking state generally has better current modulation characteristics than in its free running state.

6.3.5 Generation of High Frequency Chaotic Oscillation by Strong Optical Injection

Chaotic carrier frequency in a semiconductor laser system has almost the same or nearly the relaxation oscillation frequency of the solitary laser. For example, the oscillation very close to the relaxation oscillation frequency is at first excited in a semiconductor laser with optical feedback for the increase of the feedback strength. For a further increase of the feedback, the laser typically shows chaotic oscillations via period-doubling or quasi-period-doubling routes. Therefore, the relaxation oscillation frequency of the laser is the measure of chaotic oscillations and it plays a crucial role in the chaotic dynamics. Especially, the bandwidth of the chaotic signal is important in the chaotic secure communications discussed in Chap. 13. In such chaotic communications, the generation of a fast chaotic carrier signal is essential for a message transmission with higher bit rate. We here consider the generation

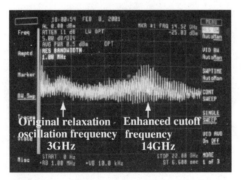

Fig. 6.17. Experimentally obtained chaotic power spectrum with enhanced cutoff frequency in a DFB semiconductor laser subjected to both optical feedback and strong optical injection. The frequency detuning between the two lasers is -3.44 GHz and the injection ratio is -5.61 dBm. The *left* and *right arrows* indicate the peaks for the chaotic carrier frequencies without and with optical injection, respectively

of fast chaotic signals in semiconductor lasers both subjected to optical feedback and optical injection. Figure 6.17 is a typical power spectrum of chaotic
oscillations obtained from a semiconductor laser which has both external optical feedback and strong optical injection. Without a strong optical injection,
the laser shows chaotic oscillation due to the external optical feedback and
the spectral peak (though it has a broad peak) is about 3 GHz, which is
comparable with the relaxation oscillation of the solitary laser. The original
relaxation oscillation frequency is shown by the arrow (left arrow). On the
other hand, the frequency of the maximum chaotic oscillation is increased
to 14 GHz (right arrow) by a strong optical injection, that is an increase by
a factor of 4.6. In this example, the frequency detuning between the master
and slave lasers is -3.44 GHz and the optical injection ratio is -5.61 dBm.

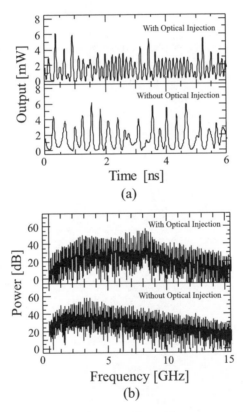

(a)

(b)

Fig. 6.18. Numerically calculated time series and power spectra of chaotic oscillations in a semiconductor laser with strong optical injection. Time series **a** with
and **b** without optical injection. Power spectra **c** with and **d** without optical injection corresponding to Figs. 6.18a and b, respectively. The parameter conditions are
$J_{T,R} = 1.3J_{th}$, $\tau_{in} = 8.0$ ps, $\kappa_{inj} = 0.143$, $\kappa_T = 0.186$, $\kappa_{T,R} = 0.186$, $\tau = 6$ ns, and
$\Delta\nu_T = \Delta\nu_R = -4$ GHz

This condition corresponds to the ordinary stable injection-locking operation in the absence of external optical feedback.

The enhancement of the chaotic carrier frequency is also numerically calculated based on the rate equations. Figure 6.18 is the result (Takiguchi et al. 2003). Figures 6.18a and b are the time series of chaotic oscillations with and without optical injection, respectively. The relaxation oscillation frequency of the solitary laser is about 2.7 GHz. Figures 6.18c and d are the corresponding rf spectra to Figs. 6.18a and b, respectively. The feedback fraction to the slave laser is taken to be a large value of $\kappa/\tau_{in} = 2.33 \times 10^{10}\,s^{-1}$ to destabilize the strongly injection-locked laser. In Fig. 6.18d, the chaotic oscillation rapidly decays out over the relaxation oscillation frequency without optical injection. On the other hand, the spectrum in Fig. 6.18c shows a bandwidth-enhanced chaotic oscillation in the presence of a strong optical injection. As can easily be seen both from the time series and the spectrum, the chaotic carrier frequency is greatly expanded up to about 8 GHz by the strong optical injection, which is as much as three times that without optical injection in Fig. 6.18d. As shown in the figure, the chaotic carrier frequency is also greatly enhanced by a strong optical injection. However, it is difficult to calculate analytically the exact enhanced bandwidth of the chaotic carrier frequency.

7 Dynamics of Semiconductor Lasers with Optoelectronic Feedback and Modulation

One of the characteristics of semiconductor lasers that is different from other lasers is the direct pump modulation. The disturbance to the injection current is an additional degree of freedom. Therefore, the disturbance to the injection current induces instabilities to semiconductor lasers and instabilities and chaotic behaviors are indeed observed by the injection current modulation. Another example of disturbances to the injection current is optoelectronic feedback, in which the output from the laser is once detected by a photodetector and, then, the detected photocurrent is fed back into the injection current. A typical feature of instabilities induced by disturbances to the injection current are the regular and irregular pulsations in the laser output. In this chapter, we discuss instabilities and chaotic dynamics of optoelectornic feedback and injection current modulation in semiconductor lasers.

7.1 Theory of Optoelectronic Feedback

7.1.1 Optoelectronic Feedback Systems

Optoelectronic feedback is one of the perturbations to the injection current in semiconductor lasers that induces instabilities. In optical feedback to semiconductor lasers, the phase sensitivity plays a crucial role in the laser dynamics. However, different from optical feedback, we do not need to consider the phase effect in optoelectronic feedback systems, since the phase information is once eliminated by a photodetection in the feedback process. Stable or unstable operations of semiconductor lasers are flexibly and reliably controlled through the injection current (Giacomelli et al. 1989, and Loiko and Samson 1992). The dynamics of semiconductor lasers with optoelectronic feedback can be described only by the two equations of the photon number and the carrier density. Therefore, optoelectronic feedback shows different dynamics from those of optical feedback and optical injection in semiconductor lasers.

For optoelectronic feedback through the injection current, there are two categories; one is positive feedback and the other negative feedback. They have different mechanisms for driving the dynamics of the laser. In negative feedback, the feedback current is deducted from the bias injection current and it induces the sharpening of the relaxation oscillation (Lee and Shin 1993).

On the other hand, the feedback current is added to the bias injection current and, as a result, the gain switching tends to drive the laser into pulsing states in the output power (Damen and Duguay 1980). Therefore, regular pulsing states induced by positive optoelectronic feedback are used as a light source for periodic pulse trains with short pico-second duration (Paulus et al. 1987). Compared with short-pulse generations by passive mode locking with a saturable absorber (typically 1 ps pulse), Q switching, gain switching (typically 10 ps pulse), or active mode locking with external modulation (typically 10 ps pulse), stable and fast pulses with variable pulse-width can be easily obtained by setting appropriately the external parameters in the optoelectronic feedback system (Inaba 1982 and van der Ziel 1985). The repetition rate of the pulses is found to be an integral multiple of the inverse of the feedback-loop delay time that is closest to the relaxation resonance frequency of the laser. Otherwise, when the time for the feedback loop is not coincident with the period of the laser relaxation oscillation or its integral multiples, the competition of the incommensurate periods results in chaotic dynamics in the laser that strongly depend on the feedback time. The chaotic pulses in this system have both chaotic peak intensities and chaotic pulse intervals.

The model of optoelectronic feedback in semiconductor lasers is the same as that treated as the injection current modulation discussed in Sect. 3.6 (Lin and Liu 2003a). Figure 7.1 shows a schematic diagram of optoelectronic feedback in a semiconductor laser. The light emitted from a semiconductor laser is detected by a photodetector and the detected photocurrent is fed back through a bias Tee circuit. The feedback may be positive or negative depending on the polarity of the output of the amplifier in the circuit. In optoelectronic feedback, the modulation is not for the complex field but for the carrier density through the disturbance to the injection current. Therefore, we use the rate equation of the photon number instead of the complex amplitude. Using (3.73) and (3.75), the rate equations for the optoelectronic feedback system is written by

$$\frac{dS(t)}{dt} = [G_n\{n(t) - n_{th}\}]S(t) \tag{7.1}$$

$$\frac{dn(t)}{dt} = \frac{J(t)}{ed}\left\{1 + \xi\frac{S(t-\tau) - S_{offset}}{S_s}\right\} - \frac{n(t)}{\tau_s} - G_n\{n(t) - n_0\}S(t) \tag{7.2}$$

where ξ is the feedback strength. The system is positive feedback for a positive value of ξ, while it is negative feedback for a negative value. S_s is the steady-state value for the photon number and S_{offset} is the constant offset in the feedback loop. The value of the offset may be zero. τ is the feedback time including time responses of the detector and the electronic circuits.

A feedback circuit with time response of 100 pico-seconds is easily available at present and the response is sufficient to follow chaotic variations in semiconductor lasers. However, when the time response of the electronic feedback circuit has the same order as the relaxation oscillation of the laser, the

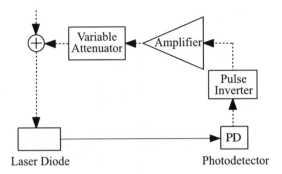

Fig. 7.1. Schematic diagram of the optoelectronic feedback system

effect of the finite response must be taken into account and another differential equation for the feedback current must be introduced besides the photon number and carrier density equations (Lee et al. 1988). We will discuss this point later. Contrary to the case of optical feedback in semiconductor lasers, photon noises induced in the optoelectronic feedback system have less effect, since the feedback is to the carrier density. This is due to the difference of the time scales of the lifetimes of carrier density and photon; the photon lifetime is typically on the order of pico-seconds, while the carrier lifetime is much longer than that and it is around nano-seconds. As noted, we do not consider the phase equation and the phase sensitivity is eliminated in the optoelectronic feedback system.

7.1.2 Pulsation Oscillations in Optoelectronic Feedback Systems

The typical feature of optoelectronic feedback in semiconductor lasers is pulsation oscillations. Depending on the feedback conditions, the laser shows various dynamic states of periodic and chaotic oscillations. However, the dynamics strongly depend on whether the system has a positive or negative feedback-loop. The details of the dynamics will be given later. Instead, we first present experimental results of pulsation oscillations in optoelectronic feedback systems. Figure 7.2 shows such examples (Tang and Liu 2001c). The laser outputs (left) and its rf spectra (right) are obtained for certain conditions of the feedback delay (the delay time including the flight of light in the loop of the laser, the detector, and the electric circuit). From Figs. 7.2a to 7.2c, the laser evolves a regular pulsing state of a constant peak intensity into a quasi-periodic pulsing state. Only one fundamental frequency at 650 MHz is excited in the spectrum in Figure 7.2a, which has the pulsing frequency of the regular pulses seen in the corresponding time series. However, ripples with small amplitude are included in the spectrum. The laser is not completely stabilized to a periodic state, although the level of the ripples is below 40 dB. The frequency of 140 MHz corresponding to the delay

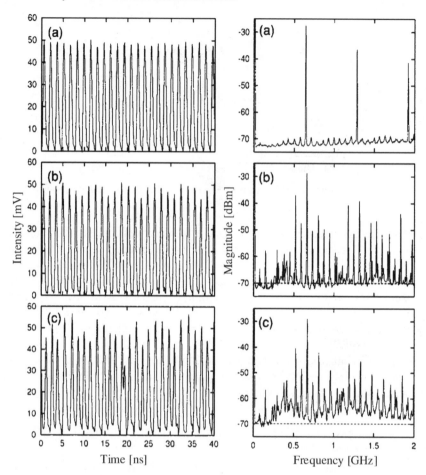

Fig. 7.2. Experimental results of time series and power spectra of various pulsing states for different delay times. **a** Regular pulsing at $\tau = 7.47$ ns. **b** Two-frequency quasi-periodic pulsing at $\tau = 7.09$ ns. **c** Chaotic pulsing at $\tau = 6.92$ ns. The laser used is a single mode DFB laser with a wavelength of 1.33 μm and the bias injection current is $J = 1.34 J_{\text{th}}$ (after Tang S, Liu JM (2001c); ©2001 IEEE)

time of the loop is excited as the second oscillation frequency in Figure 7.2b. The two frequencies are incommensurate. Therefore, the laser oscillates at the two fundamental frequencies and the oscillation is in a two-frequency quasi-periodic pulsing state. In Figure 7.2c, the broadband background is much higher than in Figure 7.2b, indicated by the dashed reference lines at −70 dBm. All over the spectrum is much broadened and the peak heights of the pulsation oscillation vary irregularly, corresponding to chaotic oscillation. Such chaotic evolution is well reproduced by numerical simulations based on the rate equations.

The feedback delay time and the feedback ratio play crucial roles in the dynamics of optoelectronic feedback systems, as is the case for optical feedback. A map of the dynamics in the phase space of the delay time and the feedback strength is calculated from the numerical simulation for the rate equations. Figure 7.3 shows the map of chaotic routes in the space for the normalized delay of $\hat{\tau} = \tau \nu_R$ and the feedback strength ξ at a fixed bias injection current (Lin and Liu 2003a). The positive value of ξ is the case for positive feedback and the negative value of ξ for negative feedback. We find that these systems share the same route to chaos and most of the dynamic states, but the frequency-locked pulsing states are clearly observed only in the negative feedback system. On the other hand, the region of the lock-

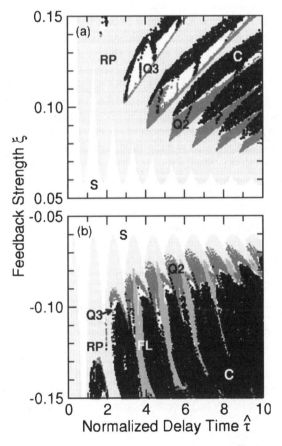

Fig. 7.3. Mappings of dynamic states of **a** positive optoelectronic feedback and **b** negative optoelectronic feedback systems at $J = 1.33 J_{\text{th}}$. S: steady-states, RP: regular pulsing, Q2: two-frequency quasi-periodic pulsing, Q3: three-frequency quasi-periodic pulsing, PL: frequency-locked pulsing, C: chaotic pulsing (after Lin FY, Liu JM (2003a); ©2003 IEEE)

ing range of the frequency-locked states is too narrow to be observable in the positive feedback. Therefore, small fluctuations in the laser parameters would drive the system into the neighboring quasi-periodic or chaotic states. Although frequency-locked states should generally exist in a two-frequency system, it is very difficult to observe experimentally the locking range of the frequency-locked states in positive optoelectronic feedback systems and they are rarely observable in real experiments.

With the increases of the normalized delay time $\hat{\tau}$ or the feedback strength ξ, the laser output shows very complicated dynamics. The laser evolves into chaos states through a quasi-periodic route following regular pulsing (RP), two-frequency quasi-periodic pulsing (Q2), three-frequency quasi-periodic pulsing (Q3), and finally chaotic pulsing states (C). Quasi-periodic states are typically observed for strong feedback strength in the positive feedback system. On the other hand, such states are limited within small feedback strength in the negative feedback system. In the mappings, RP, Q2, and Q3 states spread over large areas in the positive feedback system, while chaotic states have large areas in the negative feedback system. Therefore, we could easily obtain chaotic states in the negative feedback system, when we adjust the controllable parameters. Especially, we could observe remarkable chaotic pulsing states at large feedback strengths and long delay times. Another important difference between these two systems is the regions of the frequency-locked pulsing state (FL). In the positive optoelectronic feedback system, the states that separate the chaos islands are the regular pulsing (RP) states. However, in the negative optoelectronic feedback system, the states that separate the chaos islands are the frequency-locked pulsing (FL) states instead. Thus, the dynamics of optoelectronic feedback systems strongly depends on the positive or negative feedback even for the same delay time and the same strength.

7.2 Linear Stability Analysis for Optoelectronic Feedback Systems

7.2.1 Linear Stability Analysis

The solutions of stability and instability for the parameters in optoelectronic feedback systems are calculated by the linear stability analysis (Pieroux et al. 1994 and Grigorieva et al. 1999). Grigorieva et al. investigated the local dynamics for class B lasers with optoelectronic feedback. Applying the linear stability analysis for (7.1) and (7.2), they obtained the characteristic equation for the system as

$$\gamma^3 + \varepsilon(c^2 + 1)\gamma + c_q^2 \left\{ 1 + \gamma_0 \exp\left(-\frac{\lambda\tau}{\varepsilon_\tau} \right) \right\} = 0 \qquad (7.3)$$

where the normalized parameters are defined by $\gamma_0 = \xi J/edS_s$, $q = \tau_s J/edG_n n_s$, $c_q = \sqrt{(q-1)/(1+\gamma_0)}$, and $\varepsilon_\tau = \sqrt{\tau_{ph}/\tau_s}$. From the characteristic equation, we can calculate the boundary for stable and unstable oscillations of the laser. Figure 7.4 is the result. The stable region is located around the zero feedback coefficient ($\gamma_0 = 0$) and the left and right curves are the stability boundaries. The left curve is the stable boundary for the negative feedback, while the right curve is for the positive feedback. These curves correspond to the stable and unstable boundaries in Fig. 7.3, which is obtained from the numerical calculation for the rate equations.

The minimal values of the feedback coefficients for the stable boundaries are the same both for negative and positive feedback and the approximate form is given by

$$\gamma_{0\min} = \varepsilon_\tau \frac{q}{c_{q0}} \tag{7.4}$$

Namely, we require the above minimal value of the injection current to destabilize the laser output in optoelectronic feedback systems, whether it is a positive or negative feedback. The minimal points at $\gamma_{0\min}$ locate periodically for the increase or decrease of the delay time. For a positive feedback, the locations of the minimal points at the delay time are shifted by one quarter from the multiple period of the relaxation oscillation at a free running laser and the delay is given by

$$\tau_k = \varepsilon_\tau \frac{2\pi}{c_{q0}} \left(m + \frac{1}{4} \right) \quad m = 1, 2, 3 \ldots \tag{7.5}$$

On the other hand, the delay for a negative feedback is given by

$$\tau_k = \varepsilon_\tau \frac{2\pi}{c_{q0}} \left(m + \frac{3}{4} \right) \quad m = 1, 2, 3 \ldots \tag{7.6}$$

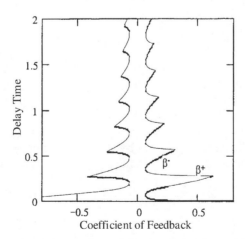

Fig. 7.4. Boundary of stability and instability in phase space of feedback coefficient and delay time at $\varepsilon_\tau^2 = 1000$ and $q = 1.5$ (after Grigorieva EV, Haken H, Kaschenko SA (1999); © 1999 Elsevier)

where $c_{q0} = \sqrt{q-1}$. For a positive feedback, the relaxation oscillation frequency decreases due to the increase of the carrier density and it is calculated as

$$\nu_R' = \frac{c_{q0}}{2\pi\varepsilon_\tau}\left(1 - \frac{\gamma_{0min}}{2}\right) \qquad (7.7)$$

The second term in the parenthesis on the right hand side of the equation is the effect of the feedback. In the case of a negative feedback, the relaxation oscillation frequency has the same form as (7.7), but the sign of the minimal value is negative and the relaxation oscillation frequency increases due to the reduction of the carrier density.

In the above discussion, we ignore the response time of the electronic circuits in the feedback system. However, we must take the transient effect into consideration, when we cannot neglect the finite response time of the photodetector. Lee et al. (1988) investigated the dynamics of semiconductor lasers with optoelectronic feedback when the photodetector has a finite response and derived the conditions for pulsation oscillations in the laser output power. They introduced the response for the feedback current as a differential equation (Kressel and Butler 1977):

$$\frac{di}{dt} = \frac{1}{\tau_i}(c_1 S_s - i) \qquad (7.8)$$

where $i = G_n \tau_s \tau_p J/ed$ is the normalized feedback current, τ_i is the 3-dB bandwidth of the feedback circuits, c_1 is the conversion efficiency of the photon density to the normalized current through the photodetector including the coupling efficiency. Taking into account the saturation effect of the current, the actual feedback current i_f through an amplifier is modeled by the following equation:

$$i_f = \frac{i_s}{1 + \exp\left\{-\frac{4(i-i_a)G_A}{i_s}\right\}} \qquad (7.9)$$

where i_s, i_a, and G_A are the saturation current, the detected photocurrent, and the gain in the circuit, respectively.

Combining (7.1) and (7.2) with (7.8), a linear stability analysis can also be applied to a negative optoelectronic feedback system. The loop gain for pulsation oscillations is calculated from the real part of the characteristic equation in the linear stability analysis. The resonance angular frequency taking into consideration of finite response of the feedback circuit is given by

$$\omega_{R,feedback}^2 = \omega_R'^2 + \frac{1}{\tau_i}\left\{\frac{1}{\tau_s} + G_n S_s + \frac{1}{\tau_{ph}}\left(\varepsilon' S_s + \beta_{sp}\frac{G_n\tau_{ph}n_0+1}{G_n\tau_s S_s}\right)\right\} \qquad (7.10)$$

where ω_R' is the angular frequency of the relaxation oscillation for the infinite time response of the feedback circuit and ε' is the effect of the nonlinear gain $\varepsilon' = \varepsilon_s G_n \tau$. The second term on the right hand side of the equation is the correction of the resonance angular frequency for the finite time response. The equation includes the effects of the nonlinear gain saturation ε_s and the spontaneous emission of light (coefficient β_{sp}). When the value of τ_i is approximated as infinity, namely, the time response is assumed to be instant, the second term in (7.10) is neglected and, then, the resonance frequency becomes exactly equal to the relaxation oscillation frequency at the infinite time response. For a finite time response of the circuit, the resonance frequency increases due to negative optoelectronic feedback. It is noted that the spontaneous emission of light induces the increase of the frequency of the relaxation oscillation as can be easily understood from (7.10). Also, the nonlinear gain shows the same tendency as the effect of spontaneous emission.

7.2.2 Characteristics of Semiconductor Lasers with Optoelectronic Feedback

Taking an example for negative optoelectronic feedback, we here present some results of the characteristics in the systems derived from the linear stability analysis. Figure 7.5a shows light-injection current (L-I) characteristics for

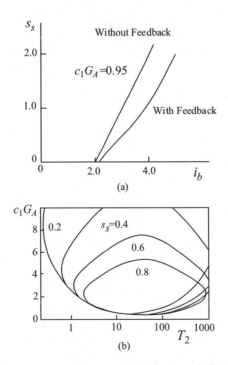

(a)

(b)

Fig. 7.5. Effects of negative optoelectronic feedback. **a** L-I characteristics. **b** Dependence of pulsing area on photon number or equivalently bias injection current, $T_2 = \tau_s/\tau_i$ and $s_s = G_n\tau_s S_s$. The parameters are $T_1 = \tau_s/\tau_{ph} = 2000$, $i_s = 2.5$, $\beta_{sp} = 10^{-5}$, and $\varepsilon' = 10^{-3}$. (after Lee CH, Shin SY, Lee SY (1988); ©1988 OSA)

a negative optoelectronic feedback (Lee et al. 1988). Since the carrier number in the active layer in negative optoelectronic feedback is reduced from that at the free running state, the laser threshold increases. As a result, we require a larger bias injection current for the same laser output power as that at the solitary oscillation. At the same time, the slope efficiency is reduced. In this model, the saturation effect for the current in the feedback loop is taken into account and, therefore, the slope of the L-I characteristic at a higher injection current is the same as that of the solitary oscillation, although the L-I curve is shifted toward the higher injection current.

Pulsation oscillations in optoelectronic feedback in semiconductor lasers strongly depend on the feedback gain G_A and the response time of the feedback circuit τ_i. The map of pulsing states in the phase space of the feedback gain and the response time is calculated from the linear stability analysis and is plotted in Fig. 7.5b. In Fig. 7.5b, the normalized coordinates of $c_1 G_A$ and τ_s/τ_i are used instead of the feedback gain and the response time. The unstable region of pulsing is inside the curves. When the feedback gain is as small as $c_1 A \ll 1$, the laser oscillates stably. For larger feedback gain, unstable pulsing starts to appear. However, the unstable region shrinks and finally disappears at a certain level of the feedback gain. The laser is stabilized for a higher bias injection current and the increase of the photon number (normalized photon number $s_s = G_n \tau_s S_s$) results in the shrinkage of the unstable region. As can be easily understood from the figure, the required closed-loop gain becomes minimum when $1/\tau_i$ is equal to the relaxation oscillation frequency. In this simulation, only the unstable pulsing oscillation boundary is investigated from the linear stability analysis. However, it is noted that various unstable dynamics and routes to chaos, not only the pulsing states but also unstable oscillations and chaotic states, are observable depending on the device and system parameters.

Figure 7.6 shows the waveforms of pulsing states for the feedback current i ($i = G_n \tau_s \tau_{ph} J/ed$), the carrier density n' ($n' = G_n \tau_{ph} n$), and the photon number s ($s = G_n \tau_{ph} S$) at certain parameter conditions (Lee et al. 1988). The excitation of pulsation oscillations in negative optoelectronic feedback is explained as follows: the detected photocurrent rom the laser is fed back into the bias injection current with a transient response of the photodetector after the laser oscillation starts at a certain time. The feedback current induces the reduction of the injection current. The injection current may become below the threshold and the laser oscillation stops for a large feedback current. Therefore, the laser oscillates at only the beginning of the rise time of the electronic feedback circuit and emits a very short pulse. When the laser oscillation stops, the feedback current decreases and the injection current to the laser increases. The injection current well exceeds the threshold, and then the next pulse is generated. These processes may be regarded as sharpening the extraction of the first spike of the relaxation oscillation. The feedback sharpens the falling edge of the first spike and suppresses the subsequent spikes. Thus, strong pulsation oscillation is generated at a high frequency

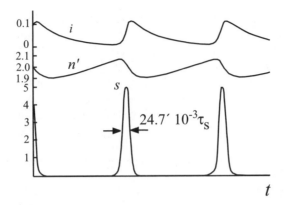

Fig. 7.6. Calculated waveform for current i, carrier density n', and photon number s with $s_s = 0.4$, $i_s = 2.5$, $c_1 G_A = 2.0$, $T_1 = 2000$, $T_2 = 10$, $\beta_{sp} = 10^{-5}$, and $\varepsilon' = 10^{-3}$ (after Lee CH, Shin SY, Lee SY (1988); © 1988 OSA)

of the order of the relaxation oscillation frequency (the frequency is shifted to a higher side due to the feedback). The pulse width is easily adjusted by appropriately choosing the feedback gain. Also the pulse repetition rate can be controlled by the bandwidth $1/\tau_i$ of the amplifier.

7.3 Dynamics and Chaos in Semiconductor Lasers with Optoelectronic Feedback

7.3.1 Chaotic Dynamics in Negative Optoelectronic Feedback

In the previous section, we demonstrated regular pulsing states in semiconductor lasers induced by optoelectronic feedback. In this section, we show in detail unstable and chaotic pulsation oscillations in such systems. As has already been discussed, there are two types of optoelectronic feedback; one is positive and the other is negative. For both cases, we can expect regular pulsing, quasi-periodic pulsing, and chaotic pulsing states in the laser output power. However, the region of the frequency-locked state in the parameter space is very narrow for positive optoelectronic feedback and it is only observable for negative optoelectronic feedback in actual experimental systems (Giacomelli et al. 1989). Similar locking states have been observed in a modulated external cavity injection laser (Lee and Shin 1993) and a modulated self-pulsing semiconductor laser (Lee et al. 1988). These frequency-locked pulsing states are experimentally observed to exhibit a harmonic frequency-locking phenomenon, where the pulsing frequency is locked to a harmonic of the delay loop frequency instead of the delay loop frequency itself.

Numerical simulations for stable and unstable pulsation oscillations for negative optoelectronic feedback were conducted and various pulsing states were observed. Here, we show the experimental results (Lin and Liu 2003a). In Fig. 7.7, sequential pulsing states including regular pulsing (RP), two-frequency quasi-periodic (Q2), quasi-periodic frequency locking (Q-FL), frequency locking (FL), and chaotic (C) states are observed by changing the frequency of the feedback loop from $f_{\text{loop}} = 49$ to $68\,\text{MHz}$. On the way to chaotic evolutions, generations of various frequency components are observed, such as excitations of the relaxation oscillation, the second funda-

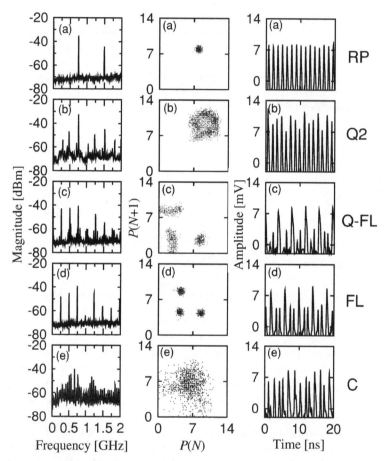

Fig. 7.7. Experimentally measured power spectra, phase portraits, and time series of different pulsing states in negative optoelectronic feedback systems. **a** RP: regular pulsing, **b** Q2: two-frequency quasi-periodic pulsing, **c** Q-FL: quasi-periodic frequency locking, **d** FL: frequency locking, and **e** C: chaotic states, where the loop frequencies are 68, 64, 51, 50, and 49 MHz, respectively. The laser used is a DFB laser of the wavelength of 1.299 μm. The bias injection current is $J = 1.17J_{\text{th}}$ and the relaxation oscillation frequency at that current is 1.5 GHz (after Lin FY, Liu JM (2003a); © 2003 IEEE)

mental frequency (the loop frequency), their harmonics, and their beats. It is also demonstrated that the rotation numbers of these frequency-locked pulsing states show a Devil's staircase structure. Three-frequency quasi-periodic pulsing (Q3) states in which an extra frequency component is added to the two-frequency quasi-periodic state (the relaxation oscillation and the loop frequencies) are observed in the numerical simulations. However, Q3 states are rarely observed in experiments. In this experiment, Q3 states are not observed, since the areas of Q3 states are so small that they are easily shifted to different neighbor states due to disturbances such as noises in the electronic circuit. The other oscillation states were well reproduced in the numerical simulations.

Figure 7.7a is a regular pulsing state for the loop frequency of $f_{\text{loop}} = 68\,\text{MHz}$. The pulsing frequency is 750 MHz. It is noted that the pulsing frequency of the laser is about one half of the resonance frequency $f_{\text{p}} \sim \nu_{\text{R}}/2$. Such sub-harmonic oscillation has also been found in a driven van der Pol oscillator (Gilbert and Gammon 2000). When the loop frequency f_{loop} becomes as small as 64 MHz, Q2 states appear. In this state, the pulsation frequency f_{p} at 760 MHz beats with a frequency at 320 MHz. The latter frequency is about five times that of f_{loop}. The frequency f_{p} is beating with the fifth harmonic of f_{loop}. Although the noise and the digitization error make the ring expected for a Q2 state diffused in the phase portrait, a ring-like distribution can still be seen (Fig. 7.7b). A Q-FL state between quasi-periodic pulsing and frequency-locked pulsing is observed at the increased loop frequency of 51 MHz as shown in Fig. 7.7c, where three smeared dots can be seen in the phase portrait. We can see the three frequency components of f_{p}, f_{loop}, and the fifth harmonic of f_{loop}. Figure 7.7d is an example of a frequency locking state. At the lower frequency of f_{loop} equal to 50 MHz, f_{loop} and f_{p} at 750 MHz become exactly commensurate, and frequency locking occurs. The state is a 5:15 (1:3) frequency locking, where f_{lock} is the fifth harmonic of f_{loop}, $f_{\text{lock}} = 5f_{\text{loop}}$, and the pulsing frequency is $f_{\text{p}} = 15f_{\text{loop}}$. The laser evolves into a chaotic state for the decrease of f_{loop} to 49 MHz as shown in Fig. 7.7e. The chaotic state (C) is identified by the spread dots in the phase portrait and the random intensity pulses seen in the time series. Hence, the experimental results bear out the simulation study, where the same sequence of states including the RP, Q2, Q-FL, FL, and C states are found and a quasi-periodic route to chaos is verified. Lin and Liu (2003) further investigated the phenomenon of frequency locking states with various rotation numbers and they obtained a Devil's staircase (which is a typical structure reconstructed from excited frequencies of chaotic oscillation in nonlinear systems) for the relation between the rotation number and the locking frequency.

7.3.2 Chaotic Dynamics in Positive Optoelectronic Feedback

Self-sustained pulsations are observed both for negative and positive optoelectronic feedback in semiconductor lasers. However, their dynamics are not

always the same. For example, frequency locking states are the typical feature
in negative optoelectronic feedback, while they are rarely observed in positive
optoelectronic feedback. The extent of chaotic regions in the parameter space
varies. Here, we discuss chaotic evolutions of pulsing states in positive optoelectronic feedback systems and demonstrate that not only the pulse height
but also the jitter of pulse sequences shows chaotic behaviors. Figure 7.8 is
a numerical result of pulsing states for the variations of the delay time in an

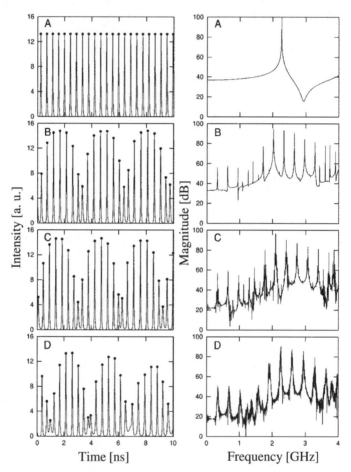

Fig. 7.8. Time series and power spectra of different pulsing states. **A**: regular
pulsing at $\hat{\tau} = 7.47$. **B**: two-frequency quasi-periodic pulsing at $\hat{\tau} = 7.25$. **C**: three-
frequency quasi-periodic pulsing at $\hat{\tau} = 7.00$. **D**: chaotic pulsing at $\hat{\tau} = 6.48$. The
pulse peak intensities are marked by the *filled circles* in the time series. The calculated spectra have relative magnitudes with decibel increment. In the numerical
simulations, the relaxation oscillation frequency is assumed to be $\nu_R = 2.5$ GHz at
$J = 1.33J_{th}$ (after Tang S, Liu JM (2001c); © 2001 IEEE)

optoelectronic feedback system (Tang and Liu 2001c). The normalized delay time is used in the figure and it is defined by $\hat{\tau} = \nu_R \tau$. With $\hat{\tau} = 7.47$ in A of Fig. 7.8, the time series shows a sequence of regular pulses with a constant pulsing intensity and interval. The corresponding power spectrum has only one fundamental pulsing frequency at $f_1 \sim 2.3\,\text{GHz}$, which is close to the resonance frequency of the laser. When the delay time is decreased to $\hat{\tau} = 7.25$ in B, the laser enters a two-frequency quasi-periodic pulsing state with the intensity modulated at a certain frequency f_2. The pulses are clearly modulated and the new modulation frequency is read to be $f_2 \sim 320\,\text{MHz}$. This f_2 is close to, but slightly less than, the inverse of the delay time of the feedback loop. The appearance of two incommensurate frequencies, f_1 and f_2, is the indication of quasi-periodicity. When the delay time is further decreased to $\hat{\tau} = 7.00$ in C, the laser enters a three-frequency quasi-periodic pulsing state as a third frequency at $f_3 \sim 23\,\text{MHz}$ shows up. The component of this frequency f_3 is very small and the frequency is incommensurate with f_1 and f_2. In other delayed feedback systems, such as semiconductor lasers with weak optical feedback (Ritter and Haug 1993a, b) and semiconductor lasers with negative optoelectronic feedback controlling the pump current (Grigorieva et al. 1999), a similar third frequency has also been found. Tang and Liu (2001c) indicated that f_3 is the result of nonlinear interaction between the laser relaxation oscillation and the delayed feedback. Finally, when $\hat{\tau} = 6.48$, the laser enters a chaotic pulsing state, as shown in D. In the chaotic states, not only the pulse height but also the separation becomes chaotic (jitter) and the corresponding spectrum is much broadened.

Phase portraits and spectra of the peak series in the Poincaré sections constructed from the corresponding time series of Fig. 7.8 are calculated and plotted in Fig. 7.9. In the phase portraits, $P(N)$ are the successive peak intensity values of the laser pulses in a given time series, while $I(N)$ are the corresponding time intervals of the pulses. The range of the rf spectra is expanded and the plots show only lower frequency components. In a regular pulsing state in A, the peak series consists of a single constant value, therefore the phase portrait is a single fixed point. In a two-frequency quasi-periodic pulsing state in B, the phase portrait shows a clear circle, since the pulsing frequency f_1 in the power spectrum of the original time series has been eliminated from the data set of the peak series under the Poincaré section and the successive peaks corresponding to f_2 are sampled. In the three-frequency quasi-periodic pulsing state in C, the two incommensurate frequencies f_2 and f_3 are both clearly shown in the spectrum of the peak series and the phase portrait is characterized by a torus. Finally, when the laser enters a chaotic pulsing state as in D, the phase portrait spreads out over a wide range. The phase portraits of $I(N + 1)$ versus $I(N)$ have the same characteristics as those of $P(N + 1)$ versus $P(N)$ at different delay times as shown in A to D of Fig. 7.9. From Fig. 7.8 and 7.9, a positive optoelectronic feedback system fits in a Rulle–Takens–Newhouse three-frequency quasi-periodic scenario to chaos (Tang and Liu 2001c).

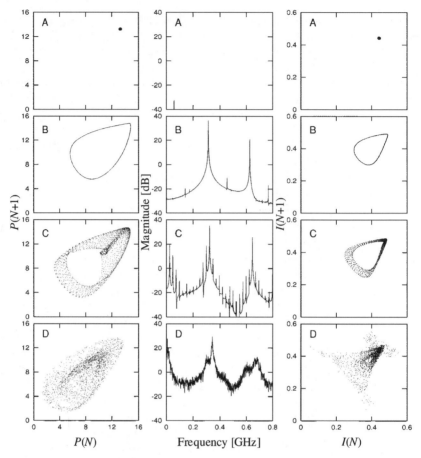

Fig. 7.9. Phase portraits and spectra for peak series extracted from the waveforms in Fig. 7.8. *Left:* phase portraits of peak intensities $P(N+1)$ versus $P(N)$. *Middle:* spectra of peak series. *Right:* phase portraits of time intervals of the pulses $I(N+1)$ versus $I(N)$. A–D indicate the corresponding pulsing states in Fig. 7.8. In A, the single dot in the phase portrait is enlarged for visibility (after Tang S, Liu JM (2001c); © 2001 IEEE)

To visualize the chaotic route in positive optoelectronic feedback in a semiconductor laser, the bifurcation diagrams are numerically calculated. Bifurcation diagrams of the extrema of the peak series versus delay time corresponding to the previous figures (Fig. 7.8 and 7.9) are plotted in Fig. 7.10. Figure 7.10a is the bifurcation diagram with the normalized delay time $\hat{\tau}$ varying from 0 to 10. For a small delay time, the effect of the feedback on the dynamics is not distinct and the feedback increases the laser output slightly over that in the free-running condition. With the increase of the delay time, the laser output shows regular pulsing with constant pulse peak intensity

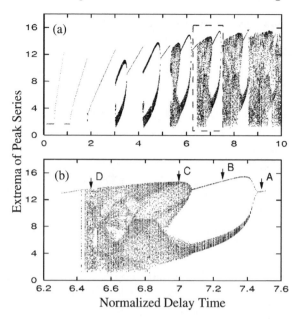

Fig. 7.10. a Bifurcation diagram of extrema of peak series with normalized delay
time $\hat{\tau}$ varying from 0 to 10. **b** Enlargement of a small region indicated by *dashed
rectangle* in **a** with identification of various pulsing states. A–D indicate the corresponding pulsing states, as in Fig. 7.8 (after Tang S, Liu JM (2001c); © 2001 IEEE)

and interval. For a larger delay, the laser follows a quasi-periodic route into
chaotic pulsing states. Fig. 7.10b is the enlarged bifurcation diagram of a part
in Fig. 7.10a denoted by the broken rectangle. A-D in the figure can be compared with the pulsing states in Fig. 7.8. The different pulsing states are
indicated by the corresponding arrows, where A is regular pulsing, B is two-
frequency quasi-periodic pulsing, C is three-frequency quasi-periodic pulsing,
and D is chaotic pulsing.

7.4 Optoelectronic Feedback with Wavelength Filter

7.4.1 System of Optoelectronic Feedback with Wavelength Filter

The frequency of a semiconductor laser changes as the increase or decrease
of the pump for the laser, i.e. the change of the bias injection current in
accordance with (5.5). The order of the frequency change is ∼mA/GHz in
usual semiconductor lasers. Therefore we can observe the dynamics induced
by the nonlinearity in a semiconductor laser with optoelectronic feedback
through a wavelength filter (frequency filter) having a bandwidth of ∼GHz.
In such systems, the nonlinearity is greatly enhanced by the introduction of

a compulsive time delay longer than the response time of the circuit inserted in the electronic feedback loop. The systems show a rich variety of dynamics, including stable and unstable oscillations and chaos, depending on the delay and other control parameters. Several chaotic systems of optoelectronic feedback with wavelength filter have been proposed in semiconductor lasers. One of them is an optoelectronic system where the emitted light passes through a Fabry-Perot resonator and the output from the Fabry-Perot is electronically detected and fed back to the laser injection current (Ohtsu 1996). The second example is the use of a birefringent plate in the optical path with the combination of optoelectronic feedback circuits to a semiconductor laser (Goedgebuer et al. 1998a and Larger et al. 1998a). Another system of optoelectronic filtered feedback in a semiconductor laser is an electro-optic (EO) modulation system, in which the emitted light from a semiconductor laser is detected by a photodetector and the signal drives an EO modulator. The light from a different laser goes through the EO module. The light is then detected by a second photodetector and the electric signal with time delay is added to the bias injection current, thus closing the loop (Goedgebuer et al. 2002). This is also an example of filtered optoelectronic systems. These examples are categorized into filtered feedback systems discussed in Sect. 4.6. These systems are used for light sources of chaotic generators in optical secure communications, which will be discussed in Chap. 13. Chaotic communications of encoding and decoding rather lower frequency signals ranging from several kHz to 100 MHz have been demonstrated using chaotic transmitter and receiver systems with optoelectronic filtered feedback as discussed here (Goedgebuer et al. 1998b, 2002; Larger et al. 1998b). In this chapter, we describe a system and dynamics of filtered optoelectroinc feedback based on a birefringent plate.

Figure 7.11 shows an example of generator of chaos in wavelength using a wavelength tunable semiconductor laser with a nonlinear feedback loop (Larger et al. 1998a). The nonlinearity is induced by the birefringent plate wet between two crossed polarizers. In a two-electrode semiconductor laser, the wavelength is fixed at a given value λ_0 by adjusting a couple of bias currents (I_0, I_1) on each of the electrodes and can be tuned electronically around λ_0 by varying current I_0 by i, while keeping I_1 constant. The feedback loop consists of a birefringent plate whose fast and slow axes are at 45° to two crossed polarizers PL_1 and PL_2, a photodetector (PD) having a time response τ with gain, and a delay line with a retardation time T. The signal is converted to the tuning current i, which is superimposed to the bias current I_0. Hence the emitted wavelength of the laser is given by $\lambda(t) = \lambda_0 + \Delta\lambda(t)$, where the additional change of the wavelength is proportional to the injection current i and is given by $\Delta\lambda(t) = (d\lambda/di)i(t)$. The power spectrum density of light at the output of PL_2 is a channeled spectrum, which is expressed as

$$NL(\lambda) = \sin^2\left(\frac{\pi D}{\lambda}\right) = \sin\left(\frac{\pi D}{\lambda_0^2}\lambda - \phi_0\right) \tag{7.11}$$

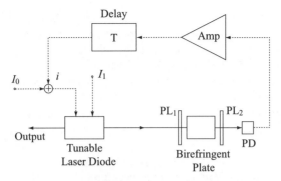

Fig. 7.11. Generator of chaos in wavelength using a wavelength tunable semiconductor laser with a nonlinear feedback loop. PL: polarizer, PD: photo-detecotr, Amp: amplifier. *Solid lines* are the optical paths and *dotted lines* are the electronic connections

where D is the optical-path difference through the birefringent plate and $\phi_0 = \pi D/\lambda_0$. Here, we assume that the additional change of the wavelength $\Delta\lambda$ is much smaller than the wavelength λ_0. Then the output wavelength of the system is ruled by the following delay differential equation

$$\lambda(t) + \tau\frac{\mathrm{d}\lambda(t)}{\mathrm{d}t} = \beta_\lambda\sin^2\left\{\frac{\pi D}{\lambda_0^2}\lambda(t-T) - \phi_0\right\} \qquad (7.12)$$

where $\beta_\lambda = PK\mathrm{d}\lambda/\mathrm{d}i$, P and K being the detected optical power by the photodetector and the gain of the amplifier. Since we consider a long delay compared with the laser response (on the order of nanoseconds) and the delay time T is around milliseconds, we need not take into account the dynamics related to the relaxation oscillation. Therefore, equation (7.12) is enough to describe the dynamics of wavelength in the tunable semiconductor laser.

7.4.2 Dynamics of Optoelectronic Feedback with Wavelength Filter

In this section, we present experimental results of optoelectronic feedback with the wavelength filter that is described in the previous section (Larger et al. 1998a). The laser is a tunable double-electrode distributed Bragg reflector (DBR) semiconductor laser whose wavelength can be tuned continuously without mode hopping. The center wavelength of the laser is $\lambda_0 = 1.550\,\mu\mathrm{m}$ and the total tuning range is $1.5\,\mathrm{nm}$ (corresponding frequency range is $200\,\mathrm{GHz}$). The wavelength is tuned by a DBR-section injection current I_0 and the wavelength tunability is $\mathrm{d}\lambda/\mathrm{d}i = 0.2\,\mathrm{nm/mA}$. The birefringent plate was a calcite slab of thickness $d_\mathrm{b} = 6\,\mathrm{cm}$, yielding an optical path difference $D = |n_\mathrm{e} - n_o|d_\mathrm{b} = 11\,\mathrm{mm}$ between its fast and slow axes, where $n_\mathrm{e} = 1.477$ and $n_o = 1.634$ are the extraordinary and ordinary refractive indices, respectively. Then its spectral transmission curve, which is also the function expressed by (7.11), exhibits seven sinusoidal peaks centered at 1.550 μm inside the 1.5 nm tuning range of the laser diode. The linewidth of the laser diode was 10 MHz. The response time of the electronic circuit is $\tau = 8.6\,\mu s$, while the delay time is much slower than the response time and is set to be $T = 0.51\,ms$.

In the wavelength filtered optical feedback, one can observe period-doubling cascade rout to chaos is obtained as the increase of the bifurcation parameter $\beta(\beta = \beta_\lambda \pi D/\lambda_0^2)$. Figure 7.12 shows experimental dynamics for different values of the bifurcation parameter β and the corresponding power spectrum. The bifurcation parameter β is tuned by changing electronically the photodetector gain K in the feedback loop. Period-2 in Fig. 7.12a is a typical square waveform. The wavelength emitted by the laser diode oscillates periodically between two states spaced by 41 pm (6.3 GHz). The spectrum in Fig. 7.12b shows odd harmonics whose amplitude decreases with increasing the odd number. The fundamental frequency is $\{2(T + \tau)\}^{-1} = 970\,Hz$. Four wavelength states (located at $\lambda_1 = 59\,pm$, $\lambda_2 = 66\,pm$, $\lambda_3 = 110\,pm$, $\lambda_4 = 114\,pm$) are obtained in the period-4 regime, as shown in Fig. 7.12c. The spectrum of period-4 in Fig. 7.12d exhibits subharmonic peaks at frequencies $(2m - 1)/\{4(T + \tau)\}$ with a positive integer m. The dc background in the spectrum is slightly increased compared with that of the period-2 regime. As β exceeds 2.11 in Figs. 7.12e and f, the peak-to-peak amplitude of these small oscillations increases, yielding a high noise-like level in the spectrum. These oscillations may be regarded as a starting chaotic regime at each of the levels of the period-4 cycle. The time evolution of Fig. 7.12e is termed period-4 chaos. When slightly increasing β from 2.12 to 2.18, the laser enters a period-2 chaos, as shown in Fig. 7.12g and h. The background level in the spectrum is relatively high, but still features harmonic peaks. The two successive regimes period-4 chaos and period-2 chaos illustrate the inverse cascade in the bifurcation diagram. These non-periodic but chaotic regimes with a growing complexity are also in very good agreement with the behaviors observed in previous experimental investigations (Gibbs 1985). Another example of periodic regime is illustrated in Figs. 7.12i and j; a complicated high-frequency periodic oscillation is observed, which is named as higher harmonic synchronization. For values of β over 2.5 in Figs. 7.12k and l, fully developed chaos is clearly obtained, without any remaining influence of the higher harmonic synchronization. The spectrum is similar to that of a white noise in the bandwidth of the dynamical system.

Figure 7.13 shows the experimental plot of the bifurcation diagram obtained as $\phi_0 = 0.3$ corresponding to Fig. 7.12. The vertical axis in the figure is the wavelength and the horizontal axis is the bifurcation parameter. The experimental bifurcation diagram quite agrees with the calculated diagram by the delay differential equation in (7.12). The period-doubling cascade (period-2 T_2 and period-4 T_4) route to chaos is obtained as the bifurcation parameter

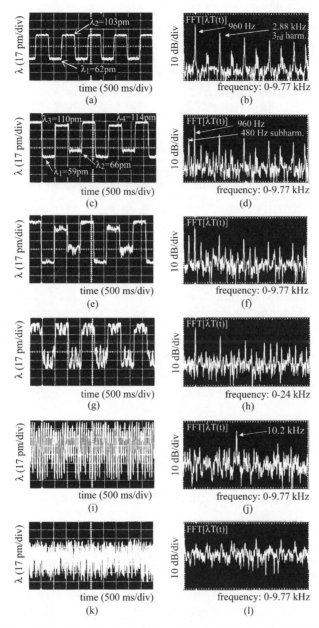

Fig. 7.12. Experimentally obtained time evolution of the wavelength (*left* column) and its FFT spectrum (*right* column) at $\phi_0 = 0.3$ for different values of the bifurcation parameter β. **a, b**: $\beta = 1.93$, period-2, **c, d**: $\beta = 2.04$, period-4, **e, f**: $\beta = 2.12$, period-4 chaos, **g, h**: $\beta = 2.18$, period-2 chaos, **i, j**: $\beta = 2.26$, higher harmonic synchronization, and **k, l**: $\beta = 2.52$, fully developed chaos (after Larger L, Goedgebuer JP, Merolla JM (1998a); © 1998 IEEE)

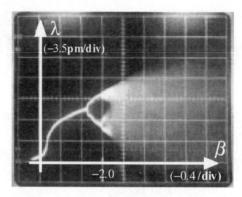

Fig. 7.13. Experimental bifurcation diagram for $\phi_0 = 3$ (after Larger L, Goedge-buer JP, Merolla JM (1998a); © 1998 IEEE)

is increased from 0.7 to 2.1. To observe further higher periodic oscillations such as T_8 and T_{16} states, we require a fine tuning of the bifurcation parameter β, but it is very difficult due to existing noises and the limitation for the resolutions of the experimental equipment and also the narrow ranges and the low stabilities of T_8 and higher periods. In usual fact, observations of higher periodic oscillations over period-4 states are rare and difficult to be observed in the real experiment. When considering the general shape of the doubling cascade, it is also in good agreement with other experimental results (Gibbs 1985).

7.5 Chaotic Dynamics of Semiconductor Lasers Induced by Injection Current Modulation

7.5.1 Instabilities of a Modulated Semiconductor Laser

The output from a semiconductor laser faithfully follows an injection current modulation as far as the modulation is small. On the other hand, for strong injection current modulation, the laser output power clearly exhibits a number of nonlinear characteristics, i.e., harmonic distortion, pulsation different from the modulation frequency, bistability, and period-doubling and quasi-period-doubling routes to chaos. The routes to chaos in semiconductor lasers have also been intensively studied not only for theoretical interest, but also for practical purposes, especially in the area of analogue modulation in optical fiber communications. Chaos in pump-modulated lasers is not the only unique feature in semiconductor lasers. It has also been observed experimentally in solid-state lasers (Klische et al. 1984) and Nd-doped fiber lasers (Phillips et al. 1987). Chaotic phenomena in class B lasers including semiconductor lasers critically depend both on the spontaneous emission (Lee et al.

1985) and the gain saturation (Chen et al. 1985). The spontaneous emission factor and the gain saturation coefficient, which act as damping factors in the small signal analysis, suppress instabilitities of semiconductor lasers. Spiky behavior of the photon density by the injection current modulation in semiconductor lasers is used for a light source of pico-second optical pulse generation in various applications (Ito et al. 1981 and O'Gorman et al. 1989). When a laser is modulated at a frequency above the intrinsic resonance frequency, the resonance peak shifts towards lower frequency as the modulation current is increased. When the frequency of the noise peak is moved to about a half of the drive frequency, the noise peak begins to be sharpened. That is, the first subharmonic is about to be generated. When the modulation current is increased further, the oscillation gradually becomes period-1. Observation of period-4 oscillations in a weakly-pulsating laser was reported (Chen et al. 1985), where self-pulsation was induced by optical damage. The observation of a period-doubling route to chaos in a Nd-doped fiber laser is also attributed to a very small spontaneous emission factor and gain saturation coefficient (Phillips et al. 1987).

Figure 7.14 is an example of pulsation oscillations by strong modulations for the injection current in a semiconductor laser (Hemery et al. 1990). In the figure, the modulation frequency and index, and the bias injection current are changed. The left plot shows the experimental results and the right shows the theoretical ones. A finite response photodetector is used in the experiment, so that the detector response is also displayed in the middle. In

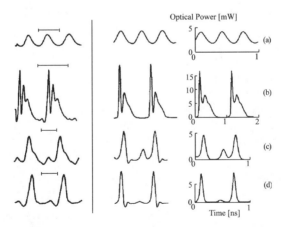

Fig. 7.14. Laser outputs by strong injection current modulation. Experimental (*left*) and theoretical (*right*). *Middle curves* account for the photodetector response in the experiment. The modulation conditions are **a** $f_m = 3.00$ GHz, $m = 0.21$, $J = 2.14 J_{\text{th}}$, **b** $f_m = 1.00$ GHz, $m = 1.40$, $J = 2.14 J_{\text{th}}$, **c** $f_m = 3.4$ GHz, $m = 2.30$, $J = 1.40 J_{\text{th}}$, and **d** $f_m = 3.33$ GHz, $m = 3.60$, $J = 1.40 J_{\text{th}}$. In each case, the *horizontal bar* indicates the period of current modulation (after Hemery H, Chusseau L, Lourtioz JM (1990); © 1990 IEEE)

the numerical simulations, the filter effect of the photodetector is taken into account, therefore the theoretical waveforms are compatible with those of the experiment. The laser is a single mode Fabry-Perot laser of a InGaAsP-InP buried heterostructure oscillating at a wavelength of 1.30 μm. The relaxation oscillation frequency depends on the bias injection current and it is less than 3 GHz under the operating conditions. For a high modulation frequency at $f_m = 3.00$ GHz but a moderate modulation index of $m = 0.21$, a waveform similar to the modulation is faithfully reproduced (Fig. 7.14a). For a low modulation frequency and a large modulation index, the laser response still has the same periodicity as the modulation, but several spikes are seen in a single period as shown in Fig. 7.14b. When both the modulation frequency and index are large, period-2 oscillations merge into the laser output power as shown in Fig. 7.14c. However, further large modulation index in Fig. 7.14d, subpeaks as observed in Fig. 7.14c vanish. Similar results of such period-doubling routes have also been reported in semiconductor lasers (Chen et al. 1985, Gallagher et al. 1987, Hori et al. 1988, and Hemery et al. 1990). These behaviors are explained as follows; the laser exhibits intense pulses and the carrier number is abruptly decreased after the oscillations of intense pulses when the injection current is strongly modulated. If the modulation frequency exceeds the relaxation oscillation frequency, a longer recovery time than the modulation period is required to restore the carrier number enough for the laser oscillation. Then, the laser may oscillate at a frequency beyond the period of the modulation and result in period-2 oscillations.

7.5.2 Linear Stability Analysis

In optoelectronic feedback systems, the feedback term is applied to the injection current as an additional perturbation as shown in (7.2). In the injection current modulation, the feedback term is replaced by the modulation as discussed in Sect. 3.6. As usual, the modulation may be a sinusoidal one with a modulation frequency f_m and a modulation index m. Again, the rate equation for the carrier density in the presence of sinusoidal injection current modulation is given by

$$\frac{dn(t)}{dt} = \frac{J(t)}{ed}\{1 + m\sin(2\pi f_m t)\} - \frac{n(t)}{\tau_s} - G_n\{n(t) - n_0\}S(t) \qquad (7.13)$$

The rate equation for the photon density remains the same as that in (7.1). Yoon et al. (1989) conducted the linear stability analysis paying attention to the nonlinear gain saturation coefficient ε_s and the spontaneous emission factor β_{sp} and discussed stability and instability for the injection current modulation. They showed resonance frequency shift, bistability, and period-doubling bifurcation in directly modulated semiconductor lasers in the laser output power. The dynamics is strongly dependent on spontaneous emission and gain saturation. As a result, it is shown that the resonance peak in the frequency response shifts toward lower frequencies as the modulation current is increased, and that it may accompany a hysteresis phenomenon. Similar results have been reported by Harth (1973). His result was obtained by neglecting both the spontaneous emission factor and the gain saturation coefficient.

In the following, we show some results obtained from the linear stability analysis (Yoon et al. 1989). In the linear stability analysis based on the parametric forced oscillation to the rate equations, we define the normalized frequency detuning δ/ω_R due to the modulation and put the laser frequency in the presence of the injection current modulation as an approximate form

$$\nu \approx \nu_R \left(1 + \frac{\delta}{\omega_R}\right) \tag{7.14}$$

After some calculations, we obtain the relation between the modulated laser output power a and the detuning δ as

$$a^2 \left\{ \left(\delta + \frac{a^2}{24\omega_R}\right) + (\tau_s \Gamma_R)^2 \right\} = \frac{(T i_b m)^2}{4} \left(1 + \frac{\delta}{\omega_R}\right) \tag{7.15}$$

where $T = \tau_s/\tau_{ph}$ and $i_b = G_n \tau_s \tau_{ph} J/ed$ is the normalized bias injection current. The power a is normalized and defined with the use of the change of photon number δS as

$$a = \frac{G_n \tau_s^2}{\tau_{ph}} \delta S \tag{7.16}$$

From these relations, the peak amplitude a_p of pulsation oscillations reads

$$a_p^2 = \frac{a_{pl}^2}{1 + \frac{a_{pl}^2}{24\omega_R^2}} \tag{7.17}$$

For small modulation index m, we can approximate the amplitude as

$$a_p \approx a_{pl} = \frac{T i_b m}{2\tau_s \Gamma_R} \tag{7.18}$$

Using the amplitude a_p, the relaxation oscillation frequency subjected to the injection current modulation is given by

$$\nu' = \nu_R \left(1 - \frac{a_p^2}{24\omega_R^2}\right) \tag{7.19}$$

The reduction of the relaxation oscillation frequency by the injection current modulation has been already discussed in Fig. 7.14.

Figure 7.15 shows the plot of the calculated frequency response of the laser output power a as a function of the modulation frequency (Yoon et al.

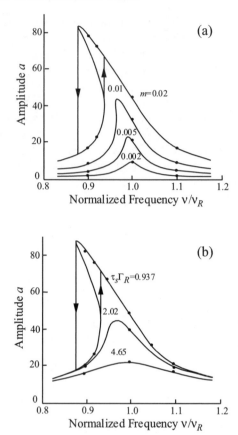

Fig. 7.15. Frequency response of laser output power a at $g_0 \tau_{\rm ph} n_s = 1$, $T = 3000$, and $i_{\rm b} = g_0 \tau_s \tau_{\rm ph} J/ed = 2.8$. Dependence on **a** modulation index m with $\tau_s \Gamma_{\rm R} = 0.937$ and **b** normalized damping factor $\tau_s \Gamma_{\rm R}$ with $m = 0.02$. Dots represent numerical results obtained from the rate equations and the *solid lines* the linear stability analysis (after Yoon TH, Lee CH, Shin SY (1989); © 1989 IEEE)

1989). Figure 7.15 a shows the response for the variations of the modulation index m at a fixed normalized damping factor of $\tau_s \Gamma_{\rm R} = 0.937$ and Fig. 7.15b for the variations of the normalized damping factor $\tau_s \Gamma_{\rm R}$ at a fixed modulation index of $m = 0.02$. Dots represent numerical results directly calculated from the rate equations, while solid lines represent analytical results by the linear stability analysis. As the modulation index is increased, the resonance peak shifts toward lower frequencies, as predicted from the calculation. It accompanies hysteresis (bistabiltiy) for $m > m_{\rm min}$. With the increase of the normalized damping constant $\tau_s \Gamma_{\rm R}$, the hysteresis region narrows and disappears. It shows that the magnitude of the normalized damping factor $\tau_s \Gamma_{\rm R}$, which is strongly related to the spontaneous emission factor $\beta_{\rm sp}$ and the gain saturation coefficient ε_s, is crucial to the dynamic behavior of the semiconductor lasers. Thus, under a strong injection current modulation, bistability is observed for the laser output power in a semiconductor laser with a large damping constant. This induces instability in the laser output and finally results in chaotic oscillations at larger injection current modulation. It is noted that the resonance frequency shift has been observed experimentally (Chen et al. 1985).

It is well-known that generation of a period-2 solution or the first period-doubling bifurcation in a directly modulated semiconductor laser is possible by subharmonic resonance occurring when the frequency of the driving current is close to twice the small signal resonance frequency. In the same manner as the previous discussion, using the normalized frequency detuning as $\delta/2\omega_R$ and putting the laser frequency as

$$\nu \approx 2\nu_R \left(1 + \frac{\delta}{2\omega_R}\right) \tag{7.20}$$

we perform the linear stability analysis for the rate equations. Analytical solutions are also obtained in the same manner as in the previous discussion, however, here, only the results are shown in Fig. 7.16. At the frequency slightly less than $2\omega_R$, there are hystereses for the increase or decrease of the frequency. It is not explicitly shown in the graph, but coexistence states of two chaotic attractors exist in the region of periodic states (period-1 and period-2) at certain modulation indexes. With an increase of the modulation index from zero, there is no subharmonic resonance until $m = m_1$. At that point, the subharmonic amplitude suddenly takes a finite value jumping to the upper branch of the amplitude. With decreasing the modulation index on the upper branch, the amplitude of the subharmonics decreases and it falls abruptly to zero at $m = m_2$. The existence of a subharmonic solution depends

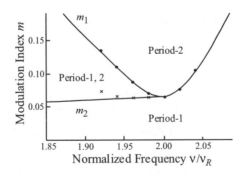

Fig. 7.16. Domain for first subharmonic bifurcation at $\tau_s \Gamma_R = 0.937$. Parameters used are the same as in Fig. 7.15. Dots and crosses represent the numerical results and the *solid lines* represent analytical results (after Yoon TH, Lee CH, Shin SY (1989); © 1989 IEEE)

on history. Thus, within the region of $m_2 < m < m_1$, two attractors coexist; one is the period-1 and the other is the period-2 (Arecchi et al. 1982). m_1 and m_2 are proportional to the damping constant $\tau_s \Gamma_R$ that increases with the increase of both β_{sp} and ε_s, and is inversely proportional to T. Experimental observations of the period-2 solution in pump-modulated lasers have been made (Siemsen 1978 and Paoli 1981). In those papers, the authors showed that period-doubling bifurcations via period-4 and period-8 oscillations are excited by the increase of the modulation as parametric resonance in the laser.

7.5.3 Chaotic Dynamics in Modulated Semiconductor Lasers

Kao et al. (1992 and 1993) investigated the stability and instability for the injection current modulation in the space of the modulation frequency and index. They numerically calculated the maps for the models of Fabry-Perot and DFB lasers. Figure 7.17 is the plot of the map for a small damping factor of $\Gamma_R/\omega_R = 0.0469$ with zero nonlinear gain $\varepsilon_s = 0$. Curve HS_m is the boundary of the hysteresis jump of the m-th spiking state; the section with broken line denotes the downward jump. Curves PD_m and PF_m are the boundaries of period-2 and period-4 of the m-th spiking state. The laser oscillation becomes spiky when the modulation current is increased to reach the minimum of the current swing approach to the threshold current level

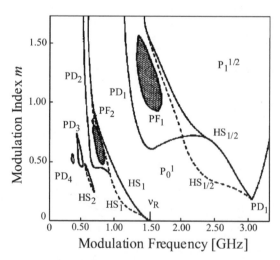

Fig. 7.17. Phase diagram with modulation frequency and index at $J = 1.5 J_{th}$. The relaxation oscillation frequency of the solitary laser at that frequency is 1.512 GHz. HS_m is the boundary of the hysteresis jump of the m-th spiking state. The section with *broken line* denotes downward jump. PD_m and PF_m are the boundaries of period-2 and period-4 oscillations of the m-th spiking state (after Kao YH, Lin HT (1993); © 1993 IEEE)

around $m = 0.5$. The laser output shows multiple spikes for $\nu < \nu_R$ and submultiple spikes for $\nu > \nu_R$. A minimum modulation current is required for the threshold PD_1 of period-2 state with frequency variation from ν_R to $2\nu_R$. The threshold $HS_{1/2}$ of hysteresis, meanwhile, overlaps with the PD_1 and two types of period-doubling can be observed. Both of them contain half-subharmonic components in the frequency spectrum and cannot be solely differentiated from the spectra. One is the normal type of period-doubling and is unrelated to the hysteresis. As discussed in the previous sections, the parameters of the nonlinear gain ε_s and the spontaneous emission coefficient β_{sp} play a crucial role for the dynamics.

Not all lasers always show unstable pulsation oscillations for the injection current modulation. In the actual case, some lasers show instabilities for strong injection current modulation, but others not. For example, the dynamics strongly depends on device structures such as the Fabry-Perot (FP) laser or the distributed feedback (DFB) laser, or the device constants of lasers even for the same laser structures. Bennet et al. (1997) observed period-2 states in a FP laser, while they identified both period-2 and period-3 states in the DFB laser. Figure 7.18 is an example. The FP laser used in the experiments has a wavelength of $1.56\,\mu m$ biased at $J = 5.88 J_{th}$ with a relaxation oscillation frequency of 13.3 GHz. The DFB laser used in the experiments has a wavelength of $1.53\,\mu m$ biased at $J = 5.26 J_{th}$ with the relaxation oscillation

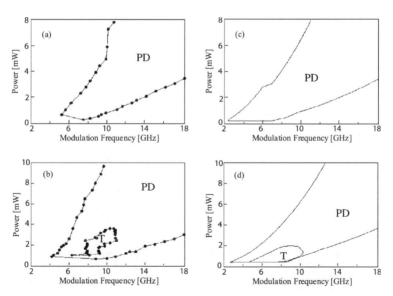

Fig. 7.18. Regions where unstable behaviors are observed. Experimental results for **a** FP laser and **b** DFB laser. **c** and **d** are the corresponding numerical simulations for **a** and **b**, respectively. PD: period-2 state, T: period-3 state (after Bennett S, Snowden CM, Iezekiel S (1997); © 1997 IEEE)

frequency of 11.3 GHz. The important parameters for the nonlinear dynamics are the nonlinear gain saturation factor ε_s and the spontaneous emission coefficient β_{sp}. The nonlinear gain saturation factor ε_s is almost the same for those two lasers and the factor of DFB lasers is generally 1.5 times larger than that of Fabry-Perot lasers. On the other hand, the difference of the spontaneous emission coefficient β_{sp} is very large. In general, the coefficient β_{sp} of Fabry-Perot lasers is 50 times larger than that of DFB lasers. Therefore, a Fabry-Perot laser is greatly affected by spontaneous emission of light. This results in less effect for unstable oscillations for the injection current modulation. Figures 7.18a and b are the plots of the map for the regions where unstable behaviors are observed in the experiments. Figures 7.18c and d are the corresponding maps for Figs. 7.18a and b, respectively, obtained by the numerical calculations from the rate equations. PD denotes the regions where period-2 states are recorded and T is the region where a period-3 state is observed. In the FP laser, there is a range of power levels and modulation frequencies over which the period-2 state is observed. In the DFB laser, there are both regions of period-2 and period-3 states. In the region where period-3 oscillation occurs, as the modulation power is increased, the route to period-3 is always via period-doubling. Thus, the dynamics of unstable pulsation oscillations depend on the device parameters, especially on the spontaneous emission coefficient. The theoretical analysis of bifurcation in semiconductor lasers by Yoon et al. (1989) found that lasers with low relaxation damping were more susceptible to bifurcation. This was also observed in the experiments. Liu and Ngai (1993) suggest that the promotion of period-3 in their bulk DFB laser was due to the unusually low value of β_{sp} measured in their laser. The observed period-3 in a DFB laser can be attributed to the low value of β_{sp} measured in that laser.

7.6 Nonlinear Dynamics of Various Combinations of External Perturbations

7.6.1 Optically Injected Semiconductor Laser Subject to Optoelectronic Feedback

Though the semiconductor laser itself is a stable laser categorized as a class B, the laser is easily destabilized by a single external disturbance, such as optical feedback, optical injection, optoelectronic feedback, or injection current modulation as has already been studied. Other possible perturbations exist. Semiconductor lasers are easily destabilized by mixed perturbations with two or more of the above external disturbances and show more complex dynamics. Indeed, such dynamics have been studied because of their importance in practical applications. For laser control to reduce chaotic noises and chaotic communications in semiconductor laser systems, composite external perturbations play essential roles in the systems. In the following, as examples, we discuss the dynamics of semiconductor lasers subjected both to optical injection and optoelectronic feedback, and both to optical feedback and injection current modulation.

At first, we consider the dynamics of semiconductor lasers subjected to optical injection and optoelectronic feedback. In the dynamics in optoelectronic feedback alone, we can observe quasi-periodic routes of chaotic pulsing (Tang and Liu 2001c) and chaotic pulsing states are generated depending on the delay time. On the other hand, a period-doubling route to chaos is identified (Simpson et al. 1994) and the frequency of chaotic oscillations is much enhanced by large optical injection compared with the free running state (Simpson et al. 1995b). However, the chaotic states can only be observed for a rather small optical injection strength with and without frequency detuning, where careful adjustment of both the injection strength and the frequency detuning is needed to operate the laser in such states (Liu et al. 1997 and Simpson et al. 1997).

In the mixed perturbations both of optical injection and optoelectronic feedback, the dynamic states including stable locking, periodic oscillation, chaotic oscillation, regular pulsing, quasi-periodic pulsing, and chaotic pulsing seen in semiconductor lasers subject to either optical injection or delayed optoelectronic feedback alone, can all be found. By controlling the optoelectronic feedback parameters, the bifurcation can be suppressed or inverted in such a manner that high-order periodic or chaotic dynamics are reduced to low-order periodic motions. In a semiconductor laser subject both to optical injection and optoelectronic feedback, coherence collapse is found to be suppressed and the bandwidth broadening effect is observed (Lawrence and Kane 1999). In this system, chaotic evolution can be observed for the change of the optical injection strength. However, there are two ways of chaotic routes, either a period-doubling route to chaotic oscillation states or a quasi-periodic route to chaotic pulsing states, even transient states in-between are identified. A notable expansion and shifting of the chaos region in the parameter space is observed in the system due to a strong optoelectronic feedback effect. The bandwidth of the chaotic state can be also greatly broadened by the introduction of optical injection. This phenomenon of chaos-region broadening and shifting together with the bandwidth enhancement is very important for increasing the bit rate of chaotic communications, which is currently limited by the bandwidth of chaos and the relaxation oscillation frequency of the laser (Tang and Liu 2001b, c, and Kusumoto and Ohtsubo 2002).

The mappings of dynamic states of a semiconductor laser subject to optical injection and optoelectronic feedback are numerically calculated and displayed in Fig. 7.19 (Lin and Liu 2003b). The mappings show stable and unstable regions as a function of the normalized optical injection strength of $\kappa'_{\mathrm{nor}} = \kappa_{\mathrm{inj}}(A_{\mathrm{m}}/\tau_{\mathrm{in}})/(A_{0\mathrm{s}}/\tau_{\mathrm{ph}})$ and the frequency detuning between the master and the slave lasers. The mappings a–d are the plots calculated for different optoelectronic feedback strengths of κ'_{nor}. In the figure, each symbol means S: steady-states, P1: period-1, P2: period-2, CO: chaotic oscillation,

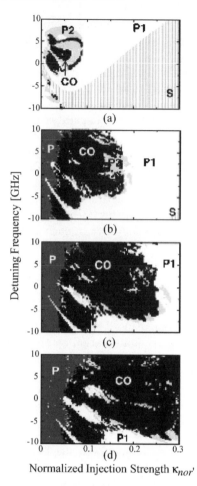

Fig. 7.19. Mappings of dynamic states of the hybrid system where the normalized feedback delay is $\tau\nu_R = 6.8$ and the optoelectronic feedback ratio ξ of **a** 0 (without feedback), **b** 0.16, **c** 0.20, and **d** 0.22, respectively. S: steady-states, P1: period-1, P2: period-2, CO: chaotic oscillation, P: pulsing states, which include regular pulsing, quasi-periodic pulsing, and chaotic pulsing. The relaxation oscillation frequency is set to 2.86 GHz (after Lin FY, Liu JM (2003b); © 2003 Elsevier)

and P: pulsing states, which include regular pulsing, quasi-periodic pulsing, and chaotic pulsing. With the increase of optoelectronic feedback strength κ'_{nor} regions of chaotic oscillations greatly expand in the map. When a laser is subjected to optical injection alone, the chaotic region is only restricted for small frequency detuning. On the other hand, when opotelectronic feedback is added to the system, the laser still behaves like a system of optical injection alone at a high injection rate, but pulsing states like optoelectronic feedback alone prevail at a lower injection rate as shown in Fig. 7.19b. In

between these two states of pulsations and chaotic oscillations, chaotic pulsing states are observed. As the feedback strength is further increased, this
expanding and shifting effect is further enhanced as can be seen in Figs. 7.19c
and d. Because of the broadening and the shifting caused by the optoelectronic feedback, this hybrid system is able to operate at chaotic states in
the strong injection region, where bandwidth enhancement of the chaotic
states is expected (Liu et al. 1997). The chaotic states (CO) of this system
in the strong injection region have bandwidths much broader than the one
that can be obtained in a laser subject to optical injection alone. This hybrid
system can be used for high-speed chaotic communications, spread-spectrum
spectroscopy, or other applications demanding broadband chaos.

7.6.2 Semiconductor Lasers
with Optical Feedback and Modulation

As shown in the previous section, when one of the mixed perturbations is
small, the behavior of the laser is dominated by the other perturbations.
As another example, we here discuss the dynamics of semiconductor lasers
subjected to optical feedback and modulation. Lawrence and Kane (2002a)
experimentally examined the chaotic dynamics for such systems, however
they considered two modulation methods for the laser power; one is the direct injection current modulation through a bias Tee and the other is the use
of an electro-optic modulator put into the path of the optical feedback loop.
For the electro-optic modulation, light emitted from the laser is modulated
by a lithium niobate electro-optic (EO) modulator. Figure 7.20 presents the
results for the map of the dynamic states in the phase space of the modulation frequency and index at a fixed external optical feedback. In the experiment, the laser used is a single mode quantum-well index-guided laser
with a wavelength of 850 nm and it is biased at $J = 1.4 J_{th}$. Under the operation condition, the relaxation oscillation frequency of the solitary laser is
about 2 GHz. The external feedback mirror is positioned at 317 mm (external
cavity resonant frequency 473 MHz) from the front facet of the laser. The
external feedback ratio is 0.22, which corresponds to strong optical external
feedback. For such strong optical feedback, the laser is generally stabilized
and the linewidth of the laser is much narrowed by the feedback. However,
the laser is no more stable with the additional strong modulation. Figure 7.20
shows the output states examined for a wide range of modulation frequencies
(from 50 MHz to 1 GHz) relative to the external cavity resonant frequency.
In the direct injection current modulation in Fig. 7.20a, a rich variety of dynamics can be observed compared with the case for the EO modulation in
Fig. 7.20b. When the modulation is small, the laser output powers for both
cases show stable oscillations except for regions corresponding to the external
cavity frequency and the relaxation oscillation frequency. Large regions of FM
and multimode output are observed at modulation frequencies close to multiples of the external cavity resonant frequency. The dominant state for the

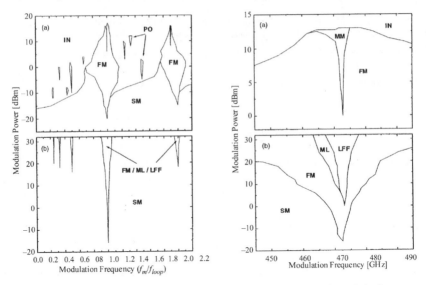

Fig. 7.20. Mappings of dynamic states in parameter space of modulation power and modulation frequency from 50 MHz to 1 GHz $(0.1 - 2.0 f_{\text{loop}})$ for **a** direct injection current modulation and **b** electro-optic modulation. Enlarged mappings for **c** direct injection current modulation and **d** electro-optic modulation corresponding to the dynamic states in Fig. 7.20 **a** and **b** close to the external cavity resonance frequency (473 MHz), respectively. IN: temporal instability, MM: multimode instability, FM: frequency modulation, LFF: low-frequency fluctuation, ML: mode-locking, SM: single-mode small-signal modulation, PO: periodic orbit (after Lawrence JS, Kane DM (2002a); © 2002 IEEE)

injection current is a temporal instability. At the temporal instability states, we can see distinct regions of periodic oscillation. In the EO modulation in Fig. 7.20b, the laser shows much different features from the case of the direct injection current modulation. FM lasing, mode-locking, and low-frequency fluctuations (LFFs) are observed for small regions centered at sub-harmonics and multiples of the external cavity resonant frequency.

Figures 7.20c and d show the enlarged plots of the dynamics concerning the modulation frequency corresponding to those of Figs. 7.20a and b, respectively. The center frequency of the plots is the external feedback frequency. For the direct modulation with low modulation powers (0 dBm) in Fig. 7.20c, the output is typical of an FM laser. As the modulation frequency is tuned close to the external cavity resonant frequency, the power modulation depth remains constant but the frequency modulation index increases. This enhancement occurs at a resonant frequency near the external cavity resonant frequency, dependent on the feedback level and the external cavity length (Lucero et al. 1988). For increased modulation power, there is a region of multimode output (labeled MM in the figure) centered at a detuning of

2 MHz from the external cavity resonant frequency. This region is characterized by an oscillation on multiple external cavity modes (Schremer et al. 1988). Such an optical state is similar to the coherence-collapsed state observed in semiconductor lasers operated with optical feedback in regime IV.

When electro-optic modulation is applied, the system behaves somewhat differently as shown in Fig. 7.20d. At modulation frequencies near the external cavity resonance, four distinct states are observed. For low modulation powers, or large detuning, the output is a single frequency. This represents very small signal modulation (labeled SM). As the modulation power is increased or the detuning from the external cavity resonant frequency is decreased, an FM lasing region is observed. This results in lower asymmetry in the FM sideband pairs. A further increase in the modulation power or decrease in the frequency detuning results in a multimode instability. However, this instability is an intermittent state similar to the LFF state observed in semiconductor lasers with optical feedback at low injection currents. For the electro-optic case, the instability close to the external cavity resonant frequency is predicted to occur theoretically using an iterative model based on a perturbation approach. It is attributed to mode competitions between external cavity modes and coupled cavity modes. The theory predicts a dynamic state consisting of large quasi-periodic amplitude modulation with irregular intermittent behavior over longer time scales, similar to the coherence collapse state in semiconductor lasers with optical feedback but with greater periodic structure than is generally observed for the case of optical feedback alone (Spencer et al. 1999).

8 Instability and Chaos
in Various Laser Structures

Edge-emitting narrow-stripe structure is not the only one for semiconductor lasers. Other than these, various kinds of laser structures of semiconductor lasers have been proposed and some of them are now in practical use. For example, self-pulsating semiconductor lasers are used for light sources of optical mass data storage systems (digital versatile disc (DVD) systems), vertical-cavity surface-emitting semiconductor lasers (VCSELs) are expected as the next generation laser light sources for communications and optical memory systems, and broad-area semiconductor lasers are promising light sources for high power laser systems. They have their own unique characteristic properties. Here, we do not discuss the details of each device structure and its characteristics, but we introduce the rate equations for such lasers and present their dynamic properties. These new laser structures have extra degrees of freedom and show instabilities and chaotic dynamics without any introduction of external perturbations. In this chapter, we discuss the dynamics of these new lasers both for solitary oscillations and external perturbations.

8.1 Multimode Lasers

8.1.1 Multimode Operation of Semiconductor Lasers

Before discussing the dynamics of semiconductor lasers with various structures different from edge-emitting lasers, we show the rate equations for multimode operating narrow-stripe edge-emitting lasers. In the preceding chapters, semiconductor lasers were assumed to be operating at a single longitudinal mode, however the laser sometimes oscillates at multimode due to noises when it is biased at a low injection current close to the threshold, otherwise it oscillates at multimode originating from their device structures even for a higher bias injection current. It intrinsically oscillates at multimode by the introduction of external perturbations such as optical external feedback. Figure 8.1 shows a multimode spectrum of a Fabry-Perot semiconductor laser operating close to the threshold. Even for a laser operating at a single mode with suppressed side mode, the side mode does not damp out by optical injection or optical feedback, and the laser dynamics are much affected by the

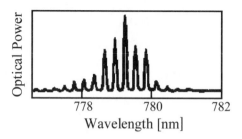

Fig. 8.1. Optical spectrum of a multimode Fabry-Perot semiconductor laser

mode behaviors as shown in Chap 6. Semiconductor lasers at solitary condition may be operated with multimode as a nature of the device structure, since the laser has a broad gain bandwidth. The separation of the longitudinal modes of an ordinary semiconductor laser is more than 100 GHz (the corresponding wavelength separation is ~1 nm) due to a short internal cavity length. However, the gain profile is as large as 20 nm or more and it has the possibility of multimodal oscillations with several oscillation lines. In semiconductor lasers of various device structures different from edge-emitting lasers, they usually operate at multimode without any external perturbations, since they originally include extra variables (extra degrees of freedom) besides those for edge-emitting lasers. We will discuss these dynamics later. Here, we first discuss the dynamics of narrow-stripe edge-emitting semiconductor lasers operating at multimode.

Semiconductor lasers have broad gain bandwidth, which is a unique feature different from other lasers. Therefore, there exist many possible oscillation lines within the band width. The gain curve is sometimes assumed as a parabolic function, although, in actual fact, the gain profile has asymmetry. The gain for shorter wavelength (higher energy of carrier) tends to be large by the band-filling effects under the condition of constant temperature and carrier injection. Also semiconductor lasers with multimode operation are much affected by this effect (Petermann 1988). The optical powers are equally distributed to respective modes below the laser threshold. However, the transfer of the power to the side modes is much restrained well above the threshold and the power of the main mode grows up resulting at a single mode oscillation. This phenomenon is well reproduced from the calculations of the rate equations for taking into consideration the multimode effects and the theory and experiments show good coincidence. The effect of side mode suppression is strongly dependent on spontaneous emission of light and the large spontaneous emission coefficient β_{sp} forces the excitation of side modes.

8.1.2 Theoretical Model of Multimode Lasers

The rate equations for the photon number and the carrier density for multimode semiconductor lasers are written by the following equations (Petermann

1988):

$$\frac{dS_j(t)}{dt} = [G_{n,j}\{n(t) - n_{th,j}\}]S_j(t) + R_{sp}(\omega_j) \tag{8.1}$$

$$\frac{dn(t)}{dt} = \frac{J(t)}{ed} - \frac{n(t)}{\tau_s} - \sum_{j=-M}^{M} G_{n,j}\{n(t) - n_0\}S_j(t) \tag{8.2}$$

where subscript j is for the j-th mode and $2M+1$ is the total mode number. $j = 0$ is the main mode. The final term in (8.1) is the effect of spontaneous emission of light. It is noted that it is not only a function of time but also a function of optical frequency. For incoherent rate equations, we do not need to consider the phase equation, since it does not couple with the other equations. On the other hand, in an actual situation, we must consider the nonlinear saturation effect for the gain and the cross-saturation effect in the photon number rate equation. Further, we must use the complex field equation instead of the photon number rate equation, when we consider coherent effects such as optical feedback. For coherent phenomena, the phase equation plays crucial role for the dynamics as has already been discussed.

At first, we study the side mode suppression ratio (MSR) in a multimode semiconductor laser. Assuming $2M+1$ oscillation lines within the gain profile and approximating the gain as a parabolic curve, the gain is written by

$$G_{n,j} = G_n \left\{ 1 - \left(\frac{j}{M}\right)^2 \right\} \tag{8.3}$$

where $M = \Delta\nu_g/\Delta\nu_l$, $\Delta\nu_g$ is the frequency width of the gain profile and $\Delta\nu_l$ the frequency of the longitudinal mode spacing. The laser output power of the j-th mode for the steady-state solution is calculated from the rate equations (8.1) and (8.2) as (Petermann 1988)

$$S_j \approx \frac{R_{sp}(\omega_j)}{\frac{1}{\tau_{ph}} - G_{n,j}} = \frac{\tau_{ph}R_{sp}(\omega_j)}{G_n} \frac{1}{\delta + \left(\frac{j}{M}\right)^2} \tag{8.4}$$

where $\delta = (1 - G_n)/G_n$. From this relation, we obtain the side mode suppression ratio as

$$\text{MSR} = \frac{S_0}{S_1} = 1 + \frac{1}{\delta M^2} = 1 + \frac{S_0}{\tau_{ph}R_{sp}} \left(\frac{\Delta\nu_l}{\Delta\nu_g}\right)^2 \tag{8.5}$$

For a semiconductor laser which is assumed to be a single mode operation, the value of MSR is larger than 20.

A single mode operation at a high bias injection current is expected for index-guided semiconductor lasers, since the spontaneous emission coefficient is as small as less than $\beta_{sp} \sim 10^{-4}$. However, the side mode suppression is weak for gain-guided semiconductor lasers with a larger spontaneous emission coefficient of $\beta_{\rm sp} \sim 10^{-3}$ and the laser tends to be oscillated with multimode. As a primary effect, a side mode is suppressed for the increase of the injection current above the threshold and the optical power is concentrated to the main mode. In actual laser oscillation, there are the effects of spatial hole burning due to standing-wave nature along the laser propagation and spectral hole burning due to the broadening of the gain profile for the increase of the optical power. As results, the side mode is suppressed and the optical power is transferred to the main mode (Agrawal and Dutta 1993). These effects cause instabilities for the main mode and play an important role for chaotic dynamics in semiconductor lasers. Similar effects are observed for optical injection to semiconductor lasers as shown in Fig. 6.5 Other effects to destabilize the main mode and enhance the side modes are the beating between the main and side modes and the four-wave mixing in the oscillation modes. The effects are strongly dependent on laser types, materials, and confinement of light in the active layer.

We have derived the multimode rate equations for semiconductor lasers in the incoherent case. However, we must take into account the effects of the nonlinear gain saturation and cross-gain saturation. Also, we must use the coherent rate equations for a semiconductor laser when the laser is subjected to external optical feedback or optical injection. When a semiconductor laser oscillates at a multimode with $2M + 1$ oscillation lines, the rate equations for the complex field and the carrier density are written by (Ryan et al. 1994)

$$\frac{\mathrm{d}E_j(t)}{\mathrm{d}t} = \frac{1}{2}(1 - i\alpha)G_{\mathrm{n},j}\{n(t) - n_{\mathrm{th},j}\}E_j(t)$$

$$- \frac{1}{2}\left(\varepsilon_{sj}|E_j(t)|^2 + \sum_{m=-M}^{M}\theta_{mj}|E_m(t)|^2\right)E_j(t) \qquad (8.6)$$

$$\frac{\mathrm{d}n(t)}{\mathrm{d}t} = \frac{J}{ed} - \frac{n(t)}{\tau_{\mathrm{s}}} - \sum_{m=-M}^{M}[K_m G_{\mathrm{n},m}\{n(t) - n_0\}]|E_m(t)|^2 \qquad (8.7)$$

where $E_j(t)$ is the field of the j-th mode, ε_{sj} and $\theta_{mj}(m \neq j)$ are the nonlinear self- and cross-saturation coefficients, respectively, and K_m is the mode gain coefficient. We omitted the Langevin noise terms in (8.6), however it may be added where necessary.

The nonlinear saturation coefficient $\alpha' = -(\partial\,\mathrm{Re}[\chi]/\partial S)/(\partial\,\mathrm{Im}[\chi]/\partial S)$ for the photon number discussed in Sect. 6.2 is ignored in the above equation. The saturation coefficients ε_{sj} and θ_{mj} do not have large values and they are in the order of $10^4\,\mathrm{s}^{-1}$. In the carrier density equation, we also ignore the mode interferences for the carrier recombination because it has small effects. A semiconductor laser operating at multimode is an unstable laser and it is easily affected by external perturbations. A multimode semiconductor laser shows mode competitions and mode switching induced by the nonlinear interactions among the modes. Mode partition noise is one of the dominant effects in multimode oscillating lasers and it is a non-negligible effect (Ahamed and

Table 8.1. Characteristic device parameters for a multimode semiconductor laser

Symbol	Parameter	Value
G_n	gain coefficient	$2.05 \times 10^{-13}\,\mathrm{m^3\,s^{-1}}$
α	linewidth enhancement factor	4.00
r_0	facet reflectivity	0.556
n_th	carrier density at threshold	$4.00 \times 10^{24}\,\mathrm{m^{-3}}$
n_0	carrier density at transparency	$1.40 \times 10^{24}\,\mathrm{m^{-3}}$
τ_s	lifetime of carrier	$2.00\,\mathrm{ns}$
τ_ph	lifetime of photon	$1.88\,\mathrm{ps}$
τ_in	round trip time in laser cavity	$6.00\,\mathrm{ps}$
V	volume of active region	$1.25 \times 10^{-16}\,\mathrm{m^3}$
ε_{ij}	nonlinear self-saturation coefficient	$1.70 \times 10^4\,\mathrm{s^{-1}}$
θ_{mj}	nonlinear cross-saturation coefficient	$1.60 \times 10^3\,\mathrm{s^{-1}}$

Yamada 2002). Each oscillation mode includes a very large relative intensity noise (RIN) and it sometimes causes problems in actual use. However, as a total intensity, a multimode semiconductor laser with partition noise has the same order of RIN as a single mode laser, since the partition noises are averaged out (Petermann 1988). Using (8.6), the rate equation for the complex field in a multimode semiconductor laser with optical feedback is given by

$$
\frac{\mathrm{d}E_j(t)}{\mathrm{d}t} = \frac{1}{2}(1 - \mathrm{i}\alpha)G_{\mathrm{n},j}\{n(t) - n_{\mathrm{th},j}\}E_j(t)
$$

$$
- \frac{1}{2}\left(\varepsilon_j|E_j(t)|^2 + \sum_{m=-M}^{M}\theta_{mj}|E_m(t)|^2\right)E_j(t)
$$

$$
+ \frac{\kappa_j}{\tau_\mathrm{in}}E_j(t - \tau)\exp(\mathrm{i}\omega_j\tau) \tag{8.8}
$$

In the following, the dynamics of multimode semiconductor lasers subjected to external optical feedback are numerically investigated. Table 8.1 shows typical values of device parameters for a multimode semiconductor laser frequently used in numerical simulations (Ryan et al. 1994).

8.1.3 Dynamics of Multimode Semiconductor Lasers with Optical Feedback

When the frequency corresponding to the round trip time of light in the external cavity is small enough compared with the relaxation oscillation frequency, the difference of the dynamics between single mode and multimode semiconductor lasers is not distinct. However, if the external cavity length becomes short and the corresponding frequency exceeds the relaxation oscillation frequency, they show different dynamics. Figure 8.2 shows an example of the difference. Bifurcation diagrams in the phase space of the carrier density and the external feedback rate are calculated both for single mode and

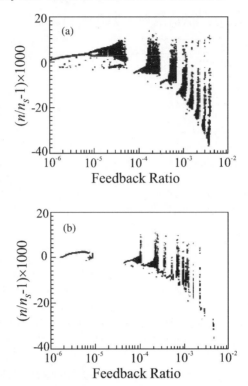

Fig. 8.2. Bifurcation diagrams of carrier density for optical feedback rate. **a** Single mode laser. **b** Multimode laser with five oscillation lines. The external cavity length is $L = 10$ cm, which corresponds to a frequency of 1.5 GHz. The relaxation oscillation frequency of the laser is assumed as 0.7 GHz. The same device parameters are assumed for both cases (after Ryan A, Agrawal GP, Gray GR, Gage EC (1994); © 1994 IEEE)

multimode semiconductor lasers (Ryan et al. 1994). For the multimode laser, five modes are assumed. In Figs. 8.2a and b, the multimode laser is stable compared with the single mode laser and a feedback level of ten times larger than the solitary case is required to destabilize the laser. Namely, the multimode semiconductor laser is less sensitive to optical feedback than the single mode semiconductor laser as far as the conditions of the device parameters are the same. This result can be understood qualitatively by noting that all modes contribute to the damping of relaxation oscillations. Even though an individual mode may be unstable in solitary oscillation, simultaneous lasing of all modes preserves the steady-state over a large range of the external feedback. The situation is quite similar to the effect of averaged relative intensity noise (RIN) in multimode semiconductor lasers (Petermann 1988). Also chaotic regions are much thinner compared to the single mode case.

At chaotic oscillations of a multimode semiconductor laser, the bifurcation diagrams cannot tell us whether all of the modes simultaneously oscillate, or whether one of the modes, or small number of them is the dominant oscillation modes. Figure 8.3 shows the simulation result for waveforms of each oscillation mode in a multimode semiconductor laser (Ryan et al. 1994). Figure 8.3 is the calculations of waveforms without and with optical feedback. Each waveform is averaged for a 10 ns time window to clearly show the difference. Switching among the modes is not distinct for the solitary oscillation, although the change of the main mode for the time evolution is visible as shown in Figure 8.3a. The solitary laser exhibits mode partition fluctuations, but it remains multi-moded most of the time. The total optical power at the solitary oscillation is almost constant for time. However, in Fig. 8.3b, only one of the modes is dominant for a certain time duration and the other

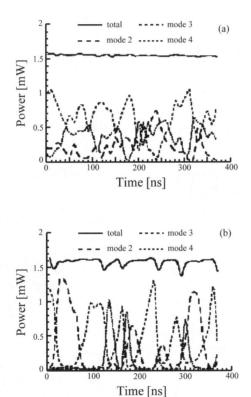

Fig. 8.3. Waveform of each mode in a multimode semiconductor laser corresponding to Fig. 8.2. Three modes are displayed. **a** Almost steady-state oscillation without optical feedback and **b** chaotic oscillation with optical feedback. The feedback ratio is 5×10^{-4} (after Ryan A, Agrawal GP, Gray GR, Gage EC (1994); © 1994 IEEE)

mode is suppressed, when the laser shows chaotic dynamics induced by the optical feedback. The main mode changes to the other mode after a certain time duration with random manner. At the moment of mode hopping, the total output power sustains irregular spikes. Switching among modes is a typical feature of multimode semiconductor lasers subjected to optical feedback. From the detailed study of the dynamics, it is understood that only one or few of the possible oscillation modes becomes the dominant mode for a certain time duration and the modes alternately switch with random manner. When the mode number is small, only one mode tends to oscillate with chaotic manner and oscillations of the other modes are suppressed. Especially, a mode of shorter wavelength tends to show chaotic oscillations for small optical feedback due to the asymmetry of the gain profile. Indeed, this mode switching has been experimentally observed (Ikuma and Ohtsubo 1998).

8.2 Self-Pulsating Lasers

8.2.1 Theory of Self-Pulsating Lasers

Self-pulsating semiconductor lasers are currently used for light sources of digital versatile disks (DVD) in optical data storage systems, since noises (actually chaotic irregular oscillations) induced by optical feedback from a disk surface are greatly suppressed by self-pulsations. In the early days, a technique of high frequency injection current modulation in CW operating semiconductor lasers was used to reduce optical feedback noises in optical disk systems. Self-pulsating semiconductor lasers are fabricated to reduce feedback noises using pulsation oscillations originating from their device structures without external control circuits. The pulsation frequency depends on each device and the bias injection current, and ranges typically from several hundreds of MHz to the order of GHz. The self-pulsation semiconductor laser itself is unstable and it sometimes shows instability for a certain region of the bias injection current even without external perturbations. The structure of self-pulsation lasers is almost the same as edge-emitting lasers except for saturable absorbing regions adjacent to the active layer as shown in Fig. 8.4. The width of the active region is usually the same size as that of edge-emitting semiconductor lasers. However, this is not the only structure of self-pulsating semiconductor lasers. The other example is a type of weak index guide (WIG) and the saturable absorbing layer is installed above the active region. The type of adjacent saturable absorbing layers (SALs) shown in Fig. 8.4 is assumed in the following discussion. However, the results are straightforwardly applicable to the WIG model.

In Fig. 8.4a the cross-section of the front facet of a self-pulsating semiconductor lasers is represented. The saturable absorbing regions are installed

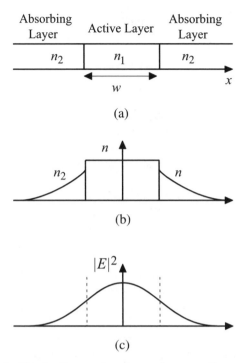

Fig. 8.4. Model of self-pulsating semiconductor lasers. **a** Cross-section of the front facet of the laser, **b** carrier distribution, and **c** optical profile. The center is the active region and both sides of the active layer are the carrier absorbing regions (after Ryan A, Agrawal GP, Gray GR, Gage EC (1994); © 1994 IEEE)

at both sides of the active layer. Figure 8.4b is the carrier distribution along the active layer. Carriers injected into the active region rapidly decay out to the regions of the saturable absorbers.

Figure 8.4c is the profile of the output power at lasing oscillation. The whole distributions of the carriers and the optical power from the active layer to the saturable absorbing regions are determined by the boundary conditions in the same manner as the treatment of the light transmission in an optical wave guide (Yamada 1993). The time constant τ_{12} of the carrier diffusion from the active layer to the absorbers is given by

$$\tau_{12} = \frac{w^2}{2D} \tag{8.9}$$

where w is the width of the active layer and D is the diffusion constant. At the same time, the carrier diffusion also occurs from the saturable absorbing regions to the active layer. Since the total number of the electrons $V_1 n_1 + V_2 n_2$ should be unchanged, the time constant τ_{21} of the carrier diffusion from the absorbers to the active layer is given by

$$\tau_{21} = \frac{V_2}{V_1}\tau_{12} \tag{8.10}$$

where V_1 and V_2 are the volumes of the active and absorbing regions, and n_1 and n_2 are the carrier densities at the active and absorbing regions, respectively. The linearized gains for the active and absorbing regions have different values, therefore we must use appropriate gains for them in the numerical simulations. Also the carrier densities above which the lasing gain becomes positive, i.e., transparent, differ from each other.

Before introducing the rate equations for self-pulsating semiconductor lasers, the mechanism of self-sustained pulsating oscillations in a SAL type laser is briefly explained. At the carrier number less than the laser threshold, carriers are accumulated in the active region by the carrier injection. When the carrier number exceeds the laser threshold, the laser oscillation starts. The carriers are rapidly absorbed by the diffusion to the saturable absorbing regions. The carrier diffusion is reduced due to the increase of the carrier number in the saturable absorbing regions, and this results in the increase of the photon number. However, the increase of the photon number causes the depletion of carriers in the active region and the decrease of the carrier number is accelerated by the diffusion of the carriers to the saturable absorbing regions. Then, the carrier number falls below the laser threshold and, finally, the laser oscillation stops. After the halt of the laser oscillation, the carrier number again increases and the next pulsating oscillation starts. This process repeats again and again and the laser shows self-sustained pulsation oscillations. The accumulation time of carriers to show lasing oscillation is typically about 1 ns (corresponding to a pulsating oscillation frequency of 1 GHz) and the width of the pulses is ~100 ps.

In the following, we assume a single mode oscillation for a self-pulsating semiconductor laser, however actual lasers are more or less multimode oscillations. Therefore, the dynamics derived from the theory do not always coincide well with the experimental results unlike in the cases of edge-emitting lasers. However, we can discuss approximate characteristics of self-pulsating semiconductor lasers, such as pulsating oscillations, pulsing frequency, and L-I characteristics. Several theoretical models have been proposed and some of them are listed in the reference (Carr and Erneux 2001), although the fundamental idea of the models is the same. Here, we assume a single mode model for a self-pulsating semiconductor laser and we introduce an additional carrier density equation for the saturable absorbing regions of the rate equations of an edge-emitting semiconductor laser (Yamada 1993 and 1996). Due to the presence of the saturable absorbing regions, carriers in the active region rapidly decay toward the absorbing regions and pulsations occur in the laser output. In such a structure, we must take into account the carrier density equations in the absorbing regions.

The rate equations for the complex amplitude $E(t)$ and the carrier densities $n_1(t)$ and $n_2(t)$ for the active and absorbing regions describing self-pulsation semiconductor lasers are written by

$$\frac{dE(t)}{dt} = \frac{1}{2}(1 - i\alpha)\left[G_{n1}\{n_1(t) - n_{th1}\} + G_{n2}\{n_2(t) - n_{th2}\}\right]E(t) \quad (8.11)$$

$$\frac{dn_1(t)}{dt} = \frac{J}{ed} - \frac{n_1(t)}{\tau_{s1}} - \frac{n_1(t) - n_2(t)}{\tau_{12}} - G_{n1}\{n_1(t) - n_{01}\}|E(t)|^2 \quad (8.12)$$

$$\frac{dn_2(t)}{dt} = -\frac{n_2(t)}{\tau_{s2}} - \frac{n_2(t) - n_1(t)}{\tau_{21}} - G_{n2}\{n_2(t) - n_{02}\}|E(t)|^2 \quad (8.13)$$

Here, subscripts 1 and 2 denote the quantities for the active and absorbing regions, respectively, and τ_{12} and τ_{21} are the carrier diffusions from the regions 1 to 2 and vice versa, respectively, as has already been defined. In the field equation, we ignored the nonlinear gain saturation effect. However, it may play an important role in self-pulsating semiconductor lasers, since the photon density becomes large due to pulsating oscillations even for a short time duration. We take into account the nonlinear gain in such a case as discussed in Sect. 3.3.4. However, we can simulate approximate characteristics of self-pulsating semiconductor lasers without considering the gain saturation effect and the term is sometimes omitted. Self-pulsation semiconductor lasers were originally aimed to reduce the effect of optical feedback noises, however the RIN is sometimes enhanced under certain conditions of the feedback. Furthermore, they are essentially unstable lasers and they sometimes show unstable or chaotic oscillations under certain ranges of the bias injection current even at solitary oscillations.

8.2.2 Instabilities at Solitary Oscillations

The self-pulsating semiconductor laser itself is an unstable laser and regular pulsating oscillation is considered as a kind of period-1 state on the way to chaotic evolution. Typical features of chaotic states in a self-pulsating semiconductor laser are pulsing oscillations with irregular pulse amplitude and jitters. The characteristics of self-pulsating semiconductor lasers are strongly dependent on the device structure and parameters. Figure 8.5 shows an example of an experimental light-injection current (L-I) characteristic of a self-pulsating semiconductor laser at solitary oscillation. The laser is a SAL type and has a maximum output power of 5 mW with an oscillation wavelength of 650 nm. The oscillation above the threshold is divided into unstable and stable regions for the bias injection current. The threshold current is about 70 mA, which is much higher than that of ordinary edge-emitting semiconductor lasers, because the strong carrier dissipation exists due to the presence of the saturable absorbing regions. Another difference is the vague threshold. The laser power does not linearly increase for the bias injection current close to the threshold, but it has a hysteresis. As demonstrated later, bistabilty is reproduced by the numerical simulations from the rate equations.

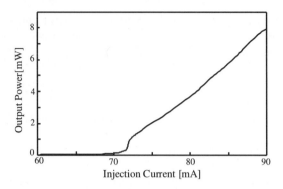

Fig. 8.5. L-I characteristic of a self-pulsating semiconductor laser. The laser is a AlGaInP multi-quantum well laser operating at a wavelength of 650 nm and a maximum output power of 5 mW

In the L-I characteristic, the laser output shows the bistable state above the threshold. In this region, the laser exhibits pulsating oscillation, but it is unstable. A waveform and its rf spectrum in this region are plotted in Fig. 8.6a. The pulse peak changes irregularly and the waveform shows a broad chaotic spectrum. Well above the laser threshold, the laser shows regular pulsing states with constant peak and separation as shown in Fig. 8.6b. However, even for such stabilized operations at solitary mode, the laser may be destabilized by optical feedback. Stability or instability of the laser operations for optical feedback is discussed in the next section. It is noted that every self-pulsating semiconductor laser does not always show the same L-I charac-

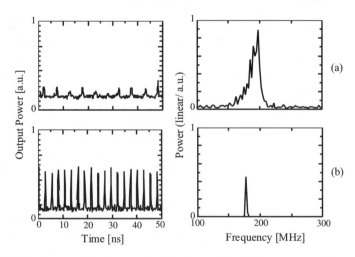

Fig. 8.6. Waveforms and rf spectra in Fig. 8.5. **a** Unstable region at a bias injection current of 72 mA and **b** stable regular pulsating oscillation at 78 mA

teristics as in Fig. 8.5. A laser with regular pulsing states is used for a light source of DVD systems.

The L-I characteristic is numerically calculated from the rate equations (8.11)–(8.13). Typical parameter values of red light self-pulsating semiconductor lasers are listed in Table 8.2. Because of the pulsation characteristics of laser oscillations, the values of the gain coefficients are larger than those of edge-emitting semiconductor lasers. Figure 8.7 is an example of the calculated L-I characteristic (van Tartwijk and San Miguel 1996). After the lasing oscillations, the laser shows bistability for the bias injection current between 48 and 58 mA and chaotic pulsating oscillations are observable in this region. For the bias injection current from 58 to 125 mA, the laser oscillates at stable regular pulsing states. Over the bias injection current of 125 mA, the laser shows stable continuous-wave (CW) oscillations. In the numerical simulations, noises induced by spontaneous emission strongly affect the pulsing frequency. At regular pulsing states without considering noises, the pulsing frequency smoothly increases with the increase of the bias injection current. In the presence of noises, a kink is observed in the characteristic curve of the pulsing frequency and the bias injection current (van Tartwijk and San Miguel 1996, and Mirasso et al. 1999).

The theoretical calculation in Fig. 8.7 well reproduces the behaviors of unstable and stable pulsating oscillations. However, the CW operation of self-pulsating lasers is not observable in experiments. The CW operation in Fig. 8.7 is achieved at a high bias injection current and such a high bias injection current may damage the laser. Another example of discrepancy between the theory and the experiment is the pulse width of the waveform.

Table 8.2. Characteristic device parameters for a self-pulsating semiconductor laser at an oscillation wavelength of 650 nm (AlGaInP laser)

Symbol	Parameter	Value
G_{n1}	gain coefficient in active region	$3.08 \times 10^{-12} \, \mathrm{m^3 \, s^{-1}}$
G_{n1}	gain coefficient in saturable absorbing region	$1.24 \times 10^{-13} \, \mathrm{m^3 \, s^{-1}}$
α	linewidth enhancement factor	4.00
n_{01}	carrier density at transparency in active region	$1.40 \times 10^{24} \, \mathrm{m^{-3}}$
n_{02}	carrier density at transparency in satuarble absorbing region	$1.60 \times 10^{24} \, \mathrm{m^{-3}}$
τ_{s1}	lifetime of carrier in active region	2.49 ns
τ_{s2}	lifetime of carrier in saturable absorbing region	1.25 ns
τ_{12}	diffusion time	2.65 ns
τ_{ph}	lifetime of photon	2.72 ps
V_1	volume of active region	$0.72 \times 10^{-16} \, \mathrm{m^3}$
V_2	volume of saturable absorbing region	$0.46 \times 10^{-16} \, \mathrm{m^3}$

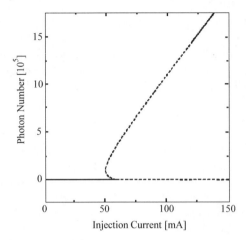

Fig. 8.7. Theoretically calculated L-I characteristic of a self-pulsating semiconductor laser (after van Tartwijk GHM, San Miguel M (1996); © 1996 IEEE)

The theoretically calculated pulse width is much smaller than the actual width. For example, the calculated pulse width is typically 10 ps, but the observed pulse usually has a width of around 100 ps. As has already been mentioned, the improvement of the theoretical model of the rate equations is still required to explain well the experimental data (Yamada 1998b). The other reason for the discrepancy is the assumption of a single mode operation for self-pulsating lasers. It is well-known that self-pulsating semiconductor lasers oscillate at multimode with many oscillation lines.

Characteristics of InGaN self-pulsating semiconductor lasers with an operating wavelength of 395 nm have been investigated (Tronciu et al. 2003). Since the difference between the carrier lifetimes of the active and saturable absorbing layers is much greater than that of red light self-pulsating semiconductor lasers, the lasers show quite different dynamics from red self-pulsating lasers. For example, the carrier lifetime of the active layer is 2.0 ns, while that of the saturable absorbing layer is only 0.1 ns. The rate equations are fundamentally the same as those in (8.11)–(8.13), but Tronciu et al. took into account of the effects of the outer regions besides the active and absorbing layers in the numerical simulations. Figure 8.8 shows some numerical and experimental results. They used the model of a WIG type laser. Close to the threshold, the laser oscillates at the CW operation without hysteresis like in Fig. 8.8a, which is quite different from the operation of red self-pulsating lasers. At a certain bias injection current, the laser at first shows self-pulsating oscillation. However, the laser recovers stable states for a bias injection current above 200 mA. The dynamics is strongly dependent on the laser cavity length. The self-pulsation range (SP) for the laser cavity length is investigated in Fig. 8.8b. We can see the agreement between the theoretical and experimental results.

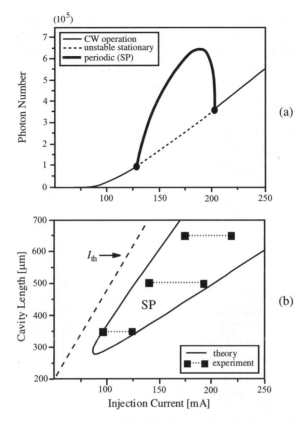

Fig. 8.8. Characteristics of an InGaN self-pulsating semiconductor laser with an oscillation wavelength of 395 nm. **a** Calculated bifurcation diagram for a laser cavity length of 500 µm. The self-pulsation region is observable from 125 to 200 mA. The other range of the injection current is a stable CW operation. **b** Self-pulsation range in the plane of laser cavity length versus injection current. Experimentally obtained ranges of the self-pulsation are indicated with *dotted lines* (after Tronciu VZ, Yamada M, Ohno T, Ito S, Kawakami T, Taneya M (2003); © 2003 IEEE)

8.2.3 Instability and Chaos by Optical Feedback

The self-pulsating semiconductor laser is fabricated as a low noise light source in optical data storage systems. However, the reduction of feedback noise is not always achieved for every feedback condition. The self-pulsating semiconductor laser has a periodic pulsation with a frequency ranging from several hundreds MHz to GHz depending on the bias injection current. There is a congenial range of optical feedback lengths to suppress feedback noise. The self-pulsating semiconductor laser sustains a forced oscillation due to the device structure and the feedback noise is reduced for a wide range of the feedback conditions. On the other hand, outside this region, the laser undergoes more noises than that of edge-emitting semiconductor lasers. Also, the laser shows chaotic behaviors for the modulation under certain bias injection current ranges. At first, we consider feedback induced noises in a self-pulsating semiconductor laser. There are two schemes of optical feedback in self-pulsating semiconductor lasers; coherent and incoherent feedback depending on the relation between the pulse separation and the feedback length. When the feedback length is small enough compared with the pulsing separation and close or less than the pulse width, the effect is coherent. For the condition $T_\mathrm{p} > \tau$ (T_p being the pulse separation), the rate equation for the complex field is given by

$$\frac{\mathrm{d}E(t)}{\mathrm{d}t} = \frac{1}{2}(1 - \mathrm{i}\alpha)\left[G_{\mathrm{n}1}\{n_1(t) - n_{th1}\} + G_{\mathrm{n}2}\{n_2(t) - n_{th2}\}\right]E(t)$$
$$+ \frac{\kappa}{\tau_{\mathrm{in}}}E(t - \tau)\exp(\mathrm{i}\omega_0\tau) \tag{8.14}$$

The equation is the same as that for the coherent case of edge-emitting semiconductor lasers except for the effects of the gain term in the saturable absorbing regions. The rate equations for the carrier densities under coherent feedback remain unchanged as in (8.12) and (8.13).

On the other hand, the rate equation for the complex field is written the same as (8.11) for incoherent optical feedback, however the carrier density equation for the active layer in (8.12) must be changed and the incoherent feedback term is added to this equation, when $T_\mathrm{p} < \tau$. The rate equation for the carrier density is given by

$$\frac{\mathrm{d}n_1(t)}{\mathrm{d}t} = \frac{J}{ed} - \frac{n_1(t)}{\tau_{\mathrm{s}1}} - \frac{n_1(t) - n_2(t)}{\tau_{12}} - G_{\mathrm{n}1}\{n_1(t) - n_{01}\}$$
$$\times \{|E(t)|^2 + \kappa_{\mathrm{i}}(1 - R_0^2)R|E(t - \tau)|^2\} \tag{8.15}$$

where R_0 and R are the intensity reflectivities of the front facet of the laser and the external reflector, respectively, and κ_{i} is the intensity coupling coefficient to the active layer.

We show some numerical results for the dynamics of self-pulsating semiconductor lasers subjected to coherent optical feedback. Figure 8.9 is an example (Yamada 1998a, b). Figure 8.9a shows the time series of the laser oscillation without optical feedback. In the figure, n_1 and n_2 are the carrier densities of the active and saturable absorbing regions, respectively, and S is the photon number. n_4 is the carrier density at the current blocking region installed above the saturable absorbing layer. The laser oscillates at the regular pulsing state and the pulsing frequency is 1.29 GHz. The calculated RIN at the solitary oscillation is less than $-130\,\mathrm{dB/Hz}$ for the lower frequency component. Figure 8.9b shows the waveforms for the same variables under optical feedback. The feedback ratio is set to be 3.3% in the average field amplitude. In the presence of optical feedback, the laser still shows a pulsating oscillation. However, the laser output power is disturbed by the feedback

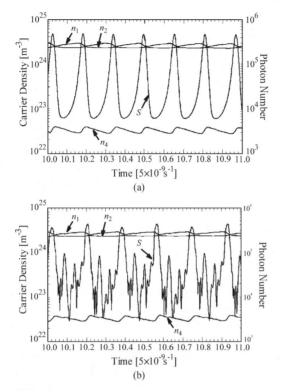

Fig. 8.9. Times series of the variables in a self-pulsating semiconductor laser at $J = 1.69J_{\text{th}}$. **a** Without optical feedback and **b** with optical feedback of 3.3% of the average field amplitude. The external cavity length is $L = 4$ cm. n_1: carrier density in active region, n_2: carrier density in saturable absorbing region, n_4: carrier density in current blocking region, and S: photon number (after Yamada M (1998a); © 1998 IEICE)

and the laser shows a chaotic oscillation. The pulse period is larger than that of the solitary oscillation and its average frequency is 1.07 GHz. It is noted that the pulse height also fluctuates and the RIN is greatly enhanced up to −90 dB/Hz for the lower frequency component.

Tartwijk et al. (1996) numerically studied the effects of optical feedback in self-pulsating semiconductor lasers and calculated pulse periods and jitters (the standard deviation from the average period) without and with optical feedback. As results, the pulse period decreases with the increase of the bias injection current and, at the same time, the jitter becomes small. With increasing the feedback time, the pulse period increases and, then, a sudden jump-down is observed at a certain feedback distance. The jump in the pulse period must be attributed to a switch of the locked-pulse frequency to a neighboring compound cavity mode, i.e., the resulting resonance frequency of the laser mode with relaxation oscillations and one of the external cavity resonance frequency. A similar trend can be observed for the pulse jitter. There is also phase sensitivity of coherent optical feedback in a self-pulsating semiconductor laser as expected from (8.14). When the feedback is small, the pulse period keeps almost the same value as that of the solitary oscillation. However, the jitter has the minimum value at a certain small feedback ratio. The increase in jitter after the optimum value manifests itself by a multi-peaked, very broad pulse period distribution. The effects agree well with the period-doubling route to chaos (Kuzentsov et al. 1986). The dynamics of coherent optical feedback is extensively studied in self-pulsating semiconductor lasers, because the lasers are used as light sources for DVD systems in which the optical feedback length is typically within several centimeters. On the other hand, the typical feedback length is several tens of centimeters to meters when the lasers are used as light sources for optical measurements. The effects for this range are incoherent.

8.2.4 Instability and Chaos by Injection Current Modulation

A few studies have been reported for the modulation properties of self-pulsating semiconductor lasers. The lasers show unstable oscillations by the modulation to the bias injection current, and also exhibit chaotic behaviors at large modulation index. For the variations of irregular pulse peaks, we can see similar chaotic bifurcations to those in ordinary edge-emitting semiconductor lasers such as discussed in Chap. 7 (Winful et al. 1986, Juang et al. 1999, 2000, and Jones et al. 2001). It is shown that the occurrence of chaotic oscillations is critically dependent on the modulation frequency. Periodic bands of chaotic dynamics are found to exist at multiples of the relaxation oscillation frequency. Not only stable pulsations resonance to the modulation frequency (locking oscillation), but also unstable pulsations and unique frequency-locked pulsations in which multiple spikes appear within some modulation period are found.

Fukushima et al. (2002) investigated experimentally and theoretically the dynamics of self-pulsating semiconductor lasers with injection current modulated and observed chaotic bifurcations for the pulsation frequency and the pulse height. Figure 8.10 shows some typical output waveforms observed by the experiments. The scales in the figure denote the period of modulation. The modulation index for the bias injection current is set to be $m = 1.06$. Each periodic state is defined as P_k^l, where k stands for the ratio of the fundamental period of the pulse train to the modulation period and l stands for the number of spikes in the fundamental period of the pulse train. Without rf modulation, the laser shows stable pulsation at a self-pulsation frequency of 245.7 MHz, as shown in Fig. 8.10a. Under rf modulation, as the modulation frequency f_m increases, the output optical pulse train is locked to the modulation frequency (P_1^1 pulsation) or its subharmonics (P_2^1 and P_3^1 pulsations) as shown in Figs 8.10d, h, and i. In the boundary regions of these frequency-locked pulsations, unstable pulsation occurs as shown in Fig. 8.10e.

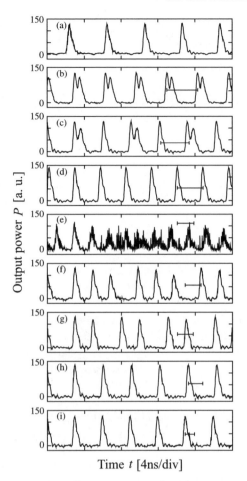

Fig. 8.10. Experimentally observed temporal waveforms of the output optical pulses at a bias injection current of 59.5 mA and a modulation index of $m = 1.06$. **a** Without rf modulation, **b** P_1^2 pulsation at $f_m = 300.0$ MHz, **c** P_2^3 pulsation at $f_m = 327.0$ MHz, **d** P_1^1 pulsation at $f_m = 360.0$ MHz, **e** unstable pulsation at $f_m = 562.0$ MHz, **f** P_4^3 pulsation at $f_m = 584.0$ MHz, **g** P_3^2 pulsation at $f_m = 592.0$ MHz, **h**+P_2^1 pulsation at $f_m = 650.0$ MHz, and **i** P_3^1 pulsation at $f_m = 1000.0$ MHz. The bar in each plot is the fundamental period of the modulation (after Fukushima T, Miyazaki H, Ando T, Tanaka T, Sakamoto T (2002); © 2002 JSAP)

It is speculated that the unstable region contains both quasi-periodic pulsation and chaotic pulsation. In the boundary regions, unique frequency-locked pulsations are also observed. One is P_1^2 pulsation in which two spikes appear within one modulation period, as shown in Fig. 8.10b. Another is P_2^3 pulsation in which two spikes and a single spike appear alternately. The others are P_4^3 and P_3^2 pulsations in which three or two spikes appear within four or three modulation periods as shown in Figs. 8.10f and g, respectively.

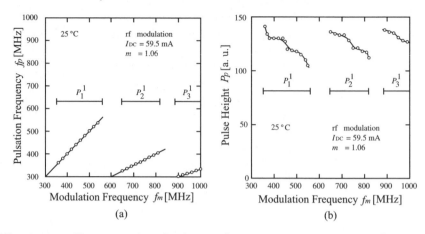

Fig. 8.11. a Characteristics of pulsation frequency versus modulation frequency under rf modulation. **b** Dependence of the pulse height on the modulation frequency under the frequency-locked pulsations of P_1^1, P_2^1, and P_3^1 (after Fukushima T, Miyazaki H, Ando T, Tanaka T, Sakamoto T (2002); © 2002 JSAP)

The experimental results are summarized in Fig. 8.11. Figure 8.11a shows the characteristics of pulsation frequency versus modulation frequency. The figure shows the regions of frequency-locked states for the modulation frequency. In between the frequency-locked states for the modulation frequency, we can observe unstable and chaotic oscillations as shown in Fig. 8.11e. Figure 8.11b shows the dependence of the pulse height on the modulation frequency under P_1^1, P_2^1, and P_3^1 pulsations. Under P_1^1 pulsation, the pulse height decreases gradually as the modulation frequency increases and eventually the pulsation becomes unstable at the modulation frequency of 560 MHz. Then P_2^1 pulsation occurs at 647 MHz. Here, the pulse height again returns to a high level. The same phenomenon is observed under P_2^1 and P_3^1 pulsations. The modulation index used in this experiment is rather high. However, similar trends in those results were obtained for lower modulation index, although each region of frequency-locked state shifts for lower modulation frequency. Fukushima et al. (2002) also compared the theoretical results with those experiments and demonstrated that the theory based on the rate equations of (8.11–8.13) with injection current modulation well explains their experiments.

8.3 Vertical-Cavity Surface-Emitting Lasers (VCSELs)

8.3.1 Theoretical Model of Vertical-Cavity Surface-Emitting Lasers

Vertical-cavity surface-emitting lasers (VCSELs) are promising devices of light sources for optical information processing and communications. Currently, VCSELs from visible wavelengths to near infrared (1.5 μm) are fabricated and their output powers reach as high as several tens of milliwatts. Also, a device that has a high modulation bandwidth of over 10 GHz with a low RIN of less than −140 dB is fabricated. A VCSEL has a disk structure with light coming out from the top or bottom of the surface. Various types of device structures have been proposed. Index- and gain-guiding structures are used for the confinements of carrier and light in VCSELs such as those for edge-emitting semiconductor lasers. Each guiding structure has merits and demerits in the laser oscillations, but the differences of the stable and unstable effects between those device structures are usually small compared with those in edge-emitting semiconductor lasers, since the length of the laser cavity is much smaller than that of edge-emitting semiconductor lasers. The details of VCSEL structures can be found in the book by Li and Iga (2002). Here, we do not discuss the details of device structures and device characteristics, but discuss the dynamics. As an example, the distributed Bragg reflector (DBR) VCSEL is shown in Fig. 8.12. The thickness of the active layer is approximately equal to the wavelength of light λ. The top view of the laser looks like a disk and its diameter is several to tens of μm. For special use, a disk diameter over 100 μm has been fabricated. In these devices, the reflectivity of the bottom surface is almost 100% and the top reflectivity of the DBR structure is around 99%. The laser light comes out from the top. Though the internal reflectivity is very high compared with edge-emitting semiconductor lasers (about 10%), VCSELs are also sensitive to optical feedback and optical injection. The photon number in the active volume is much less than that of edge-emitting lasers, and a few external photons would cause instabilities in the laser oscillations. VCSELs even for different device structures are described by the same or similar rate equations for the field and the carrier density.

There are many advantages of VCSELs for practical purposes. Since the VCSEL has a symmetric space structure, we can expect a circular beam as its output, while the beam profile of the edge-emitting laser has astigmatism.

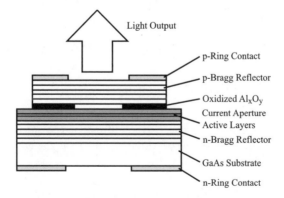

Fig. 8.12. Distributed Bragg reflector VCSEL structure

Due to a short cavity length compatible with the wavelength of light and very high reflectivity of light in the internal cavity, the laser is a very low threshold device, as low as $\sim\mu$A. From this same reason, we can produce a stabilized oscillation with a single mode that has a large mode separation ($\sim$40 THz). Another merit of VCSELs is the easiness of devising laser arrays because of the surface-emitting structure. However, VCSELs have unstable features for their operations even without any external perturbations. In addition to the time dependent phenomena, the space structures and the polarization modes give rise to instability and chaotic dynamics in VCSELs. Spatial hole burning and multi-transverse mode oscillations are often observed in the laser output to cause instabilities such as mode and polarization switching. Therefore, the VCSEL itself is an unstable laser.

The rate equations for VCSELs are similar to those for the edge-emitting laser except for the spatial terms. For a certain polarization mode, the field equation is given by (Valle et al. 1995a, b, and Law and Agrawal 1997a, b)

$$\frac{dE_j(t)}{dt} = \frac{1}{2}(1 - i\alpha)G_{nj}\{n(r, \phi, t) - n_{th,j}\}E_j(t) \tag{8.16}$$

where $n(r, \varphi, z, t)$ is the space dependent carrier density for the radial coordinates (r, φ, z) and $n_{th,j}$ is the threshold carrier density for the j-th mode. E_j is the field amplitude for the laser oscillation of the j-th spatial component, and the total complex amplitude from a VCSEL is written by

$$E_{total}(r, \phi, z, t) = \frac{1}{2}\sum_{j=1}^{M}\hat{e}_j E_j(t)\psi_j(r, \phi)A_0\sin(\beta_z z)\exp(-i\omega_{th,j}t) + c.c.$$

$$\tag{8.17}$$

where M is the total number of the spatial modes, $\hat{e}_j$ is the polarization vector for the j-th mode, ψ_j is the eigen-function for the j-th mode, β_z is the propagation constant for the z direction, and A_0 is the normalization coefficient. Since the carrier diffusion in the radial direction must be taken into account for the VCSEL oscillation, the rate equation for the carrier density is written by

$$\frac{d}{dt}n(r, \phi, t) = D\nabla_T^2 n(r, \phi, t) + \frac{J(r, \phi)}{ed} - \frac{n(r, \phi, t)}{\tau_s}$$

$$-\frac{\Gamma_d}{d}\sum_{j=1}^{M}G_{nj}\{n(r, \phi, t) - n_0\}|E_j(t)\psi_j(r, \phi)|^2 \tag{8.18}$$

where D is the coefficient for the carrier diffusion, the subscript T denotes the operation for the transverse coordinates, and Γ_d is the confinement factor for the longitudinal direction in the active layer given by

$$\Gamma_d = \int_0^d |A\sin(\beta z)|^2 dz \tag{8.19}$$

Here, the thickness of the active layer d is smaller than the total length of the laser cavity L, thus $\Gamma_d < 1$. In the derivation of the carrier density equation (8.18), we must consider the depletion of carriers for laser emission and take into account the interference terms for the external product of the vector polarizations. However, the frequency difference of the modes is usually of the order of several tens of GHz to one hundred GHz. As a result, the beating of these i and j terms, $\exp\{-i(\omega_l - \omega_j)t\}$, has a high frequency and the carrier cannot follow the oscillation. Therefore, we can neglect this effect and the equation (8.18) becomes a good approximation for the dynamics of the carrier density.

The eigen-function for the j-th mode ψ_j is a function of the polar coordinate calculated for a particular structure of the VCSEL. For example, for a weak index-guide cylindrical structure with two polarization states corresponding to the spatial LP_{01} mode, it is written by the Bessel function of the first kind $J_0(z)$ and the modified Bessel function of the second kind $K_0(z)$ and has the following form

$$\psi_j(r,\phi) = \begin{cases} \dfrac{J_0(u_{1j}r/R_a)}{J_0(u_{1j})} & \text{for} \quad r \le R_a \\[2ex] \dfrac{K_0(w_{1j}r/R_a)}{K_0(w_{1j})} & \text{for} \quad r > R_a \end{cases} \tag{8.20}$$

where R_a is the radius of the active area and u_{1j} and w_{1j} are the first roots of the eigen-value equation for the j-th polarization mode

$$\frac{u_j J_1(u_j)}{J_0(u_j)} = \frac{w_j K_1(w_j)}{K_0(w_j)} \tag{8.21}$$

They have a relation of $u_{1j}^2 + w_{1j}^2 = V_j^2$. V_j is the normalized frequency defined by

$$V_j = \frac{2\pi R_a \sqrt{\eta_{1j}^2 - \eta_j^2}}{\lambda} \tag{8.22}$$

where λ is the wavelength of light in vacuum and η_{1j} and η_j are the refractive indices for the j-th mode in the active area and the clad region.

8.3.2 Spin-Flip Model

In the derivation of the rate equations in the previous section, we consider the polarization effects in VCSELs in (8.17). However, in the physical terms, we take into account the effects of electron spin associated with light emission in the polarization dynamics in VCSELs. Specifically, left and right circularly polarizations of laser light emission are related to spin states in the conduction and valence bands. This is the origin of the polarization oscillations in VCSELs, and results in a rich variety of polarization dynamics including polarization switching frequently observed in VCSELs. San Miguel et al. proposed a different model for the rate equations of VCSELs by taking into account spin dynamics (San Miguel et al. 1995; Martin-Regalado et al. 1997; Sciamanna et al. 2002a,b). The model couples the polarization state of the electric field to the semiconductor medium by including the magnetic sublevels of the conduction and valence bands (the angular momentum numbers of electron) in quantum well devices. It is shown that laser dynamics depend significantly on the value of the relaxation rate. The polarization switching is included by the assumption of the population difference between the carrier densities with positive- and negative-spin values. From these equations, the dynamics of the laser oscillations for the lower order spatial mode can be easily explained, and the results are entirely coincident with the model discussed in the previous subsection. Although the dynamics of polarization dynamics in VCSELs can be well defined by the model, these rate equations are only applicable to the lowest spatial mode oscillation. For the treatments of dynamics related to higher spatial modes, we need the rate equations discussed in the previous section. In the following, we derive the expressions for the rate equations based on the spin dynamics.

Figure 8.13 shows the four-level model for polarization dynamics in quantum-well VCSELs. In the spin dynamics model of laser transitions, the magnetic quantum numbers in the lower edge of the conduction band have $J_z = \pm 1/2$ in accordance with up and down spin states. On the other hand, the magnetic quantum numbers for heavy holes in the upper valence band have values of $J_z = \pm 3/2$, since we can neglect the effect of light holes in quantum-well VCSELs. In the quantum state numbers, the same sign corresponds to the same spin state. Photon emitted from $+$ spin state corresponds to left circular polarization, while photon from $-$ spin state to right circular polarization. In the model, the decay rate γ_J accounts for the mixing of

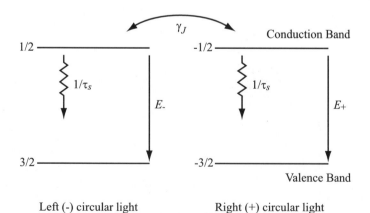

Fig. 8.13. Four-level model for polarization dynamics in quantum-well VCSEL

the populations with opposite value of J_z. This parameter is introduced to model spin-flip relaxation processes. For simplicity, we assume the same decay rate γ_J in the conduction and valence bands (San Miguel et al. 1995 and Martin-Regalado et al. 1997).

We next consider the effects of cavity anisotropies, which can be modeled in the two equations for the time evolution of the field amplitudes with two circular polarizations by replacing the linear loss rate by a matrix whose Hermitian part is associated with amplitude losses and whose anti-Hermitian part gives linear and circular phase anisotropies (also known as birefringence and circular dichroism, respectively). For VCSELs, it is known that there are two preferred modes of linear polarization that coincide with the crystal axes. These two modes have a frequency splitting associated with the birefringence of the medium. This can be modeled by a linear phase anisotropy given by a parameter γ_p, which represents the effect of different indexes of refraction for the orthogonal linearly polarized modes. In addition, the two modes may have a slightly different gain-to-loss ratio that can be related to the anisotropic gain properties of the crystal; the slightly different position of the frequencies of the modes with respect to the gain versus frequency curve, and different cavity geometries for the differently polarized modes. These effects can be modeled by an amplitude anisotropy with parameter γ_a. We assume here for simplicity that the directions of linear phase and amplitude anisotropy coincide, so that both are diagonalized by the same basis states.

We now derive the rate equations for double crossed linearly polarized light. Electron spin is associated with photon spin. We start the description of circularly polarized light for the fields of photons in such case. The laser light field is coupled to two population inversion variables; n is the sum of the upper state populations minus the sum of the lower state populations and n_J is the number of the difference between the population inversions (upper and lower state population difference) on the two distinct channels with positive or negative value of J_z (quantum spin number). Namely, in semiconductor lasers, n represents the total carrier number in excess of its value at transparency. Then the rate equations for the fields with right $(+)$ and left $(-)$ circularly polarized states are given by

$$\frac{dE_\pm(t)}{dt} = \frac{1}{2}(1 - i\alpha)G_n \left\{ n_\pm(t) - n_{th} \right\} E_\pm(t) - (\gamma_a + i\gamma_p)E_\pm(t) \qquad (8.23)$$

where $n_\pm(t) = n(t) \pm n_J(t)$, and γ_a and γ_p are the linear anisotropies representing dichroism and birefringence discussed above. Note that signs between the spin states and the subscripts of the fields are opposite.

The rate equations for the carrier density with spin-down and spin-up, $n_\pm$, are written by

$$\frac{dn_\pm(t)}{dt} = \frac{J}{ed} - \frac{n_\pm(t)}{\tau_s} - 2G_n\{n_\pm(t) - n_0\}|E_\pm(t)|^2 - \gamma_J\{n_\pm(t) - n_\mp(t)\}$$

$$(8.24)$$

where γ_J is the spin-flip rate. Using the conversion relations between linearly polarized lights and circularly polarized lights, $E_x = (E_+ + E_-)/\sqrt{2}$ and $E_y = (E_+ - E_-)/i\sqrt{2}$, and also the relations of the numbers of spin-down and spin-up with the numbers of the carrier density n and the spin-state difference n_J, $n_+ = n + n_J$ and $n_- = n - n_J$, the equations for the linearly polarized fields are reduced to be

$$\frac{dE_x(t)}{dt} = \frac{1}{2}(1 - i\alpha)G_n\{n(t) - n_{th}\}E_x(t)$$
$$+ \frac{i}{2}(1 - i\alpha)G_n n_J(t)E_y(t) - (\gamma_a + i\gamma_p)E_x(t)$$

(8.25)

$$\frac{dE_y(t)}{dt} = \frac{1}{2}(1 - i\alpha)G_n\{n(t) - n_{th}\}E_y(t)$$
$$- \frac{i}{2}(1 - i\alpha)G_n n_J(t)E_x(t) + (\gamma_a + i\gamma_p)E_y(t)$$

(8.26)

The rate γ_p reads to a frequency difference of $2\gamma_p$ between the x- and y-polarized solutions (the x-polarized solution having the lower frequency when γ_p is positive). The decay rate γ_a reads to different thresholds for these two linearly polarized solutions, with the y-polarized solution having the lower threshold when γ_a is positive. The difference of the gain between the two polarization modes varies as $G_0(1 - J/J_{sw})$ for the change of the bias injection current and the decay rate γ_a is defined by $\gamma_a = G_0 n_0(1 - J_{th}/J_{sw})$, where G_0, n_0, and J_{sw} are the intrinsic gain difference between the two polarization modes, the carrier density at transparency, and the injection current at polarization switching, respectively. Similarly, the equations for the carrier density and the difference of the spin states are calculated as

$$\frac{dn(t)}{dt} = \frac{J}{ed} - \frac{n(t)}{\tau_s} - G_n\{n(t) - n_0\}\left\{|E_x(t)|^2 + |E_y(t)|^2\right\}$$
$$+ iG_n n_J(t)\left\{E_x(t)E_y^*(t) - E_x^*(t)E_y(t)\right\}$$

(8.27)

$$\frac{dn_J(t)}{dt} = -\left(\frac{1}{\tau_s + 2\gamma_J}\right)n_J(t) - G_n n_J(t)\left\{|E_x(t)|^2 + |E_y(t)|^2\right\}$$
$$- iG_n\{n(t) - n_0\}\left\{E_x(t)E_y^*(t) - E_x^*(t)E_y(t)\right\}$$

(8.28)

The population difference n has a decay rate $1/\tau_s$ associated with spontaneous decay, while n_J has a decay rate $\gamma_J' = 1/\tau_s + 2\gamma_J$. The decay rate γ_J' accounts for the mixing of the populations with opposite value of J_z, which was introduced to model spin-flip relaxation processed and assumed to have the same value for the conduction and valence bands.

Several spin relaxation processes for electrons and holes have been identified in semiconductors, such as scattering by defects, exchange interactions between electrons and holes, and exciton-exciton exchange interactions (Martin-Regalado et al. 1997). From experimental measurements of spin relaxation times in quantum wells, the relaxation time is of the order of tens of picoseconds. Since typically $\tau_s \sim 1\,\mathrm{ns}$, and $\tau_{\mathrm{ph}} \sim 1\,\mathrm{ps}$, the spin mixing γ_J occurs on an intermediate time scale between that of the field decay and that of the total carrier population difference decay. Hence, the dynamics of the difference in spin states n_J cannot be adiabatically eliminated for the time scales of the interest. The rate equations including the magnetic sublevels of the conduction and valence bands are applied to analyses for the dynamics of VCSEL polarizations such as polarization switching and polarization insta-bilities (Sciamanna et al. 2003b, c, Sciamanna and Panajotov 2005, 2006, and Masoller et al. 2006). In particular, fruitful results are obtained for explana-tions of the dynamics for orthogonal optical injection and stabilization in VCSELs. In this book, we use mainly the theoretical treatments in the pre-vious section, which includes the spatial effects. However, the model of the magnetic sublevels is also used for VCSEL dynamics when necessary.

8.3.3 Characteristics of VCSELs in Solitary Oscillations

Even in the absence of external perturbations, VCSELs sometimes show un-stable behaviors depending on the bias injection current. Spatial and po-larization modes play important roles in the dynamic behavior of VCSELs. Higher spatial modes are easily excited for a higher bias injection current. Figure 8.14 shows the beam profiles for the lowest three spatial modes along the radial direction of a VCSEL (LP_{01}, LP_{11}, and LP_{21} modes). These modes are calculated from (8.20)–(8.22). Due to the spatial hole burning effects, the carrier distribution has a dip at the center of the disk in a VCSEL and the higher spatial modes tend to oscillate for a large bias injection current (Law and Agrawal 1997b). Figure 8.15 shows the experimentally obtained near-field images of the oscillation modes (Degen et al. 1999). The images are obtained by changing the bias injection current. Higher spatial modes are excited for the increase of the bias injection current. The excitation of higher

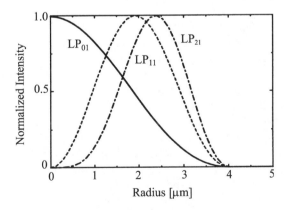

Fig. 8.14. Spatial mode distributions for LP_{01}, LP_{11}, and LP_{21} modes in VCSELs. The radius of the disk is $4\,\mu m$

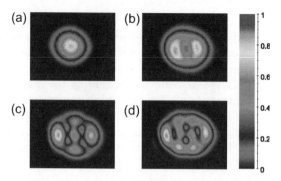

Fig. 8.15. Experimentally obtained near field images of VCSEL with 6 μm diameter at an injection current of **a** 3.0, **b** 6.2, **c** 14.7, and **d** 18 mA (after Degen C, Fischer I, Elsäßer W (1999); © 1999 OSA)

modes strongly depends on the disk diameter. For ordinary applications of VCSELs, a circular Gaussian beam of the lowest mode is desirable. To obtain such a clean beam, the diameter of a VCSEL must be small, but it is difficult to attain a high power operation at the same time.

In semiconductor materials, there exists the difference of the refractive indices between the components for the principal axis and the orthogonal axis to it because of the distortion and birefringence of the materials. The difference between the indices is very small and it is $10^{-3} \sim 10^{-4}$. For ordinary edge-emitting semiconductor lasers, the difference can be ignored due to a large asymmetric configuration for the TE and TM modes in the active layer and the laser operates at only TE mode. However, the difference plays a crucial role for the operations of VCSEL, since it has a circular disk structure of the light-emitting facet. Then, there is an ambiguity for the polarization direction of the laser oscillation. A VCSEL usually oscillates at a polarization mode along the optic axis of the material (we refer to this mode as the x-mode in the following) when the laser is biased at a low injection current. However, the polarization mode may switch from this mode to the orthogonal one for the increase of the bias injection current. This switching is mainly induced by the distortion or the birefringence of the laser material as discussed above. Therefore, we take into account the effect of the polarization modes in the field equation in (8.17). It is note that the main oscillation mode is sometimes referred to as the y-polarization mode, in accordance with the crystal axis of semiconductor laser materials. However, the main mode is here defined as x-polarization. Figure 8.16a shows the L-I characteristic of a VCSEL experimentally obtained for a disk diameter of 6 μm. At a low bias injection current, the fundamental transverse mode (higher frequency mode) starts to lase, then the orthogonal mode (lower frequency mode) grows up after the polarization switching point. Thus the main oscillation mode switches from x to y polarization mode well above the switching point. Usually, the polar-

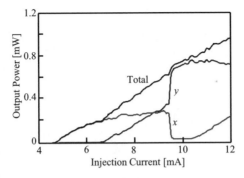

Fig. 8.16. Experimental L-I characteristic of a 6 μm diameter VCSEL at free running state with wavelength of 780 nm. x polarization mode is the oscillation for the optic axis and y polarization mode is the component perpendicular to the x component. The total power is the addition of the x and y components

ization switching has a hysteresis for the increase or the decrease in the bias injection current. In this example, clear switching of the polarization modes from x-mode to y-mode is visible at the bias injection current of 9.5 mA. Whether we can observe clear switching of the polarization modes or not strongly depends on the characteristics of the laser materials and the device structures. Some VSCELs do not show clear polarization switching for the bias injection current.

Taking into consideration the birefringence of laser materials, the polarization switching is well reproduced by the numerical simulations from (8.16)–(8.18) (Giudici et al. 1999 and Danckaerta et al. 2002). At a low bias injection current, the carrier density has a maximum value at the center of the disk in the active area and the carrier density smoothly decreases toward the edge of the disk. However, for a large bias injection current, hole burning of carriers occurs at the center of the disk. Then, the carrier density takes the maximum value a little away from the center of the disk. This induces the excitation of the orthogonal mode and the suppression of the original mode, since, for example, the hole burning due to the birefringence causes the transfer of the optical energy from the x-mode to the y-mode. Then, the laser oscillation is switched from the x-mode to the y-mode. The effects are distinct for VCSELs with large birefringence and small disk size.

Even in solitary oscillations, VCSELs show dynamic characteristics. One such type of dynamics is the anti-phase irregular oscillation of the optical power between the two polarization modes (Fujiwara and Ohtsubo 2004). Figure 8.17 shows an experimental example of anti-phase oscillations of the x and y polarization modes in a VCSEL. Unstable pulsations and bistability are sometimes observed at the switching point of the two polarization modes (Tang et al. 1997). However, not only at the switching point of the two polarization modes but also at certain bias injection currents different

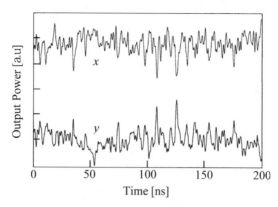

Fig. 8.17. Anti-phase oscillations of x- and y-modes in VCSEL at solitary oscillation at $J = 1.23J_{\text{th}}$. The laser is the same as in Fig. 8.16

from the switching point, does the laser show fast unstable oscillations and the two polarization modes oscillate at anti-phase manner in time. When the output power of the x polarization mode goes down, the output power of the y-mode grows up, and vice versa. This anti-phase oscillation is frequently observed in chaotic VCSELs subjected to optical feedback and injection current modulation (Besnard et al. 1997 and 1999).

8.3.4 Spatio-Temporal Dynamics in VCSELs

Lasers with spatial structures such as VCSELs and broad-area lasers have spatio-temporal dynamics induced by diffraction of light and hole-burning of carriers in the laser cavity. The cavity length of VCSELs is short, so that the effect of the diffraction of light can be neglected. However, the effect of carrier hole-burning plays a crucial role for transverse-mode oscillations. The polarization dynamics of VCSEL are strongly related to carrier hole-burning, and the laser shows pico-second instabilities even at solitary oscillations. Time averaged polarization dynamics in VCSELs, such as dynamics of near-field patterns for the bias injection current, were extensively studied, while a few studies of spatio-temporal dynamics were reported (Mulet and Balle 2002 and Barchanski et al. 2003). In this section, we present some experimental results for the spatio-temporally resolved polarization dynamics and discuss the underlining mechanism. Figure 8.18 shows the experimental results of spatio-temporal dynamics of a VCSEL obtained by a temporally resolved imaging by differential analysis, which allows us to extract the full two-dimensional evolution of the near-field intensity on timescales of 10 ps (Barchanski et al. 2003). The laser used is an oxidized VCSEL with an oscillation wavelength of 852 nm and a maximum output power of 3.8 mW. The VCSEL has a disc diameter of 14 μm.

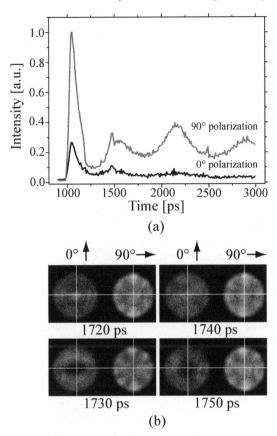

(a)

(b)

Fig. 8.18. a Evolution of the relaxation oscillations extracted from the differential images of the near-field polarization-resolved patterns. *Gray curve*: 90° polarization mode (dominant oscillation mode), and *black curve*: 0° polarization mode. **b** Snapshots for the evolution of near-field intensity of the VCSEL after the second relaxation oscillation peak. In each snapshot, the left shows the 0° polarization mode and the right 90° polarization mode. The crossing point of the overlaid crosshair is the center of the laser (Barchanski A, Gensty T, Degen C, Fischer I, Elsäßer W (2003); © IEEE 2003)

Figure 8.18a shows the laser outputs of the two polarization modes for the time evolution after a step impulse, in which each polarization mode shows relaxation oscillation after the switch-on. A period of 10 μs of the pulse prevents the occurrence of thermal effects in the experiments. The laser is biased at $2.3J_{\text{th}}$. At this bias injection current, the frequency of the relaxation oscillation is about 3 GHz. In this figure, the 90° polarization mode is the main oscillation mode. After about 3 ns from switch-on, the laser settles down to steady-state oscillation. Figure 8.18b shows snapshots for the evolution of the near-field intensity in the VCSEL after the second relaxation oscillation peak. In order to allow measurements, a displacement prism was used in the detection path to separate the laser beam into two orthogonal polarization patterns. The near-field polarized patterns of the laser was detected as an integrated image of 36,000 events by a CCD camera with a fast shutter time ~200 ps using gating triggers synchronized with the driving pulses for the bias injection current. The images in Fig. 8.18b give evidence for a rich dynamical behavior in the emission profile in both polarization directions. In the 0° polarization mode, the intensity change of the center of the disc aperture is clearly seen, namely, the center is filled with a bright spot at 1720 ps, however the center of the aperture remains mostly dark at 10 ps later. The intensity at 1740 ps is nearly uniform over the whole aperture, and then the center is again dark at 1750 ps. In the near-field patterns of the 90° polarization mode, we can see clear rotational flicker of the bright spots of the laser oscillations. For example, bright spots of the intensity rotate by 5° counterclockwise from 1720 ps to 1730 ps, while they rotate by 5° clockwise from 1740 ps to 1750 ps. This kind of rotational flicker is attributed to the spatial hole burning of the carrier density in the laser cavity. Spatial hole burning has been identified as a very important effect contributing to the observed dynamics on the examined time scales, especially for lasers having a spatial structure. Similar nonlinear dynamical behaviors has already been found and well investigated in the field of broad-area semiconductor lasers as known filamentation effects, which will be discussed in the following section.

Every transverse mode in a VCSEL corresponds to a different wavelength. Therefore, by spectrally resolving the near-field emission intensity, it is possible to investigate the dynamics of each mode separately. Figure 8.19 shows spectrally resolved near-field patterns of the VCSELs. In the observations, a spectrometer was used in the detection path to obtain spectrally resolved near-field patterns of the laser oscillations. The horizontal axis represents the spatial co-ordinate, the vertical axis is a combination of both spatial and spectral coordinates, with the wavelength increasing from the bottom to top of each snapshot. The large birefringence splitting, which is the spectral spacing among the fundamental Gaussian modes in orthogonal polarization directions, is quite noticeable. The estimation of the birefringence splitting, performed with an optical spectrum analyzer having a maximal resolution of 0.05 nm, provides a value of approximately 0.11 nm, or 50 GHz, respectively. The first image at 1430 ps, within the second relaxation oscillation peak, shows two more mode orders than during the first relaxation-oscillation peak, though the near-field pattern at the first relaxation oscillation peak is not shown here. At 1930 ps, about ten modes occur in the near-field intensity within the observed area. While the relative intensity among the 90° polarization modes mostly remains uniform, there is a drastic change in the modal behavior in the 0° polarization mode. Throughout the evolution, the four-lobed mode, which is the second oscillation mode, remains as an oscillation mode and all other modes of the 0° polarization direction show a relative smaller intensity. The fundamental Gaussian mode

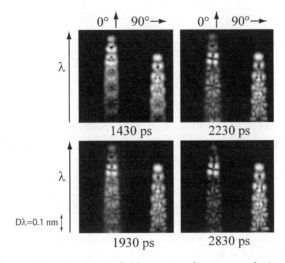

Fig. 8.19. Spectrally resolved near-field patterns for time evolution. In each snapshot, the *left* shows the 0° polarization mode and the *right* 90° polarization mode. The wavelength increases from bottom to top of each snapshot. The snapshots show the intensity change during 100 ps (Barchanski A, Gensty T, Degen C, Fischer I, Elsäßer W (2003); © IEEE 2003)

in the 90° polarization mode is spectrally aligned at the same position as the bright four-lobed mode in the 0° polarization mode. This spectral alignment implies the importance of spectral interactions related to the spatial carrier hole burning effects. Namely, the fundamental mode, observed in the 90° polarization direction, has its intensity concentrated in the center of the aperture. In contrast, the intensity of the four-lobed mode is concentrated in the periphery of the aperture, resulting in a minimal spatial overlap of both modes. The complementary oscillation between the two polarization modes originates from a competition of both polarization directions for the available gain in the active medium. The dynamics of VCSELs even at solitary oscillation still remains an interesting research field, providing further insight into the fundamental physics of semiconductor lasers and promising device optimization.

8.3.5 Feedback Effects in VCSELs

VCSELs have a high reflective mirror of the Bragg reflector within the cavity as much as the internal reflectivity of higher than 99% to realize a low laser threshold. However, the total photon number within the cavity is much smaller than that of the edge-emitting semiconductor laser and the laser is also affected by a small number of photons from an external reflector. For a small optical feedback, (8.16) is modified and the rate equations for the complex field are written by

$$\frac{dE_j(t)}{dt} = \frac{1}{2}(1 - i\alpha)G_{nj}\{n(r, \phi, t) - n_{th,j}\}E_j(t)$$
$$+ \frac{\kappa}{\tau_{in}}E_j(t - \tau)\exp(i\omega_0\tau) \qquad (8.29)$$

The other equations for the total complex field amplitude and the carrier density remain unchanged. Equation (8.29) looks like the same form as that for edge-emitting semiconductor lasers. However, the rate equations of the total field and the carrier density are the functions not only of time but also space and, as a result, the laser shows complicated behaviors compared with edge-emitting semiconductor lasers. Mutual interactions between the two polarization modes also affect the dynamics. Spatial mode competitions may also play an important role for the dynamics under a large bias injection current. Therefore, we must take into account the essential terms of the spatial polarization modes for numerical calculations of the dynamics.

Since the laser cavity length of a VCSEL is less than the optical wavelength and much smaller than those of other semiconductor lasers, the separations both for the longitudinal and transverse modes are much larger than for other lasers. Therefore, we can well apply the approximation for a single longitudinal mode operation as far as the laser has a small disk size or is biased at a modest injection current. However, the competition among spatial and polarization modes arises at a higher bias injection current. In the following, we show some characteristics of optical feedback effects in VCSELs.

Unstable oscillations of VCSELs induced by optical feedback have been numerically calculated (Law and Agrawal 1998). Table 8.3 is a typical set of

Table 8.3. Characteristic device parameters for a VCSEL at an oscillation wavelength of 850 nm

Symbol	Parameter	Value
G_n	gain coefficient	$2.90 \times 10^{-12}\,\mathrm{m^3\,s^{-1}}$
Γ_d	confinement factor	0.1
α	linewidth enhancement factor	3.80
r_1	output facet reflectivity	0.9975
r_1	bottom reflectivity	0.9995
n_{th}	carrier density at threshold	$3.80 \times 10^{24}\,\mathrm{m^{-3}}$
n_0	carrier density at transparency	$1.75 \times 10^{24}\,\mathrm{m^{-3}}$
τ_s	lifetime of carrier	1.00 ns
τ_{ph}	lifetime of photon	3.30 ps
τ_{in}	round trip time in laser cavity	22.6 fs
D	diffusion constant	$30\,\mathrm{cm^2\,s^{-1}}$
l	cavity length	1.00 μm
d	active layer thickness	0.20 μm
R_a	radius of active layer	4.00 μm

parameters used in the numerical simulations. Figure 8.20 shows a time series of the laser output for the change of the external feedback ratio. The diameter of the laser disk is 4 µm and the laser is oscillated with the lowest single mode or the lowest two spatial modes (LP_{01} and LP_{11} mode). A single polarization mode is assumed. From Figs. 8.20a–d, the two modes show period-doubling like evolutions to chaotic states. However, once they become a fixed oscillation in Fig. 8.20e and again evolve into fully chaotic oscillations in Fig. 8.20f. Spontaneous emission of light is ignored in these calculations, however it strongly affects the dynamics when the laser oscillates at chaotic states. Some such effects are the increase of noise floor and the broadening of the chaotic carrier frequency.

As dynamics of edge-emitting semiconductor lasers, low-frequency fluctuations (LFFs) have been observed. LFFs are not only the typical features of edge-emitting semiconductor lasers, but also they are observed in various types of semiconductor lasers. Fujiwara et al. (2003a) have experimentally observed LFFs in VCSELs with optical feedback from a distant reflector. Similar LFF characteristics to those of edge-emitting semiconductor lasers, sudden power dropout and gradual power recovery, are observed for the x polarization mode with the lowest spatial mode of LP_{01} at a low bias injection current. Under a LFF oscillation for the x polarization mode, the output

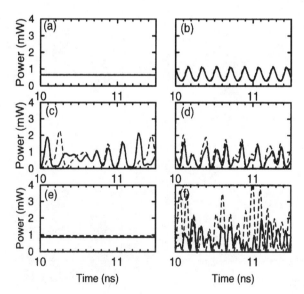

Fig. 8.20. Temporal evolutions of output power under two-mode operation with 4 µm disc contact (weak coupling) VCSEL for several feedback levels. *Solid* and *dashed* curves represent the LP_{01} and LP_{11} modes, respectively. External feedback rate κ of **a** 0, **b** 1.6×10^{-4}, **c** 5×10^{-4}, **d** 8.9×10^{-4}, **e** 1.6×10^{-3}, and **f** 2.8×10^{-3} (after Law JY, Agrawal GP (1998); © 1998 OSA)

power of the orthogonal y polarization mode also shows synchronous waveforms of LFFs with the x mode, but it is an anti-phase oscillation. Similarly
to the case of edge-emitting semiconductor lasers, the coherence of the laser
is fairly collapsed at LFF states. However, the laser still holds a single mode
operation, because of the large separation of the cavity modes (Von Lehamen
et al. 1991). The dynamic properties of LFFs in VCSELs have also been
demonstrated by numerical simulations using the model of the population
difference between the carrier densities with positive and negative spin values (Masoller and Abraham 1999 and Sciamanna et al. 2003).

Figure 8.21 shows time-averaged effects of polarization-selected optical
feedback in a VCSEL. A VCSEL used in the experiments has the disc diameter of 16 μm and the oscillation wavelength of 780 nm. The external mirror is
located at 90 cm away from the front facet of the laser. For a reference, the L-I
characteristics together with near-field oscillation patterns at 7.0, 12.0, 18.0,
and 24.0 mA are displayed in Fig. 8.21a. For steady-state oscillations, the
complementary features between the two polarization modes are clear, especially at higher bias injection current, which is the typical nature of VCSELs
either for fast or slow dynamics. The laser has no clear polarization switching between the crossed polarization modes. As noted, L-I characteristics of

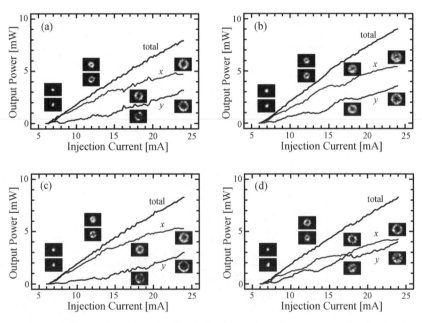

Fig. 8.21. Effects of optical feedback in VCSELs. Experimentally obtained L-I
characteristics and near-field patterns for **a** solitary oscillations, **b** total optical
feedback (intensity feedback of 9.8%), **c** x-polarization feedback (feedback of 6.1%),
and **d** y-polarization feedback (feedback of 6.0%). The *upper* mode patterns in the
insets are for x-polarization mode, while the *lower* ones for y-polarization mode

VCSELs strongly depend on device structures and used materials. Though the total L-I characteristic shows an almost linear relation for the increase of the bias injection current, each polarization component has quite different features depending on respective VCSEL structures. For example, the L-I characteristics of x- and y-polarization modes are quite different from those for one in Fig. 8.16 that shows a clear polarization switching. As a result, the near-field pattern for each VCSEL is also dependent on the device structures. For the case of total optical feedback in Fig. 8.21b, the threshold of the x-polarization mode slightly reduces from 6.4 mA to 6.2 mA, while the threshold of the y-polarization mode increases from 6.4 mA to 6.8 mA. The slop efficiency increases due to the optical feedback, but only slight changes of the L-I curves are visible. However, the spatial modes are affected by the optical feedback, especially in higher bias injection current. In the case of x-polarization optical feedback, the trends of the threshold reduction are almost the same as for the case of the total optical feedback. However, the crossed polarization component (y-mode) is greatly suppressed. On the other hand, two modes compete with each other in the case of x-polarization feedback. As a result, the laser becomes less stable at this amount of the optical feedback and it is oscillated with lower spatial orders. Polarization selective optical feedback provides quite interesting dynamics of VCSELs, and it is very important from a viewpoint of laser control.

VCSELs have been newly developed and they themselves show various dynamics under solitary oscillations and external perturbations. Therefore, the dynamics have not been well understood yet and the studies are still undergoing. As another example of dynamics in VCSELs, we here show self-oscillation properties when a portion of the laser output power is injected back into the laser after having rotated its polarization by 90 degrees with respect to the initial laser polarization state (Jiang et al. 1993). When the two polarization components of the lights are orthogonally returned to the laser from a reflector with short distance, the x component coherently couples with the y component and the y component also coherently couples with the x component in as far as the corresponding oscillation wavelengths are assumed to be the same. Figure 8.22 shows the experimental waveforms of self-modulations in VCSELs. In this experiment, a laser that shows a clear polarization switching is used. Self-modulations are observed both for the bias injection currents above and below the polarization switching point. As square wave is observed for a long external feedback, while sinusoidal waveforms are observed for a short external feedback. As can easily be recognized, the waveforms of the x and y components show anti-phase oscillations. The frequency of the oscillation is half of the frequency for the external cavity length. The polarization rotation feedback induces polarization injection locking in the VCSEL and leads to a switching of the polarization state, then self-modulation occurs in its output. The phenomenon is quite similar to injection locking in a regenerative amplifier, where very weakly injected light is sufficient to lock the laser to the incident frequency. Under the same config-

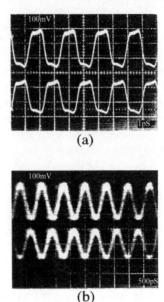

Fig. 8.22. Waveforms of polarization self-modulation signals corresponding to extended cavity lengths of **a** 16.5 cm and **b** 5.3 cm at a bias injection current of $J = 1.51 J_{\text{th}}$. The bias injection current is above the point of the polarization switching. *Upper trace:* x polarization, *lower trace:* y polarization (after Jiang S, Pan Z, Dagenais M, Morgan RA, Kojima K (1993); © 1993 AIP)

uration, a self-modulation with a frequency of 6 GHz is obtained for a short cavity length of 1 cm. These self-modulation oscillations can be used as light sources for high-speed pulse sequences. Masoller and Abraham (1999a, b) presented the numerical simulations for the model considering the population difference between the carrier densities with positive and negative spin values in VCSELs and obtained the generations of self-modulation square waves.

8.3.6 Short Optical Feedback in VCSELs

In Sect. 5.4, we discussed typical regular pulse package dynamics with LFFs induced in short cavity optical feedback in edge-emitting semiconductor lasers. Regular pulse package dynamics are also observed in VCSELs with short cavity optical feedback. However, VCSEL has complex dynamics of orthogonal polarization modes and the pulse package dynamics are substantially affected by the polarization dynamics. Thus, we cannot observe exact regular pulse oscillations as in the case of edge-emitting semiconductor lasers, due to the competitions of the crossed-polarization modes. The definition of short cavity is that the cavity length is within the length corresponding to the relaxation oscillation frequency, which is usually less than centimeters. For example, in a VCSEL with a very short external cavity condition ($\sim$10 μm), the laser and the external cavity conform a composite cavity and one polarization mode shows a periodic undulation of the output power for a period of $\lambda/2$ with the change of the external cavity length. In this state, the other mode is also excited alternately to the orthogonal mode, thus showing antiphase oscillation for the external cavity length (Arteaga et al. 2006).

Tabaka et al. (2006) investigated the pulse package dynamics with LFFs including polarization modes in a short external cavity VCSEL. As a result, for the increase of the injection current, switching from one polarization mode to the other with orthogonal polarization direction is observed. The existence of the two polarization modes in VCSELs can give rise to an additional polarization mode competition dynamics in the presence of feedback. Figure 8.23 shows the experimental results of the polarization resolved dynamics. The VCSEL with an oscillation wavelength of 986 nm has the solitary threshold of $I_{th} = 3.7$ mA and shows polarization switching at the bias injection current of $I = 4.2$ mA. The external cavity length is 6.5 cm and the mirror reflectivity is 0.3, which results in a threshold reduction of 22% from the solitary laser

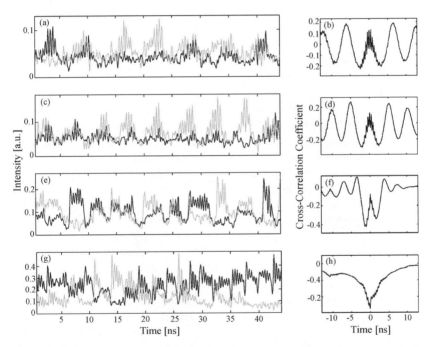

Fig. 8.23. Polarization resolved dynamics of a VCSEL in the pulse package regime for the injection current. *Left* column is time series and *right* column the corresponding cross-collation function. **a** $I = 3.2$ mA, **c** $I = 3.4$ mA, **e** $I = 3.8$ mA, and **g** $I = 5.0$ mA. *Gray plot* corresponds x polarization and *black plot* y polarization (after Tabaka A, Peil M, Sciamanna M, Fischer I, Elsäßer W, Thienpont H, Veretennicoff I, Panajotov K (2006); © 2006 AIP)

oscillation. The relaxation oscillation frequency at $I = 5.2\,\text{mA}$ corresponds to the external cavity length of 6.5 cm. Therefore, the observed dynamics in Fig. 8.23 satisfy the short cavity condition.

In Fig. 8.23a the amplitude of the peaks is still small and the shape of the single pulse package envelope is not very regular. However, the envelope of the packages can be clearly identified, which indicates that the pulse packages in the two polarization modes are almost periodic with a characteristic frequency. The pulse package dynamics in the two polarization modes can be much better recognized at $I = 3.4\,\text{mA}$, in Fig. 8.23c. In actual fact, the total intensity shows a quite regular pulse package oscillation, but the polarization resolved pulse package dynamics is not as regular as for the total intensity. The reason for this is that we observe polarization mode competition, underlying the pulse package dynamics, reducing the regularity of the pulse package dynamics in each polarization mode. This mechanism becomes more relevant at a higher injection current, approaching the polarization switching point. A gradual loss of the regularity in the pulse package dynamics as the bias injection current is increased from 3.2 to 3.8 mA. In the time series in Fig. 8.23, the pulse package dynamics temporarily take place in one of the polarization modes only in some cases and the second mode is almost turned off. In other cases the pulse package dynamics take place in the two polarization modes simultaneously. The first case of dynamics, in which the pulses are emitted in one polarization mode only, is referred to type I pulse packages. The second case of dynamics, in which the pulse package dynamics take place in the two polarization modes simultaneously, is called type II pulse packages. Similar interplay of the feedback induced complex dynamics and polarization mode competition in VCSELs has been found numerically by for LFF dynamics in the long external cavity regime (Sciamanna et al. 2003a) and experimentally confirmed (Naumenko et al. 2003b and Sondermann et al. 2003).

The cross-correlation functions corresponding to left column are shown in the right column of the figure. By increasing the injection current, we observe a continuous decrease of the modulation amplitude at the time scale of the multiples of the pulse package envelope until the peaks completely vanish, which we demonstrate in Fig. 8.23h. At higher levels of the bias injection current, the laser first emits pulses with high amplitudes while the amplitude of the following pulses progressively decreases. Moreover, in the regime of high injection currents, well above the polarization switching point, the cross-correlation function becomes negative for all time lags. This substantial change in the shape of the cross-correlation function can be associated with a remarkable change of the pulse package dynamics, reflecting a gradual transition from type II pulse package to type I one.

8.3.7 Orthogonal Optical Injection Dynamics in VCSEL

We have discussed optical injection phenomena in edge-emitting semiconductor lasers in Chap. 6. The technique is developed for frequency-locking and

stabilizing injected lasers, but the lasers are sometimes destabilized by optical injection and show a rich variety of chaotic dynamics for certain ranges
of the injection parameters as has already been discussed. In the case of
edge-emitting semiconductor laser, the laser usually emits a light with a linear polarization (TE mode) and the same polarization is used as an injection light. Once in a while, polarization-rotated optical injection (TM mode
injection) is applied to obtain a chaotic light source in edge-emitting semiconductor lasers. Normally, the excitation of the orthogonal mode is very
small in ordinary edge-emitting semiconductor lasers. However, the situation
completely changes in VCSELs, since the lasers have the ambiguity of oscillations for polarization directions. In VCSELs, optical injection including
the polarization direction plays a crucial role in the dynamics of the laser
oscillations even if the laser is oscillated at a certain fixed polarization with
a solitary mode. Depending on the injection conditions, the laser shows a rich
variety of dynamics; stabile and unstable injection locking, and even chaotic
oscillations by the optical injection. The typical feature of the dynamics is
the polarization switching between the two orthogonal polarization modes.
Further, the polarization is greatly affected by small changes of the bias injection current or the device temperature may result in a polarization switching
between the two linearly polarized modes. Control of the VCSEL polarization is a major issue in telecommunication applications. For well polarization
controlled VCSELs, polarization switching may be interesting for the development of all-optical switches. In optical injection to a VCSEL, we can obtain
a similar injection map of Fig. 6.7 or Fig. 6.8 in Sect. 6.2.3 as far as the injection polarization direction is the same as the oscillation mode of the VCSEL
(Li et al. 1996 and Ryvkinn et al. 2004). However, polarization switching dynamics encounter for an orthogonal polarization injection and the laser shows
a rich variety of dynamics in its output power and polarizations.

In order to represent the richness of the polarization dynamics in VCSELs
with orthogonal optical injection, the map of the boundaries of different dynamics is drawn as usual in the phase space of the frequency detuning and
the injection power in Fig. 8.24 (Altés et al. 2006 and Gatare et al. 2006).
The laser is under the y polarization oscillation with a single spatial mode
(the fundamental transverse mode) above the polarization switching point,
hence the main x polarization mode is suppressed. The injection power in
the horizontal axis in Fig. 8.24 is normalized to the solitary oscillation power
at this bias point. The VCSEL is externally injected by the linear polarization light with x mode and the y polarization mode dynamics of the laser is
investigated. The thin solid and the gray lines are the polarization switching
boundaries (switch-on points) for the increase of the bias injection current.
While the dashed and the thick solid lines are polarization switching boundaries (switch-off points) for the decrease of the bias injection current. In the
regions S1 and S2, the frequency of VCSEL emission is locked to the master laser. However, in the case of S2, it is the first order transverse mode
and not the fundamental transverse mode that locks to the master laser,

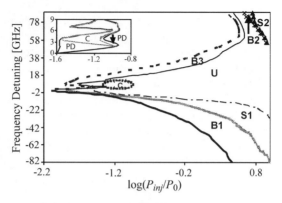

Fig. 8.24. Mapping of the dynamics of a VCSEL subject to optical injection in the phase space of the frequency detuning and the optical injection power. The laser is oscillated at the y polarization mode above the polarization switching point. The injected linear polarization light is set to be x mode. *Thin solid line* and *gray line* are polarization switching boundaries for the increase of the bias injection current. *Dashed line* and *thick solid line* are polarization switching boundaries for the decrease of the bias injection current. B1, B2, B3: bistable regions, S1, S2: stable locking regions, U: unlocking region, C: chaotic region. The inset shows period doubling (PD) dynamics around the region of instabilities C (after Altés JB, Gatare I, Panajotov K, Thienpont H, Sciamanna M (2006); IEEE 2006)

the fundamental transverse mode then being suppressed when crossing the dark green line. The unlocking of the first order transverse mode happens at smaller values of the injection power, describing bistable region B2 between the fundamental and the first order transverse mode both with the same polarization.

In Fig. 8.24, two polarization bistable regions are observed in a regime of fundamental mode emission, which correspond to two different ways of polarization switching. The first one is with frequency locking in region Bl and is confined between the gray and the thick solid lines. The second polarization bistable region of B3 is confined between the dash and the thin solid lines where the polarization switching happens without frequency locking. The two bistable regions are connected at a detuning of 2 GHz, which coincides with the birefringence frequency splitting between the two VCSEL linear polarization modes. This means that when the master laser is biased at the frequency of the x VCSEL mode (the suppressed mode under the bias injection current) a dramatic change of dynamics occurs; from polarization switching with injection locking to polarization switching without locking. For larger positive or negative detunings, the switching power is larger, and moreover the switching power is larger for a negative than for a positive detuning value. This experimental feature agrees with theoretical results on a VCSEL rate equation model (Sciamanna and Panajotov 2005). It is noted that the widths of the injection locking regions S1 and S2 and of the bistability region B1 increase with the detuning. On the other hand, the width of the bistability region B3 remains approximately constant when changing the frequency detuning. This bistable region B3 is also strongly influenced by the locking of the first order linear polarization mode (S2). For small positive detunings ranging from about $0 \sim 10\,\text{GHz}$, complicated dynamics like wave mixing, subharmonic resonance, sustained limit cycle oscillation, period-doubling and chaotic regimes (C) are observed as shown in the inset in Fig. 8.24. The example shown here is the polarization dynamics of orthogonally injected VCSELs at the bias injection current above the polarization switching point. Similar but somewhat different dynamics can be found for the bias injection current below the polarization switching point (x mode oscillations) and y polarized optical injection (Sciamanna and Panajotov 2005, 2006)

8.4 Broad Area Lasers

8.4.1 Theoretical Model of Broad Area Lasers

The high power semiconductor laser is a promising laser device for various industrial applications of high-energy optical sources, since the power conversion efficiency from electricity to light in those semiconductor lasers is much higher than in other lasers (the efficiency is more than 50%). Such high power and high efficiency lasers can be used for light sources of laser welding, pumps for solid state lasers, and laser fusion. Currently, a high power semiconductor laser over 1 kW of output is deviced by stacking lasers as arrays. One of the technologies for high power semiconductor lasers is a broad-area laser that has a broad stripe width ($\sim$100 μm which is about twenty times larger than for ordinary edge-emitting semiconductor lasers). The broad-area semiconductor laser has a broad stripe width of the active region as its name suggests. Therefore, the effects of the carrier diffusion and the diffraction of light in the active region are essential for such a structure (Diehl 2000 and Gehrig and Hess 2003). Other than that, the broad-area semiconductor laser has the same structure as ordinary edge-emitting semiconductor lasers. Figure 8.25 is an example of the device structures. The thickness of the active layer is larger than that of ordinary edge-emitting semiconductor lasers, but the oscillation of the TE mode is usually expected. However, under special installation of the device structures such as stress-induced anisotropy for the device, a broad-area semiconductor laser may oscillate at the TM mode. The internal cavity length is of the same order as for the edge-emitting laser or several times larger than that. The longitudinal dimension is typically 1 mm. The output power of a broad-area laser is more than 100 mW.

Except for the advantage of high power operation, the qualities of the laser beam show rather poor performances. For example, broad-area semiconductor lasers usually operate at multimode both for the longitudinal and

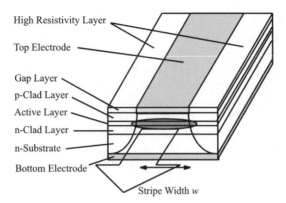

Fig. 8.25. Device structure of a broad-area semiconductor laser. The stripe width w of the active layer is as broad as $\sim$100 μm

transverse modes. The far-field pattern of a broad-area laser typically has a twin-peak. There exists a carrier hole burning effect in the active region along the stripe width at high bias injection current. The positions of the hole burning change and fluctuate with time and this gives rise to pulsating oscillations with pico-second and fast spatio-temporal filamentations (Hess et al. 1995 and Marciante and Agrawal 1998). Filamentation of broad-area semiconductor lasers, which shows zigzag motions of high intensity peaks along the internal cavity (typically the time size is several tens of pico-seconds and the spatial size several micron-meters), is one of the typical features of broad-area semiconductor lasers and it much deteriorates the laser performance. The broad-area semiconductor laser is also sensitive to external perturbations. In the following, we discuss the dynamics both without and with external perturbations.

The broad-area semiconductor laser itself is also an unstable device due to the spatial dependence in the laser oscillations (i.e., the spatial variation is an additional degree of freedom). Broad area lasers usually oscillate with multimode, however we assume a single longitudinal mode operation for simplicity. Starting from the Helmholtz equation for the complex laser field $E(x,t)$ (x is the coordinate perpendicular to the laser thickness in the active layer, i.e. the direction along the laser stripe width), the rate equation is calculated as (Rahman and Winful 1994, Merbach et al. 1995, and Levy and Hardy 1997)

$$\frac{\partial E(x,t)}{\partial t} = iD_e\frac{\partial^2 E(x,t)}{\partial x^2} + \frac{1}{2}(1 - i\alpha)G_n\{n(x,t) - n_{th}\}E(x,t) \qquad (8.30)$$

where $D_e = c/2k_0\eta^2$ is the diffraction coefficient of light (k_0 being the wavenumber in vacuum). The first term on the right hand side of the equation is the diffraction effect due to the broad active area. The diffusion effect must also be included in the rate equation for the carrier density and it is written by

$$\frac{\partial n(x,t)}{\partial t} = D_n \frac{\partial^2 n(x,t)}{\partial x^2} + \frac{J}{ed} - \frac{n(x,t)}{\tau_s} - G_n\{n(x,t) - n_0\}|E(x,t)|^2 \quad (8.31)$$

where D_n is the diffusion coefficient of the carrier and it is defined by $D_n = l_d^2/\tau_s$ (l_d is the diffusion length). In reality, the injection current is a function not only of time but also of the x coordinate. The dynamics of broad-area semiconductor lasers at solitary oscillations are numerically simulated from (8.30) and (8.31). Table 8.4 is an example of characteristic parameters of broad-area semiconductor lasers.

The field amplitude and carrier density in the active region of a broad-area semiconductor laser fluctuate in time and space. To analyze the internal local field and carrier density, we must take into account the internal field for the propagating and counter propagating waves and the polarization of the matter. Then, the electromagnetic field equation is numerically solved by using a finite difference time domain (FDTD) method (Adachihara et al. 1993, Hess and Kuhn 1996a, b, and Simmendinger et al. 1999a). Therefore, such a model is sometimes used. In the model, the field equation for the forward and backward propagations $E^+(x, z, t)$ and $E^-(x, z, t)$ along the z direction in the internal active region is given by

$$\pm\frac{\partial E^\pm}{\partial z} + \frac{\eta}{c}\frac{\partial E^\pm}{\partial t} = \frac{i}{2k}\frac{\partial^2 E^\pm}{\partial x^2} - \left(\frac{\alpha_s}{2} + i\gamma_w\right)E^\pm + \frac{i}{2}\frac{\Gamma(x)}{\eta^2\varepsilon_0 l}P_N^\pm \quad (8.32)$$

Table 8.4. Characteristic device parameters for broad-area semiconductor lasers at an oscillation wavelength of 780 nm

Symbol	Parameter	Value
G_n	gain coefficient	$2.00 \times 10^{-13}\,\mathrm{m^3\,s^{-1}}$
α	linewidth enhancement factor	3.00
r_1	front facet reflectivity	0.30
r_2	back facet reflectivity	0.95
n_{th}	carrier density at threshold	$5.11 \times 10^{24}\,\mathrm{m^{-3}}$
n_0	carrier density at transparency	$1.30 \times 10^{24}\,\mathrm{m^{-3}}$
τ_s	lifetime of carrier	$3.00\,\mathrm{ns}$
τ_{ph}	lifetime of photon	$1.88\,\mathrm{ps}$
τ_{in}	round trip time in laser cavity	$6.00\,\mathrm{ps}$
D_e	diffraction coefficient	$1.44\,\mathrm{m^2\,s^{-1}}$
D_n	carrier diffusion coefficient	$30\,\mathrm{cm^2 s^{-1}}$
l	cavity length	$500\,\mathrm{\mu m}$
w	stripe width	$50\,\mathrm{\mu m}$
d	thickness of active layer	$0.05\,\mathrm{\mu m}$

and the carrier density $n(x,z,t)$ reads

$$\frac{\partial n}{\partial t} = D_{\mathrm{n}} \left(\frac{\partial^2 n}{\partial x^2} + \frac{\partial^2 n}{\partial z^2} \right) + \frac{J(x,z)}{ed} - \frac{n}{\tau_{\mathrm{s}}} - G_{\mathrm{E,P}} \tag{8.33}$$

where l is the internal cavity length of the laser, α_{s} is the linear absorption term, γ_{w} is the parameter related to transverse and vertical variations of the refractive index due to the waveguide structure, and $\Gamma(x)$ is the confinement factor. $P_{\mathrm{N}}^{\pm}(x, z, t)$ is the nonlinear polarization of the matter accompanying the laser oscillation and is written by

$$P_{\mathrm{N}}^{\pm} = \frac{2}{V} \sum_k d_{\mathrm{cv}}(k) p_{\mathrm{r}}^{\pm}(k) \tag{8.34}$$

where $d_{\mathrm{cv}}(k)$ is the optical dipole matrix element and $p_{\mathrm{r}}^{\pm}(k)$ is the microscopic polarization function. The macroscopic generation rate $G_{\mathrm{E,P}}$ in (8.33) is given by

$$G_{\mathrm{E,P}} = -\chi'' \frac{\varepsilon_0}{2\hbar} (|E^+|^2 + |E^-|^2) + \left\{ \frac{-\mathrm{i}}{2\hbar} (E^+ P^{+*} - E^- P^{-*}) + c.c. \right\} \tag{8.35}$$

where χ'' is the imaginary part of the susceptibility. When the variables for the z direction in (8.32) and (8.33) change slowly in time, the equations reduce to the rate equations at the exit face of the laser given by (8.30) and (8.31).

8.4.2 Dynamics of Broad Area Semiconductor Lasers at Solitary Oscillations

In this section, the dynamics of broad-area semiconductor lasers at solitary oscillations are described. The L-I characteristic is the same as that for narrow-stripe edge-emitting semiconductor lasers, but the threshold current is much higher because of the broad stripe width and the high carrier injection rate. The laser threshold current is usually larger than $100\,\mathrm{mA}$. In the previous theoretical model, we assume a single mode oscillation for a broad-area semiconductor laser, however most of the actual broad-area semiconductor laser oscillates with multi-longitudinal-mode. Therefore, we must use multimode equations for the real laser model to compare with experiments. However, the single mode model can reproduce well the characteristics of broad-area semiconductor lasers. The following examples of the numerical simulations are the results for a single longitudinal mode assumption.

As discussed, the rate equations depend on the x coordinate and the spatial modes play a crucial role for the dynamics. Then, we cannot ignore the spatial dependence and must take into account the higher transverse modes. Further, the laser undergoes spatial and temporal complex dynamics due to the self-focusing effects induced by the hole burning of carriers and the diffraction of light. We will describe the dynamics later and, instead, we here discuss the time-averaged far-field profile of a laser oscillation. The output profile of broad-area semiconductor lasers has a significant wavefront distortion and the effect is remarkable for the laser of the gain-guided structure which is easy to fabricate. The far-field pattern of a broad-area semiconductor laser typically has a twin-peak profile. For multi-transverse-mode lasers, the beam quality factor is introduced to evaluate the beam quality. The beam quality factor M^2 of a far-field pattern for a laser is defined by (Hodgson and Weber 1997)

$$M^2 = \left(\frac{D_m \theta_m}{d_0 \theta_0} \right) \approx \left(\frac{D_m}{d_0} \right)^2 \tag{8.36}$$

where D_m and d_0 are the diameters of the ideal Gaussian beam and the observed beam, and θ_m and θ_0 are the divergence angles for the ideal and observed beams. The value of the beam quality factor M^2 is unity for the ideal beam, but the value for broad-area semiconductor lasers usually ranges from $10 \sim 50$ depending on the bias injection current and the stripe width.

Figure 8.26 shows a plot of far-field patterns of a broad-area semiconductor laser for a change of the bias injection current. The profile is spatially averaged. At a lower bias injection current, the laser profile has a single lobe, while the laser shows a typical twin-peak pattern for a higher injection current. The extent of the divergence of the beam in the far field is roughly determined by the average particle size of the filamentation. From this relation, the spatial size of the filamentation at the exit facet of the laser is given by (Hülsewede et al. 2001)

$$\sigma = \frac{4\lambda}{\pi\theta} \tag{8.37}$$

where θ is the diffraction angle at the far field plane. In Fig. 8.26, the divergence angle is $\theta = 0.17$ degrees and the corresponding size of the filamentation is estimated as 6 μm.

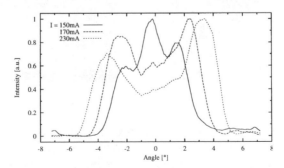

Fig. 8.26. Experimental far-field beam profile of a broad-area semiconductor laser. The laser oscillates at 780 nm and has a stripe width of 50 μm. The threshold current is 140 mA

A laser with a non-negligible spatial structure shows instabilities without any external perturbations. Another example is VCSELs, as we have already discussed. Next, we show a typical example of spatio-temporal dynamics in broad-area semiconductor lasers. Figure 8.27 is a filamentation observed in a near-field output of a broad-area semiconductor laser. We can see that bright spot particles moves back and forth in zigzag manner along the stripe width. This coil like pattern is called filamentation and it is a typical structure in broad-area semiconductor lasers (Fischer et al. 1996b and Burkhard et al. 1999). The width of migrating filaments is typically around 10 μm and it takes them about several pico-seconds to migrate from one edge of the active region to the other. Figure 8.27a is the numerical simulation for the experiment for Fig. 8.27b. Though the model is a single mode, the calculated filamentation is quite similar to the experimental one. Filamentation is universally observed not only for wide stripe lasers but also for semiconductor laser arrays.

The origin of dynamic filamentation in broad-area semiconductor lasers is not fully understood yet. However, the phenomena can be related to the effects of self-focusing, diffraction, and spatial hole burning, which depends on spatial carrier diffusion as the relevant physical mechanisms (Hess et al. 1995, and Hess and Kuhn 1996b). The self-focusing tends to guide high intensity regions resulting in a decrease of the optical gain. Thus, in the neighboring regions, the gain is higher. In addition, diffraction couples light into this neighboring region so that the spot of high intensity starts to migrate. At the edges of the active area, coupling via diffraction occurs only to one side, leading to a change of direction of migration. Figure 8.28 shows the intensity distribution of the internal cavity calculated from Maxwell-Bloch equations that include both the space dependence and the momentum dependence of

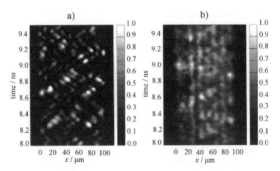

Fig. 8.27. Near-field pattern of filamentation in a broad-area laser. **a** Numerical simulation of filamentation. **b** Experimentally observed filamentation by streak camera. The bias injection current is $J = 2.0 J_{th}$. The laser has a stripe width of 100 μm and the oscillation wavelength is 814 nm. The parameters of the theoretical result are compatible with those of the experiment. The *horizontal axis* corresponds to the exit face of the active region and the *vertical axis* is the time evolution (after Fischer I, Hess O, Elsäßer W, Göbel E (1996b); © 1996 EDP Sciences)

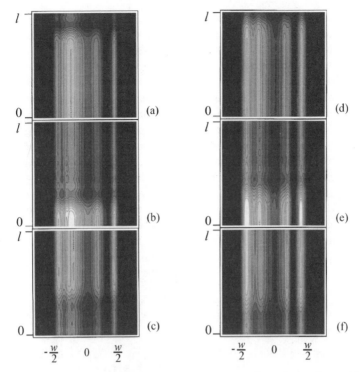

Fig. 8.28. Propagation of filamentary structures in a broad-area laser. The individual plots display snapshots showing the equi-intensity regions of the intracavity intensity. **a** At time $t = t_0$, **b** $t_0 + 1$ ps, **c** $t_0 + 2$ ps, **d** $t_0 + 2.8$ ps, **e** $t_0 + 3.5$ ps , and **f** $t_0 + 4.5$ ps. *Dark shading* corresponds to low intensity and *bright colors* to areas of high intensity. The out-coupling facet (mirror reflectivity of 0.33) is located at the lower edge of each square. The highly reflecting back-coupling mirror (mirror reflectivity of 0.99) is at the upper edge. The longitudinal extension corresponds to 250 μm; the total transverse width ($w = 50$ μm) is 70 μm (after Hess O, Kuhn T (1996b); © 1996 APS)

the charge carriers and the polarization of matter (Hess and Kuhn 1996b). The bottom of each plot is the front facet with a lower internal intensity reflectivity of 0.33 and the top is the back facet with a higher reflectivity of 0.99. Bright spots of filamentation move with the time evolutions. Other than broad-area semiconductor lasers with constant stripe width, various types of broad-area lasers have been proposed (Levy and Hardy 1997 and Fukushima 2000). To control and reduce the effect of filamentation, a flared laser having a tapered cavity has been used. In such lasers, the filamentation has been reduced but different complex spatio-temporal dynamics have been encountered.

We discussed the dependences of the laser dynamics on index- and gain-guide structures in ordinary edge-emitting semiconductor lasers in Sect. 3.7.1

and the differences between them. The dynamics strongly depend on the structures not only for solitary oscillations but also oscillations under external perturbations. However, the differences are reflected only to the parameter values in the rate equations and particular time-dependent dynamics only changes for the ranges of the parameter values, whether the laser is an index- or gain-guide structures. Nevertheless, lasers with gain-guide structure show unstable oscillations from dynamics point of view. On the other hand, broad-area semiconductor laser has a spatial structure along the stripe width (index- and gain-guided structures) and the differences between the structures give rise to large differences to the spatio-temporal dynamics. The dynamics of filamentations, which are the typical fast dynamics in broad-area semiconductor lasers, are strongly affected by the waveguiding structures.

Figure 8.29 shows the results of numerical simulations for filamentations in index- and gain-guide structures in broad-area semiconductor lasers, which have the same stripe width of $100\,\mu m$ and the same bias injection current of $1.5 J_{\mathrm{th}}$. In the numerical simulations, the same form of the rate equa-

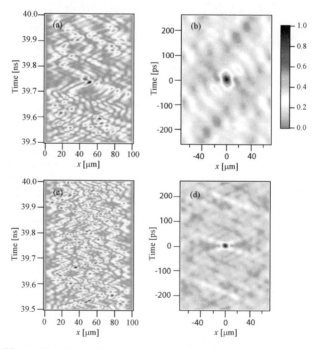

Fig. 8.29. Numerical simulations for near-field patterns in broad-area lasers (*left* column) and their spatio-temporal correlations (*right* column) at $J = 1.5 J_{\mathrm{th}}$. **a** and **b** Index-guide semiconductor laser, and **c** and **d** gain-guide semiconductor laser. Both lasers have the same stripe widths of $100\,\mu m$. Parabolic profiles for the gain, the refractive index, the injection current distribution along the stripe width are assumed in the gain-guide semiconductor laser

tions (8.30) and (8.31) is assumed, but, in the gain-guide laser, it is assumed that the gain, the refractive index, and the injection current distribution have appropriate parabolic spatial distributions along the stripe width. Figure 8.29a and b shows the near-field pattern at the exit face of the laser and its correlation function. The time and spatial sizes of the filaments in the index guide laser calculated from the correlation function are 4.1 μm and 27 ps, respectively, while those for the gain-guide laser are 2.9 μm and 12 ps, respectively. Namely, the gain-guide laser is less stable laser than the index-guide laser and the filaments strongly migrate back and forth along the active layer in the gain-guide laser as far as the values of the device parameters both for the two structures are the same. For the increase in the bias injection current, the spatio-temporal size of filaments shrinks and filaments shows strong zig-zag motions along the active layer, as a result, the both lasers become less stable.

This fluctuation of microscopic filaments reflects the performances of macroscopic laser oscillations. Figure 8.30 shows experimental results of near-field patterns for the time-averaged intensity distributions at the exit faces of index- and gain-guide broad-area semiconductor lasers. The horizontal axis points to the exit face of the laser and the intensity distribution at each bias

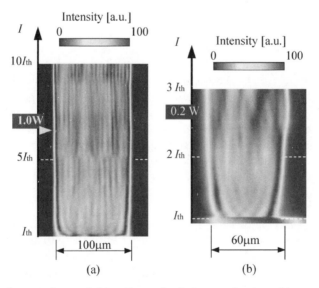

(a)   (b)

Fig. 8.30. Averaged near-field patterns for index- and gain-guide semiconductor lasers for the increase of the bias injection current. **a** Near-field pattern for an AlGaAs index-guide laser having a stripe width of 100 μm and oscillating at the wavelength of 808 nm. The threshold current is $I_{\mathrm{th}} = 160$ mA. **b** Near-field pattern for an AlGaInP gain-guide laser having a stripe width of 60 μm and oscillating at the wavelength of 642 nm. The threshold current is $I_{\mathrm{th}} = 191$ mA. Courtesy of SONY Cooperation

injection current is normalized. Each streak along the bias injection current corresponds to averaged filamentations. We can see different dynamics for several levels of the bias injection current. Although the two lasers have different stripe width, values of device parameters, and oscillation frequencies, we can recognize that the gain-guide laser is less stable laser (Asatsuma et al. 2006).

8.4.3 Feedback Effects in Broad Area Semiconductor Lasers

Instabilities in broad-area semiconductor lasers are enhanced by external perturbations. In this section, we present some instabilities and chaotic dynamics in broad-area semiconductor lasers subjected to optical feedback. The field equation in the presence of optical feedback is given by

$$\frac{\partial E(x,t)}{\partial t} = iD_e \frac{\partial^2 E(x,t)}{\partial x^2} + \frac{1}{2}(1 - i\alpha)G_n\{n(x,t) - n_{th}\}E(x,t)$$
$$+ \frac{\kappa}{\tau_{in}}E(x, t - \tau)\exp(i\omega_0\tau) \tag{8.38}$$

In the above equation, the feedback light is always returned to the original position in the active area, however the assumption may not always be true in experimental situations. The light is intentionally fed back to a different position to control the oscillation and beam profile. In that case, we must introduce the term for the space dependent optical feedback. The spatial coupling plays an important role in the laser dynamics and a locking of the laser oscillations can be expected. The beam quality inevitably deteriorate due to the broad stripe width, however fabrication of high power laser is at present the primary interest for the development for broad-area lasers and few studies have been reported for the enhancement of beam qualities. However, now the beam quality becomes the important issue for the applications of broad-area semiconductor lasers, for example, a light source for the second harmonic generation of solid state lasers and laser welding and cutting. Thus, a beam with good quality is expected. The beam quality is much improved by the introduction of optical feedback.

Dynamics similar to those of narrow-stripe edge-emitting semiconductor laser have been experimentally observed by optical feedback to broad-area semiconductor lasers. We show here one of the chaotic evolutions: the evolution of intermittent oscillations to regular chaotic states for the increase of the injection current. Figure 8.31 is an example of chaotic evolutions for the bias injection current. The threshold of the used laser is about 140 mA. Fig. 8.31a is the laser output power at the free running state just above the threshold. With the optical feedback, the reduction of the threshold is also observed in the broad-area semiconductor laser and the reduction rate is 13.9% for the external feedback rate of 6%. When the laser is biased at a low injection current, LFFs are observed. In the power recovery process

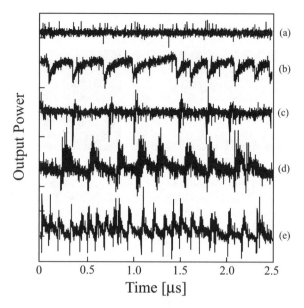

Fig. 8.31. Experimentally observed chaotic evolution for bias injection current in broad-area semiconductor laser. **a** Solitary oscillation at 140 mA. Optical feedback at bias injection currents **b** 150, **c** 162, **d** 170, and **e** 190 mA. The external cavity length is $L = 30$ cm and the external feedback strength is 6% in intensity. The laser used is the same as in Fig. 8.26

after the power dropout, the time scale of each step is also the same as the time calculated from the external cavity length. With the increase of the bias injection current, the frequency of LFFs changes and the inverse LFFs in which power jump-ups instead of power dropouts appear are observed at the bias injection current of 170 mA. Around this bias injection current, there is a kink of the L-I characteristic and the phase of the laser oscillation changes to a different state. At a further increase of the bias injection current, the laser behaves with normal chaotic oscillations, although the waveform still shows LFF-like oscillation (not fully chaotic oscillation in ordinary sense). Edge-emitting semiconductor lasers are very sensitive to the phase. Phase sensitivity also exists in broad-area semiconductor lasers subjected to optical feedback and the dynamics are much affected by the absolute phase of the external cavity (Martín-Regalado et al. 1996a, b). We return the subject of optical feedback in broad-area semiconductor lasers from the viewpoint of laser control in Chap. 10.

One of important issues for the practical applications of broad-area semiconductor lasers, such as laser cutting, is catastrophic optical damage(COD) induced by optical feedback from a target. In ordinary edge-emitting semiconductor lasers, catastrophic optical damage is also a serious problem when the laser is biased at a high injection current. The performance of the laser oscilla-

Exit Face   Near Field Pattern

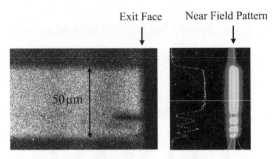

Fig. 8.32. Catastrophic optical damage (COD) induced by optical feedback in an index-guided broad-area semiconductor laser with a 50 μm stripe width and an oscillation wavelength of 808 nm. *Left*: top view of the front facet. COD can be seen in the lower part of the laser. *Right*: intensity profile of the cathode luminescence image at the front facet. Dips of light emissions corresponding to the front facet can be seen. Courtesy of SONY Cooperation

tions is significantly degraded by catastrophic optical damage and, worst case, the laser oscillation stops by the damage. The catastrophic optical damage is also critical problem in laser cutting using high power broad-area semiconductor lasers. Takiguchi et al. (2006) investigated the conditions for the occurrence of catastrophic optical damage in broad-area semiconductor lasers from a viewpoint of laser dynamics. Figure 8.32 shows an example of catastrophic optical damage observed in an AlGaAs index-guide broad-area semiconductor laser having a stripe width of 50 μm. Under filamentation oscillations in broad-area semiconductor laser, a large power is concentrated to a filament within a short time, and this effect together with a large optical feedback intensity may damage the laser. Thus, microscopic filamentations greatly affect the catastrophic optical damage, but the detailed study for catastrophic optical damage with the relation of laser dynamics has not been fully understood yet. The study is very important to prevent fatal catastrophic optical damage in broad-area semiconductor lasers as a practical issue.

8.5 Laser Arrays

Semiconductor laser arrays are also important devices for light sources with high power radiation. The laser may be composed of arrays of broad-area lasers to make an extremely high power laser device. However, here we assume that the arrays consist of ordinary edge-emitting lasers with narrow stripe width and consider the interaction among the laser elements. When the separation between the laser arrays is very small, each laser interferes and instability sometimes occurs in the total laser output. In a strict sense, we must consider all the effects of the diffraction and the carrier diffusion as already discussed in Sect. 8.4 (Münkel et al. 1996). However, we consider the situation that the coupling of lights among arrays is a dominant effect and that it is more important than those of the diffraction and the carrier diffusion. We also assume that the coupling only between the neighborhood lasers is strong, as is often the case. Thus, the rate equations for the field amplitude and the carrier density of the j-th element are given by (Winful and Rahman 1990 and Winful 1992)

$$\frac{dE_j(t)}{dt} = \frac{1}{2}(1 - i\alpha)G_n\{n_j(t) - n_{th}\}E_j(t)$$
$$- i\frac{\kappa_a}{\tau_{in}}\{E_{j+1}(t) + E_{j-1}(t)\} \tag{8.39}$$

$$\frac{dn_j(t)}{dt} = \frac{J}{ed} - \frac{n_j(t)}{\tau_s} - G_n\{n_j(t) - n_0\}|E_j(t)|^2 \tag{8.40}$$

where κ_a is the coupling ratio between the neighborhood laser elements. The spontaneous emission term is neglected in the above equations. For the numerical calculation of the rate equations, the number of laser arrays is $N + 1$ and the boundary condition is $E_0 = E_N = 0$. The rate equation for the field amplitude has the same form as the well-known equation of the coupled map lattice (CML). The CML shows typical spatio-temporal instabilities and chaos. Therefore, semiconductor laser arrays are essentially chaotic systems. Winful et al. (1992) investigated chaotic dynamics and synchronization of laser arrays based on this model.

As a different approach for the analysis of semiconductor laser arrays, the model of periodic carrier confinement and injection is proposed by extending the theory of broad-area semiconductor lasers (Merbach et al. 1995 and Martín-Regalado et al. 1996a, b). In multi-stripe laser arrays, the rate equations remain the same as (8.30) and (8.31). However, the laser is assumed to have a discrete multi-stripe structure along the x-direction as shown in Fig. 8.33. The confinement of the gain arises periodically in the active region. Then, we introduce a periodic confinement factor $\Gamma(x)$ in the wave-

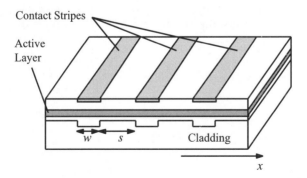

Fig. 8.33. Model of a multi-stripe semiconductor laser. Only three stripes are displayed

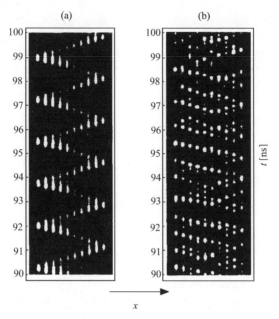

Fig. 8.34. Numerical plots of spatio-temporal output power of a ten-stripe laser array for two different injection currents. **a** $I = 34\,\text{mA}$: periodic state and **b** $I = 44\,\text{mA}$: chaotic state. The threshold current for each stripe is $I_{\text{th}} = 36\,\text{mA}$. The stripe width is $w = 5.0\,\mu\text{m}$ and the stripe separation is $s = 5.8\,\mu\text{m}$ (after Merbach D, Hess O, Herzel H, Schöll E (1995); © 1995 APS)

guide. Also, the bias injection current $J(x)$ is assumed as a periodic function with the same period. Using these assumptions, the dynamics of the multi-stripe laser are numerically investigated. For the laser arrays, filamentations are also observed. Figure 8.34 shows an example of numerically calculated filamentations (Merbach et al. 1995). The laser has a ten-stripe. We can see filamentations among laser arrays and the dynamics strongly depends on the bias injection current. In Chap. 10, we also return to the control of unstable operation of laser arrays subjected to optical feedback from the viewpoint of laser control.

9 Chaos Control and Applications

We cannot foresee the future of chaotic evolutions, since a small deviation of the initial condition in a nonlinear system results in a completely different solution of the system output. However, chaos can be controllable. Of course, a nonlinear system does not always show unstable oscillations. Or the system can be controlled and stabilized to a steady-state by appropriately shifting the parameters even when the system originally outputs irregular chaotic oscillations. In such control, the perturbations for the system may be large and the system is switched to another state from the original oscillation by the parameter shifts. However, the idea of chaos control is completely different from ordinary control methods. The perturbation for a nonlinear system is very small and little affects the state of the system. In chaos control, the system is controlled to a nearby unstable periodic or fixed orbit (saddle node point of the system). In this chapter, we discuss the method of chaos control and give some applications in semiconductor laser systems. We also demonstrate suppression of unstable oscillations in semiconductor lasers.

9.1 General Methods of Chaos Control

9.1.1 OGY Method

The physical models we encounter in real situations have more or less nonlinear characteristics, nevertheless the techniques of linearization for the systems are frequently applied and only their linear parts are used for convenience, especially in engineering. Therefore, chaos induced by nonlinear effects is usually an unfavorable phenomenon and we keep away from such irregular oscillations in practical applications. However, chaos is controllable and the first paper on chaos control was published by Ott, Grebogi, and Yorke in 1990. A nonlinear system includes various parameters and irregular oscillations of the output are generated under certain conditions in multi-dimensional parameter space. However, the operating point is not always embedded into the chaotic sea on every side. As usual, under a parameter condition, there is a possibility of the existence of unstable periodic orbits (UPO) close to the operating point of the system. Or luckily enough, a fixed point may exist near the point. Ott, Grebogi, and Yorke proposed an algorithm (OGY method) that applies appropriately estimated minute perturbations to an accessible system parameter to select and stabilize a certain nearby periodic orbit (unstable periodic orbit). Whether the method works well or fails depends on the availability of the possible nearby unstable saddle node points and the extent of the basin of unstable orbits. This idea indicates that a chaotic system can be turned into a system with multi-purpose flexibility, meaning that one can obtain various desired orbits in a simple system without dramatically modifying the configuration of the system. This method is called chaos control. The details of chaos control proposed by Ott, Grebagi, and Yorke can be found in Appendix A.3. Here, we briefly describe the idea of chaos control based on their proposal.

The application of the OGY method requires the full mathematical description of a nonlinear model. We need the attractors or the Poincaré map in advance to analyze and control the system. Based on this information, the parameter is perturbed by the mathematical method and the system is forced to fall down onto an unstable periodic orbit. Therefore, the OGY algorithm is difficult to apply to real experimental systems. The method comes from a rather mathematical basis and can only be applicable for experimental situations where one knows explicitly the exact parameter values in the dynamical system, since the parameter values are important for the calculations of unstable periodic orbits. Although the method is difficult to apply in actual situations of chaos control, it is modified and new techniques applicable to actual experimental systems are proposed by taking over the essence of the OGY method. It is noted that the idea of chaos control is a small perturbation to a nonlinear system and the original chaotic attractor is affected little by the perturbation. If the chaotic attractor is disturbed by the perturbation and the state is switched to another one, the method is not called chaos control in a strict sense. It may be a control of chaos, but not 'chaos control'.

9.1.2 Continuous Control Method

A fundamental problem existing in the OGY algorithm is the applicability for the control of high-dimensional chaotic systems. Although some attempts were made to adapt this technique to the experimental control of high dimensional dynamics, the requirements for the knowledge about the attractors and their calculations obstruct the application of the algorithm to the real time control of high-dimensional chaos. As alternative methods of chaos control for the applications to real systems, several chaos control techniques have been proposed. One of them is a continuous control proposed by Pyragas (1992). Figure 9.1 shows the schematic diagram for the control method. In the Pyragas method, which is called the continuous control method, a part of the output in a nonlinear system is detected with a delay τ_e that has an intrinsic time of the period of the chaotic attractor. The difference between the present $y(t)$ and delayed outputs $y(t-\tau_e)$ is fed back into the original sys-

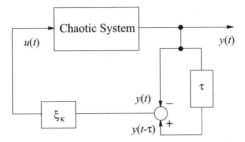

Fig. 9.1. Model of the continuous chaos control method. $y(t)$: output, τ: feedback delay, ξ_K: feedback gain, $u(t)$: feedback signal

tem. This characteristic time τ_e can be calculated either from the theoretical model or from the experimental estimation. The feedback signal is given by (Pyragas 1992, 1993, and 2001, Kittel et al. 1994, and Naumenko et al. 1998)

$$u(t) = \xi_\text{K}\{y(t - \tau_\text{e}) - y(t)\} \tag{9.1}$$

where ξ_K is an appropriate feedback gain. An example of the continuous control is given in Appendix A.3. As mentioned above, τ_e is the delay time and is chosen to be near or equal to the response time of the system. When the system is controlled, the amount of the feedback reduces to zero and the output is stabilized to a periodic or fixed state that corresponds to one of the unstable periodic orbits involved in the nonlinear system. The necessary condition in the continuous control method is only the delay time τ_e and the appropriate set of the feedback parameter enables the system to fall down to a periodic or a fixed state even when it originally outputs chaotic oscillations. It is easy to implement the method by using electronic circuits when the system response is not so fast. The method is also applicable to chaotic semiconductor laser systems.

9.1.3 Occasional Proportional Method

The other powerful method for the application in experimental systems is the occasional proportional feedback (OPF) method (Hunt 1991, Roy et al. 1992, and Liu and Ohtsubo 1994a). The OPF method modifies the OGY algorithm. The OPF method also perturbs one of the system control parameters by carefully feeding back a part of the output signal. It creates only small alterations of the attractor and pushes the system so as to stabilize it to the periodic orbit. Digital and analogue electronic circuits, such as comparator and sample/hold circuits, are required for the implementation of the method and periodic components of the attractor are extracted from the chaotic output of the system. Then, the system is stabilized to a periodic orbit by appropriately setting a synchronous signal. The seeding synchronous signal is estimated from the system parameters. Therefore, we require the information only for the characteristic time of the system, such as the delay time, in advance. However, we do not need the exact characteristic time but only a rough estimate of it. When the parameters are set within certain ranges of the appropriate values, a signal for the chaos control is autonomously output. The control signal is a pulse-like one much smaller than the chaotic oscillations and the level is also small enough for the assumption of minute perturbation for the system in chaos control. It is noted that the control signal is continuously generated in the OPF method even after the control has succeeded. This is different from the continuous control in which the control signal is eliminated after the success of the control.

Since the OPF technique is implemented using analogue and digital electronic circuits, the control can be performed very fast and is applicable to a variety of nonlinear systems. Different from the continuous control method, the electronic circuits used in the OPF method are not simple and the implementation may not be easy for a nonlinear system with fast response over nano-second oscillations. It is also pointed out that the OPF method is essentially a limited case of the OGY algorithm, when the contracting direction of the chaos attractor is infinite in strength. In laser systems, the controls of chaotic oscillations have been successfully performed based on the OPF method. The OPF method has been applied for stabilization and control for class B lasers with slow relaxation oscillations, such as solid state lasers, fiber lasers, and CO_2 lasers. In semiconductor lasers, the method is also applied to the control in optoelectronic hybrid systems. Such an example is demonstrated in Chap. 11.

9.1.4 Sinusoidal Modulation Method

Chaotic oscillations in semiconductor lasers related to the relaxation oscillations are usually on the order of nano-second and close attention must be paid to making the control circuits even when the control is possible. Here, we discuss a different method of chaos control that is very simple and suitable for practical use in systems with fast chaotic oscillations. As discussed in Chap. 4, there exist unstable periodic orbits (modes), for example, in a semiconductor laser with optical feedback. The accompanying frequencies for the modes are numerically calculated from the linear stability analysis for the system. Some of them have periodic solutions with a negative damping coefficient and they are the candidates for unstable periodic orbits (periodic solutions) close to the operating point of the system. Therefore, we think of the control of the system applying a sinusoidal modulation to one of the accessible parameters. Thus, one may expect to stabilize a chaotic oscillation to a periodic state by applying a small perturbation for a chaos parameter and a new category of chaos control is established as far as the perturbation is small enough and the original state is not so far from the periodic state (Liu et al. 1995 and Kikuchi et al. 1997).

In the nonlinear system to be controlled, we apply this method of sinus-oidal modulation to one of the accessible chaos parameters. Choosing an appropriate frequency f_0 from the mode analysis, the modulation with a small modulation index m for the parameter is given by

$$u(t) = u_o\{1 + m\sin(2\pi f_0 t)\} \qquad (9.2)$$

However, modulation frequencies derived from the linear stability analysis do not always work for the stabilization. Whether or not the control goes well depends on the extent of the basin of the attractor for the control frequency and the modulation depth. The method is an alternative one for the OPF control, since the periodic synchronous perturbation generated from the control circuits of the OPF method is replaced by a simple sinusoidal periodic signal. The sinusoidal modulation control is very easy to apply and the fast modulation is easily attained through the injection current modulation. Therefore, the method is frequently used in semiconductor laser systems.

9.2 Chaos Control in Semiconductor Lasers

9.2.1 Continuous Control

The continuous control method is suited for a system with fast chaotic oscillations in semiconductor lasers. In this section, we show an example of stabilization for chaotic oscillations induced by optical feedback in semiconductor lasers based on the continuous control method. The chaotic output from a semiconductor laser is once detected by a photodetector and electronically fed back into the bias injection current of the laser. The control signal is the difference between the detected intensities from the laser at present and before the time τ_e. Using the feedback gain ξ_K, the term for the injection current in the carrier density equation in(4.7) is replaced by

$$J = J_b[1 + \xi_K\{S(t - \tau_e) - S(t)\}] \qquad (9.3)$$

where J_b is the bias injection current without control. The second term in the parenthesis in the right-hand side equation is the control signal corresponding to $u(t)$ in (9.1). The delay time τ_e for the control is usually chosen to be the same as the optical feedback time. However, there is an optimum time delay for the control depending on the parameter conditions and it is not always the same value as the optical feedback time τ. The feedback in (9.3) looks like similar to the optoelectronic feedback discussed in Chap. 7, however it is not the same. After the continuous control has succeeded and the output of the laser is forced to a periodic or fixed oscillation, the control signal is eliminated ($S(t - \tau_e) = S(t)$) and the original state is little affected by the control.

Little work has been reported on the implementation of continuous chaos control in semiconductor laser systems. Instead of showing continuous chaos control, we here show a numerical example of dynamics states in chaotic semiconductor lasers subjected to optical feedback based on (9.3). Figure 9.2 shows the numerical example of bifurcation diagrams for a change of the control parameters (Turovets et al. 1997). Chaotic oscillations under this

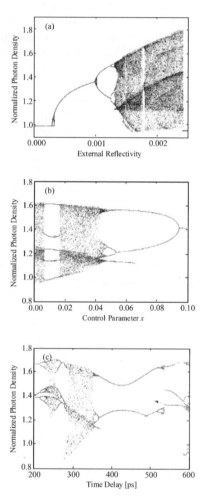

Fig. 9.2. Numerical continuous chaos control at $\xi_K = 0.05/S_0$ (S_0 is the average photon number) and $J_b = 2.0J_{th}$. The relaxation oscillation frequency at the bias injection current is about 3 GHz. **a** Bifurcation diagram for a change of external reflectivity without control at $\tau = 0.23$ ns. **b** Bifurcation diagram with chaos control for a change of feedback coefficient $x = \xi_K S_0$ at $r^2 = 0.0016$ and $\tau_e = 0.36$ ns. **c** Bifurcation diagram with control for a change of delay time τ_e at $x = \xi_K S_0 = 0.05$ and $r^2 = 0.0016$ (after Turovets SI, Dellunde J, Shore KA (1997); © 1997 OSA)

condition are not LFFs but irregular fast oscillations. Figure 9.2a shows the bifurcation diagram without feedback signal. We can see period-doubling chaotic bifurcation for the increase of the external reflectivity. Figure 9.2b shows a bifurcation diagram for the change of the feedback coefficient $x = \xi_K S_0$. For the increase of the feedback coefficient, we can see an inverse period-doubling bifurcation. The chaotic states are stabilized at periodic states or even at fixed states for the increase of the control parameter. The delay time of the optoelectronic feedback circuit is set to $\tau_e = 0.36$ ns and the delay is almost equal to the delay time in the optical feedback loop. The bifurcations are much dependent on the delay time τ_e. For the original chaotic state with $x = 0$, the laser output is stabilized to, for example, a periodic state by the change of the control parameter. However, the change may not be small and there exists a residual of the control signal in this case after the state is shifted to a stable oscillation. Figure 9.2c shows the bifurcation diagram for the change of the delay time τ_e at a fixed control parameter where the original state of the laser output is a period-2 oscillation.

In actual fact, the delay τ_e is electronically generated by an analogue circuit and the control signal is fed back into the bias injection current of the laser. Therefore, we must design a fast response circuit for fast laser oscillations. As discussed in Chap. 7, the finite response of the electronic circuits, including photodetector and amplifier, is always encountered besides of the setting of the delay τ_e. For real systems, we must take into account these effects. It is again noted that a perturbation for the chaos control must be very small and the system is scarcely affected by the control (Naumenko et al. 1998). Therefore, it is not easy to realize laser stabilization in the strict sense of chaos control. Nevertheless, the analysis of finding stable points in the bifurcation diagram gives rise to a good indication for the control of irregular oscillations of the system even if the control signal does not have small value and the residual of the control is not small.

9.2.2 Occasional Proportional Feedback Control

The OPF technique is effective for the control of rather slow response laser systems such as solid-state lasers, but it is not easy to apply the method for fast oscillating lasers. Therefore, the OPF control in chaotic semiconductor lasers has not been reported to date. Here we show an example of the system construction of OPF control in a chaotic semiconductor laser subjected to optical feedback. Figure 9.3 shows a schematic diagram of the control. A chaotic output from the laser is detected by a photodetector (PD) and is converted into a time dependent electric signal $x(t)$. A variable offset is added to $x(t)$ to bring it within a window of adjustable width. When the waveform transits within the window, the window comparator outputs a pulse, with the pulse width coincident with the length of transition. Meanwhile a synchronous signal is generated by a pulse generating circuit. The frequency of the synchronous signal is related to the delay and relaxation times of the

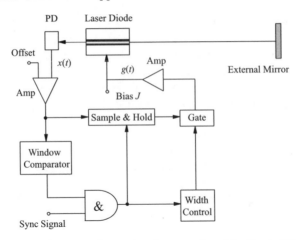

Fig. 9.3. Schematic diagram of occasional proportional feedback (OPF) control in a semiconductor laser with optical feedback

system. When the synchronous signal is coincident with the pulse from the window comparator, the sample/hold circuit is activated and acquires the waveform voltage. A gate with adjustable time width is employed to select only a part of the sampled signal. An amplifier with variable gain, offset, and polarity delivers the control signal $g(t)$ to the drive current of the semiconductor laser diode and perturbs the injection current.

The control signal $g(t)$ must be less than several percent of the bias injection current and it is a pulse like signal with a pulse width that is much shorter than the delay time of the optical feedback. By the control, the chaotic state is always pushed to a periodic or fixed orbit close to the initial point, but the control signal is continuously generated and never dies out. This is different from the result of the continuous control method. Various periodic states can be designed not only for chaotic oscillations but also for stable oscillation states by a small control signal generated by the OPF system adjusting the offset for the amplifier and the window comparator, or appropriately setting the synchronous signal of the AND gate. For design of the OPF control, we require the information about the delay in the nonlinear system in advance. Then, the synchronous signal is appropriately set in the control circuit.

9.2.3 Sinusoidal Modulation Control

The chaos control methods discussed above more or less require the detection of the chaotic signal, the processing of the post-detection signal, and feedback of it to the laser. Therefore, it is sometimes difficult to implement the methods for fast response nonlinear laser systems in practical applications. There is a simple way to realize chaos control suitable for systems with fast chaotic oscillations. We performed the linear stability analysis for the steady-

states in chaotic semiconductor laser systems in Sect. 4.2.2. The real parts
of the solutions for the characteristic equation derived from the linear stability analysis represent the damping rate of the oscillations and the imaginary
parts denote the accompanying frequencies. The frequencies obtained are the
candidates for periodic oscillations of unstable saddle node points which are
embedded into the system close to the initial operating point. In accordance
with this fact, the chaotic system can be controlled to a periodic or fixed
state by modulating the accessible parameter with one of the frequencies obtained by the linear stability analysis. The control method works indeed as
long as unstable periodic orbits are not far away from the operating point.
In a semiconductor laser with optical feedback, a sinusoidal injection current modulation is the easiest way to perform the control. The modulation is
applied to the bias injection current in (4.7) as

$$J = J_\mathrm{b}\{1 + m \, \sin(2\pi f_0 t)\} \tag{9.4}$$

where f_0 is the modulation frequency calculated from the linear stability
analysis and m is the modulation index with a small amplitude. There is an
allowable range for the parameter values of f_0 and m for successful control.
The robustness for the control depends on the extent of the attractor and
the basin of each possible linear mode.

In the following, the sinusoidal modulation control to the injection current in semiconductor lasers with optical feedback is described. Figure 9.4
shows the plot of the mode distribution calculated from the linear stability
analysis for the system (Liu et al. 1995 and Kikuchi et al. 1997). Under the

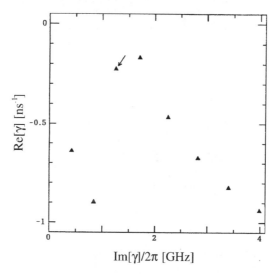

Fig. 9.4. Linear mode distribution for chaotic output in a semiconductor laser with
optical feedback at $J = 1.1J_\mathrm{th}$, $r = 1.5\%$, and $L = 25.5$ cm. Under the operation
condition, the laser originally shows chaotic oscillation

operating condition, the laser shows chaotic oscillations. All the calculated modes within this region have negative real values, therefore the calculated modes are candidates for unstable periodic solutions. However, the stability of the modes and the robustness for the control are different from one mode to another and must be investigated by using bifurcation diagrams for the control parameters. We focus on the mode indicated by the arrow in the figure and use this mode as the sinusoidal modulation control.

Figure 9.5 shows an example of chaos control using a sinusoidal modulation to the injection current in a semiconductor laser with optical feedback. The sinusoidal modulation control is performed by modulating the injection current with the modulation frequency of $f_0 = 1.251\,\mathrm{GHz}$ and the modulation index of $m = 2.1\%$ of the bias injection current. The modulation index is sufficiently small for satisfying the assumption of little effect for the original chaotic state to the system. The chaotic waveform in Fig. 9.5a is controlled to a period-1 oscillation by the modulation of one of the mode frequencies as

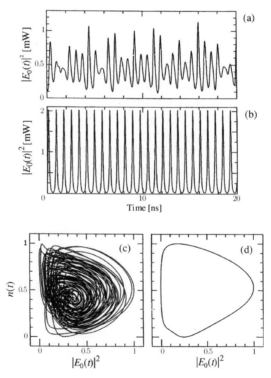

Fig. 9.5. Sinusoidal modulation control using the parameters in Fig. 9.4. **a** Original chaotic oscillation of laser output without control. **b** Controlled period-1 oscillation using the mode frequency indicated by arrow in Fig. 9.4. The modulation depth is 2.1% of the bias injection current. **c** and **d** are attractors corresponding to **a** and **b**, respectively

shown in Fig. 9.5b. Figures 9.5c and d are the attractors for Figs. 9.5a and b, respectively, in the phase space of the laser output power and the carrier density. The chaotic attractor in Fig. 9.5c is controlled to the periodic state in Fig. 9.5d. The robustness of the method for parameter variations is an important issue. In this method, there is a finite modulation range of the parameters for effective control, for example, successful control is achieved within the range of several tens to a hundred MHz centered at the exact mode frequency. However, the extent of the attractor after the control is slightly deformed by the modulation. One of the important issues of chaos control is the response time after the control signal is switched on. The time required for successful control using sinusoidal modulation has been studied by Uchida et al. (1999). According to their results, the time required for reaching the stabilization has statistical distributions for each trial, but the characteristic time is roughly ten times the laser relaxation oscillation (equivalently ten times the typical time scale of chaotic oscillations).

9.2.4 Optical Control

Injection current modulation is not the only technique to modulate accessible parameters for the implementation of chaos control in semiconductor lasers. As an alternative modulation method for chaos control, we here show an example of the introduction of an extra mirror in the feedback loop to control chaotic oscillations in a semiconductor laser (Liu et al. 1996, Liu and Ohtsubo 1999, and Ruiz-Oliveras and Pisarchik 2006). Figure 9.6 shows the optical chaos control system using double external mirrors. One of the mirrors is the external mirror that gives rise to chaotic oscillations in the semiconductor laser and the second mirror is used for the control. In this system, a beat signal induced by the mixing of lights from the two mirrors plays the same role for a sinusoidal modulation as the bias injection current modulation in the previous section. We have already performed the linear stability analysis for a single external mirror in Sect. 4.2. In the same manner as in the single mirror case, the linear stability analysis is applied to the rate equations with double

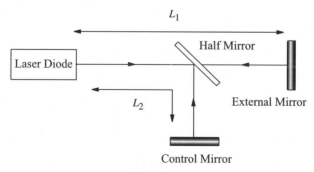

Fig. 9.6. Schematic diagram of optical sinusoidal modulation control

external mirrors. Depending on the feedback times from the two mirrors, the solutions for the optical oscillation frequency are determined from the following equation:

$$\omega_{\rm s} - \omega_{\rm th} = -\frac{\kappa_1}{\tau_{\rm in}}\{\alpha\cos(\omega_{\rm s}\tau_1) + \sin(\omega_{\rm s}\tau_1)\}$$
$$-\frac{\kappa_2}{\tau_{\rm in}}\{\alpha\cos(\omega_{\rm s}\tau_2) + \sin(\omega_{\rm s}\tau_2)\} \tag{9.5}$$

where $\omega_{\rm s}$ is the angular frequency for the steady-state oscillation, and subscripts 1 and 2 are for the first (external) and second (control) mirrors, respectively. The solution of the above equation is added to the steady-state phase and this additional term has the effect of a sinusoidal modulation for the control of the system.

In the strict sense of chaos control, the additional optical feedback term in (9.5) must be small not to disturb the original feedback system. However, the system is usually affected strongly by the second feedback mirror and the system might not fall into a nearby unstable periodic orbit by the control, but it is pulled apart from an accompanying anti-mode by the interference and is forced to another periodic or fixed state far from the original state. The system after the control may have a different attractor from the original one and, therefore, such a control is not categorized into chaos control with small perturbation in the sense of OGY control. However, the technique is sometimes effective for stabilizing chaotic irregular oscillations. In the following, we demonstrate the control method for stabilizing LFF oscillations in semiconductor lasers with optical feedback as an example of optical control. When LFFs occur, the oscillation mode always accompanies an unstable anti-mode as explained in Sect. 5.3. The oscillation mode is a stable mode close to or at the maximum gain mode of the laser oscillation. During chaotic itinerary due to the perturbation, the state is once trapped into the anti-mode and this induces low-frequency fluctuations (LFFs). If the system is separated apart from the anti-mode by the control, the laser is forced to a stable oscillation. The steady-state solutions for the photon number and the carrier density in the double-cavity system are written by

$$A_{\rm s}^2 = \frac{J/ed - n_{\rm s}/\tau_{\rm s}}{G_{\rm n}(n_{\rm s} - n_0)} \tag{9.6}$$

$$n_{\rm s} = n_{\rm th} - \frac{2\kappa_1}{\tau_{\rm in}G_{\rm n}}\cos(\omega_{\rm s}\tau_1) - \frac{2\kappa_2}{\tau_{\rm in}G_{\rm n}}\cos(\omega_{\rm s}\tau_2) \tag{9.7}$$

To realize the control, we make the condition where the accompanying anti-mode of only the first mirror is separated from the maximum gain mode by the introduction of the second mirror according to the steady-state solutions in (9.6) and (9.7).

Figure 9.7 shows the phase diagram for the carrier density $n(t)$ and the phase $\psi(t) = \phi(t) - \phi(t - \tau) + \omega_0\tau$ ($\omega_0 = \omega_{\rm th}$) in the presence of the control

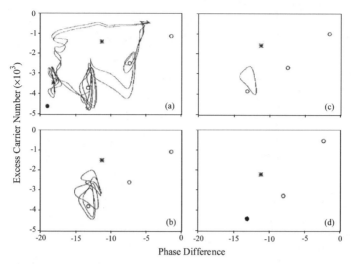

Fig. 9.7. Phase trajectories observed in space $[\phi(t) - \phi(t - \tau_1) + \omega_0\tau, n(t)]$. *Asterisks:* anti-modes. *Open circles:* unstable external cavity modes. *Filled circles:* stable external cavity modes. **a** LFF for $\kappa_1\tau_{\mathrm{ph}}/\tau_{\mathrm{in}} = 0$. **b** Quasi-periodic behavior with frequency locking for $\kappa_1\tau_{\mathrm{ph}}/\tau_{\mathrm{in}} = 0.5\times10^{-3}$. **c** Limit cycle corresponding to periodic behavior for $\kappa_1\tau_{\mathrm{ph}}/\tau_{\mathrm{in}} = 1.0\times10^{-3}$. **d** Stationary behavior for $\kappa_1\tau_{\mathrm{ph}}/\tau_{\mathrm{in}} = 4.0\times10^{-3}$. The laser locks into a stable external cavity mode. The feedback ratio of the first mirror is $\kappa_1\tau_{\mathrm{ph}}/\tau_{\mathrm{in}} = 4.6 \times 10^{-3}$ in all cases. The delay times are $\tau_1 = 1\,\mathrm{ns}$ and $\tau_2 = 0.2\,\mathrm{ns}$ (after Rosigter F, Mégret P, Deparis O, Blondel M, Erneux T (1999); © 1999 OSA)

(Rogister et al. 1999). Without the second mirror in Fig. 9.7a, the laser is trapped in one of the anti-modes (denoted by asterisks) close to the maximum gain mode (black dot) and shows chaotic oscillation (chaotic trajectory of LFFs in the figure). By the introduction of the second mirror in Fig. 9.7b, anti-modes are separated far away from the oscillation modes (open circles) and the laser does not show catastrophic oscillations like LFFs. In Fig. 9.7b, the attractor still looks chaotic but with a compact attractor and the laser oscillates at a quasi-periodic state near one of the external modes. With the increase of the feedback level from the second mirror, the laser is stabilized at a periodic state with an attractor of a close loop in Fig. 9.7c and finally forced to a fixed state at the maximum gain mode (black dot) in Fig. 9.7d. Figure 9.8 is an experimental result for the control of LFFs (Rogister et al. 2000). The figure shows optical spectra observed by a Fabry-Perot spectrometer for the change of the external feedback level. The feedback strength of the second mirror is estimated from the threshold reduction rate by the feedback and it is written on the right-hand side of the figure. Without the control by the second mirror, the coherence of the laser oscillation collapses due to LFF oscillations as shown in Fig. 9.8a. In Figs. 9.8b and f, the laser output power shows

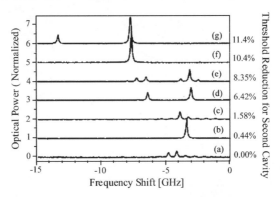

Fig. 9.8. Experimental optical spectra measured in double cavity configuration as a function of threshold reduction owing to the feedback strength of the second mirror. **a** LFFs in the absence of second feedback. The increase of the feedback strength of the second mirror leads to stabilization (**b** and **f**) interspersed with unstable regions (**c–e** and **g**). The optical spectra have been normalized with respect to the maximum of trace **b**. Single mode oscillation of the laser is guaranteed by a grating mirror put into the path of the optical feedback loop for the second mirror. The feedback fraction is estimated by the reduction rate of the threshold current due to the feedback. The feedback strength of the first mirror is fixed at 7.1% counted by the threshold reduction in all cases. The bias injection current is just above the threshold and the lengths of the external mirrors are $L_1 = 21\,\mathrm{cm}$ and $L_2 = 19\,\mathrm{cm}$ (after Rogister F, Sukow DW, Gavrielides A, Mégret P, Deparis O, Blondel M (2000); © 2000 OSA)

fixed constant states and the laser oscillates at single mode. The laser shows a period-1 state with the frequency of the relaxation oscillation in Fig. 9.8d. Thus, the laser is controlled to a periodic or fixed state by appropriately choosing the feedback strength of the second mirror. The noise for the lower frequency component is also much suppressed by the control (Rogister et al. 1999).

9.3 Controlling Chaos and Noise Suppression

9.3.1 Noise Suppression by Sinusoidal Modulation

The main noise source in free running semiconductor lasers is the spontaneous emission of photons in laser media. Noises in semiconductor lasers are greatly enhanced by optical feedback. The detailed definition and descriptions for the noise characteristics in semiconductor lasers can be found in the book by Petermann (1988). The general description of noise effects and relative intensity noise (RIN) in semiconductor lasers are also given in Sect. 3.5 in this book. We again define the RIN in relation to the noise of the optical

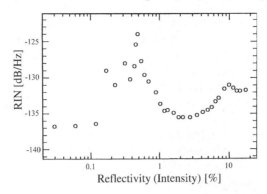

Fig. 9.9. Experimentally obtained RIN in a semiconductor laser with optical feedback. The laser is a Fabry-Perot type with an oscillation wavelength of 780 nm at the maximum power of 5 mW. The bias injection current is $J = 1.5J_{\text{th}}$ and the external cavity length is $L = 20$ cm

power δS to the mean power $\langle S \rangle$ according to (see (3.87))

$$\text{RIN} = \frac{\langle \delta S^2 \rangle}{\langle S \rangle^2}, \tag{9.8}$$

where the optical output power from the laser is defined by $S(t) = \langle S \rangle + \delta S(t)$. In actuality, the feedback induced irregular intensity fluctuation is not a noise but a chaotic fluctuation. However, the effects of the phenomena are similar to noises in free running semiconductor lasers. Therefore, we use the same notation for the feedback induced irregular fluctuations.

Figure 9.9 shows an experimentally obtained RIN in the presence of optical feedback in a Fabry-Perot semiconductor laser. The RIN is plotted for the external feedback strength (intensity), however the actual feedback to the active layer is much less than the externally calculated reflectivity due to the losses and absorption of light through the optical components and the diffraction loss of the focusing lens. At an external reflectivity around 0.1% (compatible with 0.001% in theoretical calculation), the RIN has almost the same level as the solitary oscillation. Above this level, the RIN abruptly increases and reaches the maximum value at the external reflectivity of 1% (the corresponding theoretical feedback strength is about 0.1 to 0.01%). Without optical feedback, RIN for ordinary edge-emitting semiconductor lasers under solitary oscillations is less than -140 dB/Hz. However, in the presence of optical feedback, the RIN increases to much higher than -125 dB/Hz in the feedback regions of III $\sim$ V (chaotic and coherence collapse regimes) as discussed in Sect. 4.1. In these regions, we can observe the broadening of the oscillation linewidth, chaotic behaviors of the laser output, and the coherence collapse state of the laser. A laser with a RIN above -125 dB/Hz cannot be used as a light source for optical data storage systems because of the increase of bit-rate errors. Even though the feedback level in this regime is very small, it affects the performance of the laser operation. To know the dynamics in this feedback regions are also very important from the viewpoint of practical applications, such as the use for optical data storage systems, since the returned light from the disk surface in those systems is almost of the same order. On the other hand, the RIN decreases with further increase of the feedback level. The noise characteristics are dependent not only on the reflectivity of the external mirror, but also on other system parameters such as the bias injection current and the external mirror position.

Unstable oscillation of semiconductor lasers subjected to optical feedback was stabilized by the introduction of a sinusoidal modulation to the injection current as shown in the previous section. Figure 9.10 shows the numerical result of the noise suppression by the sinusoidal modulation method (Kikuchi et al. 1997). The RIN of the solitary laser is $\sim -140\,\mathrm{dB/Hz}$ in the lower frequency region (solid line). A frequency peak at about 3 GHz is the relaxation oscillation component. In the presence of optical feedback, the noise is extremely enhanced and it is about $-120\,\mathrm{dB/Hz}$ in the lower frequency region as is shown as the broken line in the figure. The noise level exceeds the allowed criterion for a light source of the optical data storage system. The dotted line shows the result of the control. One of the mode frequencies (in this case, it is 2.38 GHz) is chosen as a control frequency and the laser is modulated by this frequency through the injection current modulation. By the modulation, the laser shows synchronous oscillation (period-1) and the RIN in the lower frequency region is reduced to $-130\,\mathrm{dB/Hz}$. The modulation amplitude is small and $m = 0.15$. The noise level attained by the control is enough for the requirement of a light source for the optical data storage system.

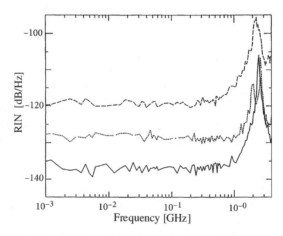

Fig. 9.10. RIN for sinusoidal modulation control based on numerical simulation. *Solid line:* RIN of solitary oscillation. *Broken line:* RIN for optical feedback without control. *Dotted line:* RIN for sinusoidal control. The modulation frequency is 2.38 GHz and the modulation index is 0.15

The modulation index of $m = 0.15$ used in Fig. 9.10 may be still larger from the viewpoint of chaos control. The modulation more or less affects the laser oscillation and a closer investigation shows deviations of the chaos attractor from the original one. The modulation index must be less than about 0.01 for the ideal chaos control. Nevertheless, the laser is stabilized by a rather small signal compared with ordinary forced sinusoidal control without optimized frequency. In optical data storage systems, the suppression of optical feedback noise is an important issue. The light is reflected from a disk surface and is returned into the laser active layer. The feedback light induces much noise in the laser and causes serious problems for the performance of the data read-out. In actual optical data storage systems, a high frequency injection current modulation on the order of several hundreds MHz to one GHz has been employed to suppress feedback-induced noises (Arimoto et al. 1986 and Gray et al. 1993). In such optical data storage systems, the modulation index over $m = 1.0$ was frequently used. The modulation depth is much larger than that of the chaos control and a laser is sometimes brought below the threshold by the modulation.

It is known from the empirical basis that there is a best modulation frequency for each optical data storage system. From the viewpoint of chaos control in semiconductor laser systems, the high frequency modulation technique is closely related to the sinusoidal modulation control in chaos control, in which the chaotic orbit is stabilized to a periodic oscillation, though the modulation depth is much larger than the expected chaos control. The strong modulation technique used for optical compact disk systems is considered as a forced oscillation to the light source. However, the method does not always work for every selected modulation frequency. As we have discussed in Chap. 7, strong modulation to a semiconductor laser sometimes gives rise to unstable chaotic oscillations. Therefore, the optimized modulation frequency has some relation to the sinusoidal modulation control of chaos, even though the modulation frequency is selected on the empirical basis. Recently, a self-pulsation laser was used for light sources of DVD in optical data storage systems. The chaos control algorithm introduced here may give us important information for the design of such devices and systems. The essence of chaos control is that the control does not change the original dynamics of the nonlinear system. However, the original dynamics may be changed due to a very small but non-negligible modulation amplitude. In that case, the idea of chaos control is still effective for the control of an existing unstable periodic orbit as far as the modulation is small.

To decide the optimum frequency in the sinusoidal modulation control, we must establish the model for a real system and estimate the frequency. However, it is not easy to obtain all the parameter values of the system in advance. Nevertheless, we can guess the frequencies. The relaxation oscillation frequency of the laser is one of the candidates for the optimum frequency. Another one is the external cavity frequency and its higher harmonics. From the linear stability analysis, the mode frequencies of the optical feedback system are not always equal to the exact external cavity frequencies, but there exist mode frequencies close to the external cavity frequency. As has already been discussed, there is a certain extent for the tolerance of the modulation frequency for a successful chaos control. In reality, which frequency is the best for the control and suppression of noises must be examined and tested by using each possible frequency. Noise induced by optical feedback is much suppressed by the best selection of the modulation frequency and, thus, the laser is stabilized with a small modulation power.

9.3.2 Stability and Instability of LFFs by Injection Current Modulation

Low-frequency fluctuation (LFF) is one of the typical instabilities of semi-conductor lasers with optical feedback. Even when the laser seems to be oscillated at a stable state under optical feedback, the laser becomes un-stable by the injection current modulation and shows LFF oscillations for certain ranges of the modulation frequency. However, the laser shows stable oscillations for other frequency ranges. On the contrary, the laser oscillating at the LFF state due to optical feedback can be stabilized by the injection current modulation with appropriate frequencies. There exist typical time scales of the laser oscillation in optical feedback systems and the external mode is usually very close to one of the linear modes derived from the linear stability analysis. Here, we show the control of LFFs by the injection current modulation with a frequency close to the external cavity mode. For a detuned modulation from the external mode frequency, the laser shows LFFs under certain parameter conditions. However, the laser output with LFFs is experi-mentally stabilized to a synchronous oscillation by a modulation frequency close to the external mode (Takiguchi et al. 1998 and 1999b). The modulation frequencies are not exactly equal to the external mode frequency calculated from the external cavity length. In the experimental study by Taguchi et al. (1998, 1999), LFF oscillations from a semiconductor laser subjected to op-tical feedback were suppressed by the injection current modulation with the resonance frequency with small modulation index (about $-10\,$dB). By the modulation, the laser oscillation is locked to the control frequency, but the locking range is very narrow. For example, the modulation frequency for the stable oscillation is several hundreds of MHz at the external cavity length of the order of ten centimeters, but the range of the modulation frequency for stable oscillations is at most several MHz. Stabilization of LFFs has also been reproduced by numerical simulations. The stabilization of LFFs by the injection current modulation is considered to be the same technique as the sinusoidal modulation control discussed above.

Figure 9.11 shows the stabilization of LFFs by the injection current modu-lation (Takiguchi et al. 2002). Figure 9.11a is an example of LFFs by the injection current modulation with the frequency far away from the external cavity frequency. The laser is originally a single mode laser, but it oscillates

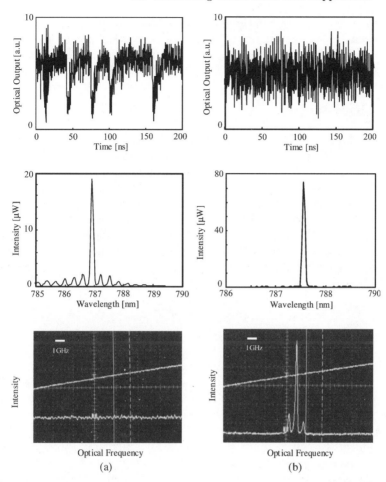

Fig. 9.11. Experimental results of modulation induced LFFs and stabilization in a semiconductor laser with optical feedback. The external cavity length is $L = 27$ cm. Top to bottom: waveform of LFFs, optical spectrum observed by a spectrum analyzer, and optical spectrum observed by a Fabry-Perot spectrometer. **a** LFFs for a modulation frequency of 390 MHz and a modulation depth of -4 dBm. **b** Stabilized oscillation for a modulation frequency of 548 MHz and a modulation depth of -4 dBm. The external cavity frequency is 556 MHz

at multimode and the coherence of the laser is completely destroyed, as is observed from the spectrum in a Fabry-Perot spectrometer. On the other hand, the laser is stabilized and oscillates at a single mode for the frequency close to the external cavity mode (note that it is not exactly equal to the external cavity frequency). The laser oscillates at single mode and also recovers its coherence as can be seen from the optical spectra both of the spectrum analyzer and the Fabry-Perot spectrometer. The small spectral subpeaks appeared at the main mode in the Fabry-Perot spectrometer correspond to the components of the modulation. The laser is also stabilized by the injection current modulation with frequencies of the higher harmonics of the fundamental frequency of the external cavity mode.

9.3.3 Chaos Targeting

We have discussed the optical control method in Sect. 9.2.4 in which the laser oscillation was fixed to the maximum gain mode by the introduction of the second mirror. A similar technique can be applied to stabilize feedback induced chaotic oscillations so as to lower the state onto the maximum gain mode by adjusting the optical phase. The control method is called dynamic targeting (Wieland et al. 1997 and Hohl and Gavrielides 1998). The dynamic targeting is generally realized by compensating the parameter values to shift and stabilize the laser oscillation when the system becomes an unstable state. From (4.10), the condition of the stability, i.e., the maximum gain mode condition is given by

$$\omega_0 \tau = \frac{\kappa_1}{\tau_{\text{in}}} \tau \alpha \;\; \text{mod} \; 2\pi \tag{9.9}$$

When the system does not satisfy this condition, the laser becomes unstable and shows LFFs and chaotic oscillations. Therefore, we could adjust the parameters to stabilize the laser when the system deviates from this condition. The laser is forced to a fixed state by controlling the bias injection current to satisfy the condition of (9.9), since the optical frequency ν_0 can be changed by the bias injection current according to the relation of (5.1).

Figure 9.12 is an example of the results for dynamic targeting control (Hohl and Gavrielides 1998). The bifurcation diagram in Fig. 9.12a is a calculated bifurcation cascade of a semiconductor laser subjected to optical feedback at the pump of $S = \frac{2\tau_{\text{ph}}}{G_n \tau_s}\left(\frac{J}{ed} - \frac{n_{\text{th}}}{\tau_s}\right) = 0.001$ for the change of the feedback fraction $\kappa \tau_{\text{ph}}/\tau_{\text{in}}$, where κ is the feedback parameter defined previously. For the change of the optical feedback level, the laser shows a bifurcation cascade from stable state, period-doubling, to chaotic oscillations. The dashed curve denotes the stabilized laser intensity when the feedback phase is adjusted by means of the optical frequency ν_0 as the feedback strength is increased. For the adjustment of the phase x, the rate equation of the field is written by

$$\frac{dE(t)}{dt} = \frac{1}{2}(1 - i\alpha)G_n\{n(t) - n_{\text{th}}\}E(t)$$
$$+ \frac{\kappa}{\tau_{\text{in}}}E(t - \tau)\exp\{i(\omega_0 - x)\tau\} \tag{9.10}$$

By the extra phase $x\tau$ in (9.10), the laser oscillation is fixed to the maximum gain mode. In actuality, the laser output decreases due to the decrease of the bias injection current by the adjustment. The laser output decreases for

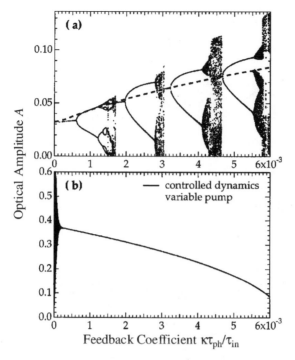

Fig. 9.12. a Computed bifurcation cascade for a semiconductor laser with optical feedback pumped 0.001 above threshold. The *dashed curve* shows the stabilized laser intensity as the feedback phase is adjusted by means of ω_0 as the feedback strength is increased. **b** Stabilization of laser intensity by varying the injection current by $\Delta S = 0.17\Delta\omega_0$ (corresponding to 1.1 GHz/mA). The intensity is stabilized and falls off as the injection current is decreased. The parameter conditions are $L = 15\,\mathrm{cm}$, $\omega_0\tau = -1.45\,\mathrm{rad}$, and $\alpha = 4$. The feedback ratio here is defined by $\kappa\tau_{\mathrm{ph}}/\tau_{\mathrm{in}}$. The conversion relation from the injection current to the optical frequency $\Delta J = 0.085(\tau_{\mathrm{s}}/\tau_{\mathrm{ph}})G_n ed\Delta\omega_0$ is used (after Hohl A, Gavrielides A (1998); © 1998 OSA)

the increase of the feedback strength but stabilizes to a fixed state as shown in Fig. 9.12b. Thus, the system under irregular oscillations can always be pulled back by dynamic targeting control. It is again noted that the dynamic targeting is not categorized as chaos control in the sense of the OGY method, since the system is always changed to a different chaotic state by dynamic targeting.

10 Stabilization of Semiconductor Lasers

Chaos and instabilities are not only the effects in semiconductor lasers with external perturbations. Semiconductor lasers are also strongly stabilized by external perturbations under appropriate parameter conditions. Optical injection from a different laser is a typical example of laser stabilizations. Other examples are weak or strong optical feedback, phase-conjugate optical feedback, grating optical feedback, and optoelectronic feedback. The longitudinal and transverse modes, frequency, power, and polarizations of semiconductor lasers are stabilized by external perturbations. Stabilizations are especially important in newly developed semiconductor lasers (VCSELs, broad-area semiconductor lasers, laser arrays, and so on), since these lasers involve instabilities even in their solitary oscillations. In this chapter, stabilization and control of semiconductor lasers are discussed based on rather simple configurations of external perturbations. Other than these examples, we present laser stabilizations using photonic structures and quantum-dot structures for newly developed lasers. The control methods introduced here may not be chaos control discussed in the previous chapter, but they are closely related to the ideas of dynamic and chaos controls in semiconductor lasers.

10.1 Linewidth Narrowing by Optical Feedback

10.1.1 Linewidth Narrowing by Strong Optical Feedback

The spectral line of edge-emitting semiconductor laser usually has a width of several to several tens of MHz even its solitary oscillation, which is 100 times larger than ordinary lasers. Such broad spectral linewidth mainly originates from the existence of a non-zero linewidth enhancement factor known as the α parameter, which is discussed in Sect. 3.5.5. For applications for coherent optical communications and coherent optical measurements, it is essential to use frequency-stabilized and narrow-linewidth light sources. The linewidth of a solitary edge-emitting semiconductor laser is determined by the device parameters and given by (3.102) as $\Delta \nu = R_{\mathrm{sp}}(1 + \alpha^2)/(4\pi S)$. Therefore, we can reduce the linewidth by appropriate designs for a semiconductor laser; the use of semiconductor materials with small α parameter, lower spontaneous emission rate, and higher bias injection current (higher photon rate). As a different technique, the linewidth of semiconductor lasers can be reduced by weak optical feedback as discussed in Sect. 5.1.3. The linewidth narrowing induced by optical feedback is realized within the ranges of feedback regimes II $\sim$ III and the linewidth is reduced by a factor of two or three from that of the solitary laser. However, with the increase of optical feedback level, the semiconductor laser is destabilized by the feedback and shows instabilities in feedback regime IV. As a result, the linewidth is much broadened of the order of GHz and the laser finally reaches coherence collapse state. On the contrary, further increase of the optical feedback level over several percents in amplitude, the laser is again stabilized by strong optical feedback and the linewidth is greatly reduced compared with the solitary oscillation linewidth. This effect can be used for laser stabilization. Many studies for linewidth narrowing induced by strong optical feedback in semiconductor lasers have been reported previously (Fleming and Mooradian 1981b, Patzak et al. 1983, Wyatt and Devlin 1983, Chraplyvy et al. 1986, Kazarinov and Henry 1987, and Tromborg et al. 1987). By using a complex feedback system and careful tuning of the system, Stoehr et al. (2006) achieved ultra-narrow linewidth control as small as 1 Hz. They used a complex system consisting of a semiconductor laser locked in a single stage to a stable high-finesse reference cavity. The phase locking technique with the combination of optical feedback and injection were frequently used for realizing ultra-narrow linewidth and the method is promising for applications in optical communications. However, we discuss here rather simple optical feedback systems for linewidth narrowing in semiconductor lasers.

First, we discuss the linewidth narrowing in semiconductor lasers with strong optical feedback (regime V) from a plain reflector. For narrowing the linewidth of a semiconductor laser, the combination of a low front facet reflectivity and a high back facet reflectivity, which are achieved by anti-reflection (AR) coating techniques, is frequently used to perform the effective reduction of the linewidth. Strong optical feedback is applied to such an AR coated semiconductor laser. Therefore, we assume different internal reflectivities for the front and back facets as r_1 and r_2, respectively. Then the composite reflectivity at the front facet together with the external reflector, which was already discussed in (4.25), can be written by

$$r_{\text{eff}} = \frac{r_2 + r \exp(i\omega_0\tau)}{1 + r_2 r \exp(i\omega_0\tau)} \tag{10.1}$$

For a small external reflectivity r, (10.1) reduces to (4.26) corresponding to a weak optical feedback case. According to (4.32), the linewidth in the presence of optical feedback is given by

$$\Delta\nu_{\text{ex}} = \frac{\Delta\nu}{F^2} = \frac{R_{\text{sp}}(1+\alpha^2)}{4\pi S_{\text{s}} F^2} \tag{10.2}$$

where S_{s} represents the steady-state solution for the laser intensity in the presence of optical feedback. The linewidth narrowing factor F is calculated from (Tromborg et al. 1987)

$$F = 1 + \frac{\alpha}{\tau_{\text{in}}} Re \left[\frac{d \ln r_{\text{eff}}}{d\omega_0} \right] - \frac{1}{\tau_{\text{in}}} \text{Im} \left[\frac{d \ln r_{\text{eff}}}{d\omega_0} \right] \qquad (10.3)$$

The above equations can be applied to general cases of optical feedback level and the factor F reduces (4.33) for a weak optical feedback in regimes II and III. However, for strong optical feedback, we must use the reflectivity r_{eff} in (10.1). It is noted that the above relations are only valid for stable oscillations of semiconductor lasers and they are not applicable to chaotic oscillations in feedback regimes III to IV. As discussed above, a low front facet reflectivity is sometimes used for the linewidth reduction. In the limit of $r_2 = 0$, the composite reflectivity reads $r_{\text{eff}} = r \exp(i\omega_0 \tau)$ and the resulting linewidth is given by

$$\Delta\nu_{\text{ex,min}} = \frac{\Delta\nu}{\left(1 + \frac{\tau}{\tau_{\text{in}}}\right)^2} \qquad (10.4)$$

For $\tau/\tau_{\text{in}} \gg 1$, the above equation is approximated as

$$\Delta\nu_{\text{ex,min}} = \frac{\Delta\nu}{\left(\frac{\tau}{\tau_{\text{in}}}\right)^2} \qquad (10.5)$$

For example, the linewidth of a solitary semiconductor laser can be reduced to 1/1000 by appropriately choosing the optical feedback level, the external cavity length, and the front facet reflectivity.

Figure 10.1 shows the instance of linewidth narrowing by strong external optical feedback (Atzak et al. 1983). The solid lines represents the numerical results for the linewdith as a function of R_2 for different external reflectivities of $R = r^2$ normalized at that for $R_2 = r_2^2 = 0$. The laser assumed is an InGaAsP laser with an oscillation wavelength of $1.5\,\mu\text{m}$ and the internal cavity length of $190\,\mu\text{m}$. The external cavity length is set to be $20\,\text{cm}$. The minimum linewidth of $\Delta\nu_{\text{ex,min}}$ is $2\,\text{kHz}$. The solid circle in the figure is the experimental result for a strong optical feedback in the reference (Wyatt and Devlin 1983). The experimental conditions of the used laser system are $R_2 = 0.04$ and $R = 0.04$. The reduced linewidth of $10\,\text{kHz}$ in the experiment is well coincident with the theoretical expectation. The linewidth of the solitary laser is $\sim10\,\text{MHz}$ and the linewidth is reduced at a factor of three by the strong optical feedback. As another example, Chraplyvy et al. (1986) also conducted an experiment for linewidth reduction by strong optical feedback for an InGaAsP DFB laser with an oscillation wavelength of $1.5\,\mu\text{m}$ and internal cavity length of $250\,\mu\text{m}$. Under the conditions of the internal reflectivity of 0.04, the external cavity length of $18\,\text{cm}$, and the external reflectivity of 0.16, they obtained a linewidth of $40\,\text{kHz}$ for the original oscillation linewidth of $\sim10\,\text{MHz}$. The reduction of the linewidth by a factor of three is typical in this method.

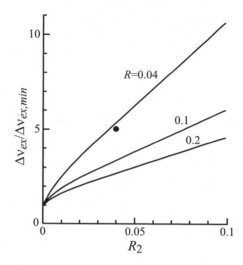

Fig. 10.1. Numerical simulations for normalized linewidth of a semiconductor laser as a function of the facet reflectivity $R_2 = r_2^2$. $\Delta\nu_{\mathrm{ex,min}} = 2$ kHz is the minimum linewidth for $R_1 = r_1^2 = 0.31$, $R_2 = r_2^2 = 0.04$, and $R = r^2 = 0.04$. *Solid circle* is the experimental result (Patzak E, Sugimura A, Satio S, Mukai T, Olesen H (1983); © IEEE 1983)

10.1.2 Linewidth Narrowing by Grating Feedback

Grating optical feedback is used for the purpose of the selection of longitudinal modes and the stabilization for the selected mode in semiconductor lasers. We have already introduced the dynamic properties induced by grating optical feedback in semiconductor lasers in Sect. 5.5. In the previous discussion, we presented the theoretical background of the dynamics of grating feedback and showed some numerical results. Here, we show some experimental results for grating feedback in semiconductor lasers from a viewpoint of linewidth narrowing. The essential effects in grating feedback are the same as those of optical feedback from a plain mirror. However, grating feedback enables desirable frequency selection for the laser oscillations, side-mode suppressions, and lowering the frequency jitter. Figure 10.2 shows the example of linewidth narrowing by grating feedback in a semiconductor laser for different three external cavity lengths of $L = 25$, 50, and 100 cm (Olesen et al. 1983). The used laser is a single-mode InGaAsP semiconductor laser of cleaved facets with the oscillation wavelength of 1.55 µm and the internal cavity length of 310 µm. The laser stably oscillates in a single mode less than the bias injection current of $1.8 J_{\mathrm{th}}$, but it oscillates at a multi-mode over that injection current. The linewidth of the free-running laser is rather large and it is 45 MHz. The grating mirror is a diffraction grating of 600 grooves/mm and the blaze wavelength of 1.25 µm, arranged in a Littrow configuration. The external cavity length (external phase) is finely tuned over a few micron-meter using a piezo-

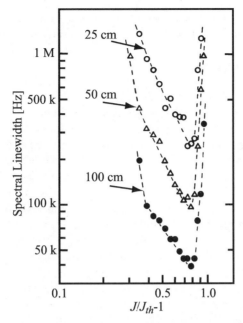

Fig. 10.2. Linewidth reduction for bias injection current in a semiconductor laser with grating optical feedback for the external cavity lengths of 25, 50, and 100 cm (after Olesen H, Saito S, Mukai T, Saitoh T, Mikami O (1993); © JSAP 1983)

electric element in the optical feedback loop. The estimated optical feedback fraction is around 10^{-4}, so that the feedback corresponds to weak to moderate optical feedback. In Fig. 10.2, the linewidth inversely proportional to the bias injection current, which is consistent with the relation in (10.2). The linewidth is also reduced by the increase in the external cavity length in accordance with the increase of the C parameter. Over the bias injection current of $1.8J_{\rm th}$, the linewidth of the laser rapidly increases due to multi-mode oscillations. In this example, the minimum linewidth of 40 kHz is achieved by the grating optical feedback under the conditions of the external cavity length of $L = 100$ cm and the bias injection current at $1.79J_{\rm th}$. Except for the conventional plain grating mirror, fiber Bragg grating and volume holographic grating are used as alternative feedback mirrors (Naumenko et al. 2003a and Ewald et al. 2005). Ewald et al. (2005) demonstrated linewidth reduction up to 7 kHz using optical feedback from a volume holographic grating for a no special AR-coated semiconductor laser operating at the oscillation wavelength of 852 nm.

10.1.3 Linewidth Narrowing
by Phase-Conjugate Optical Feedback

Phase-conjugate optical feedback discussed in Chaps. 4 and 5 can be also used as the linewidth control of semiconductor lasers. For phase-conjugate mirrors, they are categorized into fast or slow response natures with respect to the response time of semiconductor lasers (mainly carrier response time of semiconductor lasers). The dynamics of respective phase-conjugate effects are different with each other. However, in either case, a spatial phase-conjugate wave is generated from a phase-conjugate mirror, so that the method has the advantages of self-alignment and aberration correction through the optical components. It is expected that these properties will result in good coupling back into the laser and give rise to the active use for the laser stabilization. In the strict sense, instantaneous phase-conjugate mirror is not available and we must more or less consider the effects of finite response of phase-conjugate mirror. Therefore, slow response phase-conjugate mirror, such as $BaTiO_3$ photorefractive crystal, is used for the purpose of alignment free optical feedback due to the generation of only a spatial phase-conjugate wave. Many studies for the stabilization of semiconductor lasers with phase-conjugate optical feedback were reported (Vahala et al. 1986, Kürz and Mukai 1996, Kürz et al. 1996, Liby and Statman 1996, and Vainio 2006).

First, we discuss the linewidth narrowing induced by fast response phase-conjugate mirror in semiconductor lasers. Kürz et al. (Kürz and Mukai 1996, and Kürz et al. 1996) employed a broad-area semiconductor laser as a fast phase-conjugate device. Figure 10.3 shows the experimental setup of the system. An AlGaAs broad-area semiconductor, which is biased at less than but very close to the threshold, is pumped by a frequency-stabilized tunable semiconductor laser and a probe beam from a semiconductor laser to be stabilized is injected to the broad-area laser. The phase-conjugated wave is generated from the broad-area laser and it is fed back into the probe laser. The frequency from the pump laser is appropriately chosen and it is stabilized less than the oscillation linewidth of 20 kHz. The front and back facet reflectivities of the

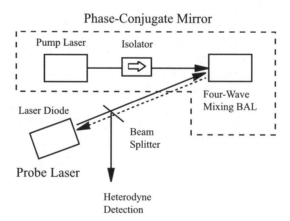

Fig. 10.3. Experimental setup for stabilizing semiconductor laser with phase-conjugate optical feedback from a four-wave mixing broad-area semiconductor laser (BAL)

broad-area laser are 0.05 and 0.95, respectively, to generate easily a phase-conjugate wave. Since the stripe width of the broad-area laser is wide and is 50 μm, the probe beam can be injected at a non-zero angle with respect to the pump beam, which causes the emission of amplified pump, probe, and phase-conjugate signals in distinct spatial direction. The pump beam, back reflected beam from the back facet, and probe beam conform to a four-wave mixing configuration and the phase-conjugate wave is generated back to the laser to be controlled. Using the broad-area semiconductor laser as a phase-conjugate material, the external incidence angle of 5.2° is possible in this configuration, which corresponds to the internal angle of 1.4° between the pump and probe beams. The response time is dependent on the carrier lifetime and the diffusion constant of the carrier density grating inside the broad-area laser. The response time of the phase-conjugate mirror is estimated around 1 ns. The probe laser, which is the stabilized laser, is an uncoated single-stripe semiconductor laser with the oscillation wavelength of 830.6 nm. The free-running laser has a linewidth of 5 MHz at the bias injection current with the mode suppression ratio of 30 dB.

Figure 10.4 shows the experimental result of the linewidth narrowing. The broad-area semiconductor laser is biased at 98% of the threshold and as small as less than 1% of the probe laser power is injected to the broad-area laser, giving a phase-conjugate reflectivity of 6%. The observed phase-conjugate feedback to the probe laser is about 1 μW (6×10^{-5} of its output power). Figure 10.4 is the power spectrum measured by a delayed self-heterodyne technique. The observed spectral linewidth is 25 kHz, which is close to the instrument resolution of 14 kHz in this measurement. The strong side-mode suppression of 50 dB is also visible in the figure, thus not only the linewidth narrowing but also the reduction of the side mode suppression ratio are attained by the phase-conjugate optical feedback. The dependence of the side-mode frequency separation $c/4L$ ($L = 2.29$ m corresponding to the side-mode

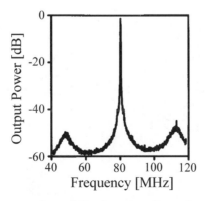

Fig. 10.4. Power spectrum of a stabilized semiconductor laser measured with a delayed self-heterodyne technique (after Kürz P, Mukai T (1996); © OSA 1996)

frequency separation of 32.7 MHz) confirmed that the origin of the spectral narrowing comes not from conventional optical feedback effect (for the conventional case, this would be $c/2L$) but phase-conjugate optical feedback effect. In spite of such long external cavity length, the laser is stabilized more than for hours due to the phase-conjugate optical feedback.

Contrary to the case of fast phase-conjugate optical feedback, the effects of spatial phase-conjugation are only dominant for the dynamics of semiconductor lasers with a slow response phase-conjugate reflector. Therefore, the linewidth of semiconductor laser with slow response phase-conjugate optical feedback is written by the similar form as the case of conventional optical feedback as

$$\Delta\nu_{\mathrm{pc}} = \frac{\Delta\nu}{\left\{1 + C\cos(\phi_{\mathrm{pc}} + \tan^{-1}\alpha)\right\}^2} \qquad (10.6)$$

where $\Delta\nu$ is again the oscillation linewidth for a solitary semiconductor laser. We assume a weak phase-conjugate optical feedback in the above equation. In conventional optical feedback from a plain reflector, the phase ϕ_{pc} in (10.6) is replaced by $\omega_{\mathrm{s}}\tau$, where ω_{s} is the steady-state value of the oscillation frequency and τ is the round-trip time of light in the feedback loop. Therefore, the linewidth from a conventional optical feedback shows a periodic change for the increase or decrease of the external mirror position. However, the phase-conjugate feedback is phase insensitive as has already been discussed in Chaps. 4 and 5, even for a slow response phase-conjugate mirror. The phase is determined by the boundary conditions of the laser and the device characteristics of the phase-conjugate mirror. We can assume the constant phase as integer multiple of 2π, i.e., $\phi_{\mathrm{pc}} = 2m\pi$. Then the linewidth monotonically decreases for the increase of the length of the phase-conjugate mirror, since the C parameter is simply proportional to the delay time τ. For $\phi_{\mathrm{pc}} = 2m\pi$, the reduced linewidth is given by

$$\Delta\nu_{\mathrm{pc}} = \frac{\Delta\nu}{\left(1 + \kappa\frac{\tau}{\tau_{\mathrm{in}}}\right)^2} \qquad (10.7)$$

In actual fact, the reflected light at the phase-conjugate mirror has a constant phase shift, so that we must take into account this effect. When the phase shift satisfies the condition as $\phi_{\mathrm{pc}} + \tan^{-1}\alpha = 2m\pi$, one obtains the minimum spectral linewidth given by

$$\Delta\nu_{\mathrm{pc}} = \frac{\Delta\nu}{(1 + C)^2} \qquad (10.8)$$

Several experimental studies for linewidth narrowing in semiconductor lasers with slow response phase-conjugate optical feedback have been reported (Vahala et al. 1986, Liby and Statman 1996, and Vainio 2006). For example, using an uncoated single-mode semiconductor laser with optical feedback from a photorefractive BaTiO$_3$ mirror, the linewidth is reduced one-tenth of that of the free-running laser (Liby and Statman 1996). In this example, the system has a four-wave mixing configuration and the power reflectivity of 0.2% is used. The linewidth reduction is not only the effect of phase-conjugate optical feedback. Another important aspect is the elimination of frequency jitter of the laser oscillations. The fact comes from the slow response of the photorefractive mirror. Any jitter from the central lasing frequency does not have enough time to write a new holographic grating. Only that central frequency is coupled back into the laser. Due to static gratings compared with the dynamics of the laser fluctuation, long term frequency stability is achieved. Vaino (2006) attained the linewidth reduction as small as 25 kHz in an anti-reflection coated semiconductor laser with strong phase-conjugate optical feedback from a Ce-doped BaTiO$_3$ photorefractive mirror. Since the front facet of the internal reflectivity is as very small as 10^{-4} in this case, strong feedback is realized with relatively small phase-conjugate signal from a self-pumped phase-conjugate configuration. In this experiment, a long life frequency stability of 1.5×10^{-8} during 100 s is observed. Thus, not only the linewidth reduction, but also stable center oscillation frequency, is attained even by slow response phase-conjugate optical feedback.

10.1.4 Linewidth Narrowing by Resonant Optical Feedback

Resonant optical feedback is one of the methods for stabilization and linewidth narrowing in semiconductor lasers (Laurent et al. 1989 and Iannelli et al. 1993). For example, the linewidth of an AlGaAs semiconductor laser oscillating at 780 nm is stabilized to 20 Hz by a resonant optical feedback using a spectral line from a rubidium atomic vapor (Hashimoto and Ohtsu 1987). Another example, which is frequently used, is optical feedback from a Fabry-Perot resonator. Here, we introduce the stabilization of the spectral line in semiconductor lasers with resonant optical feedback from a Fabry-Perot interferometer. Figure 10.5 shows the geometry of the optical feedback locking system from a Fabry-Perot resonator. The beam from a semiconductor laser is sent through a tilted confocal Fabry-Perot interferometer to avoid the optical feedback effect of the direct reflected beam. Only the transmission-like beam (type II beam shown in the figure) is coupled with the light of the internal laser cavity.

For a single mode semiconductor laser under weak optical feedback from a Fabry-Perot resonator, the rate equations for the complex field is written by (Laurent et al. 1989)

$$
\begin{aligned}
\frac{\mathrm{d}E(t)}{\mathrm{d}t} &= \frac{1}{2}(1 - \mathrm{i}\alpha)G_{\mathrm{n}}\left\{n(t) - n_{\mathrm{th}}\right\}E(t) \\
&+ \sum_{m=0}^{\infty} \frac{\kappa_m}{\tau_{\mathrm{in}}} E(t - \tau_m) \exp\left\{\mathrm{i}\omega_0 \tau_m - \mathrm{i}\phi(t - \tau_m)\right\}
\end{aligned}
\tag{10.9}
$$

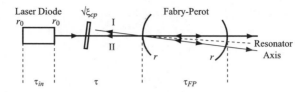

Fig. 10.5. Schematic of optical feedback locking system from a Fabry-Perot resonator. To obtain a transmission type (II) beam for the optical feedback, the resonator axis is slightly tilted, thus avoiding a reflection type (I) beam into the laser cavity

where the extra feedback term is the effect of multiple interferences of the beam in the Fabry-Perot resonator. Here, the m-th feedback coefficient is given by

$$\kappa_m = \sqrt{\xi_{\mathrm{cp}}} \frac{1 - r_0^2}{r_0} r(1 - r^2) r^{4m} \tag{10.10}$$

where ξ_{cp} denotes the power mode coupling factor, $\tau_m = \tau + (2m + 1)\tau_{\mathrm{FP}}$ is the roundtrip delay time between the laser and the Fabry-Perot resonator for each reflection from the cavity (τ_{FP} being the roundtrip time in the Fabry-Perot cavity), r_0 is the facet reflectivity of the semiconductor laser, and r is the mirror reflectivity or the Fabry-Perot resonator.

From a steady-state analysis similar to that in Sect. 4.2, we obtain the solution for the laser oscillation frequency as

$$\omega_{\mathrm{th}}\tau = \omega_{\mathrm{s}}\tau$$
$$+ C_{\mathrm{FP}} \frac{\sin\left\{\omega_{\mathrm{s}}(\tau + \tau_{\mathrm{FP}}) + \tan^{-1}\alpha\right\} - r^4 \sin\left\{\omega_{\mathrm{s}}(\tau - \tau_{\mathrm{FP}}) + \tan^{-1}\alpha\right\}}{1 + F_{\mathrm{FP}}^2 \sin^2(\omega\tau_{\mathrm{FP}})} \tag{10.11}$$

where the modified C parameter for resonant optical feedback from a Fabry-Perot resonator is given by

$$C_{\mathrm{FP}} = \frac{\sqrt{1 + \alpha^2}}{\tau_{\mathrm{in}}} \tau \sqrt{\xi_{\mathrm{FP}}} \frac{1 - r_0^2}{r_0} \frac{r(1 - r^2)}{1 - r^4} \tag{10.12}$$

and the factor F_{FP} is defined by

$$F_{\mathrm{FP}} = \frac{2r^2}{1 - r^4} \tag{10.13}$$

For a high-finesse Fabry-Perot $r \sim 1$, the modified C parameter is given by

$$C_{\mathrm{FP}} = \frac{\sqrt{1 + \alpha^2}}{\tau_{\mathrm{in}}} \tau \sqrt{\xi_{\mathrm{FP}}} \frac{1}{2} \frac{F_{\mathrm{FP}}}{F_0} \tag{10.14}$$

where $F_{FP} = \pi r^2 (1 - r^4)$ and $F_0 = \pi r_0 (1 - r_0^2)$ are the finesses of the Fabry-Perot and laser cavities, respectively. Employing a similar derivation of the linewidth for a solitary laser discussed in Sect. 3.5.5, the linewidth of the laser oscillation under optical feedback from a Fabry-Perot resonator is calculated as

$$\Delta\nu_{FP} = \frac{\Delta\nu}{(1 + 2C_{FP}\frac{\tau_{FP}}{\tau})^2} \quad (10.15)$$

where $\Delta\nu$ is the linewidth of the free-running laser given by (3.102). For a large optical feedback $C_{FP}\tau_{FP}/\tau \gg 1$, the linewidth can be simply written by

$$\Delta\nu_{FP} = \frac{\Delta\nu_0}{\xi_{FP}\left(\frac{\tau_{FP}}{\tau}\frac{F_{FP}}{F_0}\right)^2} \quad (10.16)$$

where $\Delta\nu_0 = R_{sp}/(4\pi S)$ is the Schawlow-Townes linewidth. The form of the linewidth is quite similar to (10.5), but it is important to note that this expression no longer contains the linewidth enhancement factor α. This means that the confocal Fabry-Perot cavity imposes the laser frequency and thereby can be interpreted as a decoupling of the phase noise to the amplitude noise.

Employing a resonant optical feedback from a Fabry-Perot interferometer, Laurent et al. (1989) demonstrated the reduction of the linewidth from 20 MHz of a free-running semiconductor laser to 4 kHz in a single mode GaAlAs semiconductor laser. They used a GaAlAs Fabry-Perot semiconductor laser with the oscillation wavelength of 850 nm. The conditions for the linewidth narrowing are as follows; the free-spectral-range (FSR) of 375 MHz, the finesse of the Fabry-Perot of $\sim$100, the cavity length of the Fabry-Perot resonator of 200 mm, and $\tau \approx 2\tau_{FP}$. Several studies for linewidth narrowing of the order of kHz have been reported in semiconductor lasers with Fabry-Perot optical feedback (Dahmani et al. 1987). The method of the resonant Fabry-Perot optical feedback provides not only narrowing the laser's linewidth but also the stabilization of the center-oscillation-frequency of the laser.

10.2 Linewidth Narrowing by Optoelectronic Feedback

In the applications of linewidth narrowing in semiconductor lasers, the method of optoelectronic feedback has the advantages of high stability, reproducibility, flexibility, and controllability over the optical feedback methods, since the feedback is negative and the noise and fluctuation induced by photons from the laser are averaged out due to the slower response of carriers than the photon lifetime. Also, the feedback loop can be designed accurately through a computer simulation, as is the case when designing conventional analogue feedback electronic circuits. Here, we discuss linewidth narrowing

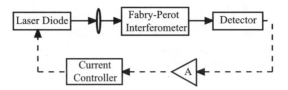

Fig. 10.6. Schematic diagram of optoelectronic feedback system for stabilization of semiconductor laser. The *solid line* corresponds to optical path and the *broken line* to electronic connection

in semiconductor lasers with negative optoelectronic feedback. Figure 10.6 shows the schematic diagram of the feedback system. The light from the laser passes through a frequency-selected Fabry-Perot resonator and detected by a photodtector. The current signal is amplified and is negatively fed back into the bias injection current of the laser. The stability of this feedback system strongly depends on the noise performance of the detector. Further, the detected signal by the photodetector fluctuates due to the fluctuation of the laser frequency, $\delta\nu(t)$, induced by the spontaneous emission in the cavity. The fluctuation is proportional to the laser output power P and the fluctuation of the transmittance of the Fabry-Perot interferometer $\delta T_{\mathrm{FP}}(\nu)$, namely, $\delta P = P\delta T_{\mathrm{FP}}(\nu)$. The stability of feedback is also affected by the noise performance of the photodetector and the linewidth is determined by taking into account these factors.

The variation of the minimum detectable power δP_{min} of the photodetector is defined from the noise theory for optical detectors (Ohtsu and Kotajima 1985). The detectors concerned here are solid-state detectors. From the theory, we can derive the relation of the signal-to-noise ratio (SNR) for a FM noise detection, $\mathrm{S/N} = \mathrm{d}P/\mathrm{d}P_{\mathrm{min}}$. For the SNR of $\mathrm{S/N} = 1$, we define the minimum detectable frequency fluctuation for $\delta\nu(t)$ as $\delta\nu_{\mathrm{min}}$, and consider that the square of $\delta\nu_{\mathrm{min}}$ corresponds to the magnitude of the Allan variance for the minimum detectable FM noise. Then, we obtain the following relation between the Allan variance R_{E} and the minimum fluctuation $\delta\nu_{\mathrm{min}}$:

$$R_{\mathrm{E}}(\tau) = \left\langle \frac{E(t)E^*(t+\tau)}{|E(t)|^2} \right\rangle = \exp\{\mathrm{i}2\pi\nu_0\tau - 2(\pi\tau)2\sigma_{\mathrm{F}}^2(\tau)\} \qquad (10.17)$$

where $\sigma_{\mathrm{F}}^2 = (\delta\nu_{\mathrm{min}})^2$ is the variance of the power spectrum of the FM feedback noise and it is usually assumed to have a Gaussian variation. The Fourier transformed spectrum has a Lorentzian shape. Using these relations, the half-width of half-maximum (FWHM) of the spectral line can be calculated and finally given by (Ohtsu and Kotajima 1985)

$$\Delta\nu_{\mathrm{FBm}} = \frac{8}{9\pi}\left(\frac{c}{\eta_{\mathrm{FP}}L_{\mathrm{FP}}}\right)^2 \frac{(1-R_{\mathrm{FP}})^2}{R_{\mathrm{FP}}} \frac{Fk_{\mathrm{BT}}T + \frac{h\nu_0}{\xi}M}{P} \qquad (10.18)$$

where the subscript FP denotes for Fabry-Perot resonator, η_{FP} is the refractive index of the resonator, L_{FP} is the internal cavity length, R_{FP} is the internal reflectivity of resonator mirrors, F is the noise figure of the detector, k_B is the Boltzmann constant, T is the temperature, M is the excess noise factor, ξ is the quantum efficiency of the laser, and P is the incident light power into the Fabry-Perot resonator.

Figure 10.7 shows the numerical result of the attainable minimum linewidth $\Delta\nu_{FBm}$ of the feedback system for the reflectance of the Fabry-Perot resonator calculated from (10.18). Typical values of the excess noise factors for an avalanche photodiode (APD) $M = 9$ and a PIN photodiode $M = 1$ are used in the numerical calculations. The minimum linewidth is reduced by the increase of the reflectance and strongly depends on the excess noise factor. From Fig. 10.7, the linewidth can be ultimately reduced to a value less than $1\,\mathrm{kHz}$ when $L_{FP} = 10\,\mathrm{mm}$ and $R_{FP} > 0.9$. This makes the electrical feedback a quite promising technique to realize an ultra-narrow linewidth. Figure 10.8 shows the experimental results of oscillation linewidths for the variation of the normalized injection current for different reflectances of the Fabry-Perot interferometer. The laser used is a single mode DFB semiconductor laser with the oscillation wavelength of $1.50\,\mu\mathrm{m}$. Two types of Fabry-Perot filters consisted of rod of fuzed silica with the cavity lengths of $10\,\mathrm{mm}$ are used and the reflectance of the interferometers are $R_{FP} = 0.90$ and 0.95. The gain and bandwidth of the amplifier in the electronic feedback circuit are $30\,\mathrm{dB}$ and $100\,\mathrm{MHz}$, respectively, and the feedback delay of the circuit is $13\,\mathrm{ns}$. Depending on the reflectivity of the Pabry-Perot mirrors, a power low is established for the normalized injection current (namely the laser output power). A linewidth narrower than $1\,\mathrm{MHz}$ is obtained by feedback and the

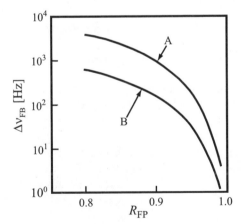

Fig. 10.7. Attainable linewidth reduction for the reflectance of the Fabry-Perot interferometer. **A**: Ge-APD detector, and **B**: Ge-PIN detector. The parameters are $T = 293\,\mathrm{K}$, $\nu_0 = 2 \times 10^{14}\,\mathrm{Hz}$, $P = 3\,\mathrm{mW}$, $n_{FP} = 1.4$, $L_{FP} = 10\,\mathrm{mm}$, $F = 25$, $\eta = 0.73$ (Ohtsu M, Kotajima S (1985); © IEEE 1985)

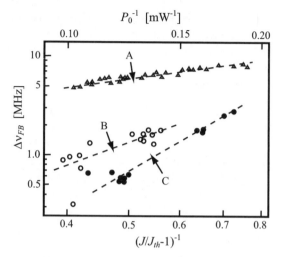

Fig. 10.8. Linewidth for the inverse of the normalized injection current for different reflectances of the Fabry-Perot interferometer. **A**: Solitary sate, **B**: $R_{FP} = 0.9$, and **C**: $R_{FP} = 0.95$ (Ohtsu M, Kotajima S (1985); © IEEE 1985)

minimum linewidth in this graph is 330 kHz. Modifying the experimental system, Ohtsu et al. (1990) demonstrated a linewidth reduction of 560 Hz for an AlGaAs laser of 830 nm wavelength by optoelectronic feedback and for the cavity length of the Fabry-Perot resonator of 30 mm. The original linewidth of the free-running laser is 3.45 MHz, so that the reduction of factor of four is achieved in this experiment.

10.3 Stabilization in Lasers with Various Structures

10.3.1 Noise Suppression in Self-Pulsation Semiconductor Laser

Noise suppression in semiconductor lasers cased by external perturbations is an important issue not only for edge-emitting lasers but also other newly developed semiconductor lasers. We showed some numerical results for the dynamics of self-pulsating semiconductor laser subjected to coherent optical feedback. The noise performance of self-pulsating semiconductor lasers with optical feedback is strongly dependent on the external cavity conditions. Figure 10.9 shows the calculated RINs under different feedback condition (Yamada 1998a). The RIN at solitary oscillation is well below the required noise level for the use of optical data storage systems and the noise is only enhanced close to the relaxation oscillation frequency. However, for the optical feedback at 3.3% in amplitude and the external reflector length of $L = 4$ cm ($\tau = 0.27$ ns), the noise level of the lower frequency component increases up to -90 dB/Hz. The laser at this noise level cannot be used for a light source

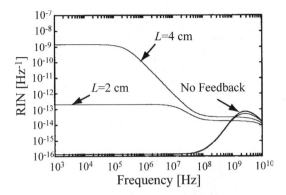

Fig. 10.9. RIN of a self-pulsating semiconductor laser in the presence of optical feedback at $J = 1.69 J_{\mathrm{th}}$. The feedback ratio is 3.33% (after Yamada M (1998a); © 1998 IEICE)

in optical data storage systems (the tolerance for external feedback noise is less than $-125\,\mathrm{dB/Hz}$). On the contrary, the feedback noise is greatly suppressed by as much as less than $-130\,\mathrm{dB/Hz}$ at the external cavity length of $L = 2\,\mathrm{cm}$ ($\tau = 0.13\,\mathrm{ns}$), which is below the noise level required in optical data storage systems. The essence of chaos control is that a chaotic state is shifted to a periodic or a fixed state (saddle node points) embedded in the chaotic sea in the parameter space by a small perturbation to the parameters. The operation point may not be always situated close to such unstable saddle node points. When the state is close enough to the unstable saddle node points, the control is successful. However, it would fail when the state is far from such points. The origin of instability induced by optical feedback in self-pulsating semiconductor laser is explained by the inappropriate combination of the operating condition of the system including the laser device parameters. Therefore, it is not easy to suppress optical feedback noise only from the design of the parameters and structure of semiconductor devices. We must consider the whole system for the optimizing operation condition for the use of self-pulsating semiconductor lasers in optical data storage systems.

10.3.2 Stabilization of VCSELs

Semiconductor laser is essentially a stable class B laser, however it is only true for edge-emitting lasers in the absence of external perturbations, which are described by simple rate equations with time development. Newly developed semiconductor lasers have additional degrees of freedom compared with ordinary edge-emitting semiconductor lasers and they are not stable lasers any more. They show unstable chaotic oscillations without any external perturbations as discussed in Sects. 8.3 and 8.4. Therefore, the stabilization in those lasers even in their solitary oscillations is an important issue in practical applications. Those lasers are either stabilized by external control or by installation of some control structures inside the device. To reduce unstable oscillations, the idea of chaos control is still effective for stabilizing those lasers. In VCSELs, the linewidth narrowing and the stabilization of the spatial modes are important and the lasers are controlled by optical feedback and optical injection in the same manner as those for edge-emitting semiconductor lasers. A VCSEL usually oscillates at the lowest fundamental spatial mode, when it is biased close to the threshold. However, depending on the device parameters, the laser may show a polarization switching between two polarization modes even for a single spatial mode oscillation. With the increase of the bias injection current, higher spatial modes are ready to oscillate and the instability of the laser increases. Therefore, the control and stabilization to the fundamental Gaussian mode with fixed polarization are essential in the applications of VCSELs at any bias injection current.

Optical feedback and optical injection are effective for the control and the stabilization of the polarization mode (Marino et al. 2003 and Romanelli et al. 2005). In a short cavity optical feedback regime, the laser is periodically stabilized with synchronizing periodic stability enhancement for the change of the external cavity length as described in Sects. 5.4 and 8.3 (Arteaga 2007). Another important issue is the stabilization of VCSELs for the injection current modulation in the presence of optical feedback, which is a possible situation for optical data storage systems. The VCSEL sometimes shows unstable oscillations due to the existence of the two perturbations. Yu (1999) conducted numerical simulations for such a model in VCSELs and discussed the stability and instability of the laser oscillations from a viewpoint of the device designs. To obtain high single-mode output power (of the order of a few milliwatts and over) within a large range of injection currents could be interesting for communication systems and optical data storage systems. For these applications, it is important to avoid multimode oscillation in large aperture devices and to control the polarization of the emitted field. This requires us to develop methods for achieving the control of the transverse and polarization modes emitted by VCSELs. Several approaches have been undertaken to increase the single-mode output power of VCSELs: increasing the cavity length, hybrid implant oxide VCSELs, surface etching, or implementing a passive anti-guide region (Mario et al. 2003). In this section, we present the control and stabilization both of the spatial and polarization modes to the single mode in a VCSEL using a frequency selective optical feedback from a grating mirror, even when the laser is biased at a higher injection current.

Figure 10.10 shows an experimental example of the controls by frequency selective optical feedback in a VCSEL. The VCSEL used has a disc diameter of 16 μm and oscillates at the wavelength of 840 nm. The laser has the threshold of 1.6 mA. At threshold, the VCSEL emits only in its fundamental transverse mode, but the two orthogonal polarizations are active. Figure 10.10a shows the free-running polarization-resolved optical spectra at the bias injection current of 1.70 mA. In Fig. 10.10, the third order trans-

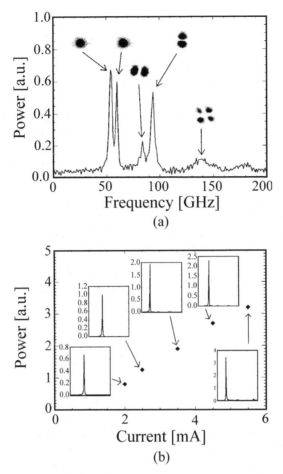

Fig. 10.10. Single-mode operation of a VCSEL induced by frequency selective optical feedback. **a** Optical spectra and frequency-resolved transverse profiles of multi spatial-mode and polarization oscillations at free-running state. **b** Result of control (after Marino F, Barland S, Balle S (2003); © IEEE 2003)

verse mode is close to the threshold and only one of the polarization modes is visible. The separation between the two polarization modes within the same spatial component is about 10 GHz, while the successive spatial mode separation is about 40 GHz. For the control, the emitted light from the VCSEL is optically fed back into the laser cavity using the first-order reflection of a diffraction grating in the grazing incidence configuration (Littman external cavity), thus frequency selective optical feedback is achieved. The external cavity length is 15 cm, corresponding to a free spectral range of 1 GHz. Using the frequency selective control, Marino et al. (2003) selected any of the two polarization components of a particular spatial mode among the spatial modes. For this purpose, the grating lines rotated at 45° with respect to the orthogonal polarization modes to obtain the grating efficiency for the both polarization components be the same.

From the point of view of the applications, it could be very important to optimize the single fundamental-mode output power and the pump current range of stable single-mode operation. Figure 10.10b shows the result of control of one of the polarization modes (main mode) with the fundamental transverse component. In this occasion, the polarization direction is taken to be perpendicular to the grating lines in order to have the strongest feedback as possible. The threshold reduction induced by the optical feedback is ~10%. During the measure, the tuning mirror angle is readjusted because of the modal frequency shift due to the current change. In Fig. 10.10b, we observe the fundamental single-mode emission until $I \sim 4\,\mathrm{mA}$ ($2.5I_{\mathrm{th}}$) the corresponding output power is 2.7 mW. At this current, the output power of the solitary laser is 1.9 mW and four transverse modes are emitting. Within this range of currents, the emission is stable and the polarization is perpendicular to the grating rulings. For higher injection currents, the output starts to loose stability and higher order transverse modes appear.

Other than the control of VCSELs by external optical feedback, optically pumped vertical-external-cavity surface-emitting lasers known as VECSEs are emerging as an important category of semiconductor lasers (Holm et al. 1999, and Fan et al. 2006). Optical pumping of these devices allows high power ($0.1 \sim 1.0\,\mathrm{W}$), circularly symmetric TEM_{00} mode laser operation to be achieved without the need for post-growth processing and at modest cost. Another example is the use of the effect of external optical feedback on the transient response of anti-resonant reflecting optical waveguide (ARROW) VCSELs (Chen and Yu 2004). The proper design of ARROW can suppress the excitation of high-order transverse leaky modes, as well as increase the critical feedback strength so that stable high-power single-mode operation of VCSELs can be obtained even under the influence of strong external optical feedback. The control of spatial modes using the installation of photonic structures in VCSELs, which is discussed in the following section, is categorized into this technique. The control and stabilization of spatial and polarization modes are still important issues for the applications of VCSELs.

10.3.3 Stabilization of Broad-Area Semiconductor Lasers

Control of the emission properties of broad-area semiconductor lasers is also an important issue, such as for pump source for solid-state lasers, spectroscopy, and material processing. Shaping beam profile and recovering laser coherence in broad-area semiconductor lasers are expected for the practical applications (Champagne et al. 1995, Simmendinger et al. 1999a, b, Wolff and Fouckhardt 2000, Raab and Menzel 2002, Lawrence and Kane 2002b, and van Voorst et al. 2006). Several schemes to control and stabilize for the emission properties of broad-area semiconductor lasers have been proposed.

Some of them are based on the improvements of device structures and the others are based on external control. Examples of the latter case include injection locking in a master-slave configuration, frequency selective optical feedback, and spatial filtering optical feedback. In broad-area semiconductor lasers, we can expect the similar techniques of laser control and stabilization as those for edge-emitting semiconductor lasers. However, we must take into account unique characteristics of the dynamic properties of broad-area semiconductor lasers originated from the large stripe width. One of the instabilities that strongly affects the oscillation properties and the beam qualities is the filamentation effect discussed in Sect. 8.4. Though the filamentation is an irregular and fast phenomenon in time, it is controllable by ordinary external perturbations to broad-area lasers. Indeed, several studies were reported for the selection and control of spatial oscillation modes using the method of optical feedback (Martin-Regalado et al. 1996a, Mandre et al. 2003, 2005, Wolff et al. 2003, and Chi et al. 2005). Even by such simple control, the filamentation of a broad-area laser is greatly suppressed and the time-averaged beam profile, the oscillation linewidth, the beam profile, and the power stability, are much improved.

First, we discuss the stabilizations of unstable oscillations induced by filamentation in broad-area semiconductor lasers based on spatially filtered optical feedback. Figure 10.11 is an experimental example of chaos control of a broad-area semiconductor laser by optical feedback (Mandre et al. 2003, 2005). The laser is a broad-area semiconductor laser with the stripe width of $100\,\mu\text{m}$ and the output power of $1\,\text{W}$ operating in pulse mode to minimize thermal effects. The output power is fed back into the active region by an external mirror with a spatial filter. The spatial filtering is very simple, and is accomplished by the intrinsic beam divergence. Depending on the shape of the spatial filter (the shape of the reflecting mirror curvature), the filamentation can be controlled and even completely eliminated for a certain condition of the filter configuration. The external mirror is positioned at $10\,\text{mm}$ from the laser facet and the reflectivity of the mirror is about 50% in intensity. The free-running laser shows unstable oscillations with filamentations like that shown in Fig. 8.27. Figure 10.11 shows a plot of near field patterns at the laser exit face observed by a streak camera. The horizontal axis is the position of the laser exit face and the vertical axis is the time development for the laser oscillation. The optical feedback is switched on at time $t = 0\,\text{ns}$. Figure 10.11a shows a plot of filamentation for a lower bias injection current at $J = 1.75J_{\text{th}}$, while Fig. 10.11b is for a higher injection at $J = 3.0J_{\text{th}}$. For the lower bias injection current, filamentation is greatly suppressed by the feedback after a certain time lapse from the switch-on (around $8\,\text{ns}$) and the averaged spatial beam profile becomes almost single. For the higher bias injection current, the filamentation is not completely eliminated, however it is enough suppressed compared with filamentations under no control compared with Fig. 8.27. In usual fact, thermal lensing due to the high carrier density concentration in the active layer may increasingly affect the beam character-

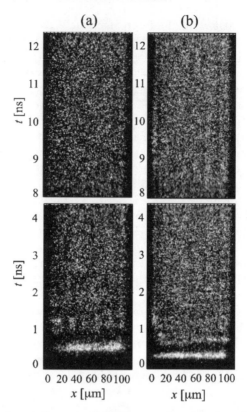

Fig. 10.11. Experimentally controlled filamentations in a broad-area semiconductor laser by spatially filtered optical feedback. Near field patterns observed by a streak camera at the bias injection currents of **a** $J = 1.75J_{th}$ and **b** $J = 3.0J_{th}$ (after Mandre SK, Fischer I, Elsäßer W (2003); © OSA 2003)

istics at high pump current, thus leading the degradation of the beam quality. Nevertheless, stabilization of the emission dynamics is still achieved by the scheme even at high operation currents.

Next, we show coexistence states of chaotic attractors in broad-area semiconductor lasers subjected to optical feedback. This indicates the possibility of the control for broad-area semiconductor lasers by the method of optical feedback (Fujita and Ohtsubo 2005). The coexistence states of chaotic attractors were already described in Sect. 6.2.4. Similar effects are also observed in edge-emitting semiconductor lasers with narrow stripe width. Figure 10.12 shows an experimental example of coexistence states of chaotic oscillations in a broad-area semiconductor lasers. The initial condition determined the convergence to one of the coexistent chaotic attractors. Once the system falls on a certain chaotic orbit, the system always goes around the same chaotic attractor. However, if the perturbation to the system is strong enough, there is a possibility of the system to switch from one chaotic orbit to another. Fig-

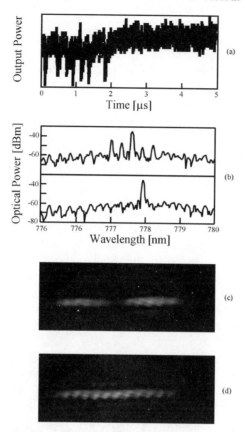

Fig. 10.12. Experimental result of coexistence states of a chaotic attractors in a broad-area semiconductor laser at $J = 1.0J_{\text{th}}$. **a** Coexistence state of LFFs and stable oscillation. Optical spectra at **b** LFF state and **c** stable oscillation. Near-field spatial patterns at **d** LFF state and **e** stable oscillation. The stripe width of the laser is 50 μm. The external cavity length is $L = 30\,\text{cm}$ and the feedback strength is 5% of the average intensity

ure 10.12a shows such a case. The laser is suddenly trapped to a steady-state oscillation with constant output from LFF oscillation. In this experiment, the laser alternately switches from stable to unstable oscillations for the period of several to several tens of milliseconds. When the laser shows a LFF oscillation, it oscillates at multimode as easily understood from the optical spectrum in Fig. 10.12b. However, the optical spectrum in Fig. 10.12c is almost single for the constant part of the output power in Fig. 10.12a. One of typical features of broad-area semiconductor lasers is a twin-peak intensity profile of the far-field pattern. Fig. 10.12d shows the twin-peak pattern at the LFF oscillation in Fig. 10.12a. On the other hand, the pattern is a single lobe when the stable oscillation is achieved. The spatial mode of the laser oscillation is also stabilized to the lowest spatial mode. The beam profile and the coherence are easily controlled and stabilized not only by grating mirror feedback but also by conventional optical feedback and phase-conjugate feedback (Lawrence and Kane 2002b, and van Voorst et al. 2006). The control of the beam qualities of broad-area semiconductor lasers is important for practical applications and is still an ongoing issue.

10.3.4 Stabilization of Laser Arrays

The control of laser arrays is important for the implementation of high power semiconductor lasers. The technique of optical feedback is also used for the control of laser arrays (Münkel et al. 1996, 1997, Martín-Regaldo et al. 1996a, and Chi et al. 2003). Narrowing the spectral linewidth of a multi-array stack of semiconductor lasers is very important in material processing and laser welding, especially for high power applications of semiconductor lasers. Stacked arrays of semiconductor lasers are also controlled by external optical feedback. Zhu et al. (2005) conducted the linewidth narrowing of stacked semiconductor laser arrays by grating optical feedback. In their experiment, the spectral power of the narrowed laser is increased up to approximately 3.5 times that of the free-running case. The method of optical feedback is also effective for the control and stabilization of the oscillation line, beam profile,

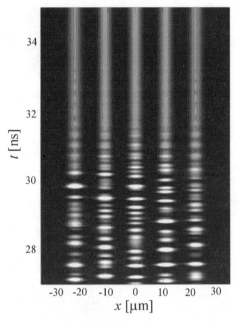

Fig. 10.13. Numerical result of spatio-temporal dynamics of a five-stripe laser at $J = 4.0 J_{\mathrm{th}}$, $\kappa = 3 \times 10^{-2}$, and $\tau = 0.0877$ ns, where the feedback is switched on at $t = 30$ ns. The stripe width of each array is 5 μm and the separation of the lasers is 6 μm (after Münkel M, Kaiser F, Hess O (1997); © 1997 APS)

and power stability in laser arrays. As discussed in Chap. 8, the same technique for calculating rate equations in broad-area semiconductor laser is used for the numerical simulations for laser arrays. Figure 10.13 shows the control of a semiconductor laser array with five elements (Münkel et al. 1997). The laser shows unstable chaotic oscillations at the free running state. When the optical feedback is applied to the array at a time of $t = 30$ ns, each laser oscillation is stabilized within $1 - 2$ ns after the switching. The lasers at the free running state behave like pulsation oscillations, but each laser oscillates at almost spatially single mode after the control. The phase sensitivity also plays the important role for the laser oscillations and the effects must be taken into account for the control. A few theoretical studies on the control of laser arrays have been reported, but little has been investigated on the control of actual laser arrays. The model described here is considered as rather strong coupling between the adjacent laser elements. The control of laser arrays with loose coupling is also an important issue as a real model, but the study is left for the future.

10.4 Controls in Nobel Structure Lasers

10.4.1 Photonic VCSELs

To control and stabilize newly developed semiconductor lasers, external control techniques have been developed. We discussed some instances in the previous section. In this section, we present topics of controls of newly developed semiconductor lasers based on installations of device structures. The first example is a photonic structure installed in adjacent to a laser exit face in VCSELs. Single mode operation of lasers is important for the diffraction limit of the output beam in various applications. VCSEL is expected to operate by stable oscillations at the fundamental single spatial mode with a fixed polarization direction. The stable single mode laser with small beam divergence angle is advantageous for reliable high-speed data transmission. Even for short-distance links, multi-mode VCSELs undergo various problems at high modulation frequencies. In addition to the single mode operation, the control and stabilization of the output polarization are essential for most applications. In spite of strong demand for single mode operations, stable single spatial mode of VCSEL is difficult to obtain with a high output power, as discussed in the preceding sections. In general, a large device size enhances the maximum output power while degrading the single-mode stability due to the thermal lens effect and carrier non-uniformity. To attain single-mode VCSELs with high output power, the improvement of device structures such as anti-guide optical confinement was proposed. However, we here discuss a different method by installing a photonic structure on top of the device as a post-processed technique, although it has some trade-offs, for example for the cavity loss or current confinement. Periodic structure with some small

defects or holey structure is a common technique for photonic VCSELs (Furukawa et al. 2004, Lee et al. 2004, and Liu et al. 2004). The excitation of
higher spatial modes is suppressed and the laser is forced to stay to a single
mode by a photonic structure installed on the top of the exit surface of the
laser cavity.

Furukawa et al. (2004) employed triangular shaped holes as a photonic
structure for the purpose of increasing the field penetration of higher order
spatial modes to the hole side. Figure 10.14 shows an example of the holey
structures, which has 12 triangular holes. The holey structure is installed
on the top of the DBR reflector of an oxide-confined VCSEL. The holes are
aligned so that they surrounded the center modal area and the tip penetrates
the oxide aperture. Figure 10.15 shows the result of the stabilizations for

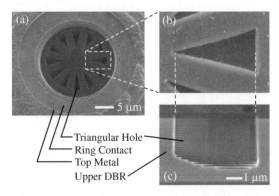

Fig. 10.14. Example of scanning electron microscope images of a triangular holey
structure with twelve holes. **a** Full top view. **b** Magnified top view of triangular
hole. **c** Cross-sectional side view of the hole formed on upper multilayer DBR (after
Furukawa A, Sasaki S, Hoshi M, Matsuzono A, Moritoh K, Baba T (2004); © AIP
2004)

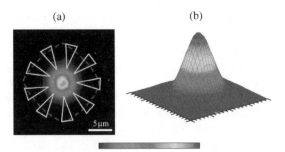

Fig. 10.15. Emission pattern from a photonic VCSEL with nine triangular holes.
a Near-field pattern at the maximum laser output power of 2 mW. The *solid line*
shows positions of triangular holes and the *dashed line* represents the oxide aperture. **b** Far-field pattern of the VCSEL (after Furukawa A, Sasaki S, Hoshi M,
Matsuzono A, Moritoh K, Baba T (2004); © AIP 2004)

another holey structure with nine triangular holes. The figure shows the near- and far-field patterns at the maximum output power. The dashed line in Fig. 10.15a denotes the boundary of the oxide aperture of a diameter of 15 μm. The holes are aligned sufficiently close together with a relatively large penetration of 4 μm to the oxide aperture. The half-maximum diameter of the near-field pattern is 3.1 μm and the half-maximum divergence angle is 6.4°, which agrees closely with the diffraction limit of the near-field pattern. The side mode suppression ratio is greatly enhanced and the ratio of 45 ∼ 50 dB is attained. There are several variations for photonic structures to control emissions in semiconductor lasers. A monolithically integrated surface grating on top of VCSELs is another example of the structures (Debernardi et al. 2005), by which the oscillation wavelength is selected to coincide with the grating frequency and the polarization of the laser oscillations is controlled to the direction perpendicular to the grating lines.

10.4.2 Quantum-Dot Broad-Area Semiconductor Lasers

The importance of beam shaping for broad-area semiconductor lasers was dis- cussed in Sect. 10.3.3. The beam quality of broad-area semiconductor lasers is strongly affected by filamentations, so that the suppression of the effects is essential in applications. It is well know that beam filamentation in semicon- ductor laser strongly depends on the linewidth enhancement factor a (Mar- ciante and Agrawal 1998). Quantum-dot devices exhibit less filamentation than quantum-well devices, since the quantum-dot structure can reduce the linewidth enhancement factor, resulting in suppression of filamentations com- pared to quantum-well lasers. In the quantum-dot laser, the strong localiza- tion of carrier inversion and the small amplitude-phase coupling enable a sig- nificant improvement of beam quality compared with quantum-well lasers of identical geometry (Gehrig and Hess 2002, Smowton et al. 2002, Ribbat et al. 2003, and Gehrig et al. 2004). For the description of a quantum-dot structure, some modifications for the rate equations of ordinary broad-area semiconduc- tor lasers are necessary to account for the characteristics of a quantum-dot laser ensemble. For modeling the dynamic, an additional term of the differen- tial term is taken into account to describe carrier escape and carrier capture into the dot from the wetting layer (Gehrig and Hess 2002 and Gehrig et al. 2004). The detail of the derivation for the quantum-dot Maxwell-Bloch equa- tions can be found in the reference (Gehring and Hess 2002). Here, we show only the results for the comparison between quantum-dot and quantum-well lasers.

Figure 10.16 shows the comparisons of the beam qualities between quan- tum-dot and quantum-well lasers obtained by numerical simulations and experiments. Quantum dots at an appropriate density are embedded into the active region of the laser cavity. In the presence of quantum dots, the dy- namic motions of filaments are greatly suppressed and the laser can emit fairly stable beam compared with quantum-well semiconductor lasers. The laser is

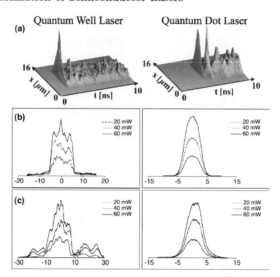

Fig. 10.16. Filamentation control by quantum-dot structure in a broad-area semi-conductor laser. *Left*: Quantum-well laser. *Right*: Quantum-dot laser. **a** Time resolved dynamic behaviors after laser switch-on (numerical simulation). Averaged near field patterns for **b** theory and **c** experiments (Gehrig E, Hess O, Ribbat C, Sellin RL, Bimberg D (2004); © AIP 2004)

an InGaAs quantum-dot laser (dot density 10^{11} cm^{-2}) of the strip-width of 6 µm at the weavelength of 1.1 µm. The laser has a rather small stripe width, but it is enough to show the dynamic properties of filamentations typically observed in broad-area semiconductor lasers. Depending on the dot density and each size of quantum dot, the laser shows inhomogeneous broadening, which results in the change of the linewidth enhancement factor. The α parameter can be even negative under a certain operating condition of the laser and the filamentation behavior of a quantum-dot laser should be drastically different from a quantum-well laser. As easily seen from Fig. 10.16a, the quantum-well laser has strong filamentations, while the quantum-dot laser is well stabilized enough after the switch-on relaxation oscillation. The averaged beam profiles both for the simulations and the experiments are shown in Fig. 10.16b and c. The averaged beam profiles for quantum-well laser contain irregular peaks in their envelopes, which is the effect of filamentations on the order of pico-second. On the other hand, the good quality beam profiles are obtained for the quantum-dot laser. The beam quality factor M^2 ($M^2 = (D/d)^2$: D and d being the diameters for the observed and ideal Gaussian beams) is much reduced in comparison with that of quantum-well lasers. For fiber coupling purposes, the beam quality of $M^2 < 2$ is desirable. The obtained beam quality of the quantum-dot laser well satisfies the demand for such application. Quantum-dot lasers have a much better beam quality compared to quantum-well lasers of the same geometry. The strong localization of the carriers in the dots in combination with the reduced amplitude phase coupling thus guarantees a good spatial quality. In VCSELs, quantum-dot structure is sometimes used to control unstable oscillations such as polarization-mode switching, higher spatial mode excitations, and mode competitions. The instabilities have been reduced by introducing physical distortions to the laser materials. The polarization mode of VCSEL is also stabilized to a single mode by embedding quantum dots into the active layer (Huffaker et al. 1988 and Yu et al. 2006). Quantum-dot structures are promising to stabilize laser oscillations not only in newly developed semiconductor lasers but also in ordinary edge-emitting semiconductor lasers.

11 Stability and Bistability in Feedback Interferometers, and Their Applications

On the way to chaotic evolution, bistability and multistability are observed in the outputs of self-mixing semiconductor lasers. In a periodic state, the laser output shows not simply periodic oscillation but also hysteresis. Novel applications have been proposed based on these phenomena, for example, a displacement measurement is performed by counting the fringes obtained from bistable self-mixing interference between the internal field and the optical feedback light in the laser cavity. The direction of the displacement is simultaneously determined from asymmetric waveforms showing hysteresis. We discuss various methods for optical metrology based on self-mixing interference effects in semiconductor lasers. This chapter does not deal with the detailed descriptions of the methods and their accuracies but with the introduction of the principles of the methods.

11.1 Optical Feedback Interferometers

11.1.1 Bistability and Multistability in Feedback Interferometers

Laser interferometry is a well established technique for the measurement of vibrations and displacement of objects. In the interferometry, for example, the displacement of the order of optical wavelength is measured from the fringe analysis of sinusoidal variations of signals from the interferometer output. We have investigated the self-mixing effects in semiconductor lasers with optical feedback in Chap. 4. We have also shown that the laser output exhibited periodic undulations with half of the optical wavelength for the change of the external cavity length under an appropriate condition of the external reflectivity. In self-mixing semiconductor lasers, the returned light from an external reflector interfers with the internal laser oscillation and the original field in the laser cavity plays a role for the reference wave. Therefore, we can conduct interferometic measurements using semiconductor lasers with optical feedback based on the same principle as the ordinary laser interferometers and we can obtain the absolute position, displacement, and vibration of the external reflector.

In this interferometric measurement, we use periodic oscillations, especially period-1 states, prior to chaotic oscillations on the way to period-doubling bifurcations. Since self-mixing in semiconductor lasers is a nonlinear effect, not only the absolute value of the displacement but also the additional information of the direction of motion (whether the object is approaching or is going away from the laser) can be easily determined from the analysis for the fringe pattern. A semiconductor laser itself plays a role not only as a light source but also as a self-mixing detector in the measurement. In commercially available semiconductor lasers, a photo-diode is usually installed within the laser package as a monitor of the laser output power and we can use it as a detector for the fringe analysis. Therefore, we can construct very a compact sensor for the interferometic measurements. Also, we do not require complex processing for the post-detection signal. However, it is noted that the technique is limited to a certain range of the reflectivity of the external reflector. For large reflectivity of the external reflector, the detected signal may not be a periodic oscillation but a chaotic irregular oscillation. We cannot apply the method for such a case of a large reflectivity of the external reflector.

In the following, we investigate the interferometric measurements using bistable states (period-1 states) of light outputs in self-mixing semiconductor lasers discussed in Chap. 4. The optical configuration is the same as that in Fig. 4.1 and the rate equations of the model are given by (4.5)–(4.7). In the presence of optical feedback in a semiconductor laser, the oscillation angular frequency changes from ω_0 (the solitary oscillation) to ω_s. The relation between the two angular frequencies is given by

$$\omega_0 \tau = \omega_s \tau + C \sin(\omega_s \tau + \tan^{-1}\alpha) \tag{11.1}$$

where $C = \kappa\tau\sqrt{1+\alpha^2}/\tau_{\text{in}}$. The dynamics in semiconductor lasers subjected to optical feedback strongly depend on the C parameter. We are interested in the parameter region of $C \sim 1$ in this chapter, where the laser shows periodic oscillations prior to the onset of chaotic oscillations. Using (4.9)–(4.11), the steady-state value of the laser output is given by

$$S_s = A_s^2 = \frac{\dfrac{\tau_s J}{ed} - n_s + \dfrac{2\kappa}{G_n \tau_{\text{in}}}\cos\omega_s\tau}{1 - \dfrac{2\kappa\tau_{\text{ph}}}{\tau_{\text{in}}}\cos\omega_s\tau}\frac{\tau_{\text{ph}}}{\tau_s} \tag{11.2}$$

Since we are considering a rather small coefficient κ of optical feedback, the difference between the laser output powers with and without optical feedback is small. Then, the difference can be approximated as follows:

$$\Delta S = S_s - S_s|_{\kappa=0} \approx \Delta S_0 \cos\omega_s\tau \tag{11.3}$$

where $\Delta S_0 = 2\kappa\tau_{\text{ph}}^2(J/ed - n_s/\tau_s)/\tau_{\text{in}}$. In actual fact, ΔS is a time dependent function because the carrier density also varies with time by the optical feedback. We introduce a normalized function $F(t)$ as $F(t) = \Delta S(t)/\Delta S_0$.

Substituting (11.3) into (11.1) and using the relation $\omega_0\tau = 2kL$, the external cavity length as a function of time t is given by (Donati, Giuliani, and Merlo 1995)

$$L(t) = \frac{1}{2k}\left[\cos^{-1}F(t) + \frac{C}{\sqrt{1+\alpha^2}}\left\{\alpha F(t) + \sqrt{1-F^2(t)}\right\} + 2m\pi\right]$$

$$\frac{dF}{dt}\cdot\frac{dL}{dt} < 0 \tag{11.4a}$$

$$L(t) = \frac{1}{2k}\left[-\cos^{-1}F(t) + \frac{C}{\sqrt{1+\alpha^2}}\left\{\alpha F(t) - \sqrt{1-F^2(t)}\right\} + 2(m+1)\pi\right]$$

$$\frac{dF}{dt}\cdot\frac{dL}{dt} > 0 \tag{11.4b}$$

where m is a non-negative integer number ($m = 0, 1, 2, \ldots$). The laser output varies for the change of the external cavity length, but the waveform has asymmetric features depending whether the external reflector moves toward or away from the laser. Then, we can determine the displacement of the external reflector and also the direction of movement in accordance with the relation in (11.4a, b).

Next, we investigate the effect of optical feedback at bistable states of the laser output power. For a small optical feedback of $C = 0.6$, the laser output power is a periodic oscillation as shown in Fig. 11.1a and the period is just half of the optical wavelength (Donati et al. 1995). The variation of the waveform is smooth, but it is not a symmetrical shape as expected from the above discussion. In this numerical simulation, it is assumed that the external mirror is moving away from the laser, i.e., the phase $\omega_0\tau$ is increasing. If the phase $\omega_0\tau$ is decreasing, the laser output shows the reversed waveform to Fig. 11.1a. Therefore, we can determine the direction of the movement from the shape of the waveform. For a large value of a C parameter of $C = 3$, the laser output power still varies with the period of $\lambda/2$, but shows hysteresis as shown in Fig. 11.1b. At this parameter value, the laser output power takes bistable states for a certain range of the phase. Therefore, we can expect a significant difference between the shapes of the waveforms for the increase or decrease of the external mirror position. With further increase of the C parameter value, skew of the waveform is enhanced and the laser output takes multi-stable states. These multi-stable states are rarely observed in actual situations and the laser behaves the chaotic oscillations under these conditions, since multi-stable states are usually "unstable" in real systems.

Figure 11.2 presents the experimental results of self-mixing signals for different optical feedback strengths (Giuliani et al. 2001). The external reflector is put on a loudspeaker and the loudspeaker is driven by a sinusoidal signal in Fig. 11.2a. When the feedback is small in Fig. 11.2b (the feedback strength in intensity is roughly estimated as 10^{-7}), the laser output power shows a periodic undulation whose period is equal to half of the optical wave-

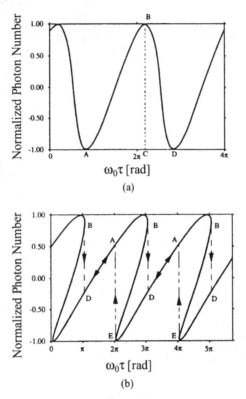

Fig. 11.1. Numerical calculation of laser output power ΔS for phase $\omega_0\tau$. **a** $C = 0.6$ and **b** $C = 3$ with hysteresis. The linewidth enhancement factor is chosen as $\alpha = 6$ (after Donati S, Giuliani G, Merlo S (1995); © 1995 IEEE)

length. In Fig. 11.2c (the feedback strength of 10^{-5}), the laser output power is quite different from the periodic state in Fig. 11.2b. The waveform is still periodic, but the waveform for the increase of the phase is completely different from that for the decrease. Then, the absolute value of the displacement of the external reflector is obtained by counting the peaks of the undulations in the waveform and the direction of the movement is clearly discriminated by examining the waveform. The feedback intensity of 10^{-5} corresponds to the periodic state just before the onset of chaotic evolution in a semiconductor laser with optical feedback. For the large feedback strength of 10^{-4} in Fig. 11.2d (corresponding to a moderate to strong feedback in regime IV), the coherence of the laser is completely destroyed and there is no interference between the internal and external lights. As a result, the laser output power exhibits a similar waveform to the driving signal. However, the signal is broadened due to the modulation of fast chaotic oscillations. It is also noted that the offset phase of the signal is generally not always equal to that of the driving signal.

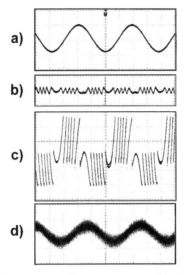

Fig. 11.2. Experimentally observed self-mixing signals for a change of external reflector. **a** Driving sinusoidal signal of external reflector. Laser output signals for external reflectivities of **b** 10^{-7} with periodic state, **c** 10^{-5} with hysteresis, **d** 10^{-4} with coherence collapse state. The driving signal corresponds to the change of the external reflector for 1.3 μm/div. The oscillation wavelength of the laser used is 800 nm. The time scale is 1 ms/div (after Giuliani G, Donati S, Passerini M, Bosch T (2001); © 2001 SPIE)

11.1.2 Interferometric Measurement
in Self-Mixing Semiconductor Lasers

We can measure the change of the external cavity length on the order of half of the optical wavelength by using the self-mixing effect in semiconductor lasers and also determine the direction of the change. Based on these principles, we here discuss the concrete methods for the measurement of displacement, velocity, vibration, and absolute position of the external reflector. Each measurement includes a particular processing algorithm for the detected signals, however the fundamental methods of signal processing for those measurements still contain the common technique (Donati et al. 1995). Before discussing each technique, we take the measurement for the displacement of an external reflector as an example and show the detection and analysis for periodic signals in the self-mixing laser output. Figure 11.3 is an example of the signal processing systems. The light reflected from a target mirror is mixed with the original laser field in the laser cavity and the mixed signal is detected by a photodetector installed in the laser package. The detected signal passes through an amplifier and a high-pass filter. Then, the up- and down-edges of the periodic signal for every $\lambda/2$ period are counted by a counter. As a result, the displacement of the target reflector including

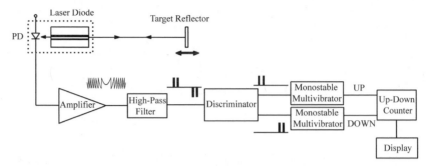

Fig. 11.3. Basic circuits of signal processing for interferometric measurements in self-mixing semiconductor lasers

the direction of the movement during the counting is calculated. The basic resolution of the measurement is $\lambda/2$ in this technique. It is noted that the SNR of the self-mixing interferometer using semiconductor lasers is limited by the efficiency of the coupling photodetector, and it is about 20 dB poorer than of conventional interferometry with the with 50/50 half mirror (Giuliani et al. 2002). However, we can construct a very simple measurement system with high flexibility by the self-mixing interferometer using semiconductor lasers. In the following discussions, we assume that the laser output due to the mixing is a periodic signal with period $\lambda/2$ without notice.

In the following, we show typical signals observed in the self-mixing semiconductor lasers. We take an example of the displacement measurement of an external target under an appropriate condition of the external optical feedback for bistability operation of $C > 1$. Figure 11.4 is an experimental self-mixing signal for a sinusoidal displacement of the object. In the figure, the upper trace is the experimental self-mixing signal for a sinusoidal target

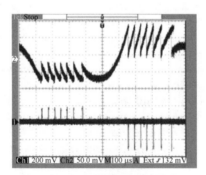

Fig. 11.4. *Upper trace*: experimental self-mixing signal obtained for a sinusoidal target displacement of 3.3 μm peak-to-peak amplitude and 1 kHz frequency, *lower trace*: analogue derivative of self-mixing signal, showing up/down pulses. The timescale is 100 μs per division (after Giuliani G, Norgia M, Donati S, Bosch T (2002); © 2002 IOP)

displacement of 3.3 µm peak-to-peak amplitude and 1 kHz frequency. The lower trace is an analogue derivative of a self-mixing signal, showing up and down pulses, where the states of up and down pulses correspond whether the target is coming towards or going away from the laser. By this approach, displacement of retroreflective target has been successfully measured over 1 m distance with an allowed maximum speed of 0.4 m/s, solely limited by electronic bandwidth. The maximum target distance is limited by the coherence length of semiconductor lasers, being usually several meters to 10 meters. For an appropriate target reflectivity satisfying the condition $C > 1$, the self-mixing signal becomes a sawtooth-like waveform and, then, an accuracy better than $\lambda/2$ can be achieved by linearization of the interferometric fringe, i.e., the function defined in (11.4a, b) is approximated by ideal sawtooth. A resolution of 65 nm has been achieved using a semiconductor laser with a wavelength of 780 nm, in which the resolution is improved by a factor of 6 with respect to conventional fringe-counting technique (Servagent et al. 1998). Residual inaccuracy is caused by the nonlinearity of the actual self-mixing waveform. In the following section, we discuss several particular examples of the self-mixing measurements in semiconductor lasers.

11.2 Applications in Feedback Interferometer

11.2.1 Displacement and Vibration Measurement

In the signal processing system in Fig. 11.3, we obtain the number N of counted pulses as the output and the number is assumed to be large enough. Then, the displacement $\Delta L(L = L_0 + \Delta L, L_0$ being the offset length) of the external reflector is given by the following relation (Donati et al. 1996 and Merlo and Donati 1997):

$$\Delta L = N\frac{\lambda}{2} + O(\lambda) \approx N\frac{\lambda}{2} \tag{11.5}$$

where $O(\lambda)$ is the residual of the counts. The direction of the displacement is determined from the total counted number of the up- and down-edges. Therefore, N has a plus or minus sign. Vibration measurement of an external reflector is also conducted by the same principle. For vibration measurement, the follow-up for time varying signals is important. When the time response of the signal processing circuits is fast enough, the measurement is limited by the response of the laser, i.e., the relaxation oscillation. Since the response of the laser is over nano-second, the total response of the measurement system with fast electronic circuits is up to nano-second and it is much faster than time variations of the ordinary mechanical vibrations we are considering.

The detection of a target displacement is the basic for interferometric measurement. We have shown an example of displacement measurement using self-mixing semiconductor lasers in the previous section. Here, we discuss vibration measurement in a self-mixing interferometer, which is the same principle as displacement measurement. When the amplitude of a target reflector under vibration is large enough (larger than the optical wavelength), we can obtain the frequency of the vibration from a Fourier transform analysis for the detected signal in the interferometer. Regardless of the optical feedback strength, the maximum frequency contained in the self-mixing signal for the case of a target vibrating at a frequency f_0 with amplitude ΔL is proportional to the product $f_0 \Delta L$. Indeed, a sinusoidal object vibration of $140\,\mathrm{Hz}$ frequency and $7.86\,\mu\mathrm{m}$ peak-to-peak amplitude is successfully measured by the method (Scalise 2002). However, only the product can be measured by the method, and we cannot obtain details of the vibration, such as the profile of the vibration amplitude.

To reconstruct a waveform of an object vibration, a closed loop technique is proposed. The principle of the measurement and the processing electronic circuits after the detection of a self-mixing signal are shown in Fig. 11.5 (Giuliani et al. 2002). At a moderate optical feedback of $C > 1$, we obtain a sawtooth-like interferometric signal as an output form the self-mixing in a semiconductor laser as has already been discussed. Figure 11.5a shows the principle of linear measurement of small target vibrations by locking the interferometer phase to half a fringe in the moderate feedback regime, where the interferometric signal can be approximated as having a triangular shape. For a moment, we consider a small amplitude object vibration. At an operating offset intensity at S_0, the self-mixing output S is linearly proportional to the vibration amplitude and the waveform of the vibration is directly observed by

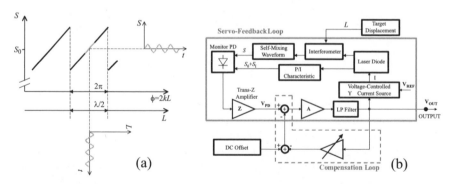

(a)

(b)

Fig. 11.5. a Principle of linear measurement of small target vibrations by locking the interferometer phase to half a fringe in the moderate feedback regime. The vertical axis represents the power emitted by the semiconductor laser, where S_0 is the power emitted by the unperturbed semiconductor laser. The horizontal axes represent interferometric phase and target displacement, respectively. b Block diagram for the self-mixing vibrometer accomplishing the phase-locking and phase-nulling techniques. The details of input and output variables and each block are explained in the text (after Giuliani G, Bozzi-Pietra S, Donati S (2003); © 2003 IOP)

an oscilloscope as far as the peak-to-peak amplitude of the object vibration is within $\lambda/2$.

By employing an additive active phase-tracking method, the maximum measurable vibration amplitude can be extended up to several hundred micron meters. The active phase-tacking system is designed so that a constant number of wavelengths are contained in the path from the semiconductor laser to the target. Figure 11.5b shows the block diagrams of the signal processing system. The blocks contained in the solid box constitute the servo-feedback loop and the blocks contained in the dashed box make up the compensation path. The main block is the self-mixing interferometer operating in the moderate feedback regime, whose phase must be kept at a constant value, corresponding to half an interferometric fringe. The target displacement ΔL acts as a perturbation to the system, and it generates a variation $\Delta\phi$ of the interferometric phase. The phase variation $\Delta\phi$ causes a proportional variation ΔS in the power emitted by the laser through the self-mixing effect given by $\Delta S = \beta_{\mathrm{tr}}\Delta\phi$ (β_{tr} being the slope coefficient of the triangular transfer characteristics of the interferometer). The power variation is detected by the monitor photodetector and converted into the voltage signal ΔV_{PD} by the transimpedance amplifier, which is given by the relation as $\Delta V_{\mathrm{PD}} = \sigma Z \Delta S$ (σ is the net efficiency of the photodiode and Z is the transresistance). This signal is then amplified by a factor A, low-pass filtered, and fed to the input of the voltage-controlled laser current source with admittance Y, thus generating a variation I of the injection current as $\Delta I = AY\Delta V_{\mathrm{PD}}$. This, in turn, gives rise to a variation $\Delta\lambda$ of the laser wavelength such that $\Delta\lambda = \Delta I d\lambda/I$. The feedback loop ensures that the phase variation generated by the laser wavelength variation is exactly opposite (at least at first order) to that caused by target displacement. The amplified error signal V_{OUT} fed to the current source is a perfect replica of the target displacement, and it constitutes the instrument output.

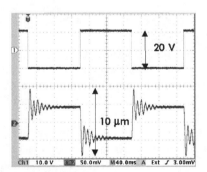

Fig. 11.6. Examples of vibration measurement. The target is a loudspeaker with a black paper surface driven by a 10 Hz square wave. *Upper traces*, loudspeaker drive signal, *lower traces*, vibrometer output signal (after Giuliani G, Bozzi-Pietra S, Donati S (2003); © 2003 IOP)

Figure 11.6 is an example of vibration measurements using the self-mixing vibrometer. The object is a loudspeaker driven by a square wave of 10 Hz. The semiconductor laser used is a commercial single- mode Fabry-Perot type with maximum power of 40 mW at the oscillation wavelength of 800 nm. The distance from the laser to the target is 80 cm. The amplitude of the vibration is much lager than $\lambda/2$, but we can obtain the full waveform of the oscillations as shown in the figure. As we can see, the damped resonance oscillations of the loudspeaker are clearly visible.

11.2.2 Velocity Measurement

Velocity measurement is considered as a continuous change of the external mirror position and is easily performed by the extension of the displacement measurement. When the external mirror moves, the detected signal changes as $\Delta S = \Delta S_0 \cos \omega_s \tau$ in accordance with (11.3). The external cavity round-trip time τ is a time dependent function and is proportional to the external cavity length L. Considering the angle θ of the motion for the optical axis, the round-trip time τ is written by (Bosch et al. 2001)

$$\tau(t) = \frac{L_0 \pm vt \cos \theta}{c} \tag{11.6}$$

where v is the speed of the external mirror. The signs of the velocity term account for the direction of the motion; the plus sign is for the object moving away from the laser and the minus sign is for moving toward the laser. The self-mixing in semiconductor lasers is of the heterodyne detection and the term related to the velocity in (11.6) corresponds to a Doppler shift component in the self-mixing (Groot and Gaillatin 1989, Shinohara et al. 1989 and Aoshima and Ohtsubo 1992). As has already been discussed in the displacement measurement, we can discriminate the direction of the movement from the shape of waveforms of the self-mixing signal. Two-dimensional velocity measurement is easily implemented by extending the one-dimensional measurement.

Figure 11.7 shows an experimental self-mixing signal in time-domain from a rough rotating disc with a small feedback fraction $C < 1$ (Giuliani et al. 2002). In Fig. 11.7a, the self-mixing amplitude is strongly deformed by speckle modulation compared with a flat reflecting surface. Therefore, it may be difficult to extract the velocity information from the time signal by the fringe counting technique, as done in the displacement measurement. However, the harmonic component corresponding to the disc velocity is easily obtained from the Fourier spectrum as shown in Fig. 11.7b, although the spectrum has a broadened peak due to the speckle effect. Using the technique of the self-mixing in semiconductor lasers, velocity measurements ranging from a rigid surface of ~ 100 m/s to a slow blood flow of $\sim$mm/s have been performed (Özdemir et al. 2000 and Giuiliani et al. 2002).

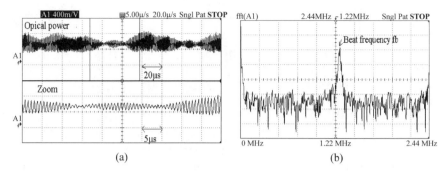

Fig. 11.7. Example of velocity measurements using self-mixing semiconductor laser. **a** Time-domain self-mixing signal for velocity measurement on a rotating diffusing target. **b** FFT spectrum of the signal. The Doppler beat frequency 1.46 MHz corresponds to a speed of 0.56 m/s (after Giuliani G, Norgia M, Donati S, Bosch T (2002); © 2002 IOP)

11.2.3 Absolute Position Measurement

When the injection current of a semiconductor laser is modulated, not only the laser output power, but also the oscillation frequency of the laser, change in accordance with the relation in (5.5). The same periodic undulation signal like in Fig. 11.2b or c is observed for the laser output under the condition of the C parameter of $C \sim 1$ when a ramp signal is applied to the bias injection current of a semiconductor laser at a fixed external mirror position. The period of the undulations is equal to $c/2L$. For the measurement of the absolute position of a target (distance), a ramp signal, which has a linear increase or decrease for the time development, is usually used. By the ramp modulation, the oscillation frequency is also linearly changed. For a change of the injection current, the wavelength of the laser oscillation varies as $\Delta\lambda$, then the change of the wavenumber Δk is written by

$$\Delta k = -2\pi \frac{\Delta\lambda}{\lambda^2} = 2\pi \frac{\Delta\nu}{c} \qquad (11.7)$$

where $\Delta\nu$ is the frequency change due to the injection current variation. For the reflecting mirror positioned at L from the laser facet, the change of the optical phase in the self-mixing interferometer due to the modulation is $\Delta\phi = \Delta k \cdot 2L$. The quantity of $\Delta\phi/2\pi$ is the number of interferometric fringes occurring from the wavelength variation $\Delta\lambda$ observed in the self-mixing interferometer, which is given by the following relation

$$\frac{\Delta k \cdot 2L}{2\pi} = N + O(N) \qquad (11.8)$$

Here, $O(N)$ represents the residual of fringe number, which corresponds to the maximum error in the distance measurement. By counting the number

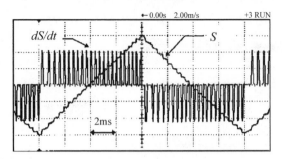

Fig. 11.8. Self-mixing signal for absolute distance measurement, obtained for a 0.8 mA current modulation in a Fabry-Perot semiconductor laser. The pulses are the analogue derivative of the laser output power, which corresponds the fringes to be counted for the distance measurement (after Giuliani G, Norgia M, Donati S, Bosch T (2002); © 2002 IOP)

of fringes, one obtains the distance of the reflector from the laser facet, and the distance L is given by

$$L = \frac{\lambda^2}{2\Delta\lambda}N = \frac{c}{2\Delta\nu}N \qquad (11.9)$$

Figure 11.8 shows the detected output power S of the laser swept by a ramp signal (Giuliani et al. 2002). Looking more closely, the signal resembles step-wise variations of the output power, although the macroscopic change of the detected signal shows a linear increase or decrease for the time development. This step-wise change is induced by the selections of successive resonance external modes by the variation of the bias injection current for the laser; thus the output power shows not a smooth change but a step-wise change. The analogue derivative of the laser output power $\mathrm{d}S/\mathrm{d}t$ becomes a train of pulse-like signals, and this corresponds to the fringe signals discussed before. Counting the number of fringes for the duration of the ramp signal, we obtain the absolute distance with the relation in (11.9). The error in this measurement is the quantization error of fringes and the error corresponds to the maximum residual of the fringe counting in (11.8). Thus the maximum error is given by $c/2\Delta\nu$. Mourat et al. (2000) conducted the distance measurement using the self-mixing interferometer of a tunable multi-electrode DBR semiconductor laser having continuous tunable range up to 375 GHz and attained the accuracy of the measurement less than 0.5 mm for the distance of the order of meters. The accuracy is quite coincident with the theoretical resolution of $c/2\Delta\nu = 0.4$ mm.

11.2.4 Angle Measurement

Self-mixing interferometry is also applied for small angle measurement. In the angle measurement, coherence collapse states like in Fig. 11.2d is used (Giuliani et al. 2001). Figure 11.9 shows the experimental setup for small

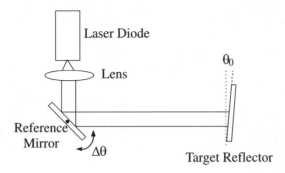

Fig. 11.9. Experimental setup for angle measurement in a self-mixing semiconductor laser. The external mirror under test is tilted with a small angle θ_0. The reference mirror is dithered with a small amplitude

angle measurement in a self-mixing semiconductor laser. An external mirror under test is tilted with a small angle θ_0 for the optical axis. In the optical setup, the direction of the illuminating light beam is changed by a reference mirror and the beam is directed to the reflector under the test. The feedback level is in regime IV and the laser output power shows coherence collapse states as shown in Fig. 11.2d when $\theta_0 = 0$. For a non-zero tilt angle, the feedback strength from the reflected light decreases with the increase of the tilt angle, but the reflected light is still fed back into the laser cavity and the tilt is such a small angle. In the measurement, the reference mirror put into the optical path is dithered with small amplitude of the tilt angle $\Delta\theta$. At the reference mirror angle for compensating the reflector tilt θ_0, the amount of the feedback light takes the maximum value. When the tilt of the reference mirror is a periodic function with time, the laser shows synchronous output with the modulation. However, the phase of the detected periodic function differs from that of the modulation due to the initial offset angle θ_0. Figure 11.10 shows

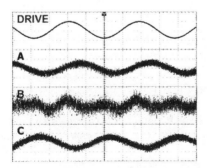

Fig. 11.10. Waveforms of self-mixing laser outputs. *Upper trace:* drive signal of reference mirror. Laser output powers **A**: with negative tilt, **B**: zero tilt, and **C**: positive tilt. The frequency of the driving signal is 180 Hz (after Giuliani G, Donati S, Passerini M, Bosch T (2001); © 2001 SPIE)

the experimental results of the laser outputs in the angle measurement. In this figure, signal B corresponds to zero tilt of the external mirror and the output power includes the second harmonic component due to rather strong optical feedback. However, the initial phases of signals A and C are shifted from that of the driving signal and the small angles are calculated from the phase shifts. The tilt angle of the external mirror has a linear relation with the detected phase shifts for a certain range of the tilt. In this technique, we need some calibration for a particular setup of the experiment. We can perform a small tilt angle detection on the order of $10^{-6} \sim 10^{-4}$ rad based on this technique.

11.2.5 Measurement of Linewidth and Linewidth Enhancement Factor

The linewidth enhancement factor α of a semiconductor laser is an important parameter in deciding its dynamics characteristics. As discussed in Chap. 3, the real and imaginary parts of the complex susceptibility in semiconductor lasers are not determined independently, but they are related. This fact gives rise to a non-zero finite value of the linewidth enhancement factor. For most lasers such as gas lasers, the value of the linewidth enhancement factor α is zero, while it is around $\alpha = 3 \sim 7$ in semiconductor lasers (see Sect. 3.3.3). As a result, the linewidth of the laser oscillation is broadened by as much as several tens of MHz to 100 MHz. On the other hand, for lasers with a linewidth enhancement factor of $\alpha = 0$, the linewidth is usually less than MHz, as discussed in Chap. 3. There are several methods of measuring the factor (Okoshi et al. 1980). It can also be measured by analyzing the laser output power for a sinusoidal modulation of the external mirror position in a self-mixing interferometer. For a certain range of optical feedback strength of an external reflector, a periodic sawtooth-like wave is observed for the change in external cavity length in the laser output power. First, we show the linewidth measurement in semiconductor lasers based on the self-mixing effects.

We assume that the oscillation frequency of a semiconductor laser is $\nu = \nu_0 + \delta\nu$, where $\delta\nu$ is the fluctuation of the laser oscillation. In the measurement of the linewidth, the position of the external mirror is modulated by a sinusoidal signal and the external mirror is vibrated with a small amplitude compatible with the order of the optical wavelength. Using the modulation for the external mirror position $l_m(t)$ with zero mean and putting the external cavity length $L = L_0 + l_m(t)$, the back-reflected field phase is given by

$$\phi = \frac{4\pi}{c}\nu L = \frac{4\pi}{c}\nu_0 L + \frac{4\pi}{c}\nu_0 l_m(t) + \frac{4\pi}{c}\delta\nu L \qquad (11.10)$$

The phase ϕ is a periodic function with period $\lambda/2$, but it is a statistical function due to random fluctuation of $\delta\nu$.

On averaging the phase and its square and calculating the covariance, the statistical root-mean-square (rms) phase related to the linewidth is given by (Giuiliani and Norgia 2000)

$$\sqrt{\langle(\Delta\phi)^2\rangle} = \sqrt{\langle\phi^2\rangle - \langle\phi\rangle^2} = \frac{4\pi}{c}L_0\overline{\delta\nu} \qquad (11.11)$$

where $\overline{\delta\nu}$ is the average of the frequency fluctuations. The average fluctuation $\overline{\delta\nu}$ is equal to $\Delta\nu$ in (3.100), which gives the relation between the phase fluctuation and the linewidth enhancement factor α. Figure 11.11 shows the experimental result of jitter in the measurement of the linewidth. Figure 11.11a shows the driving signal for the position of the external mirror and the periodic output power. Figure 11.11b shows the zoomed frame with the superposition of subsequent single-sweep acquisitions of the self-mixing signal. The periodic signal contains detailed structures, and there is jitter in up- or down-edges of the sawtooth-like wave in the laser output power. From the statistical average of the jitters, the relation between the rms phase and the linewidth is calculated according to (11.11). In the real experiment, the measurement is repeatedly conducted for different absolute positions of the external mirror and the value of $4\pi\delta\nu_0/c$ is obtained as the proportional coefficient. The amount of feedback required to achieve the self-mixing regime is moderate (i.e., around 10^{-6} in power), so that the optical feedback little affects the linewidth of the laser oscillations and the linewidth measured

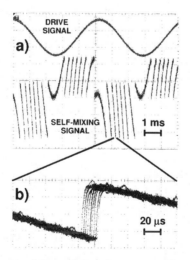

Fig. 11.11. Measurement of linewidth enhancement factor. **a** Driving signal of external mirror position (*upper trace*, 1 V/div corresponding to 1.43 µm/div target displacement) and corresponding laser output (*lower trace*. 1 ms/div time scale) in self-mixing interferometer. **b** Zoomed frame of superposition of subsequent single-sweep acquisitions of signal (time scale of 20 µs/div corresponds to the phase variation of 0.5 rad/div) (after Giuliani G, Norgia M (2000); © 2000 IEEE)

under the small perturbations remains almost the same value as the solitary oscillations.

The linewidth enhancement factor can be obtained by the above same optical system by calculating two phase separations; one is the separation between a zero-crossing phase of the approaching signal and the phase at the adjacent down-edge, and the other is the separation between a zero-crossing phase of the leaving signal and the phase at the adjacent up-edge. From the comparison between the two phase values, the linewidth enhancement factor can be calculated either graphically or numerically (Yu et al. 2004). The values of the linewidth enhancement factor from 2.2 to 4.9 are experimentally obtained for various different lasers with different oscillation wavelength. The values of the linewidth enhancement factor measured using the proposed technique are in good agreement with those obtained by using the self-heterodyne method (Okoshi et al. 1980).

11.3 Active Feedback Interferometer

11.3.1 Stability and Bistability in Active Feedback Interferometer

Another type of feedback interferometer is a system of a two-arm interferometer with optoelectronic feedback. Here, we discuss the feedback of the interference light to the bias injection current of a semiconductor laser. Such a system is considered as a kind of filtered feedback systems discussed in Sect. 4.6. For example, in a Twyman-Green interferometer, the optoelectronic feedback technique is applied to stabilize the fringe of the interferometer output from disturbances such as vibrations. Such a system opens wide applications for the fringe analysis and measurements of laser interferometer

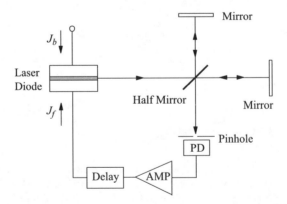

Fig. 11.12. Model of Twyman-Green active feedback interferometer. The fringe of the interferometer output is detected by a photodetector (PD) through a pinhole smaller than the fringe spacing

under various circumstances of the atmosphere (Yoshino et al. 1987). Figure 11.12 is an example of laser interferometers with optoelectronic feedback. The interferometer output is detected by a photodetector through a small pinhole. The diameter of the pinhole is assumed to be much smaller than the fringe spacing. The detected photocurrent is fed back to the bias injection current of the light source of the semiconductor laser. The principle of the stabilization of the interferometer is as follows: the detected optical power deviates when the fringe is disturbed by the external perturbation. Then, the detected photocurrent changes and the injection current to the laser is modulated. The change of the laser output power induces the optical frequency change so as to compensate and cancel the fluctuations of the fringe intensity at the detection point. The variation of the optical frequency is at most several GHz in ordinary feedback interferometers. We can ignore the effect of disturbance by the optical frequency change on the accuracy of the interferometric measurement, since the ratio of the change to the center optical frequency is only less than 10^{-5}. However, care must be taken with respect to the feedback strength. In this active interferometer, stability, multistability, and chaos appear in the laser output depending on the feedback strength and the response time of the feedback loop. In the following, we discuss the principle and behaviors of the active feedback interferometer, and applications for chaos control and signal generations.

We have investigated the effects of optoelectronic feedback in semiconductor lasers in Chap. 7. A similar treatment can be applied to the active interferometer, but the feedback signal is an interference fringe. The rate equations we use are

$$\frac{dS(t)}{dt} = G_n \{n(t) - n_{th}\} S(t) + R_{sp} \tag{11.12}$$

$$\frac{dn(t)}{dt} = \frac{1}{ed}\{J - \xi x(t)\} - \frac{n(t)}{\tau_s} - G_n\{n(t) - n_0\}S(t) \tag{11.13}$$

where $x(t)$ is the term of optoelectronic feedback. As discussed in Chap. 7, the electronic feedback circuit usually has a finite time response and the variable $x(t)$ follows a differential equation similar to (7.8). Here, we write the response of the feedback term as follows:

$$\tau_i \frac{dx(t)}{dt} = -x(t) + \frac{J_f(t)}{\xi} \tag{11.14}$$

where τ_i is again the response time of the electric circuit and J_f is the feedback current to the bias injection current. $x(t)$ is the variable of the feedback and it corresponds to the photon number as a physical quantity. Therefore, ξ is the conversion efficiency from the current density to the photon number. The feedback current of the active interferometer is easily calculated as (Ohtsubo and Liu 1990 and Liu and Ohtsubo 1992a, b)

$$J_f(t) = \xi x_b - G_A \xi x(t)[1 + b\cos\{\kappa_i x(t) - \phi_0\}] \tag{11.15}$$

where x_b is the reference signal in the feedback circuit and G_A is the gain of the circuit. The cosine term on the right hand side of (11.15) denotes the fringe in the interferometer output and b is the visibility of the fringe. ϕ_0 is an offset phase in the interferometer. The laser frequency is changed by the feedback current. The cosine term in (11.15) is the effect of the frequency change. Using the optical frequency ν_0 without feedback, the frequency $\nu(t)$ in the presence of feedback is written by $\nu(t) = \nu_0 - \beta_f x(t)$. Then, the argument of the cosine function reads

$$-\kappa_i x(t) + \phi_0 = -\frac{4\pi D_i \nu(t)}{c} = -\frac{4\pi D_i \beta_f}{c} x(t) + \frac{4\pi D_i \nu_0}{c} \qquad (11.16)$$

where D_i is the difference of the interferometer arms and β_f is the conversion efficiency from the photon number to the oscillation frequency in the semiconductor laser.

If the responses of the electronic circuits and the laser are much faster than the time-varying external disturbances for the interferometer, only (11.15) is sufficient to describe the system characteristics of the active interferomenter. We here discuss the stability and instability of the system when the response time of the feedback circuit is fast enough, $\tau_i \sim 0$. Indeed, possible mechanical vibrations for the interferometer are less than 1 kHz. Therefore, solutions of stability, bistability, and multistability of the laser output are investigated from the crossing points for the graph of $y = x(t)$ and $y = J_f(t)$ in (11.15). When the disturbance for the interferometer is small enough, we can obtain a stable solution of the interferometer. In this active interferometer, the configuration of the imbalance interferometer is essential, since the feedback signal depends on the difference according to (11.16). The interferometer is always stabilized at a certain fringe pattern as far as the deviation or distortion of the fringe pattern by the disturbance is smaller than the fringe separation. Thus, we can attain robust interferometric measurement under unfavorable conditions of disturbances and the fringe analysis is performed under such severe conditions. When the disturbance is large enough with exceeding the fringe spacing, multi-stable states appear in the laser output

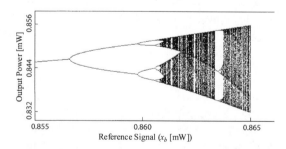

Fig. 11.13. Bifurcation diagram of laser output in active delay feedback interferometer for change of bias injection current. The parameters are $G_A = 0.05$, $\kappa_i = 32\pi$, and $\phi_0 = 0$

and hops of the optical frequency through the feedback are induced. This gives rise to chaotic behaviors in the laser output (Ohtsubo and Liu 1990, and Liu 1994). The technique of the active interferometer cannot be applied for the phase scanning interferometer, since the phase shift of the fringe is an essential technique in the phase scanning interferometry.

The active interferometer shows rich varieties of dynamics when the feedback circuit has a time delay. By the introduction of the delay, the system exhibits stability, instability, and chaotic states depending on the feedback delay and ratio. We here consider the following modified equation for (11.16) for the delayed system (Liu and Ohtsubo 1992a, b):

$$J_f(t) = \xi x_b - G_A \xi x(t - \tau_e)[1 + b\cos\{\kappa_i x(t - \tau_e) - \phi_0\}] \qquad (11.17)$$

where τ_e is the delay time in the feedback circuit. Figure 11.13 shows the calculated bifurcation diagram of the laser output for the change of the reference signal (the bias injection current). The bifurcation diagram is obtained

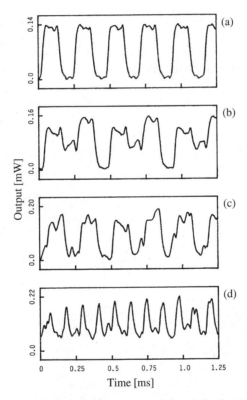

Fig. 11.14. Chaotic oscillations in an active delayed feedback interferometer at $G_A = 0.05$, $\kappa_i = 32\pi$, $\phi_0 = 0$, and $\tau = 0.1\,\mathrm{ms}$. **a** $x_b = 0.423\,\mathrm{mW}$ with period-1 state, **b** $x_b = 0.30\,\mathrm{mW}$ with period-4 state, **c** $x_b = 0.473\,\mathrm{mW}$ with period-1 chaos, and **d** $x_b = 0.573\,\mathrm{mW}$ with fully developed chaos

by assuming the difference equation described by (11.17) instead of solving the continuous rate equations. The laser output clearly shows typical chaotic evolution via period-doubling bifurcation. Figure 11.14 shows experimentally obtained waveforms in the active interferometer for the change of reference signal level. With increasing the reference signal level, the laser output evolves from a periodic oscillation into chaotic states. The period $2T$ of the period-1 oscillation in Fig. 11.14a is about $2T = 0.22$ ms and it is almost equal twice the delay time of the circuit of $\tau_e = 0.10$ ms. The difference of time $T - \tau_e = 0.01$ ms is equal to the intrinsic delay τ_i of the whole circuit except for the extra delay circuit. In this chaotic system, we can easily design periodic orbits by appropriately choosing the system parameters and generate arbitrary waveform sequences in the laser output prior to chaotic states. These higher harmonic oscillations are used for the applications of chaotic associative memory (Liu and Ohtsubo 1992b, 1993, and 1994b).

11.3.2 Chaos Control in Active Feedback Interferometers

Chaotic oscillations in active delayed feedback interferometers can be also controlled to periodic or fixed states based on the chaos control method. In this section, we describe chaos control in the active interferometer by the occasional proportional (OPF) method. The active feedback interferometer

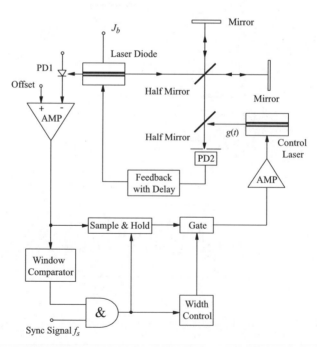

Fig. 11.15. OPF control for chaotic oscillations in an active delay feedback interferometer

was originally designed for the isolation of rather slow response mechanical vibrations. Therefore, the system is very suited of the OPF technique (Liu and Ohtsubo 1994a, b). We employ the OPF control system discussed in Fig. 9.3. Figure 11.15 shows the schematic diagram for the OPF control in the active interferometer. From the detected output power from PD1, an appropriate sampling control signal is generated in the OPF control circuit with a synchronous signal (Sync Signal) and the control signal with a small amplitude is overlapped into photodetector PD2 for the fringe detection as a small perturbation. After successful control, the chaotic output of a laser oscillation is fixed to a periodic state.

Figure 11.16 is the experimental result of the OPF control. The delay time of the circuit is $\tau_e = 2.0 \,\mathrm{ms}$. Under the experimental condition, the laser exhibits chaotic oscillation as shown in Fig. 11.16a. The typical frequency

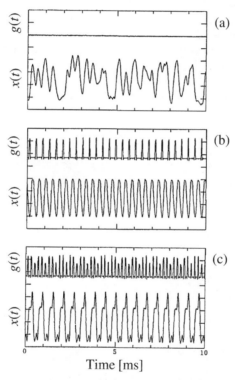

Fig. 11.16. Experimental results of chaos control in an active delay feedback interferometer. **a** Chaotic oscillation without control. **b** 11-th harmonics of fundamental period-1 orbit with control (synchronous frequency is 2.64 kHz). **c** 7-th harmonic oscillation of fundamental period-1 orbit with control (synchronous frequency is 5.04 kHz). The delay time of the circuit is $\tau_e = 2.0 \,\mathrm{ms}$. *Upper trace* of each figure is the control signal $g(t)$ (arbitrary amplitude) and *lower trace* is the controlled waveform $x(t)$

of the chaotic signal is 0.24 kHz and it is almost equal to the sum of the times τ_i and τ_e. For the frequency of the synchronous signal in the control circuit of 2.64 kHz, the system is controlled to a periodic state. In Fig. 11.16b, the controlled waveform is the 11-th harmonics of the fundamental period-1 orbit. On the other hand, the laser is controlled to the 7-th harmonics of the fundamental period-1 orbit for the synchronous frequency of 5.04 kHz in Fig. 11.16c. The corresponding sampling frequency used as a control signal $g(t)$ is 21 multiples of the fundamental frequency. In delay differential systems, we can design and generate arbitrary multi-valued waveforms (isomer signals) of higher periodic orders for the fundamental periodic oscillation by adding extra control circuits to the systems (Liu and Ohtsubo 1991 and Liu et al. 1994c). In this example, the control signal is a very small perturbation to the chaotic oscillation and its amplitude is less than 3% of the bias injection current. Therefore, the OPF control applied here is approximately considered as a category of chaos control in the meaning of the OGY algorithm. In the OGY method, the control signal is eliminated after the success of the control, but the control signal is continuously lasting with the same level. In the OPF method, the system must be always pushed by the control signal to fix a certain attractor of unstable periodic orbit.

12 Chaos Synchronization in Semiconductor Lasers

Another possibility of applications of chaotic semiconductor lasers is chaotic secure communications. The key to chaotic communications is chaos synchronization between two nonlinear systems. If two nonlinear chaotic systems operate independently, the two systems never show the same output because of the sensitivity of chaos for the initial conditions. However, when a small portion of a chaotic output from one nonlinear system is sent to the other, the two systems synchronize with each other and show the same output under certain conditions of the system parameters. This scheme is called chaos synchronization. It is very surprising that two chaotic systems share the same waveform, since chaos is sensitive to the initial conditions and its future is unpredictable. In this chapter, we overview chaos synchronization in chaotic semiconductor laser systems for the introduction of the secure chaos communications discussed in Chap. 13.

12.1 Concept of Chaos Synchronization

12.1.1 Chaos Synchronization

We cannot expect the same chaotic oscillation for two nonlinear systems even when they are the same configuration having the same parameter values, because chaos has strict sensitivity to the initial conditions of the parameters. For example, two chaotic systems with the same parameters may at first output similar signals when the difference between the initial conditions is small enough in the ordinary sense. Then, the two signals show a small difference with lapse of time and, then, the difference rapidly increases for further time development. Finally, the two systems behave in a completely different manner in as far as the difference between the initial conditions is not zero. However, there is a possibility of showing the same output in two nonlinear systems if the two systems possess a common subsystem with the same parameter values, otherwise if a small amount of the signal from one of the two systems is transmitted to the other. Under this condition, the systems output completely the same chaotic signal. The scheme is called "chaos synchronization."

The idea of chaos synchronization between two nonlinear systems was proposed by Pecora and Carroll in 1990. They used a Lorenz system with three variables for the demonstration. In their system, an output from one of the variables as a subsystem in a transmitter was sent to a receiver. Then, they showed chaotic synchronization between the transmitter and receiver systems. After their proposal, synchronization phenomena in various chaotic systems including lasers have been reported. The idea and principle of chaos synchronization are described in Appendix A.4. Chaos synchronization between two nonlinear systems is not self-evident and this is a real surprise, since we cannot expect the same output even for the same two chaotic systems as far as the two systems are isolated from each other. The origin of chaos synchronization has not been fully understood yet and the theoretical background has not been established. However, chaos synchronization has been observed by numerical simulations and experiments in various nonlinear systems. In laser systems, synchronization of chaos was experimentally demonstrated in CO_2 lasers (Sugawara et al. 1994) and solid state lasers (Roy and Thornburg 1994). After that, many theoretical and experimental researches for chaos synchronization in various laser systems including semiconductor lasers were published.

Here, we show the general idea of chaos synchronization. Figure 12.1 is a one-to-one system of chaos synchronization. The receiver of chaotic system 2 consists of the same configuration as chaotic transmitter system 1 and also has the same device characteristics as those of system 1. A small portion of the transmitter output is sent to the receiver. In Fig. 12.1a, the transmitter signal is unidirectionally coupled to the receiver and the chaotic output from the receiver synchronizes with the transmitter under an appropriate condition. In laser systems, an optical isolator is usually used to realize unidirectional coupling and the laser output from the transmitter is optically injected to the receiver laser. As a matter of fact, transmitter and receiver lasers may not be the same types as chaotic light sources, or even the transmitter may not be the same kind of laser as the receiver laser. As far as the

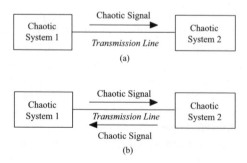

Fig. 12.1. General idea of chaos synchronization of a one-to-one transmitter-receiver system

transmitter can simulate and transmit a possible chaotic waveform of the receiver laser with the same optical frequency, successful chaos synchronization can be achieved. Indeed, a virtual chaotic waveform numerically simulated by a computer is also used as a transmitter signal to a real receiver laser for chaos synchronization. Chaos synchronization is realized for a negative value of the maximum conditional Lyapunov exponent for the difference between the transmitter and receiver signals. Figure 12.1b is a chaos synchronization system of the mutual coupling of signals. The system is not suitable for application of chaotic secure communications, however it has quite interesting characteristics as a chaos synchronization system and has been extensively studied as a chaos synchronization scheme.

Chaos synchronization is attained not only in one-to-one transmitter-receiver systems but also in the multiple transmitter-receiver systems shown in Fig. 12.2. In this system, all the transmitters and receivers may be the same system, but each transmitter laser exhibits different chaotic output from the others. In this case, the parameter values for each pair of the transmitter and receiver systems must be the same and they become the key for chaos synchronization. Otherwise, a transmitter is a different system from each other and one of the receivers may play a counterpart to the transmitter. Chaotic signals from the transmitters are sent through a single transmission line and broadcasted to each receiver. In the receiver systems, each chaotic signal from the transmitters only synchronizes with the corresponding receiver having the same system and device characteristics. Indeed, chaos synchronization has been demonstrated in a few of multiple transmitter-receiver systems (Liu and Davis 2000).

In the proposal of chaos synchronization by Pecora and Carroll, the system is divided into two subsystems. In their model, the transmitter has two subsystems, while the receiver has only one of the two subsystems (for details, the reader is referred to Appendix A.4). The chaotic signal from one of the

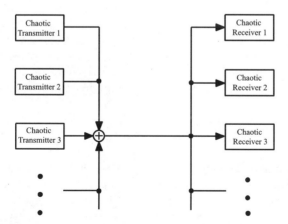

Fig. 12.2. Multiple transmitter-receiver systems of chaos synchronization

subsystems is transmitted to the receiver. Then, the receiver conforms the complete chaotic system by the signal transmission and the receiver synchronizes with the transmitter under an appropriate condition of the parameters. The idea of chaos synchronization was immediately applied in real electronic circuit systems after the proposal by Pecora and Carroll (Cumo et al. 1993). However, the method is not straightforwardly applicable to laser systems, since we cannot divide the dynamics of laser variables into subsystems.

Chaos synchronization strategies developed for most nonlinear systems, such as nonlinear circuits, cannot be directly implemented on semiconductor lasers because of a number of significant differences between semiconductor lasers and other nonlinear dynamical systems. The differences are as follows;

1) A semiconductor laser is an integrated entity that cannot be easily decomposed into subsystems.
2) For a given laser, it is not possible to arbitrarily adjust its intrinsic dynamical parameters and they can be only varied through their linear dependence on the laser power by varying the bias point of the laser.
3) One of its dynamical variables, the carrier density, is not directly accessible externally and, therefore, cannot be used to couple the transmitter and receiver lasers for synchronization.
4) When the output laser field of the transmitter laser is transmitted and coupled to the receiver laser, both its magnitude and phase are transmitted and coupled. It is not possible to only transmit and couple the magnitude but not the phase, or only the phase but not the magnitude.

By using a driving signal to link two chaotic systems, synchronization can be achieved if the difference between the outputs of the two systems possesses a stable fixed point with zero value. As an alternative technique in laser systems, the difference between certain variables in transmitter and receiver lasers can be used as control parameters for synchronization (Annovazzi-Lodi et al. 1996). In semiconductor lasers, master-slave configurations are frequently used as chaos synchronization systems suitable for chaotic secure communications. The schemes of optical feedback, optical injection, and optoelectronic feedback are used as typical chaotic generators in semiconductor lasers. Chaos synchronization in particular systems is discussed in the subsequent sections in this chapter. They are mostly numerical demonstrations of chaos synchronization, however several experimental results have been reported.

12.1.2 Generalized and Complete Chaos Synchronization

There are two different origins of chaos synchronization in nonlinear delay differential systems, such as in semiconductor laser systems of optical feedback and optoelectronic feedback. One is synchronization of chaotic signals based on optical injection phenomena. The other is complete chaos synchronization in which the two systems can be written by a set of the identical rate equations in a mathematical sense. We will discuss the two synchronization schemes in this section. In the ordinary sense, chaos synchronization occurs immediately after a receiver receives a chaotic signal from a transmitter when the transmitter and receiver are divided into several subsystems (see Appendix A.4). In this case, the time lag of the signal in the receiver system is defined by time τ_c, which is the transmission time of signal from the transmitter to the receiver. Namely, using the chaotic signals $\boldsymbol{y}(t)$ and $\boldsymbol{y}'(t)$ from the transmitter and receiver systems, respectively, the relation

$$\boldsymbol{y}'(t) = K_p \boldsymbol{y}(t - \tau_c) \tag{12.1}$$

is obtained (Ohtsubo 2002a). In (12.1), K_p is the proportional coefficient, and $\boldsymbol{y}$ and $\boldsymbol{y}$' are essentially vector variables. In laser systems, this type of chaos synchronization is achieved by optical injection locking and amplification of signals from the transmitter to the receiver. This is the well-known phenomenon of injection locking in laser systems. The receiver output is usually an amplified signal of the transmitted signal (the gain is not necessary larger than unity). Therefore, an excellent synchronized waveform is obtained in the receiver system when the amplification is faithfully achieved. However, distortions are usually introduced to the injection-locked waveforms and the correlation between the transmitter and receiver outputs is less than unity. This scheme is called generalized synchronization.

On the other hand, there exists a different scheme of chaos synchronization from the generalized one in delay differential systems. We assume a system like a delay differential system such as optical feedback or optoelectronic feedback in a semiconductor laser. The differential equation in the transmitter output $\boldsymbol{y}(t)$ is described by

$$\frac{\mathrm{d}\boldsymbol{y}(t)}{\mathrm{d}t} = f(\boldsymbol{y}(t), \mu_p) + \kappa_{p0}\boldsymbol{y}(t - \tau) \tag{12.2}$$

where μ_p is the vector of chaos parameters, κ_{p0} is the feedback coefficient in the system, τ is the delay time, and f is the nonlinear function describing the delay differential system. Assuming that a small portion of the transmitter signal is sent to the receiver, the receiver equation is written by

$$\frac{\mathrm{d}\boldsymbol{y}'(t)}{\mathrm{d}t} = f(\boldsymbol{y}'(t), \mu_p) + \kappa_{p1}\boldsymbol{y}'(t - \tau) + \kappa_{p2}\boldsymbol{y}(t - \tau_c) \tag{12.3}$$

where κ_{p1} is the feedback coefficient in the receiver system, κ_{p2} is the coupling coefficient between the transmitter and the receiver, τ_c is the transmission time of the signal from the transmitter to the receiver. From a comparison between (12.2) and (12.3), we obtain the condition for the equivalent forms of the two differential equations as

$$\boldsymbol{y}'(t) = \boldsymbol{y}(t - \Delta\tau) \tag{12.4}$$

$$\Delta t = \tau_c - \tau \tag{12.5}$$

$$\kappa_{p0} = \kappa_{p1} + \kappa_{p2} \tag{12.6}$$

Under the above conditions, the receiver system is mathematically described by the equivalent equation such as that of the transmitter system and the receiver generates completely the same output as the transmitter (not an amplified signal but a complete copy of the transmitter signal). Therefore, the synchronization scheme is called complete chaos synchronization and it is distinguished from generalized synchronization of chaotic oscillations. The above examples are of chaos synchronization for unidirectionally coupled nonlinear systems. However, we can consider mutually coupled systems for chaos synchronization. In that case, there are also two types of chaos synchronization, i.e., generalized and complete schemes. We will discuss chaos synchronization in mutually coupled laser systems later in this chapter.

The difference between the complete and generalized synchronization is clear from (12.1) and (12.4) and the scheme of chaos synchronization in

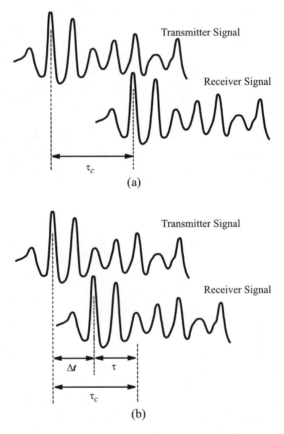

Fig. 12.3. Time lag between transmitter and receiver waveforms in chaos synchronization. Time lags **a** in generalized chaos synchronization and **b** in complete chaos synchronization. τ_c is the transmission time of the signal from the transmitter to the receiver and τ is the optical feedback time in the transmitter and receiver systems

a particular system is easily distinguished by investigating the time lag between the transmitter and receiver outputs. Figure 12.3 shows the relations of time lags in the two schemes. The receiver outputs a synchronized waveform immediately after it receives the transmitter signal in generalized chaos synchronization in Fig. 12.3a. Therefore, the time lag between the two outputs is τ_c. On the other hand, a synchronous chaotic signal in the receiver is generated in advance to receiving the transmitter signal for complete chaos synchronization as shown in Fig. 12.3b. The time lag $\Delta\tau$ in the complete chaos synchronization is less than the signal transmission between the transmitter and receiver systems. Complete chaos synchronization is sometimes called anticipating chaos synchronization due to its origin (Masoller 2001). However, it has been proved that anticipating chaos synchronization is not a unique phenomenon in delay differential systems, but also it is universally observed in differential systems. Indeed, Voss (2000) demonstrated anticipating chaos synchronization in a Rössler system that is described by a set of simple differential equations. Further, it is proved that anticipating chaos synchronization is not equivalent to complete chaos synchronization. Kusumoto and Ohtsubo (2003) observed anticipating chaos synchronization based on the injection locking phenomenon in semiconductor lasers with optical feedback. The investigation of chaos synchronization for the mathematical and physical backgrounds is still undergoing and many subjects are left for future study.

12.2 Theory of Chaos Synchronization in Semiconductor Lasers with Optical Feedback

12.2.1 Model of Synchronization Systems

There are two schemes of chaos synchronization in delay differential systems. A semiconductor laser subjected to optical feedback is a delay differential system and different dynamics like the Lorenz system are observed (see Appendix A.4). The systems of semiconductor lasers with optical feedback have been frequently used for chaotic generators in chaos synchronization and numerous reports have been published (Ohtsubo 2002b and the references therein). In the following, detailed explanations of synchronization in chaotic semiconductor lasers subjected to optical feedback are given. Examples for some other systems will be presented later.

In laser systems, a small portion of the output from one of variables (usually the laser output power or the complex field) is sent to the receiver laser instead of sharing common variables. Chaos synchronization is very sensitive to parameter mismatches between the transmitter and receiver systems. For example, even for semiconductor lasers coming from the same wafer, we cannot expect exactly the same oscillation frequencies for the transmitter and receiver lasers under the same bias injection current. That is in itself a reason why chaos synchronization is difficult to achieve in laser systems. However, laser frequency is easily tuned by changing the bias injection current and a slave laser frequency can be locked to a master laser by optical injection within a certain range of the frequency detuning. Therefore, we can achieve robust chaos synchronization using the frequency-pulling effect by carefully choosing the parameter conditions.

Chaos synchronization is achieved not only in a master-slave configuration of transmitter and receiver systems but also in a mutual coupling system (Fujino and Ohtsubo 2001 and Heil et al. 2001b). We can see complicated dynamics in mutual coupling systems compared with those in unidirectionally coupled systems. A few studies have been reported for chaos synchronization in mutual coupling systems and the study is still undergoing. Mutual coupling systems are not suited for the application to chaotic secure communications, but they are used for phase locking and control of laser arrays (Winful and Rahman 1990, Sauer and Kaiser 1998 and Garcia-Ojalvo et al. 1999). Chaos synchronization has been extensively studied in class B lasers and many experimental results have been reported. The semiconductor laser with optical feedback is the excellent model of chaos synchronization both for the theoretical and experimental studies.

We here discuss systems for chaos synchronization in semiconductor lasers with optical feedback (Ohtsubo 2002a). Figure 12.4 schematically shows the chaos synchronization systems. Figure 12.4a shows a unidirectional coupling system in which the receiver laser is isolated from the transmitter laser by an optical isolator. Both the transmitter and receiver systems have optical feedback loops and this configuration is called a closed-loop system. In Fig. 12.4b, the system is also a unidirectional coupling, but the receiver system does not have a feedback loop. This asymmetric system is called an open-loop system. The robustness and accuracy of chaos synchronization in the open-loop system are quite different from those in the closed-loop system. In chaos synchronization, the transmitter must output a chaotic signal. However, the receiver systems may or may not be chaotic without receiving the transmitter signal. Chaos synchronization is achieved by an injection of a chaotic signal. As a matter of fact, Fig. 12.4b shows a special case of Fig. 12.4a. Indeed, the system in Fig. 12.4a reduces to the system in Fig. 12.4b, when we put the reflectivity of the external reflector equal to zero. We mostly discuss chaos synchronization for the closed-loop configuration of Fig. 12.4a, but the open-loop system is implicitly included in the discussion. Figure 12.4c is a mutual coupling system. Here, the isolator in Fig. 12.4a is removed. Then, each laser behaves as a transmitter and a receiver. In this system, each laser outputs different chaotic signals or steady-state signals before coupling. After the coupling, the two lasers output the same chaotic signal. There are also two synchronization schemes (complete and generalized schemes) in the mutual coupling case.

(a) Closed-Loop System

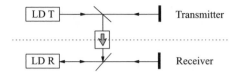

(b) Open-Loop System

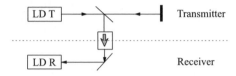

(c) Mutual Coupling System

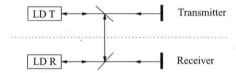

Fig. 12.4. Schematic diagram of chaos synchronization systems in semiconductor lasers with optical feedback. **a** Symmetric unidirectional coupling system, **b** asymmetric unidirectional coupling system, and **c** mutual coupling system. LD T: transmitter laser, LD R: receiver laser

12.2.2 Rate Equations in Unidirectional Coupling Systems

In this section, we investigate the theoretical treatment for chaos synchronization in the unidirectional coupling closed-loop system shown in Fig. 12.4a. The rate equations for the transmitter and receiver lasers are written by the same equations as those for the model discussed in Chap. 4 except for the light transmission term in the receiver rate equations (Ahlers et al. 1998). The rate equations for the transmitter laser are written by

$$\frac{\mathrm{d}A_{\mathrm{T}}(t)}{\mathrm{d}t} = \frac{1}{2}G_{\mathrm{n,T}}\{n_{\mathrm{T}}(t) - n_{\mathrm{th,T}}\}A_{\mathrm{T}}(t) + \frac{\kappa_{\mathrm{T}}}{\tau_{\mathrm{in,T}}}A_{\mathrm{T}}(t - \tau_{\mathrm{T}})\cos\theta_{\mathrm{T}}(t) \quad (12.7)$$

$$\frac{\mathrm{d}\phi_{\mathrm{T}}(t)}{\mathrm{d}t} = \frac{1}{2}\alpha_{\mathrm{T}}G_{\mathrm{n,T}}\{n_{\mathrm{m}}(t) - n_{\mathrm{th,T}}\} - \frac{\kappa_{\mathrm{T}}}{\tau_{\mathrm{in,T}}}\frac{A_{\mathrm{T}}(t - \tau_{\mathrm{T}})}{A_{\mathrm{T}}(t)}\sin\theta_{\mathrm{T}}(t) \quad (12.8)$$

$$\frac{\mathrm{d}n_{\mathrm{T}}(t)}{\mathrm{d}t} = \frac{J_{\mathrm{T}}}{ed} - \frac{n_{\mathrm{T}}(t)}{\tau_{\mathrm{s,T}}} - G_{\mathrm{n,T}}\{n_{\mathrm{T}}(t) - n_{\mathrm{0,T}}\}A_{\mathrm{T}}^2(t) \quad (12.9)$$

$$\theta_{\mathrm{T}}(t) = \omega_{\mathrm{0,T}}\tau + \phi_{\mathrm{T}}(t) - \phi_{\mathrm{T}}(t - \tau_{\mathrm{T}}) \quad (12.10)$$

where subscript T represents the transmitter laser. The rate equations for the receiver laser read

$$\frac{dA_R(t)}{dt} = \frac{1}{2}G_{n,R}\{n_R(t) - n_{th,R}\}A_R(t)$$
$$+ \frac{\kappa_R}{\tau_{in,R}}A_R(t - \tau_R)\cos\theta_R(t) + \frac{\kappa_{cp}}{\tau_{in,R}}A_R(t - \tau_c)\cos\xi_c(t) \quad (12.11)$$

$$\frac{d\phi_R(t)}{dt} = \frac{1}{2}\alpha_R G_{n,R}\{n_R(t) - n_{th,R}\} - \frac{\kappa_R}{\tau_{in,R}}\frac{A_R(t - \tau_R)}{A_R(t)}\sin\theta_R(t)$$
$$- \frac{\kappa_{cp}}{\tau_{in,R}}\frac{A_R(t - \tau_c)}{A_R(t)}\sin\xi_c(t) \quad (12.12)$$

$$\frac{dn_R(t)}{dt} = \frac{J_R}{ed} - \frac{n_R(t)}{\tau_{s,R}} - G_{n,R}\{n_R(t) - n_{0,R}\}E_R^2(t) \quad (12.13)$$

$$\theta_R(t) = \omega_{0,R}\tau + \phi_R(t) - \phi_R(t - \tau_R) \quad (12.14)$$

$$\xi_c(t) = \omega_{0,T}\tau_c + \phi_R(t) - \phi_T(t - \tau_c) + \Delta\omega t \quad (12.15)$$

where subscript R denotes the receiver lasers, κ_{cp} is the injection rate from the transmitter to the receiver laser, and $\Delta\omega$ is the angular frequency detuning. The last terms in (12.11) and (12.12) are the effect of the chaotic signal from the transmitter. When the external feedback is zero in the receiver system, i.e., $\kappa_R = 0$, the model reduces to the open-loop system in Fig. 12.4b.

12.2.3 Generalized Chaos Synchronization

One of the origins of chaos synchronization in a semiconductor laser with optical feedback is the injection locking and amplification phenomenon in a system modeled by delay differential equations. The condition for complete chaos synchronization, which is discussed in the next section, is very strict and most cases of chaos synchronization observed in lasers are based on the injection locking and amplification phenomenon. Therefore, experimental results of chaos synchronization in laser systems were mostly for generalized chaos synchronization. We here consider the condition for the generalized chaos synchronization in a system of a semiconductor laser with optical feedback. For generalized chaos synchronization, the average optical power injected to the receiver laser is large, as much as several tens of percent in amplitude (several percent in intensity), while it is much less than several percents for the case of complete chaos synchronization. In generalized chaos synchronization, the relation between the field amplitudes for the transmitter and receiver lasers is given by

$$A_R(t) \propto A_T(t - \tau_c) \quad (12.16)$$

Namely, the synchronized chaotic output in the receiver is generated upon receiving the transmitter signal. Therefore, the time lag between the outputs of the transmitter and receiver lasers is equal to the transmission time τ_c.

12.2.4 Complete Chaos Synchronization

Next, we consider the conditions where the two rate equations for the transmitter and receiver lasers are written by the identical set of equations, namely, the conditions for complete chaos synchronization. The model of chaotic generators for the transmitter and the receiver is also a semiconductor laser with optical feedback. We assume that the device parameters in the two lasers are the same and the two lasers oscillate at the same frequency, i.e., zero frequency detuning $\Delta\omega = \omega_{0,m} - \omega_{0,s} = 0$. Further, the lasers are biased at the same injection current and the external feedback conditions are also the same for the transmitter and receiver lasers, except for different values of the feedback coefficients, κ_T and κ_R. Under these assumptions, the conditions for complete chaos synchronization read (Ohtsubo 2002b)

$$A_R(t) = A_T(t - \Delta t) \tag{12.17}$$

$$\phi_R(t) = \phi_T(t - \Delta t) - \omega_0 \Delta t(\quad \text{mod } 2\pi) \tag{12.18}$$

$$n_R(t) = n_T(t - \Delta t) \tag{12.19}$$

$$\kappa_R = \kappa_T + \eta_c \tag{12.20}$$

$$\Delta t = \tau_c - \tau \tag{12.21}$$

The delay differential equations (12.11)–(12.13) in the receiver laser have completely identical forms to those in (12.7)–(12.9) of the transmitter laser. The scheme is called compete chaos synchronization. The receiver laser outputs the synchronous chaotic signal before receiving the transmitted signal by anticipating it in advance to the time $\tau = \tau_T = \tau_R$. The parameters in the two laser systems must be identical to satisfy the conditions for complete chaos synchronization, however there are certain ranges of tolerances for the parameter mismatches when we allow a little deterioration of the correlation between the transmitter and receiver outputs. Usually, it is not easy to achieve complete chaos synchronization in real laser systems and a few experimental studies for complete chaos synchronization have been reported (Liu et al. 2002a).

12.2.5 Mutual Coupling Systems

We discuss chaos synchronization in mutually coupled semiconductor lasers modeled in Fig. 12.4c. For simplicity, we put the reflectivities of external mirrors in the transmitter and receiver systems equal to zero without loss of generality, i.e., we remove the external mirrors. In the mutual coupling system, each laser plays a role for the virtual external mirror to the counterpart laser. Therefore, even without the optical feedback loop, the lasers can show chaotic oscillations due to mutual optical injections, as discussed in Chap. 6. Mutual coupling lasers with optical feedback is a straightforward extension of the discussion here. We can also observe both complete and generalized chaos synchronization in mutually coupled semiconductor lasers (Hohl et al. 1997 and 1999). The rate equations for one of the lasers are written by

$$\frac{dA_1(t)}{dt} = \frac{1}{2}G_{n,1}\{n_1(t) - n_{th,1}\}A_1(t) + \frac{\kappa_{inj,2}}{\tau_{in,1}}A_2(t - \tau_c)\cos\theta_1(t) \quad (12.22)$$

$$\frac{d\phi_1(t)}{dt} = \frac{1}{2}\alpha_1 G_{n,1}\{n_1(t) - n_{th,1}\} - \frac{\kappa_{inj,2}}{\tau_{in,1}}\frac{A_2(t - \tau_c)}{A_1(t)}\sin\theta_1(t) \quad (12.23)$$

$$\frac{dn_1(t)}{dt} = \frac{J_1}{ed} - \frac{n_1(t)}{\tau_{s,1}} - G_{n,1}\{n_1(t) - n_{0,1}\}A_1^2(t) \quad (12.24)$$

$$\theta_1(t) = \omega_{0,1}\tau + \phi_1(t) - \phi_2(t - \tau_c) + \Delta\omega t \quad (12.25)$$

The rate equations for the other laser are also given by symmetrical forms as

$$\frac{dA_2(t)}{dt} = \frac{1}{2}G_{n,2}\{n_2(t) - n_{th,2}\}A_2(t) + \frac{\kappa_{inj,1}}{\tau_{in,2}}A_1(t - \tau_c)\cos\theta_2(t) \quad (12.26)$$

$$\frac{d\phi_2(t)}{dt} = \frac{1}{2}\alpha_2 G_{n,2}\{n_2(t) - n_{th,2}\} - \frac{\kappa_{inj,1}}{\tau_{in,2}}\frac{A_1(t - \tau_c)}{A_2(t)}\sin\theta_2(t) \quad (12.27)$$

$$\frac{dn_2(t)}{dt} = \frac{J_2}{ed} - \frac{n_2(t)}{\tau_{s,2}} - G_{n,2}\{n_2(t) - n_{0,2}\}E_2^2(t) \quad (12.28)$$

$$\theta_2(t) = \omega_{0,2}\tau + \phi_2(t) - \phi_1(t - \tau_c) - \Delta\omega t \quad (12.29)$$

where subscripts 1 and 2 are for the respective lasers and $\Delta\omega = \omega_1 - \omega_2$ is the angular frequency detuning between the two lasers.

In the mutual coupling systems, there are two solutions of chaos synchronization: one is based on injection locking and amplification phenomena and the other is complete chaos synchronization. For the case of synchronization due to injection locking, one of the two lasers plays the role of a master laser and the other is a slave. Then, the relation between the two amplitudes is written by

$$A_2(t) \propto A_1(t - \tau_c) \quad (12.30)$$

or

$$A_2(t - \tau_c) \propto A_1(t) \quad (12.31)$$

These are not exact solutions for (12.22)–(12.29) in a mathematical sense. However, these relations are confirmed by numerical simulations and experiments. Most cases of chaos synchronization observed in real experiments in mutually coupled semiconductor lasers are based on generalized chaos synchronization. In these cases, the optical transmission power is as large as several tens of percent of the average amplitude of the chaotic variation. The percentage is almost the same as that in a unidirectional coupling system of a generalized chaos synchronization scheme. Which laser becomes master

or slave is determined by the differences of the operating conditions of the
lasers and the parameter mismatches.

On the other hand, there is an identical solution for complete chaos
synchronization in mutually coupled semiconductor lasers. The conditions
follow

$$\Delta\omega = 0 \tag{12.32}$$
$$A_2(t) = A_1(t) \tag{12.33}$$
$$\phi_2(t) = \phi_1(t) \tag{12.34}$$
$$n_2(t) = n_1(t) \tag{12.35}$$

Namely, the two lasers simultaneously output the same chaotic signals even
for a finite transmission time τ_c of light. The scheme is also considered as
anticipating chaos synchronization.

12.3 Chaos Synchronization in Semiconductor Lasers with an Optical Feedback System

12.3.1 Chaos Synchronization – Numerical Examples

We here show some numerical simulations of chaos synchronization in the
closed-loop systems shown in Fig. 12.4b. Figure 12.5 shows examples of generalized and complete chaos synchronization (Murakami and Ohtsubo 2002).
Figure 12.5a shows a chaotic signal to be transmitted. Figure 12.5b shows
the receiver output under the condition of generalized chaos synchronization and Fig. 12.5c shows the correlation plot between the waveforms of
Figs. 12.5 a and b. Figure 12.5d shows the receiver output under the condition of complete chaos synchronization and Fig. 12.5e is the correlation plot
between the waveforms of Figs. 12.5a and d. The optical transmission power
is 22% ($\kappa_{cp}/\tau_{in,R} = 74.9\,\mathrm{ns}^{-1}$) in the generalized chaos synchronization. On
the other hand, it is as small as 1.5×10^{-4}% ($\kappa_{cp}/\tau_{in,R} = 1.96\,\mathrm{ns}^{-1}$) in the
complete case. The time for the light transmission between the transmit-
ter and receiver lasers is set to zero for simplicity in this figure. Therefore,
the time lag between the two lasers is zero for generalized synchronization,
while it is -1 ns for complete chaos synchronization. An excellent correlation
between the transmitter and receiver outputs is obtained for the complete
chaos synchronization. The difference of the time is exactly equal to the theoretically expected time lag $\Delta\tau$. Thus, we can distinguish the type of chaos
synchronization by investigating the time lag between the transmission signal
and the receiver output.

The attractors in the transmitter and receiver lasers show the same orbit
under complete chaos synchronization, since the two systems follow completely the identical equations. Then, the receiver output traces the same
orbit as that of the transmitter due to injection of a small seed from the

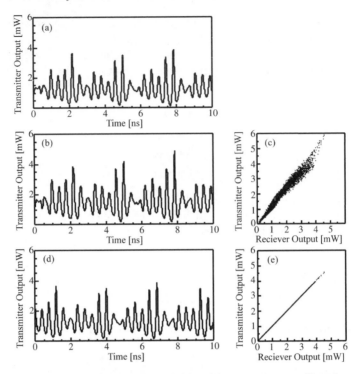

Fig. 12.5. Chaos synchronization in a closed-loop system. **a** Chaotic transmission signal, **b** receiver output of generalized chaos synchronization at $\kappa_{cp}/\tau_{in,R} = 74.9\,\mathrm{ns}^{-1}$, **c** correlation plot for **a** and **b**, **d** receiver output of complete chaos synchronization at $\kappa_{cp}/\tau_{in,R} = 1.96\,\mathrm{ns}^{-1}$, and **e** correlation plot for **a** and **d**. The conditions are $J = 1.3 J_{th}$, $\tau = 1\,\mathrm{ns}$, $\kappa_T/\tau_{in,T} = 1.96\,\mathrm{ns}^{-1}$, and $\Delta\omega = 0$

transmitter. On the other hand, the receiver output is an amplified copy of the transmitter signal in generalized chaos synchronization. Therefore, the synchronized signal almost looks the same as the waveform of the transmitter, however the chaotic attractor in the receiver laser has some deviations from that of the transmitter. Figure 12.6 shows chaotic attractors of the receiver laser in the phase space of the laser output power and the carrier density. Figure 12.6a is the chaotic attractor of the transmitter signal in Fig. 12.5a. Figure 12.6b is the chaotic attractor of the receiver output corresponding to Fig. 12.5b. The general view of the orbit is quite similar to Fig. 12.6a, but they are different. The extent of the orbit in Fig. 12.6b is slightly larger than that of Fig. 12.6 a and the receiver signal is amplified. Also, the carrier density in Fig. 12.6b is lowered to less than the threshold by the strong optical injection from the transmitter laser and this results in the reduction of the gain. For the case of complete chaos synchronization in Fig. 12.5d, the chaotic orbit of the receiver laser is the same as in Fig. 12.6a. From these facts, generalized

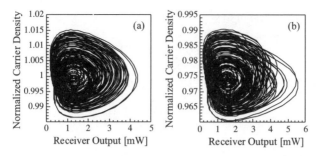

Fig. 12.6. Chaotic attractors in the phase space of laser output power and carrier density. **a** Chaotic attractors correspond to **a** Fig. 12.5b and **b** Fig. 12.5d

chaos synchronization is clearly a different phenomenon from complete chaos synchronization.

Optical injection is widely used for signal transmission from transmitter to receiver lasers in chaotic communications. When we consider optical injection, the stable and unstable map in the phase space of the injection ratio and the frequency detuning it is very useful to know the injection properties as discussed in Chap. 6. Here, we show the conditions and distributions of successful chaos synchronization using the map. Figure 12.7 presents the map of stable and unstable injection locking areas. The boundaries of stable and unstable injection locking, and unlocking for the solitary laser are shown

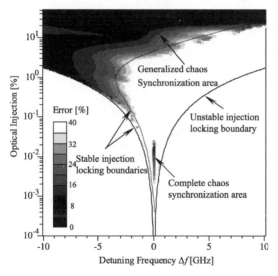

Fig. 12.7. Map of stable and unstable injection locking in receiver laser and areas of generalized and complete chaos synchronization. An open-loop optical feedback system is assumed. The reflectivity of the external optical feedback mirror in the transmitter system is $10^{-2}\%$ in intensity

as solid curves in the figure. The vertical axis is the optical injection (in intensity) from the transmitter to the receiver laser and the horizontal axis is the frequency detuning between the transmitter and receiver lasers. Excellent chaos synchronization is attained at the dark areas in the map. The error of chaos synchronization in the figure is defined by the following equation:

$$\sigma_{\text{error}} = \frac{\langle |S_{\text{T}} - S_{\text{R}}| \rangle}{\langle S_{\text{R}} \rangle} \tag{12.36}$$

where S_{T} and S_{R} are the intensities of the transmitter and receiver lasers, and $\langle \cdot \rangle$ denotes the ensemble average. Generalized chaos synchronization occurs in a wide range of the frequency detuning and the optical injection in the stable injection-locking area, while complete chaos synchronization takes place at the unstable injection-locking area. From the comparison of this map with Fig. 6.7, complete chaos synchronization is attained at chaotic states within the unstable injection-locking area in a simple optical injection-locked laser. The area of complete chaos synchronization is very narrow with zero detun-

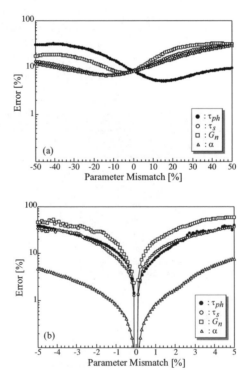

Fig. 12.8. Calculated synchronization error as a function of parameter mismatch. τ_{ph}: photon lifetime, τ_{s}: carrier lifetime, G_{n}: gain, and α: linewidth enhancement factor. **a** Generalized and **b** complete chaos synchronization. The parameter values are the same as those in Fig. 12.5

ing and small optical injection due to the requirement of strict parameter coincidence.

The effects of parameter mismatches between the transmitter and receiver systems are very important for applications of chaos synchronization to secure optical communications. Figure 12.8 shows the plots of synchronization errors for the mismatches of various laser device parameters. Figure 12.8 a shows the errors of generalized chaos synchronization. The permissible errors for the parameter mismatches are large in generalized chaos synchronization and we can expect robust chaos synchronization. However, the synchronization errors are always larger than 1% of the average laser intensity variations, since the origin of the synchronization comes from optical injection locking and amplification phenomena, and distortions of the synchronous waveform from the transmitter signal are always presented. From the investigation of the transmitter and receiver waveforms, they are still quite similar with each other when the errors for the parameter mismatches are less than a few percent. It is noted that the best synchronization is not always attained at zero parameter mismatches. Figure 12.8b is the effects of parameter mismatches in complete chaos synchronization. As expected, chaos synchronization is achieved with high accuracy at almost zero parameter mismatches and the synchronization errors rapidly increase with the increase of the parameter mismatches. Thus, strict conditions are required for successful chaos synchronization in the complete case.

12.3.2 Chaos Synchronization – Experimental Examples

Many theoretical studies on chaos synchronization in semiconductor lasers with optical feedback have been published, but only a few experimental investigations have been reported in real experimental systems (Takiguchi et al. 1999 and Fujino and Ohtsubo 2000, Fischer et al. 2000, and Sivaprakasam et al. 2000a). In this section, we show some examples of experimental results for chaos synchronization. Figure 12.9 shows the experimental results of chaos synchronization in a closed-loop system discussed in Fig. 12.4a. Figure 12.9a is the output waveforms of the transmitter and receiver lasers without signal transmission. As far as the two lasers are isolated, the output powers have no correlation as shown in Fig. 12.9b. When a fraction of the transmitter output is sent to the receiver, the synchronous waveform in Fig. 12.9c is obtained. The transmitted optical power from the transmitter to the receiver is rather strong, as much as 4.6% of the average power of the receiver laser. Therefore, the synchronization is a generalized case. In Fig. 12.9a, the two lasers show chaotic outputs. However, it is not always necessary for the receiver laser to be oscillated at a chaotic state and the receiver laser may be a steady-state oscillation even in the presence of optical feedback. The feedback level in the receiver laser is usually less than that in the transmitter laser and the receiver laser may show a synchronous chaotic oscillation after the optical injection from the transmitter. Chaos synchronization has

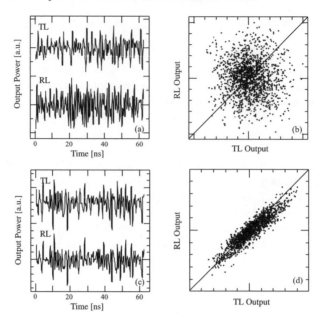

Fig. 12.9. Experimental chaos synchronization in a closed-loop system. **a** Waveforms of transmitter and receiver lasers without coupling. **b** Correlation plot for **a**. **c** Waveforms of transmitter and receiver lasers with coupling. The transmitted optical power is 4.6% of the average optical power of the receiver laser. **d** Correlation plot for **c**

been also demonstrated in an open-loop system. Of course, the receiver laser oscillates at a steady-state without coupling of the transmitter signal in that case. The robustness of chaos synchronization is much dependent on whether the system is an open- or closed-system. We will again discuss the differences in Chap. 13.

Only a few experimental studies have been reported for complete chaos synchronization, since the conditions of complete chaos synchronization are too severe to be achieved in real experiments (Sivaprakasam et al. 2001a and Liu et al. 2002a). At complete chaos synchronization, the time lag of the waveforms between the transmitter and receiver lasers is given by $\Delta\tau = \tau_c - \tau$. Liu et al. (2002a) conducted complete chaos synchronization in an open-loop system of semiconductor lasers with optical feedback. They changed the external cavity length and examined the time lag between the transmitter and receiver signals. They observed the change of the time lag proportional to the external cavity length (the proportional coefficient is negative) and showed that their schemes were for complete chaos synchronization. In their experiment, the parameters of the transmitter and receiver lasers were carefully chosen to have almost the same characteristics and the initial frequency detuning between the transmitter and receiver lasers was set to be less than several tens of MHz.

12.3.3 Anticipating Chaos Synchronization

Anticipating chaos synchronization was at first introduced as a synchronization phenomenon peculiar to nonlinear delay differential systems. Later, it was proved that anticipating chaos synchronization is also observed in low dimensional dissipative systems described by simple differential equations (Ahlers et al. 1998, Voss 2000, and Ohtsubo 2002b). Voss demonstrated that anticipating chaos synchronization is realized in a Rössler system (continuous system) that is described by three differential equations. Therefore, anticipating chaos synchronization is not only a unique feature in delay differential systems, but also it is a universal phenomenon in chaotic nonlinear systems. In chaos synchronization systems of semiconductor lasers with optical feedback, anticipating chaos synchronization was also observed outside of the parameter regions for ordinary complete chaos synchronization in the stable and unstable injection-locking map. In that case, the chaos synchronization originated from optical injection locking and amplification effects, but the time lag of the waveforms between the transmitter and receiver lasers was equal to that of anticipating synchronization.

Kusumoto and Ohtsubo (2003) conducted a detailed study of chaos synchronization in the stable injection locking area in Fig. 12.7. Figure 12.10 shows their results. Figure 12.10a plots the anticipating chaos synchronization in the stable injection locking area. The time lag corresponds to that of anticipating synchronization, but the synchronization originates from the ordinary injection locking effect. Within the white circle in the figure, the value of the correlation coefficient between waveforms of the transmitter and receiver lasers exceeds 0.94. In this open-loop system, the optical feedback ratio in the transmitter system is as high as 0.3. Complete chaos synchronization is achieved around the optical injection rate of 0.3 at zero frequency detuning (marked A). Of course, the synchronization is an anticipating one under this condition. However, the area of anticipating chaos synchronization expands over a wide region in the stable injection-locking map. For example, the synchronization at point B is still an anticipating one as a time lag of the waveforms, but the synchronization originates from the injection locking effect. Figure 12.10b plots the trajectories of the transmitter and receiver outputs corresponding to point B in Fig. 12.10a. The plot is in the phase space of the phase difference and the normalized carrier density. Black trace denotes the trajectory for the transmitter laser and gray trace is for the receiver laser. If the chaos synchronization is complete, the trajectory of the transmitter laser perfectly overlaps with that of the receiver in the map. However, the two trajectories are separated from each other in the phase space. This phase shift between the two trajectories is equal to the frequency detuning between the two lasers. Also the carrier density of the receiver laser is lowered by the optical injection. This fact proves that the phenomenon originates from optical injection locking and amplification. According to the detailed study by Peters-Flynn et al. (2006), the laser output that is cate-

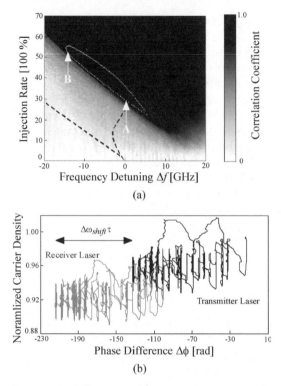

Fig. 12.10. a Anticipating chaos synchronization region in the phase space of frequency detuning Δf and optical injection rate in the open-loop system. The feedback rate in the transmitter system is 0.3. **b** Chaotic trajectories for transmitter and receiver lasers at point B in Fig. 12.10a. The trajectories are plotted in the phase space of the phase difference $\Delta\phi = \phi_j(t) - \phi_j(t - \tau_t)$ and the normalized carrier density $n_j/n_{\mathrm{th},j}$ ($j = T$ or R). The *black trace* is for the transmitter laser and the *gray trace* for the receiver laser

gorized as anticipating chaos synchronization in the stable injection locking area in Fig. 12.10 sometimes shows a mixed state of waveforms corresponding to anticipating and injection amplification signals. The two states irregularly switch in time in their numerical simulations. The details of the phenomena and the origin of the switching are not fully understood yet. Anticipating chaos synchronization is not a unique phenomenon accompanying complete chaos synchronization, but it is a universal nature in nonlinear chaotic systems.

12.3.4 Bandwidth Enhanced Chaos Synchronization

We discussed the enhancement of the cutoff frequency in a chaotic semiconductor laser by a strong optical injection in Sect. 6.3. Such semiconductor

lasers are used as light sources of chaotic generators for chaos synchronization and communications. The cutoff frequency can be varied by adjusting
the fraction of the optical injection. For example, a modulation bandwidth of
~20 GHz for the original relaxation oscillation frequency of $3 \sim 4$ GHz was attained by strong optical injection. In chaotic communications, the maximum
data transmission rate is determined by the cutoff frequency of chaotic carrier
signals and the cutoff frequency is roughly equal to the maximum modulation
bandwidth of the laser. Higher modulation bandwidth is also demanded in
various applications such as direct modulations in semiconductor lasers. Figure 12.11 shows the schematic diagram of open-loop chaos synchronization
systems with enhanced chaotic frequency (Takiguchi et al. 2003). Both the
transmitter and receiver semiconductor lasers, LD T and LD R, are strongly
injected from external semiconductor lasers, LD1 and LD2, with the same
characteristics of the device parameters. Both the transmitter and receiver
lasers oscillate at the stable injection-locked state in the absence of optical feedback. Figure 12.12 demonstrates an example of bandwidth-enhanced
chaos synchronization. The conditions are the same as those in Fig. 6.18. The
upper trace in Fig. 12.12a is a time series of the transmitter output and the
lower one is that of the receiver. The frequency detuning between the transmitter and receiver lasers is set to be zero and the observed time lag is equal
to $\Delta t = \tau_c - \tau = -6$ ns. Therefore, the synchronization scheme is for the
complete case or so called anticipating chaos synchronization. Figure 12.12b
is the correlation plot. The correlation coefficient is calculated to be 0.954
and the two lasers show good synchronization. However, we obtain a better
figure of the correlation coefficient for complete chaos synchronization in the
absence of strong optical injection. The range for small synchronization error

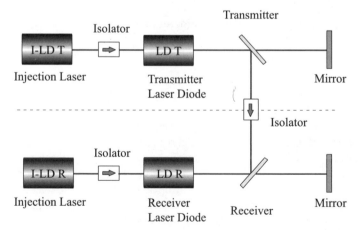

Fig. 12.11. Schematic diagram of bandwidth-enhanced chaos synchronization system. I-LD T and I-LD R are the injection lasers to the transmitter and receiver
lasers. LD T and LD R are the transmitter and receiver lasers

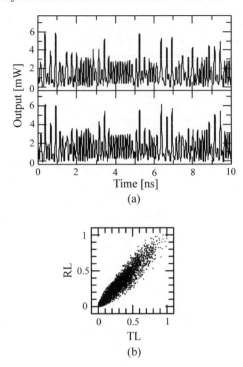

Fig. 12.12. Bandwidth-enhanced chaotic time series **a** and their correlation plot **b**. The time lag of $\Delta t = -6$ ns between the two waveforms is compensated. The conditions for the lasers are the same as those in Fig. 6.18. The synchronization scheme is complete

is very narrow for the parameter mismatches in the complete case. Even if the two lasers have the same device parameters and operate under the same conditions, chaos synchronization is realized under the limited parameter values and their ranges are usually very narrow.

12.3.5 Incoherent Synchronization Systems

The frequency detuning of the transmitter and receiver lasers plays a crucial role for the performance of chaos synchronization when the two lasers coherently couple. The difference of the frequencies must be at least within a few GHz. As discussed in Sect. 5.6, we can observe chaotic oscillations in systems of semiconductor lasers with incoherent optical feedback. Chaos synchronization is also realized in incoherent systems. We do not pay particular attention to the frequency detuning in this system. Figure 12.13 schematically shows the setup of an open-loop chaos synchronization system in semiconductor lasers with incoherent optical feedback. In the optical setup, the feedback light in the transmitter system is incoherently coupled with the internal laser

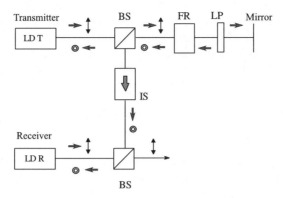

Fig. 12.13. Setup of an incoherent open-loop chaos synchronization system. The feedback light in the transmitter laser is incoherently coupled with the internal optical field of the laser. The transmitted light is also incoherently coupled to the receiver laser. RF: Faraday rotator, PL: polarizer. Symbols ⊙ and ↕ denote the two orthogonal polarization directions of light

field using the combination of a Faraday rotator (RF) and a polarizer (PL) in the optical feedback loop. The transmission light is also incoherently coupled to the receiver laser.

In incoherent chaos synchronization, we do not need to consider the rate equation for the optical phase. Therefore, the model is described by the equations for the photon number and the carrier density. For the transmitter laser, we obtain (Rogister et al. 2001)

$$\frac{\mathrm{d}S_{\mathrm{T}}(t)}{\mathrm{d}t} = G_{\mathrm{n,T}}\{n_{\mathrm{T}}(t) - n_{\mathrm{th,T}}\}S_{\mathrm{T}}(t) + R_{\mathrm{sp,T}} \tag{12.37}$$

$$\frac{\mathrm{d}n_{\mathrm{T}}(t)}{\mathrm{d}t} = \frac{J_{\mathrm{T}}}{ed} - \frac{n_{\mathrm{T}}(t)}{\tau_{\mathrm{s,T}}} - G_{\mathrm{n,T}}\{n_{\mathrm{T}}(t) - n_{0,\mathrm{T}}\}$$
$$\times \left\{ S_{\mathrm{T}}(t) + \frac{\kappa'_{\mathrm{T}}}{\tau_{\mathrm{in,s}}}S_{\mathrm{T}}(t - \tau_{\mathrm{T}}) \right\} \tag{12.38}$$

For the receiver

$$\frac{\mathrm{d}S_{\mathrm{R}}(t)}{\mathrm{d}t} = G_{\mathrm{n,R}}\{n_{\mathrm{R}}(t) - n_{\mathrm{th,R}}\}S_{\mathrm{R}}(t) + R_{\mathrm{sp,R}} \tag{12.39}$$

$$\frac{\mathrm{d}n_{\mathrm{R}}(t)}{\mathrm{d}t} = \frac{J_{\mathrm{R}}}{ed} - \frac{n_{\mathrm{R}}(t)}{\tau_{\mathrm{s,R}}} - G_{\mathrm{n,R}}\{n_{\mathrm{R}}(t) - n_{0,\mathrm{R}}\}$$
$$\times \left\{ S_{\mathrm{R}}(t) + \frac{\kappa'_{\mathrm{R}}}{\tau_{\mathrm{in,R}}}S_{\mathrm{R}}(t - \tau_{\mathrm{R}}) + \kappa_{\mathrm{cp}}S_{\mathrm{T}}(t - \tau_{\mathrm{c}}) \right\} \tag{12.40}$$

Also, the subscripts T and R are for the transmitter and receiver lasers. The feedback light is coupled to the carrier density as a delayed signal. The final term in (12.40) is the coupling of incoherent light from the transmitter. Since the coupling between the transmitter and receiver lasers is incoherent, the receiver laser is not injection-locked. However, complete chaos synchronization is also achieved under the appropriate conditions. Assuming all the device parameters of the two lasers to be the same, the conditions read

$$P_R(t) = P_T(t - \Delta t) \tag{12.41}$$
$$n_R(t) = n_T(t - \Delta t) \tag{12.42}$$

where $\Delta t = \tau_c - \tau$. Under these conditions, complete anticipating chaos synchronization is realized. As a matter of course, the condition for the frequencies of the lasers is not included.

12.3.6 Polarization Rotated Chaos Synchronization

Chaotic oscillations of semiconductor lasers are observed not only by parallel-polarization optical feedback, but also by polarization-rotated optical feedback. The system of polarization-rotated optical was described in Sects. 4.5.2 and 5.7.2. The dynamics of polarization-rotated chaos synchronization were studied theoretically and experimentally (Sukow et al. 2004, 2005, 2006; Ju 2005; Shibasaki et al. 2006). A semiconductor laser with polarization-rotated optical feedback can be used as a light source for a system of chaotic synchronization and communications. Here, we discuss chaos synchronization in a system of semiconductor lasers with polarization-rotated optical feedback. As an example, take a chaotic generator shown in Fig. 4.8b. The system we consider is shown in Fig. 12.14, in which the laser light with TE mode emitted from a transmitter laser (LD T) passes through a Faraday rotator and is reflected by a partial mirror (PM). The feedback light again passes through the Faraday rotator and fed back into the laser cavity as a TM mode light (cross-polarized component to the TE mode). While the light passed through the partial mirror goes through another Faraday rotator and is converted to a TM polarization mode. Then the crossed-polarization light is injected to the receiver laser. Thus, the polarization-rotated chaos synchronization is realized under appropriate conditions for the transmitter and receiver lasers.

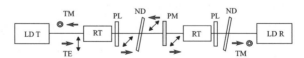

Fig. 12.14. System for polarization rotated chaos synchronization. LD T: transmitter laser, LD R: receiver laser, TE: TE polarization mode, TM: TM polarization mode, RT: polarization rotator (45° rotation), PL: polarizer, ND: neutral density filter

In the open-loop chaos synchronization system in Fig. 12.14, the rate equations for the transmitter read (Shibasaki et al. 2005)

$$\frac{dE_{\text{T,TE}}(t)}{dt} = \frac{1}{2}(1 - i\alpha_{\text{T}})G_{\text{n,T,TE}}\{n(t) - n_{\text{th,T,TE}}\}E_{\text{T,TE}}(t) \qquad (12.43)$$

$$\frac{dE_{\text{T,TM}}(t)}{dt} = \frac{1}{2}(1 - i\alpha_{\text{T}})G_{\text{n,T,TM}}\{n(t) - n_{\text{th,T,TM}}\}E_{\text{T,TM}}(t)$$
$$+ \frac{\kappa}{\tau_{\text{in}}}E_{\text{T,TE}}(t - \tau)\exp(-i\Delta\omega_{\text{TE,TM}}t$$
$$+ i\omega_{0T}\tau + \phi_{\text{T,TM}}(t) - \phi_{\text{T,TE}}(t - \tau)) \qquad (12.44)$$

$$\frac{dn(t)}{dt} = \frac{J_{\text{T}}}{ed} - \frac{n(t)}{\tau_{\text{s,T}}} - \{n(t) - n_{\text{0,T}}\}$$
$$\times \{G_{\text{n,T,TE}}|E_{\text{T,TE}}(t)|^2 + G_{\text{n,T,TM}}|E_{\text{T,TM}}(t)|^2\} \qquad (12.45)$$

For the receiver systems, the rate equations are written by

$$\frac{dE_{\text{R,TE}}(t)}{dt} = \frac{1}{2}(1 - i\alpha_{\text{R}})G_{\text{n,R,TE}}\{n(t) - n_{\text{th,R,TE}}\}E_{\text{R,TE}}(t) \qquad (12.46)$$

$$\frac{dE_{\text{R,TM}}(t)}{dt} = \frac{1}{2}(1 - i\alpha_{\text{R}})G_{\text{n,R,TM}}\{n(t) - n_{\text{th,R,TM}}\}E_{\text{R,TM}}(t)$$
$$+ \frac{\kappa_{\text{cp}}}{\tau_{\text{in,R}}}E_{\text{T,TE}}(t - \tau_{\text{c}})\exp(-i\Delta\omega t$$
$$+ i\omega_{0T}\tau + i\phi_{\text{R,TM}}(t) - i\phi_{\text{T,TE}}(t - \tau_{\text{c}})) \qquad (12.47)$$

$$\frac{dn(t)}{dt} = \frac{J_{\text{R}}}{ed} - \frac{n(t)}{\tau_{\text{s,R}}} - \{n(t) - n_{\text{0,R}}\}$$
$$\times \{G_{\text{n,R,TE}}|E_{\text{R,TE}}(t)|^2 + G_{\text{n,R,TM}}|E_{\text{R,TM}}(t)|^2\} \qquad (12.48)$$

where subscripts TE and TM stand for two crossed polarization modes, subscript T and R correspond to for the transmitter and receiver lasers, $\Delta\omega_{\text{TE,TM}}$ represents the frequency detuning between TE and TM modes in the transmitter laser, $\Delta\omega$ is the frequency detuning between the transmitted light and the TM light in the receiver laser. The other parameters are the same meaning defined in Sect. 11.2.2. The dynamic properties of the transmitter and receiver lasers in polarization-rotated chaos synchronization are numerically studied by using these coupling equations.

In transmitter and receiver systems in semiconductor lasers with polarization-rotation optical feedback, one can attain both regimes of chaos synchronization, i.e. complete and generalized cases. Figure 12.15 shows numerical examples of polarization-resolved waveforms both for complete and generalized chaos synchronization in polarization-rotated optical feedback regimes. When the injection ratio from the transmitter to the receiver lasers is equal to the optical feedback ratio in the transmitter laser, namely, $\kappa = \kappa_{\text{cp}}$, complete chaos synchronization can be achieved under appropriate parameter

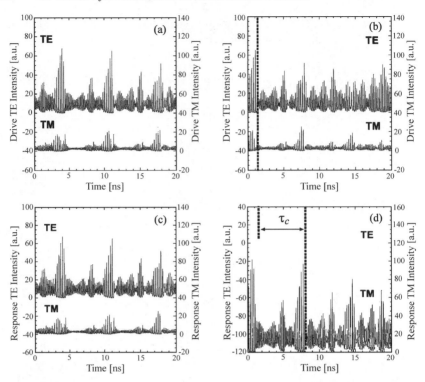

Fig. 12.15. Numerically calculated temporal waveforms of TE and TM mode intensities at synchronization. **a** and **c** Complete chaos synchronization for the transmitter and the receiver lasers, respectively, at $\kappa = \kappa_{cp}$. **b** and **d** Generalized chaos synchronization for the transmitter and the receiver lasers, respectively, at $\kappa = 3\kappa_{cp}$. (after Shibasaki N, Uchida A, Yoshimori Y, Davis P (2006); © IEEE 2006)

conditions, which is the same case for non-rotated optical feedback. For complete synchronization, the TE mode waveform of the transmitter laser completely synchronizes with the TE mode waveform of the receiver laser, and the TM mode waveform of the response laser also synchronizes with the TM mode waveform of the drive laser, as shown in Figs. 12.15a and c. Without loss of generality, the transmission time from the transmitter to the receiver lasers is equal to the delay time in the feedback loop in the transmitter, i.e, $\tau = \tau_c$. Therefore, the time lag between the synchronized signals is zero in complete chaos synchronization. In this configuration, the receiver output is a complete copy of the transmitter signal. On the other hand, for a strong optical injection $\kappa = 3\kappa_{cp}$, the receiver laser is injection-locked to the transmitter laser. Then the receiver output of TM mode is an amplified version of the transmitted chaotic signal of the TE mode oscillation from the transmitter laser as shown in Figs. 12.15b and d. The time lag between the waveforms of the TE mode in the transmitter laser and the TM mode in the receiver laser is τ_c, which is the evidence of generalized chaos synchronization. In this example, the TE mode in the receiver laser is completely suppressed and only the TM polarization component is the oscillation mode.

As in the case for synchronization in semiconductor lasers with normal (non-polarization-rotated) optical feedback, the chaos in the two regimes (complete and generalized synchronization) are distinguishable by the delay of the chaotic waveform with respect to that of the injected signal. From the detailed study for polarization-rotated chaos synchronization, it is proved that chaos synchronization can be performed even if there is a large mismatch in the optical frequencies of the lasers (Shibasaki et al. 2006). It is worth noting that synchronization can be maintained in the presence of the detuning by adjusting the injection strength. This feature is very different from the case of semiconductor lasers with non-rotated optical feedback, where complete synchronization is only very weakly robust against detuning. It should also be also pointed out that chaos synchronization can be maintained even in the presence of the detuning by an appropriate adjustment of the injection strength. Good synchronization can be maintained at the condition of positive detuning and small injection strength and at the condition of negative detuning and large injection strength. This asymmetric feature may result from the α parameter (linewidth enhancement factor) of semiconductor lasers, in the sense that chaos synchronization in semiconductor lasers with polarization-rotated optical feedback does not require strict matching of optical frequency. This feature of robustness with respect to optical frequency is particularly important for practical implementations of secure communication systems using chaos synchronization.

12.4 Chaos Synchronization in Injected Lasers

12.4.1 Theory of Chaos Synchronization in Injected Lasers

Semiconductor lasers exhibit chaotic oscillations by optical injection from a different laser as discussed in Chap. 6. We can use an optically injected laser as a light source for chaos synchronization as depicted in Fig. 6.1. However, an optical injection system is not a delay differential system, so that complete chaos synchronization is not generally realized in this system. We can also consider two types of synchronization systems: closed- and open-loop systems (Chen and Liu 2000). Both the transmitter and receiver lasers are optically injection-locked from external lasers in the closed-loop system, while only the transmitter laser is injection-locked in the case of the open-loop system. The open-loop system is also a special case of the closed-loop system. In the following, we formulate chaos synchronization in closed-loop systems of optically injected semiconductor lasers.

The optical injection-locking semiconductor laser is a coherent system. Therefore, the model must be described by the rate equations of the field, the phase, and the carrier density

$$\frac{dA_{\mathrm{T}}(t)}{dt} = \frac{1}{2}G_{\mathrm{n,T}}\{n_{\mathrm{T}}(t) - n_{\mathrm{th,T}}\}A_{\mathrm{T}}(t) + \frac{\kappa_{\mathrm{inj,T}}}{\tau_{\mathrm{in,T}}}A_{\mathrm{inj,T}}(t)\cos\theta_{\mathrm{T}}(t) \quad (12.49)$$

$$\frac{d\phi_{\mathrm{T}}(t)}{dt} = \frac{1}{2}\alpha_{\mathrm{T}}G_{\mathrm{n,T}}\{n_{\mathrm{T}}(t) - n_{\mathrm{th,T}}\} - \frac{\kappa_{\mathrm{inj,T}}}{\tau_{\mathrm{in,T}}}\frac{A_{\mathrm{inj,T}}(t)}{A_{\mathrm{T}}(t)}\sin\theta_{\mathrm{T}}(t) \quad (12.50)$$

$$\frac{dn_{\mathrm{T}}(t)}{dt} = \frac{J_{\mathrm{T}}}{ed} - \frac{n_{\mathrm{T}}(t)}{\tau_{\mathrm{s,T}}} - G_{\mathrm{n,T}}\{n_{\mathrm{T}}(t) - n_{0,\mathrm{T}}\}A_{\mathrm{T}}^2(t) \quad (12.51)$$

$$\theta_{\mathrm{T}}(t) = -\Delta\omega_{\mathrm{T}}t + \phi_{\mathrm{T}}(t) - \phi_{\mathrm{inj,T}}(t) \quad (12.52)$$

The parameters in the above equations are essentially the same as those of previous equations. $\Delta\omega_{\mathrm{T}}$ is the frequency detuning between the transmitter laser and the injection laser. The receiver driven by the transmitted signal can be described by

$$\frac{dA_{\mathrm{R}}(t)}{dt} = \frac{1}{2}G_{\mathrm{n,R}}\{n_{\mathrm{R}}(t) - n_{\mathrm{th,R}}\}A_{\mathrm{R}}(t) + \frac{\kappa_{\mathrm{inj,R}}}{\tau_{\mathrm{in,R}}}A_{\mathrm{inj,R}}(t)\cos\theta_{\mathrm{R}}(t)$$
$$+ \eta_{\mathrm{c}}A_{\mathrm{T}}(t - \tau_{\mathrm{c}})\cos\xi_{\mathrm{c}}(t) \quad (12.53)$$

$$\frac{d\phi_{\mathrm{R}}(t)}{dt} = \frac{1}{2}\alpha_{\mathrm{R}}G_{\mathrm{n,R}}\{n_{\mathrm{R}}(t) - n_{\mathrm{th,R}}\} - \frac{\kappa_{\mathrm{inj,R}}}{\tau_{\mathrm{in,R}}}\frac{A_{\mathrm{inj,R}}(t)}{A_{\mathrm{R}}(t)}\sin\theta_{\mathrm{R}}(t)$$
$$- \kappa_{\mathrm{cp}}\frac{A_{\mathrm{T}}(t - \tau_{\mathrm{c}})}{A_{\mathrm{R}}(t)}\sin\xi_{\mathrm{c}}(t) \quad (12.54)$$

$$\frac{dn_{\mathrm{R}}(t)}{dt} = \frac{J_{\mathrm{R}}}{ed} - \frac{n_{\mathrm{R}}(t)}{\tau_{\mathrm{s,R}}} - G_{\mathrm{n,R}}\{n_{\mathrm{R}}(t) - n_{0,\mathrm{R}}\}A_{\mathrm{R}}^2(t) \quad (12.55)$$

$$\theta_{\mathrm{R}}(t) = -\Delta\omega_{\mathrm{R}}t + \phi_{\mathrm{R}}(t) - \phi_{\mathrm{inj,T}}(t) \quad (12.56)$$

$$\xi_{\mathrm{c}}(t) = \omega_{0,\mathrm{T}}\tau_{\mathrm{c}} + \phi_{\mathrm{R}}(t) - \phi_{\mathrm{T}}(t - \tau_{\mathrm{c}}) + \Delta\omega t \quad (12.57)$$

where $\Delta\omega_{\mathrm{R}}$ is the detuning between the receiver and injection lasers, and $\Delta\omega$ is also the detuning between the transmitter and receiver lasers. As can be understood from these equations, chaos synchronization in the systems originates from injection locking and amplification. Under the special conditions of

$$\frac{1}{\tau_{\mathrm{ph,R}}} = \frac{1}{\tau_{\mathrm{ph,T}}} + 2\kappa_{\mathrm{cp}} \quad (12.58)$$

$$\omega_{0,\mathrm{T}}\tau_{\mathrm{c}}(\mod 2\pi) = 0 \quad (12.59)$$

$$\Delta\omega = 0 \quad (12.60)$$

$$\Delta\omega_{\mathrm{R}} = \Delta\omega_{\mathrm{T}} \quad (12.61)$$

we obtain complete chaos synchronization as (Liu et al. 2001)

$$A_R(t) = A_T(t - \tau_c) \tag{12.62}$$

$$\phi_R(t) = \phi_T(t - \tau_c) \tag{12.63}$$

The condition in (12.58) includes the internal device parameters (τ_{ph}) and the external coupling constant. Therefore, it is usually difficult to realize complete chaos synchronization in this system, since the adjustment of the parameters is extremely difficult in real experiments.

12.4.2 Examples of Chaos Synchronization in Injected Lasers

Figure 12.16 shows the experimental results of chaos synchronization using chaotic semiconductor lasers by optical injection (Liu et al. 2001a). The system used is an open-loop system and, therefore, optical injection is only presented in the transmitter system. The frequency detuning between the transmitter and injection lasers is changed to generate various chaotic states. Chaos synchronization is realized at period-1, period-2, and chaotic oscillations. Figure 12.17 plots the numerical results of synchronization errors for the parameter mismatches in the system (Chen and Liu 2000). The figure corresponds to the synchronization errors in complete chaos synchronization.

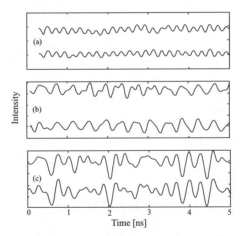

Fig. 12.16. Experimental chaos synchronization using chaotic semiconductor lasers by optical injection. The system is an open-loop. The frequency detuning between the transmitter and injection lasers is changed to generate various chaotic states at a fixed injection rate. Synchronized waveforms at **a** period-1, **b** period-2, and **c** chaotic oscillations. In each figure, the *upper trace* is the transmitter output and the *lower trace* is the receiver output. The lasers used are DFB lasers with an oscillation wavelength of 1.3 μm. The chaos synchronization is achieved under the complete condition (after Liu JM, Chen HF, Tang S (2001a); © 2001 IEEE)

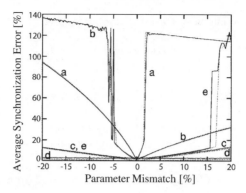

Fig. 12.17. Synchronization error (%) versus parameter mismatch (%). *Solid curves* a, b, c, d, and e correspond to parameters of cavity decay rate, spontaneous carrier decay rate, differential carrier relaxation rate, nonlinear carrier relaxation rate, and linewidth enhancement factor, respectively. The intrinsic Langevin noise is considered in the simulation. Each corresponding *dotted curve* is obtained without including the intrinsic noise (after Chen HF, Liu JM (2000); © 2000 IEEE)

The errors of synchronization are asymmetry for the parameter mismatches, which is similar to the trend for the case of the optical feedback system in Fig. 12.8b. The tolerances for the differential carrier relaxation rate, the nonlinear carrier relaxation rate, and the linewidth enhancement factor are much less effective, but the mismatch of the cavity decay rate $1/\tau_{ph}$ greatly affects the performance of chaos synchronization. In real devices, we could not access and vary the device parameters arbitrarily, therefore we must carefully select lasers with similar characteristics even if the lasers come from the same wafer. It is said that semiconductor laser devices have parameter mismatches within $5 \sim 20\%$ in industrial standards. In real lasers, we must also take noise effects into account. Therefore, the coupling coefficient κ_{cp} from the transmitter to the receiver lasers must be larger than a certain value to take on a negative value for the conditional Lyapunov exponent.

12.5 Chaos Synchronization in Optoelectronic Feedback Systems

12.5.1 Theory of Chaos Synchronization in Optoelectronic Feedback Systems

We discussed chaotic oscillations in optoelectronic feedback in semiconductor lasers in Chap. 7. We here assume the chaotic generators of optoelectronic feedback lasers depicted in Fig. 7.1. In optoelectronic feedback systems, the rate equations for the photon number and the carrier density are enough for describing the systems. Optoelectronic feedback systems in semiconductor lasers have an advantage of excellent synchronization performance over optical feedback and optical injection systems. Since the time scale for the carrier density is three figures larger than that of the photon lifetime, the performance and accuracy of chaos synchronization in optoelectronic feedback systems are different from those of optical feedback and optical injection systems. The points will be again discussed in the next chapter from the viewpoint of data transmission capability in chaotic communications. The rate equations for the photon number and the carrier density in a transmitter of optoelectronic feedback are written by

$$\frac{dS_T(t)}{dt} = G_{n,T}\{n_T(t) - n_{th,T}\}S_T(t) + R_{sp,T} \tag{12.64}$$

$$\frac{dn_T(t)}{dt} = \frac{J_T}{ed}\{1 + \xi_T S_T(t - \tau_T)\}$$
$$- \frac{n_T(t)}{\tau_{s,T}} - G_{n,T}\{n_T(t) - n_{0,T}\}S_T(t) \tag{12.65}$$

where ξ_T is the coefficient of the optoelectronic feedback circuit in the transmitter. The rate equations for the receiver laser are given by

$$\frac{dS_R(t)}{dt} = G_{n,R}\{n_R(t) - n_{th,R}\}S_R(t) + R_{sp,R} \tag{12.66}$$

$$\frac{dn_R(t)}{dt} = \frac{J_R}{ed}\{1 + \xi_R S_R(t - \tau_R) + \xi_{cp}S_T(t - \tau_c)\}$$
$$- \frac{n_R(t)}{\tau_{s,R}} - G_{n,R}\{n_R(t) - n_{0,R}\}S_R(t) \tag{12.67}$$

where ξ_R is the coefficient of the optoelectronic feedback circuit in the receiver and ξ_{cp} is the coupling coefficient from the transmitter to the receiver lasers. As discussed in Chap. 7, when the electronic feedback circuit has a finite time response, the feedback terms $s(t) = \xi_T S_T(t - \tau_T)$ and $s(t) = \xi_R S_R(t - \tau_R) + \xi_{cp}S_T(t - \tau_c)$ are replaced by the following integral equation:

$$y(t) = \int_{-\infty}^{t} f(t' - t)s(t')dt' \tag{12.68}$$

where $f(t)$ is the response function of the electronic circuit. The optoelectronic feedback system is also a delay differential system like the optical feedback system. However, the optoelectronic feedback system is quite different from optical feedback and optical injection systems. For example, the chaotic output from optoelectronic feedback is generally irregular pulsing states. The driving signal to the laser in the optoelectronic feedback system is also a chaotic signal, but the signal is not linearly proportional to the optical output power (Liu et al. 2001b). Chaos synchronization in optoelectronic feedback is generally a complete type (Tang et al. 2001).

12.5.2 Examples of Chaos Synchronization in Optoelectronic Feedback Systems

Figure 12.18 shows the results of chaos synchronization in an open-loop opto-electronic feedback system (Tang et al. 2001). As discussed in Chap. 7, the typical feature of chaotic oscillations in optoelectronic feedback systems is periodic or irregular pulsations of the laser output power. In the figure, chaos synchronization is achieved for various states of chaotic oscillations by chang-ing the feedback time τ_T in the electronic feedback circuit. The synchroniza-tion scheme is complete chaos synchronization. Figure 12.19 shows the nu-merical simulation for the model. Figure 12.19a is the bifurcation diagram for the normalized delay time $\hat{\tau} = \tau \nu_R$ in the transmitter laser. Fig. 12.19b is the maximum conditional Lyapunov exponent. Here, the parameter c_p is defined by

$$c_p = 1 - \frac{\xi_R}{\xi_T + \xi_{cp}} \tag{12.69}$$

When the system is a closed-loop, $c_p = 0$, while $c_p = 1$ for an open-loop. From this figure, the maximum conditional Lyapunov exponent in the open-loop

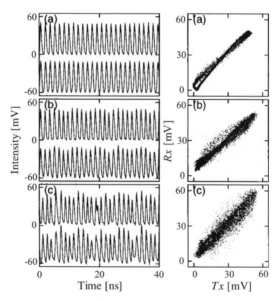

Fig. 12.18. Time series and correlation plots of synchronization at three different pulsing states under $c_p = 1$. **a** Regular pulsing at $\tau = 7.47$ ns, **b** two-frequency quasi-periodic pulsing at $\tau = 7.09$ ns, **c** chaotic pulsing at $\tau = 6.92$ ns. In **a–c** the *upper trace* is for the transmitter and the *lower trace* is for the receiver. The *left row* is the time series and the *right row* is the correlation plots. The laser is a DFB laser with the oscillation wavelength at 1.30 µm. The relaxation oscillation frequency of the laser is 2.5 GHz at the operating condition (after Tang S, Chen HF, Liu JM (2001); © 2001 OSA)

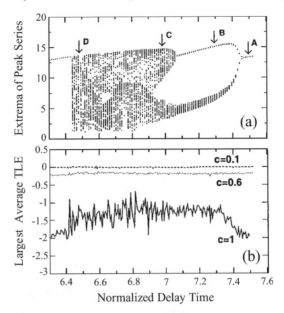

Fig. 12.19. Conditions for stable route-tracking synchronization. **a** Bifurcation diagram. **A**: regular pulsing, **B**: two-frequency quasi-periodic pulsing, **C**: three-frequency quasi-periodic pulsing, **D**: chaotic pulsing. **b** Largest average conditional Lyapunov exponent (transverse Lyapunov exponent) of coupled system for the same dynamic states as in **a** under different coupling strengths, $c_p = 0.1$, 0.6, and 1 (after Tang S, Chen HF, Liu JM (2001); © 2001 OSA)

system is smaller than that of the closed-loop system. Therefore, the open-loop system can achieve stable chaos synchronization compared with the closed-loop system. The effects of parameter mismatches for chaos synchronization have also been studied in optoelectronic feedback systems and similar results as for optical feedback systems are obtained (Abarbanel et al. 2001).

12.6 Chaos Synchronization in Injection Current Modulated Systems

Semiconductor lasers are sensitive to injection current modulation and sometimes show chaotic oscillations for certain conditions both of the device parameters and the modulation frequency and index as discussed in Chaps. 6 and 7. However, chaotic oscillations by the frequency modulation occur only under limited conditions in ordinary edge-emitting semiconductor lasers. Therefore, we briefly introduce a chaos synchronization system using a frequency modulated self-pulsating semiconductor laser as a chaotic light source. The output from a self-pulsating semiconductor laser shows regular pulsating oscillations for the ideal case. The laser is used as a light source of a digital versatile disk system. However, the laser easily exhibits chaotic oscillations (irregular pulsing states) under an appropriate combination of the modulation frequency and index for the bias injection current as discussed in Sect. 8.2. In the chaotic oscillations, pulse heights of the pulsating output irregularly fluctuate and this fluctuation is proved to be chaotic. Jones et al. (2001) demonstrated numerically chaos synchronization in symmetrical modulation systems. They used a very high frequency modulation of 3.4 GHz with a large modulation index of 0.3. Their model was a coherent coupling between the transmitter and receiver lasers. Since a self-pulsating semiconductor laser is once brought almost below the laser threshold after a pulsation, incoherent coupling must be taken into account through a long transmission line between the transmitter and receiver systems. Only a few studies have been published on chaos synchronization in modulated semiconductor lasers to date.

12.7 Chaos Synchronization in Mutually Coupled Lasers

12.7.1 Chaos Synchronization of Semiconductor Lasers with Mutual Optical Coupling

Mutually coupled oscillators are of great interest because of the important insight they provide into coupled physical, chemical, and biological systems. Mutually coupled semiconductor lasers can be used for a system of chaos synchronization. However, only a few reports have been published on the characteristics of mutually coupled systems, since they are not suited for chaotic communications, which are the main concern for the applications of chaos synchronization (Hohl et al. 1997 and 1999, Heil et al. 2001b, Mirasso et al. 2002a, and Mulet et al. 2002). Several reports have been published for chaos synchronization with mutually coupled edge-emitting semiconductor lasers (Spencer et al. 1998, Spencer and Mirasso 1999, and Fujino and Ohtsubo 2001). As an experimental example of chaos synchronization in mutually coupled lasers, we here show chaos synchronization in chaotic VCSELs in LFF regimes (Fujiwara et al. 2003b). In the experiments, mutually coupled VCSELs without external mirrors are used. Even when the lasers are stable at the free running state, they exhibit chaotic oscillations under mutual coupling. Figure 12.20 shows the results of chaos synchronization. Figure 12.20 a plots time series and their optical spectra of the x polarization components of the two lasers at LFF oscillations. Figure 12.20b shows the results for the y polarization components. The lasers are biased at low injection currents and their spatial modes are the lowest ones, i.e., LP_{01} mode. In the experiment, the two polarization modes are mutually coupled with each other. In this example, chaos synchronization occurs by the coupling with the x polarization components and one of the outputs of the orthogonal y components synchronizes to the other under the anti-correlation effect. This fact is easily understood from the observation of optical spectra in the figure. Anti-phase

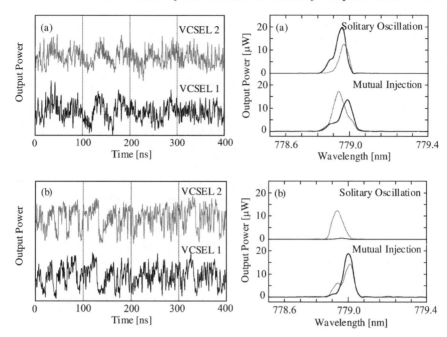

Fig. 12.20. Chaos synchronization in mutually coupled VCSELs in an LFF regime.
a x-polarization mode and **b** y-polarization mode. The *left* is the time series and the
right is the corresponding optical spectra. *Solid curves:* VCSEL 1, *dotted curves:*
VCSEL2. The coupling time of light between the two lasers is 4 ns

correlation of the two orthogonal polarization components is a unique feature in VCSELs. At the solitary oscillations, the y polarization modes are
dominant lasing modes and the laser powers are almost concentrated to the
y polarization modes. On the other hand, the x polarization modes increase
and become the dominant modes after the mutual coupling. The laser oscillation of VCSEL2 lags with respect to VCSEL1 with 4 ns (the transmission time
of light from one laser to the other), therefore the synchronization is a generalized case. Fujiwara and Ohtsubo (2004) also showed chaos synchronization
for a selective polarization mode in mutually coupled VCSELs. Though the
other mode is not coupled, the remaining modes synchronize with the anti-
phase correlation effect. Chaos synchronization in VCSELs occurs even for
the two different spatial modes as far as detuning of the oscillation frequencies
is negligibly small.

12.7.2 Chaos Synchronization of Semiconductor Lasers with Mutual Optoelectronic Coupling

The dynamics and chaos synchronization for mutually coupled systems in
semiconductor lasers with optoelectronic feedback was studied by Tang et al.

(2004). In the system, mutual coupling can act as a negative feedback to stabilize the coupled oscillators or it can increase the complexity of the system inducing a highly complex chaos depending on the operating conditions. A quasi-periodicity and period-doubling bifurcation, or a mixture of the two, is found in such a system. Also, the system exhibits a unique state of stabilizing and quenching the oscillation amplitude of two pulsating oscillators, a phenomenon known as "death by delay". Although the chaotic waveforms are very complex with broad spectra, a high quality of synchronization between the chaotic waveforms is observed. Such synchronization is achieved because of the effect of mutual coupling and the symmetric design between the two lasers. Figure 12.21 shows a schematic diagram for semiconductor lasers with mutual optoelectronic coupling. The fundamental chaotic oscillator is the same as the system of optoelectronic feedback as discussed in Sect. 12.5.1. A part of an emitted light from semiconductor laser LD 1 is once detected by photodetector PD 1 and electronically fed back into the bias injection current of the laser with delay τ_1. On the other hand, the other light is detected by photodetector PD 2 and fed into the bias injection current of semiconductor laser LD 2 with transmission time T_1. Similarly, semiconductor laser LD 2 also has an optoelectronic feedback loop with time delay τ_2. A part of an emitted light from the laser LD 2 is also fed into the first laser with transmission time T_2, thus mutual coupling of the system is attained.

The system of the mutual coupling lasers can be easily described by extending the discussion in Sect. 12.5. For semiconductor laser LD 1, one reads

$$\frac{dS_1(t)}{dt} = [G_{n1}\{n_1(t) - n_{th1}\}] S_1(t) \tag{12.70}$$

$$\frac{dn_1(t)}{dt} = \frac{J_1(t)}{ed}\{1 + \xi_{f1} S_1(t - \tau_1) + \xi_{cp1} S_2(t - T_2)\}$$
$$- \frac{n_1(t)}{\tau_{s1}} - G_{n1}\{n_1(t) - n_{01}\} S_1(t) \tag{12.71}$$

Here, we assume an instantaneous response of the electronic feedback circuit, however we can apply (12.68) for a finite response of the circuit. For the second laser, the coupling equations can be written as symmetric forms as

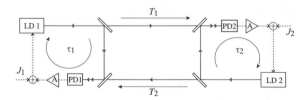

Fig. 12.21. Schematic diagram of mutually coupled semiconductor lasers with optoelectronic feedback. LD: laser diode, PD: photodiode, A: amplifier, I: bias injection current

above equations by

$$\frac{\mathrm{d}S_2(t)}{\mathrm{d}t} = [G_{\mathrm{n2}}\{n_2(t) - n_{\mathrm{th2}}\}] S_2(t) \tag{12.72}$$

$$\frac{\mathrm{d}n_2(t)}{\mathrm{d}t} = \frac{J_2(t)}{ed}\{1 + \xi_{\mathrm{f2}}S_2(t - \tau_2) + \xi_{\mathrm{cp2}}S_1(t - T_1)\}$$

$$- \frac{n_2(t)}{\tau_{\mathrm{s2}}} - G_{\mathrm{n2}}\{n_2(t) - n_{02}\} S_2(t) \tag{12.73}$$

In mutually coupled semiconductor lasers, not only that the output of one laser is coupled into the dynamics of the other laser, but also that the time delay introduced by the mutual coupling further increases the dimension of the degree of freedom in the coupled lasers. Consequently, a lot of interesting dynamics have been observed in such mutually coupled semiconductor lasers. For example, optoelectronic feedback can drive semiconductor lasers into nonlinear oscillations, such as regular pulsing, quasi-periodic pulsing, or chaotic pulsing under certain conditions of the device and feedback parameters. One of typical features in this system is a death by delay, in which two limit-cycle oscillators suddenly stop oscillating due to a time-delayed coupling between these oscillators by tuning the feedback parameters (Tang et al. 2004). The phenomenon of death by delay has been observed in many other mutually coupled limit-cycle oscillators, which do not necessarily have a delayed feedback In the mutually coupled optoelectronic feedback systems described by (12.70)–(12.73), we can observe periodic death islands of the laser oscillations at a certain coupling strength for the increase of the coupling delay time $T_1 + T_2$. In reality, there is always a bandwidth limitation from the components such as the amplifiers, the photodetectors, and even the lasers. Consequently, the mutually coupled semiconductor laser system is not only highly nonlinear but also highly dispersive. The system can have a quasi-periodic pulsing route, a period-doubling pulsing route, or a mixture of these two bifurcations to chaos. The system has very interesting properties as a viewpoint of nonlinear dynamics. However, we here focus on the synchronization properties of the system.

Experimental and theoretical studies for synchronization of mutually coupled semiconductor lasers with optoelectronic feedback were reported by Tan et al. (2004) and Chiang et al. (2005). The two semiconductor lasers are operated in states of regular oscillations or quasi-periodic oscillations under the effect of optoelectronic feedback before the mutual coupling is applied. Once the mutual coupling is applied, dramatic effects can be observed on the original nonlinear oscillations. Figure 12.22 shows an experimental example of chaos synchronization in this system. Without mutual coupling, the waveforms of the two lasers may exhibit either typical pulsing states or chaotic pulsing states at certain oscillation conditions. In this case, the waveform from PD 1 is a regular pulsing state with one fundamental frequency, while that from PD 2 is a quasi-periodic pulsing state as shown in Fig. 12.22a. With mutual coupling, highly complex chaotic outputs form the two lasers

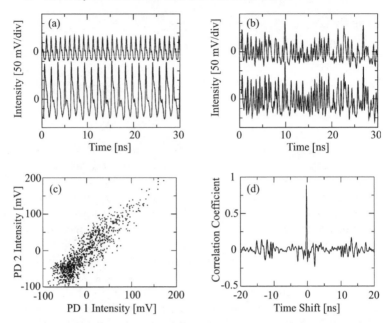

Fig. 12.22. Experimental chaos synchronization in semiconductor lasers with mutual optoelectronic coupling. $T_1 = T_2 = 15.4$ ns and $\tau_1 = \tau_2 = 15.4$ ns. **a** Time series of regular pulsing states in the two lasers before mutual coupling. **b** Chaotic time series after mutual coupling. **c** and **d** Correlation plot of the photodiode outputs after mutual coupling (after Tang S, Vicente R, Chiang MC, Mirasso CR, Liu JM (2004); © IEEE 2004)

are observed under the conditions of $T_1 = T_2 = \tau_1 = \tau_2 = 15.4$ ns as shown in Fig. 12.22b. It is noted that the two waveforms have a zero time lag and the type of synchronization is complete. Figures 12.22c and d show the correlation plot between the outputs form PD 1 and PD 2, and the time shifted correlation, respectively. The detailed properties of chaos synchronization in mutually coupled semiconductor lasers with optoelectronic feedback were reported in the references (Chiang et al. 2005, 2006).

13 Chaotic Communications in Semiconductor Lasers

Chaotic data encoding or scrambling is a technology for overcoming the difficulties of the digital methods in secure communications. Using chaotic lasers as light sources, high-speed and broadband secure communications can be established. In this chapter, we discuss cryptographic applications of chaos in semiconductor lasers. The technique we treat in this chapter is an analogue chaotic encryption and decryption. Messages to be sent are encoded into chaotic time series generated from a chaotic semiconductor laser and decoded by a chaotic laser with the same characteristics. The key for chaos communications is chaos synchronization, which we discussed in the previous chapter. In chaotic communications, a small message is embedded into a chaotic laser carrier and the total signal is sent to the receiver. Only the chaotic oscillation is reproduced based on chaos synchronization and a chaos-pass filtering effect in the receiver laser. By subtracting the receiver output from the transmission signal, the message is successfully decoded.

13.1 Message Encryption in a Chaotic Carrier and Its Decryption

13.1.1 Chaotic Communications

The development of efficient technologies for high-speed and massive data transmissions is an urgent subject in the rapidly growing information-oriented society. One of the important issues of information and communication networks is the security problem. In secure data transmissions, a message to be sent is usually encoded by computer software and the security of encoding is guaranteed by the complexity of the calculations necessary to decode the original message. However, the development of digital computer technology is so fast that the standard code for scrambling data in secure communication systems can be soon decoded by a fast computer. On the other hand, the enhancement of the complexity of calculations for encoding and decoding messages may lose real-time processing of data transmissions. In the meantime, the method of quantum computing has been developed as one of the candidates to decipher quickly encoded data in standard secure communication systems. As an alternative method, chaotic communications have been

proposed for high-speed and broadband capabilities with hardware based secure communications (Kennedy et al. 2000, and Dachselt and Schwarz 2001).

There are two techniques for chaos-based secure communications: one is
digital encoding and the other is analogue encryption. As examples of digital
techniques, the method of code scrambling based on chaotic signal generations
such as discrete-sequence (DS) optical code division multiple access (CDMA)
is used for chaos communications. The method of chaos CDMA uses long life
chaotic non-correlated data sequences embedded into the chaotic orbits as
CDMA codes. It is verified that the generated codes have an advantage over
the existing Gold series for the irregularity and non-correlation properties
(Chen et al. 2001). Another example is the technique of secure chaos key
generation instead of random numbers in ordinary secure communication
systems (Uchida et al. 2003). Another one is an analogue technique. In this
chapter, we are concerned with the analogue method, since chaos in laser systems is best suited for analogue data encryption and decryption by nature. In
the analogue technique, when a fraction of a chaotic signal from a transmitter is sent to a receiver, the two systems synchronize with each other under
certain conditions, as discussed in the previous chapter. Not only the two
system configurations but also the chaos parameters of the two systems must
be the same for perfect chaos synchronization. The merits of the use of semiconductor lasers in chaotic communications are clear, since light is the carrier
of modern basic communication channels and the generation of high-speed
and broadband signals is easily attained compared with e.g., nonlinear electronic circuits. Chaotic communications require special hardware to generate
a chaotic signal and to realize synchronization. Even if one tries to decode
messages by computing or guessing chaotic states from the signals obtained,
it is very difficult to decode messages by available techniques without knowledge of the chaos keys because they are embedded into high-dimensional
chaotic spaces (Ohtsubo 2002b).

First, we show the basic idea of analogue chaos communications. Figure 13.1 shows the model for analogue chaos communication systems. The
basics of the technique is chaos synchronization between two nonlinear systems, transmitter and receiver, as already noted. A message with small amp-

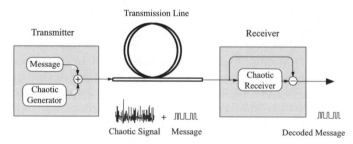

Fig. 13.1. General concept of chaos synchronization with analogue data encryption

litude (compared with the chaotic variations of the transmitter output) is embedded into a chaotic carrier in the transmitter. The chaotic carrier with the message is sent to the receiver through the communication channel. In the receiver, the system only synchronizes to the chaotic signal from the transmitter. Then, the message is decoded by subtracting the receiver output from the transmitted signal. If the amplitude of the message is small enough, we can achieve successful chaos synchronization even if the transmission signal includes the perturbation (message) to the chaotic signal. However, small signal approximation is not always necessary. For example, a message may be comparable with chaotic variations in the chaos modulation (CMO) technique as discussed later.

Chaotic data communications using laser systems are categorized into three classes depending on the techniques of message encoding and decoding (Ohtsubo 2002a,b, Liu et al. 2002b, and Ohtsubo and Davis 2005). They are chaos masking (CMA), chaos modulation (CMO), and chaos shift keying (CSK). Each method can be mathematically formulated by laser rate equations in transmitter and receiver lasers and the dynamic behaviors of the systems can be described by these rate equations. Even for optical communications, we can also generate chaotic oscillations from nonlinear electronic circuits and transmit a chaotic signal converted by optoelectronic devices through optical channels. However, the carrier of communications is light and laser chaos is best suited for such purposes. Therefore, many systems using chaos of various laser systems have been proposed for chaos synchronization and communications. In spite of existing work, we still need extensive studies about many subjects to put the systems into practical use, for example, the degree of security, the accuracy of synchronization for parameter mismatches between transmitter and receiver systems, robustness of communications, and other things.

13.1.2 Chaos Masking

Following the proposal of chaos synchronization in nonlinear systems, Pecora and Carroll (1991a, b) pointed out the possibility of secure communications based on chaos synchronization. They used electronic circuits to realize Lorenz chaos (see Appendix A.4). In their method, a chaotic signal (variable x) in a transmitter system is sent to a receiver system as a synchronous signal. At the same time, a chaotic variable z in the transmitter with a small message m was sent to the receiver and chaos synchronization between the transmitter and receiver systems could be achieved. The receiver system consisted of a subsystem of the variables y and z. After subtracting the synchronized signal z' in the receiver from the transmitted signal $z + m$, the message was successfully decoded. The technique is essentially categorized into the method of chaos masking (CMA). However, two different channels were required for the data transmission in their method. Cuomo et al. (1993)

proposed a method of chaotic communication using a single transmission channel for the same Lorenz system as that of Pecora and Carroll.

In a laser system, we cannot divide the system into subsystems as shown in Appendix A.4. Therefore, a fraction of a chaotic laser output power from a transmitter is sent to a receiver laser through a single communication channel. Figure 13.2 shows the general three schemes of optical communications in analogue chaotic systems. Figure 13.2a shows the system of chaos masking (CMA), where a small message $m(t)$ is embedded into a chaotic carrier $x(t)$ in a transmitter and, then, the signal of $x(t) + m(t)$ is sent to a receiver. The receiver system is the same as that of the transmitter and the two systems are operating at the same parameter values. Only the chaotic signal of $x(t)$ is reproduced in the receiver system if the amplitude of the message is small enough. Then, the message $m(t)$ is decoded by subtracting the receiver output $x(t)$ from the transmission signal $x(t) + m(t)$. To hide a message into

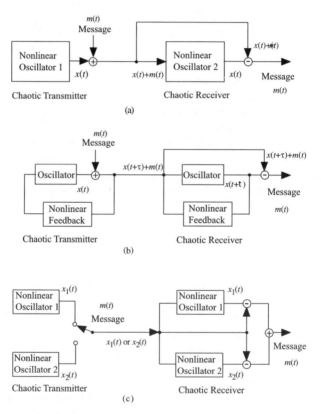

Fig. 13.2. General schemes of optical communications in analogue chaotic systems. Models of **a** chaos masking (CMA), **b** chaos modulation (CMO), and **c** chaos shift keying (CSK)

chaotic carriers securely (namely, mask the message) and to reproduce good quality of a decoded message, the amplitude of the message must be sufficiently small compared with the averaged chaotic carrier signal. Usually, the fraction is less than 1% of the average chaotic power.

13.1.3 Chaos Modulation

In the method of chaos modulation (CMO) shown in Fig. 13.2b, both a chaotic carrier and a message conform a new chaotic oscillation in the nonlinear system. Therefore, a message embedded into the chaotic carrier may not be small. The transmitter and receiver systems can be described by the equivalent mathematical differential equations. Therefore, complete chaos synchronization is achieved in the system and an excellent synchronous signal can be obtained in the receiver output. The method resembles CMA, however CMO is essentially a different technique from CMA. As shown in Fig. 13.2b, a message is mixed with a chaotic carrier in the nonlinear oscillator and the two signals conform a new chaotic state different from the original one. In CMO, a delayed feedback system is usually used as a chaotic generator. The new chaotic signal is given by $x(t + \tau) = f(x(t) + m(t))$ after the delay time τ, where f is the nonlinear function of the system. This new signal together with the message $x(t + \tau) + m(t + \tau)$ is sent to the receiver. Since the transmitter and the receiver are the same nonlinear systems, the chaotic oscillation $x(t + \tau) = f(x(t) + m(t))$ is exactly reproduced in the receiver system as chaos synchronization. By subtracting the synchronized chaotic signal $x(t + \tau)$ from the transmitted signal $x(t + \tau) + m(t + \tau)$, we can decode the message. Sometimes, the message is decoded by dividing the transmitted signal by the synchronized chaotic signal in the receiver. From the point of view of encoding and decoding message, the method of CMO has no restriction on the magnitude of the message as a secure communication, since both the chaotic carrier and the message conform new chaotic states in the nonlinear systems. However, in the optics case, the message is usually decoded as an intensity $I(t) = |E(t)|^2$. Therefore, the amplitude of the message must be small enough when we use the ordinary message decoding technique. Furthermore, the degree of security for data transmissions becomes worse when the signal level of a message is large. Therefore, the amplitude of a message in CMO should also be small.

13.1.4 Chaos Shift Keying

The signal shift keying technique, which is frequently used in ordinary communication systems, is also applicable to chaotic data transmissions. Fig. 13.2c is an example of such system diagrams. In chaos shift keying (CSK), two chaotic states $x_1(t)$ and $x_2(t)$ are generated in a transmitter system. The switching itself to send either chaotic state is a message $m(t)$. In the receiver system, each state $x_1(t)$ or $x_2(t)$ is detected by the technique of chaos synchronization. Therefore, two sets of chaotic generators are usually prepared both for the transmitter and the receiver. However, the difference between two chaotic states in the CSK system must be very small, since the message can be easily estimated from the attractors when the difference of chaotic oscillations between the two states is too large. In chaos synchronization in nonlinear systems, the time required for the synchronization between receiver and transmitter is finite for the switching of chaotic states. Therefore, we must take into account the transient and finite response of signals for practical use of the systems.

13.1.5 Chaotic Data Communications in Laser Systems

Shortly after the proposal of chaos synchronization by Pecora and Carroll, Colet and Roy (1994) demonstrated chaos synchronization in laser systems using loss modulated solid state lasers by numerical simulations and predicted the possibility of chaotic communications based on such nonlinear systems. They showed chaotic data transmission of a binary bit-sequence with a rate of 100 kbps in the system. VanWiggeren and Roy (1998a, b) demonstrated data transmission for secure communications based on chaos modulation (CMO) using laser systems. They proposed a ring fiber laser system with an optical feedback loop (delay loop) as a chaotic generator. A message to be transmitted was put into the feedback loop as a modulation, then a new chaotic oscillation was produced in the feedback system. They successfully demonstrated data transmission higher than a bit rate of 100 Mbps. Goedgebuer et al. (Goedgebuer et al. 1998b and Lagar et al. 1998b) also reported chaotic data transmission based on optoelectronic feedback using wavelength-to-current conversion systems in semiconductor lasers. Their method was also categorized into CMO. Other CMO systems were also proposed by using laser systems (Luo et al. 2000, Abarbanel et al. 2001, Liu et al. 2001c, and Tang et al. 2001). Tang and Liu (2001a) experimentally demonstrated data transmission of a pseudo-random binary bit-sequence with a 2.5 Gbps non-return-to-zero (NRZ) signal corresponding to the OC-48 standard bit rate in optoelectronic feedback semiconductor laser systems.

After the demonstration of chaotic communications based on CMO, the method of chaos masking (CMA) was widely studied theoretically and experimentally because of the ease of implementation in semiconductor laser systems (Mirasso et al. 1996, Sánches-Díaz et al. 1999, Sivaprakasam and Shore 1999 and 2000b, c, White and Moloney 1999, Jones et al. 2000, and Rogister et al. 2001). Annovazzi-Lodi et al. (1996 and 1997) proposed a method of CSK using semiconductor lasers with optical feedback. Also studies on chaotic communications based on CSK in various laser systems have been reported (Liu and Davis 2001, Davis et al. 2001, and Mirasso et al. 2002b). Almost all these systems used chaotic oscillations in class B lasers, such as solid state lasers, fiber lasers, and semiconductor lasers. In chaotic laser communications, the effects of optical feedback, optical injection from a different laser, and optoelectronic feedback have been frequently used to generate chaotic signals.

Liu et al. (2002) investigated three configurations of systems using semiconductor lasers (optical injection locking, optical feedback, and optoelectronic feedback systems) and compared the performances of data transmissions for three different techniques (CMA, CMO, and CSK). As a result, the optoelectronic feedback system with CMO showed excellent performance for data transmissions. As discussed later, chaotic carrier frequency, which is closely related to the relaxation oscillation of the solitary laser, is an important measure of the capability of data transmissions and semiconductor lasers with high frequency response are indispensable as high speed chaotic generators for that purpose. Distributed feedback (DFB) lasers of near infrared wavelength oscillations are frequently used for chaotic laser communications, since they are suitable for chaotic light sources of ordinary communication systems with high frequency response. Vertical-cavity surface-emitting lasers (VCSELs) are also promising devices for future semiconductor lasers and also chaotic lasers. Other lasers such as MQW lasers with visible oscillations may be used for chaotic light sources for short-range communications. However, we assume edge-emitting lasers as chaotic generators in the following. Even if device structures are different from each other, the system that is described by the same laser rate equations shows the same dynamics of chaotic oscillations as discussed earlier.

As stated in Chap. 12, there are two types of mechanisms of chaos synchronization: one is complete chaos synchronization and the other is synchronization of chaotic oscillation by optical injection locking and amplification. In a delay differential system, there is a solution for complete chaos synchronization where transmitter and receiver lasers can be described by mathematically identical forms of the equations. On the other hand, we can expect synchronization of chaotic oscillations based on the injection locking phenomenon in nonlinear amplifying systems in the chaotic transmitter and receiver lasers. We can use both systems for secure communications based on chaos synchronization, although the degree of security is different in the two schemes.

13.2 Cryptographic Applications in Optical Feedback Systems

13.2.1 Chaotic Communications in Optical Feedback Systems

In this section, we focus on chaotic secure communications using systems of semiconductor lasers with optical feedback. Sánches-Díaz et al. (1999) numerically studied chaotic communications based on CMA in the systems and

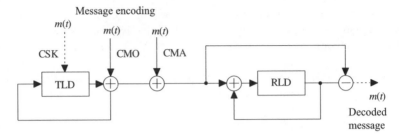

Fig. 13.3. Schematic diagram of chaotic communications in semiconductor lasers with optical feedback. $m(t)$ is a message signal to be embedded. The position for each message encoding scheme is shown in the figure. The *solid lines* are optical connections and the *broken lines* are electronic connections

demonstrated data transmissions of a bit rate of 4 Gbps. In their method, a direct modulation to the injection current in a transmitter semiconductor laser was used as the message encoding, therefore the technique was in principle CMO rather than CMA. However, it is assumed to be CMA as long as the modulation amplitude of a message is very small. Annovazz-Lodi et al. (1996) proposed a CSK system in semiconductor lasers with optical feedback using systems consisting of a single chaotic transmitter and two chaotic receivers. Mirasso et al. (2002b) also proposed a CSK system using a single chaotic generator of a semiconductor laser with optical feedback both for a transmitter and a receiver. They numerically demonstrated data transmissions of a bit rate of 2 Gbps. The technique is called ON/OFF CSK. A lot of theoretical and numerical studies on chaotic data transmissions and communications have been reported in semiconductor lasers with optical feedback. However, only a few experimental studies have been published. Experimental data transmission of binary messages in optical feedback systems with bit rate over giga-bit-per-second has not been reported to date. Instead, a few results of data transmission of a sinusoidal wave with a frequency over gigahertz have been demonstrated (Kusumoto and Ohtsubo 2002). Also few theoretical and experimental studies have been reported for CMO in systems of semiconductor lasers with optical feedback (Liu et al. 2001c). A system of a semiconductor laser with optical feedback has a merit for its simplicity, especially in CMA and CSK. However, we need extra devices for the message modulation in CMO, such as electro-optic modulators.

Using a chaotic generator of a semiconductor laser with optical feedback, we formulate the rate equations for transmitter and receiver lasers. Figure 13.3 shows the schematic diagram of chaotic communications in semiconductor lasers with optical feedback. Embedding a message into the transmitter system, the transmitter can be modeled by the following coupled equations for the complex field E and the carrier density n, according to the configurations of unidirectionally coupled semiconductor lasers with optical feedback (Ohtsubo 2002a and Liu et al. 2002b)

$$\frac{dE_T(t)}{dt} = \frac{1}{2}(1 - i\alpha_T)G_{n,T}\{n_T(t) - n_{th,T}\}E_T(t)$$

$$+ \frac{\kappa_T}{\tau_{in,T}}E_T(t - \tau_T)\exp(i\omega_{0,T}\tau_T) + \eta_{CMO}m_{CMO}(t) \qquad (13.1)$$

$$\frac{dn_T(t)}{dt} = \frac{J_T}{ed}\{1 + \eta_{CSK}m_{CSK}(t)\} - \frac{n_T(t)}{\tau_{s,T}}$$

$$- G_{n,T}\{n_T(t) - n_{0,T}\}|E_T(t)|^2 \qquad (13.2)$$

whereas the receiver driven by the transmitted signal can be described by

$$\frac{dE_R(t)}{dt} = \frac{1}{2}G_{n,R}(1 - i\alpha_R)\{n_R(t) - n_{th,R}\}E_R(t)$$

$$+ \frac{\kappa_R}{\tau_{in,R}}E_R(t - \tau_R)\exp(i\omega_{0,R}\tau_R)$$

$$+ \frac{\kappa_{cp}}{\tau_{in,R}}E_R(t - \tau_c)\exp(i\omega_{0,T}\tau_c - i\Delta\omega t)$$

$$+ \eta_{CMA}m_{CMA}(t - \tau_c) + \eta_{CMO}m_{CMO}(t - \tau_c) \qquad (13.3)$$

$$\frac{dn_R(t)}{dt} = \frac{J_R}{ed} - \frac{n_R(t)}{\tau_{s,R}} - G_{n,R}\{n_R(t) - n_{0,R}\}|E_R(t)|^2 \qquad (13.4)$$

where $\Delta\omega = \omega_{0,T} - \omega_{0,R}$, κ_{cp} is the coupling coefficient of light from the transmitter to the receiver, $m_{CMA}(t)$, $m_{CMO}(t)$, and $m_{CSK}(t)$ are the message sequences corresponding to CMA, CMO, and CSK systems, respectively, and η_{CMA}, η_{CMO}, and η_{CSK} are the actual modulation coefficients for each system. For example, when we consider a chaotic communication system with CMA, we put $\eta_{CMA} \neq 0$, and $\eta_{CMO} = \eta_{CSK} = 0$. The modulation depth of the message in chaotic secure communications is generally very small so as not to disturb the chaotic attractors. In actual systems, the perturbation due to the message encoding is not for amplitude but for optical intensity, except for the case of injection current modulation. However, we can approximately assume amplitude perturbation as far as it is very small compared with the average chaotic amplitudes. Otherwise we could formulate rate equations for intensity perturbations for the numerical simulations.

As can be easily recognized from (13.1)–(13.4), there is a condition where the rate equations in the transmitter are mathematically identical to those in the receiver in a CMO system. Therefore, complete chaos synchronization is performed in this system. On the other hand, for the other cases (CMA and CSK systems), a message always behaves as a small perturbation to each chaotic system. Therefore, the modulation coefficients η_{CMA} and η_{CSK} must be small enough to achieve good chaos synchronization. They should be usually less than 1% of the average of the chaotic fluctuations.

13.2.2 Chaos Masking in Optical Feedback Systems

In the following, we describe the particular technique for each modulation scheme in optical feedback systems. In CMA, a small message to be sent is added to a chaotic carrier from the transmitter laser. The system under consideration is the same as in Fig. 11.9a. In CMA, we set $\eta_{\mathrm{CMA}} \neq 0$ and $\eta_{\mathrm{CMO}} = \eta_{\mathrm{CSK}} = 0$ in (13.1)–(13.4). Since there is a message term on the right hand side of the receiver equation (13.3), complete chaos synchronization is not realized in this system in a strict sense. However, we can approximately observe complete chaos synchronization as long as the modulation depth of the message is small enough. The accuracy of chaos synchronization depends on the parameter mismatches for the chaos keys. For example, chaos synchronization with high accuracy is achieved when the total light input to the receiver laser both from the external reflector and the transmitter laser is almost equal to the amount of external feedback in the transmitter.

In spite of the presence of the perturbation of a message in a transmitted signal, only the chaotic carrier is reproduced in the receiver output in CMA. This phenomenon is known as chaos-pass-filtering. This fact is verified by numerical and experimental studies (Ohtsubo 2002b). The phenomenon of chaos-pass-filtering in nonlinear systems is not obvious and needs some explanation. The origin of chaos-pass filtering will be discussed in the following section (Murakami and Shore 2005). A small message is also considered as a perturbation for a chaotic system like noises in the system. It seems that noises are discriminated from intrinsic chaotic dynamics induced in the nonlinear system. As far as perturbation in a system is small, original chaotic dynamics is preserved in the transmitted signal. As is easily understood, the message in CMA is decoded by subtracting the receiver output from the transmitted signal of the chaotic carrier together with the message. The accuracy of the decoding becomes worse when the modulation depth of the message increases, though it depends on the synchronization schemes (complete or optical injection locking regime). Further, the security of data transmissions is degraded with the increase of the modulation depth, since the message may be directly visible in the transmitted signal.

In CMA, a message is added to a chaotic carrier generated from a transmitter. For example, a chaotic carrier is modulated through an electro-optic modulator, which is an intensity modulation to the chaotic carrier. However, an alternative method is frequently used for encoding a message. A message is simply added to the bias injection current and it is an easier way to modulate intensities in semiconductor lasers. Strictly speaking, it is a technique of CMO rather than CMA. However, it reduces to the method of CMA when the level of a message is small enough. Indeed, a lot of theoretical and experimental work has been published based on the same techniques of injection current modulation as message encoding in the system.

Before showing the results for data transmissions and decoding of messages in CMA, the unique phenomenon of chaos-pass-filtering is discussed.

Figure 13.4 shows an experimental example for chaos synchronization when a message is embedded into the transmitter signal (Kusmoto and Ohtsubo 2002). The message is added to a chaotic carrier as an injection current modulation in the transmitter laser and it is a sinusoidal wave with a frequency of 1.5 GHz. The relaxation oscillation frequency of the solitary laser is about

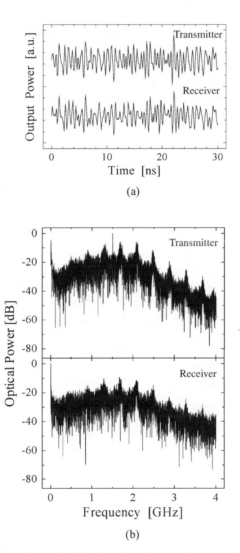

Fig. 13.4. Chaos-pass-filtering effect. **a** Waveforms of transmitted signal and receiver output in a closed-loop system of semiconductor lasers with optical feedback. A message of a sinusoidal wave of a frequency of 1.5 GHz with a modulation depth of −14 dB is included in the transmitter signal. **b** Corresponding rf spectra to Fig. 13.4a. The synchronization is based on an optical injection locking scheme

4 GHz. The two chaotic waveforms look the same as shown in Fig. 13.4a and they are synchronized with each other in spite of the presence of the message. The synchronization is also confirmed by the correlation plot.

Figure 13.4b shows the RF spectra corresponding to the waveforms in Fig. 13.4a. Besides the broad spectral peaks of the external cavity mode and its higher harmonics, a sharp spectral peak for the message of 1.5 GHz is clearly visible in the transmitter spectrum. On the other hand, the spectrum bears resemblance to that of the transmitter but no distinct spectral component for the message is present in the receiver spectrum due to a chaos-pass-filtering effect. As a result, we can extract the message simply by subtracting the reproduced chaotic signal in the receiver laser from the transmitted signal, thus chaotic communication is realized. The receiver laser generates only the intrinsic chaotic oscillations the same as that from the transmitter if the message embedded into the chaotic carrier is small enough. The effect of the insensitivity for small external perturbations to chaos can be considered as a different phenomenon such as the sensitivity of chaos for initial conditions. Chaos seems to distinguish external perturbations and the system nonlinearity.

Looking at the spectrum of the transmitted signal in Fig. 13.4, the question may arise that the message may be extracted by filtering the waveform with a narrow bandpass filter at the message frequency. Figure 13.5 shows the filtered waveforms for the decoded message as well as the transmitted signal and the receiver output. The waveforms are the results for a narrow bandpass filter of ±100 MHz centered at the message frequency of 1.5 GHz. The decoded message is a simple subtraction of the receiver output from that to the transmitter. The decoded message is reproduced as a good sinusoidal oscillation, which is almost the same signal as the original message. However, the filtered waveforms for the transmitted signal and the receiver output are not good harmonic signals and they are even not in-phase with the message signal. The degree of the security of communications must be

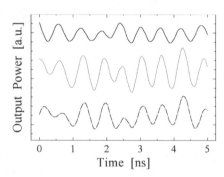

Fig. 13.5. Message decoding for signals in Fig. 13.3a with a narrow bandpass filter of ±100 MHz centered at a message frequency of 1.5 GHz. **a** Decoded message (*upper trace*) and transmitter (*middle trace*) and receiver (*bottom trace*) outputs

evaluated for actual data of binary bit-sequences. However, the results obtained in Fig. 13.5 show some of evidence for the security in the present systems.

As an example of a signal transmission of binary data in CMA, numerical results by Sánches-Díaz et al. (1999) are shown. They conducted data transmissions of pseudo-random bit-sequences of a 4 Gbps NRZ signal in a closed-loop system. The message is a small perturbation of 0.5% of an averaged chaotic oscillation in the transmitter and it is fed to the injection current to the laser as a direct modulation. Since the perturbation is sufficiently small, the method is considered as CMA. They also assumed the nonlinear effect of signal transmissions through optical fibers between the transmitter and receiver systems. Figure 13.6 shows their results. In their method, the decoded signal is obtained by the comparison of the transmitted signal $E_{\mathrm{trans}}(t)$ ($\propto E_{\mathrm{T}}(t)$) with the decoded one $E_{\mathrm{R}}(t)$ as

$$m'(t) = \sqrt{\frac{|E_{\mathrm{trans}}(t)|^2}{|E_{\mathrm{R}}|^2} - 1} \tag{13.5}$$

The fidelity of the chaotic signal after the transmission through the optical fiber may be degraded due to a nonlinear dispersion effect in the fiber. The data transmission in optical fiber is described by the following nonlinear Schrödinger equation (Ohtsubo 2002a)

$$\mathrm{i}\frac{\partial E(z,T)}{\partial z} = -\frac{\mathrm{i}}{2}\alpha_{\mathrm{f}}E(z,T) + \frac{1}{2}\beta_2\frac{\partial^2 E(z,T)}{\partial T^2}$$
$$- \gamma_{\mathrm{non}}|E(z,T)|^2 E(z,T) \tag{13.6}$$

where $E(z,T)$ is the slowly varying complex field, z is the propagation distance, and T is the time measured in the reference frame moving at the group velocity. γ_{non} is the nonlinear parameter that takes into account the optical Kerr effect, α_{f} is the fiber loss, and β_2 is the second order dispersion parameter. They also used a low-pass Fabry-Perot filter with a bandwidth of 5 GHz to obtain the final decoded message of Fig. 13.6f, though a higher order Butterworth electronic filter is usually used. Excellent chaos synchronization was attained in spite of the effect of nonlinear optical fiber transmission through 50 km and the message was successfully reconstructed.

The quality of the reproduced chaotic signal and, accordingly, the decoded message in the receiver are degraded by the transmission through the nonlinear optical fiber. The system performance at the modulation of 2 Gbps is displayed in Fig. 13.7 for different propagation length in optical fiber (Sánches-Díaz et al. 1999). Whether the message is included in the chaotic carrier or not, the synchronization becomes worse for a long fiber transmission and the quality of the decoding becomes worse accordingly. The degradation comes from both the dispersion and nonlinearities of optical fiber. In actual optical

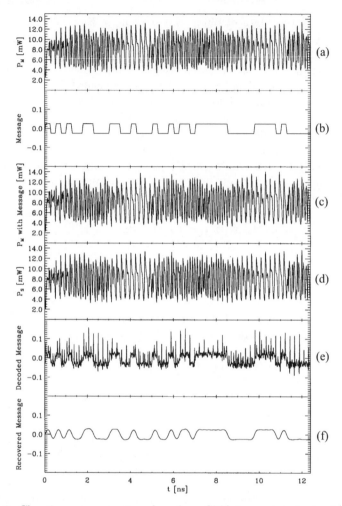

Fig. 13.6. Chaotic communications based on CMA in semiconductor lasers with optical feedback. **a** Chaotic oscillation from transmitter, **b** binary bit-sequence for data transmission, **c** chaotic waveform together with message, **d** synchronized chaotic oscillation in receiver after data transmission through a nonlinear optical fiber of 50 km, **e** decoded signal, and **f** low-pass filtered waveform for decoded signal (after Sánches-Díaz A, Mirasso CR, Colt P, García-Fernández P (1999); © 1999 IEEE)

fiber transmission, there is a loss of light through the optical fiber. Therefore, an in-line amplifier is placed at a certain distance in practical optical fiber communications. This may cause further degradation of chaotic signals due to the enhancement of the nonlinear effects after the amplification. In-line amplifiers would aggravate the negative effect of the fiber nonlinearities on the synchronization.

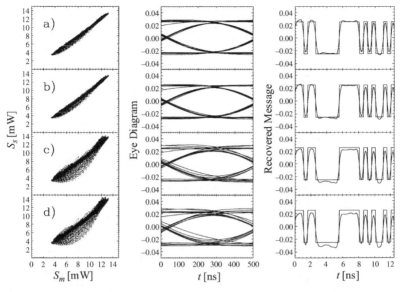

Fig. 13.7. Signal recovery for fiber transmissions of 50, 100, 150, and 200 km from **a** to **d**, respectively. The *left column* shows the correlation plots between the chaotic carrier in the transmitter without a message and the receiver output at synchronization. The *middle* shows the corresponding eye patterns. The *right* shows the decoded messages after filtering (after Sánches-Díaz A, Mirasso CR, Colt P, García-Fernández P (1999); ©1999 IEEE)

13.2.3 Chaos Modulation in Optical Feedback Systems

In CMO, we put $\eta_{CMO} \neq 0$ and $\eta_{CMA} = \eta_{CSK} = 0$ in (13.1)–(13.4). A message is added within the transmitter, for example in the feedback loop in Fig. 12.4a, and the transmission signal of $E(t + \tau) = f(E(t) + m_{CMO}(t))$ is generated and transmitted to the receiver. The original chaotic signal together with the message conforms a new chaotic signal and, therefore, the message may essentially have a large amplitude. However, care must be taken when embedding a large message signal into chaotic oscillations, since the message itself may explicitly appear in the transmission signal for the worst case. For this reason, a message with small amplitude is usually used in CMO. The method of CMO is approximately equal to CMA when an encoding message is small enough.

We present a numerical example of the CMO method. This scheme requires the achievement of complete chaos synchronization and hence is usually difficult to realize by experiment (Liu et al. 2002a). The possibility of transmission of a NRZ pseudo-random binary sequence at 2.5 Gbps using CMO was demonstrated by numerical simulation. Figure 13.8 shows the results (Liu et al. 2001c). In this example, it is assumed that an electro-optic modulator is inserted into the optical feedback loop, and is used to modulate

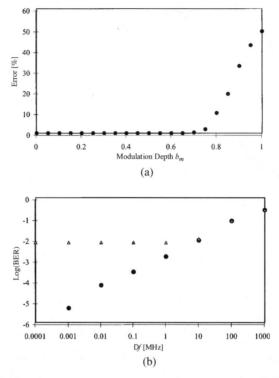

Fig. 13.8. Calculated synchronization error and bit error rate (BER) for normalized dimension less bias injection currents of 0.67, normalized feedback and injection ratios of 0.01. **a** Synchronization error versus modulation amplitude b_m. **b** BER versus frequency detuning at a modulation factor of $b_m = 0.15$. *Black dots:* no noise, *triangles:* noise at SNR= 40 dB

the feedback signal as $m_{CMO}(t) = 1 + b_m$ for "1" and $m_{CMO}(t) = 1 - b_m$ or "0", where b_m $(b_m < 1)$ is the modulation depth. Figure 13.8a shows the dependence of the synchronization error on the amplitude of the embedded signal b_m, and Fig. 13.8b is the dependence of the bit rate error (BER) on the frequency of the detuning between the transmitter and receiver lasers, both with noise added (triangles) and without noise (circles). The results show good synchronization for a wide range of relative modulation amplitudes and sensitive dependence of the BER on the detuning, which are strong features of the CMO method using complete synchronization.

13.2.4 Chaos Shift Keying in Optical Feedback Systems

A set of chaotic generators is usually required both for transmitter and receiver systems in CSK. Indeed, chaotic signals from two transmitters are switched according to the binary value of 0 or 1 of a message sequence and they are sent to the two receivers. The receivers synchronize to the corresponding transmitters and the decoding is done by the synchronization. Annovazzi-Lodi et al. (1997) used a single transmitter consisting of a semiconductor laser with optical feedback in CSK instead of two transmitters. A NRZ binary message is put into the injection current of the laser at a certain bias point. According to the message, two chaotic states are generated and they are sent to the receiver. The difference between the two chaotic attractors must be sufficiently small not to be distinguished easily for secure communications. The receiver is a set of chaotic generators and they have the same characteristics except for the injection currents. The injection currents are set at either the high or low value of the binary message. Then, each receiver synchronizes with the corresponding chaotic state and synchronous and asynchronous (chaotic bursts) signals are obtained for the time sequence from comparison between the transmitted signal and the receiver output. Figure 13.9 shows one of the receiver outputs in the CSK system. The square waveform is a message to be transmitted and the error signal is the receiver output. The other receiver laser outputs the compensating signal to the waveform in Fig. 13.9. Then, the message is decoded from these two signals.

When two nonlinear chaotic systems synchronize with each other, the receiver does not respond immediately after it receives a chaotic signal from the transmitter. Usually the receiver outputs the synchronous signal after a certain transient time. In a CSK system, the chaotic oscillation switches from one state to the other according to the ON/OFF signals. Therefore, we must consider the synchronization recovery time after the switching of signals. This limits the efficiency of the possible bit rate of the data transmission. This synchronization recovery time depends on each system configuration (open- or close-loop system) and the device and system parameters. The typical fre-

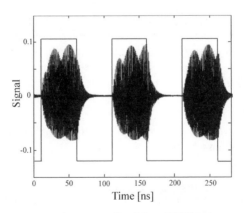

Fig. 13.9. Square message to be transmitted in a CSK system and error signal from one of the receiver outputs. The system has a single transmitter of a semiconductor laser with optical feedback and the receiver consists of two optical feedback systems (after Annovazzi-Lodi V, Donati S, Scirè A (1997); © 1997 IEEE)

quency of a chaotic carrier in semiconductor lasers with optical feedback is of the order of the relaxation oscillation frequency at the solitary oscillation and a message frequency must be less than this. The time required for the synchronization in CSK is from nano-second to several nano-seconds depending on the system parameters (Vicente et al. 2002). On the other hand, it is possible for CMA and CMO methods to achieve higher data transmissions than the relaxation oscillation frequency, since chaotic oscillations in the systems have broadband characteristics over the relaxation oscillations. In the CSK discussed here, the transmitter laser is directly modulated by a message through the injection current, so that it looks like a method of CMO. However, the original chaotic attractor is not affected by the modulation. It is distinguished from CMO as long as the amplitude of the message is small (less than 1% of the averaged chaotic oscillation). The synchronization recovery time will be discussed in Sect. 13.7.

13.2.5 Chaotic Communications in Incoherent Optical Feedback Systems

Incoherent optical feedback systems are also used for chaotic communications based on chaos synchronization as discussed in the previous chapter. In the systems, the laser output power from the transmitter is coupled with the carrier density of the receiver laser. Therefore, we do not need to consider the optical phase and tune the optical frequencies between the transmitter and receiver lasers. As a result, we can easily achieve chaos synchronization. However, the grade of the security in the communications is deteriorated, since one of the keys for secure chaotic communications is eliminated. Nevertheless, the output generated by a semiconductor laser with incoherent optical feedback is a higher dimensional chaos and the system has enough complexity for secure communications. It is still not easy to reproduce the transmitter chaos in the receiver without knowing the chaos parameters in the system. Rosigter et al. (2001) conducted chaotic data transmission based on CSK using incoherent optical feedback systems by numerical simulations. The synchronization system is the same as in Sect.12.3.5. They succeeded in the data transmission of 250 Mbps pseudo-random-bit sequences with excellent quality using ON/OFF CSK. The level of the message was only 0.3% of the bias injection level and the secure communication was achieved by hiding the data behind the chaotic signal.

13.2.6 Chaos Pass Filtering Effects

The effect of chaos-pass filtering is essential to attain successful secure communications by hiding messages behind chaotic signals in chaos synchronization systems. In accordance with the effect, the chaotic signal, which is transmitted from the transmitter laser, is only reproduced by the receiver laser even if the transmitted signal includes a message. Thus, the message is correctly decoded by subtracting the chaos signal of the receiver laser from the transmitted signal under appropriate conditions for the chaos synchronization system. However, the above expression of 'chaotic signal is only reproduced' may not be correct. As will be discussed later, the effect of the chaos-pass filtering is that the main component of the chaotic signal from the transmitter laser is closely copied in the receiver laser and the chaos transmittance is usually equal to or larger than unity, while the transmittance of the message, whose frequency component is less than the relaxation oscillation, is much less than unity. Several studies for the effects of chaos-pass filtering have been reported both theoretically and experimentally (Uchida et al. 2003b, Paul et al. 2004, and Murakami and Shore 2005, 2006). Especially, chaos-pass filtering plays a crucial role in chaotic masking systems. In the following, we discuss the effect of chaos-pass filtering in semiconductor lasers with optical feedback.

Consider a chaos masking system in semiconductor laser with optical feedback described by (13.1)–(13.4) and assume three different sinusoidal modulations for the injection current J_T in the transmitter laser as messages. Figure 13.10 shows the calculated transfer function between the transmitted signal and the receiver output. The chaos synchronization system assumed here is a type of open loop and the synchronization is generalized one. To show the effect of chaos-pass filtering explicitly, the ratio of the optical injection from the transmitter to the receiver laser is as large as 50%. A transmission rate of several percents from the transmitter to the receiver is sufficient to attain chaos synchronization in the laser system and, indeed, such a small rate is usually used in a synchronization system using chaotic semiconductor lasers. The parameters of the solitary laser without optical feedback are $J = 1.3J_{th}$ and $f_R = 2.43$ GH. The optical delay in the transmitter laser is set to be $\tau = 1$ ns and the frequency detuning between the transmitter and receiver lasers is assumed to be -0.1 GHz. Figure 13.10a shows the power spectra for the transmitter signal including the messages and the synchronized receiver output. The power spectrum of the receiver laser is vertically shifted by -15 dB to show clearly the difference. The frequency component of the chaotic signal is concentrated from $1 - 10$ GHz. In the both spectra, embedded spectral peaks of sinusoidal signals are clearly seen. From the closer look at the two spectra, the spectra in the lower frequency components in the receiver are largely suppressed. On the other hand, the signal higher than the relaxation oscillation frequency is much enhanced in the receiver side.

Figure 13.10b shows the response ratio between the transmitter signal and receiver output. Since the injection rate from the transmitter to the receiver is strong as much as 50% in this case, the gain of the chaotic signal component in the receiver laser is much lager than 0 dB. As for the message components, it is seen from the figure that the transmission rates for the lower

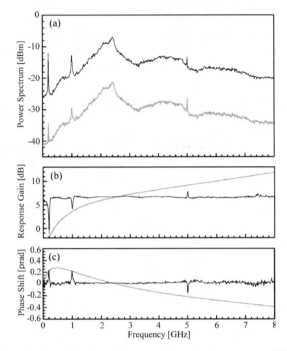

Fig. 13.10. Transfer functions in a chaos synchronization system. The transmitting signal from the transmitter is a chaotic signal together with three sinusoidal messages with modulation frequencies of 0.2, 1.0 and 5.0 GHz. The modulation indices are 0.05, 0.05, and 0.10, respectively. **a** Power spectra for transmitter signal and receiver output. *Solid line*: transmitter, and *gray line*: receiver. The receiver spectrum is vertically shifted by −15 dB. **b** Transfer function from the transmitter to the receiver. *Solid line*: transfer gain, and *gray line*: transfer function for sinusoidal modulation only (modulation index is 0.05). **c** Phase shift between the transmitter signal and the receiver output. *Solid line*: phase shift between the transmitter signal and the receiver output, and *gray line*: phase shift for sinusoidal modulation only (modulation index is 0.05) (Murakami A, Shore KA (2005); © APS 2005)

frequency components below the solitary relaxation oscillation frequency are greatly decreased compared with the transmission gain of the chaotic signal, while the transmission rate for the higher frequency component is larger than unity. The gray line in the figure shows the transfer function from the transmitter to the receiver laser for only sinusoidal modulation with a modulation index of 5% to the solitary transmitter laser. The cross-over point of the chaotic gain and the sinusoidal modulation response is exactly equal to the relaxation oscillation frequency of the semiconductor laser at solitary oscillation. Figure 13.10c shows the phase shift between the transmitter signal and the receiver laser. The solid line represents the phase shift, which is calculated from the Fourier components of the transmitter signal and the receiver output. There is no phase shift for the chaotic transmittance, while the sinusoidal signals with lower frequency components have a positive phase shift and the higher frequency component has a negative shift. The gray line shows the phase shift for only sinusoidal modulation with a modulation index of 5% to the solitary transmitter laser, which corresponds to the gray line in Fig. 13.10b. Again, the cross-over point of the chaotic phase shift and the phase shift for the sinusoidal modulation is exactly equal to the relaxation oscillation frequency. It is noted that the phase shift becomes $-\pi$ at the resonant frequency by the strong optical injection from the transmitter to the receiver laser, when only a sinusoidal modulation is transmitted (though the resonant frequency is 12.7 GHz in this case and the frequency is out of scope in this graph).

As discussed above, the frequency components of chaotic signals concentrate at and around the relaxation oscillation frequency and they typically range from 1 to 10 GHz in chaotic semiconductor lasers. Another interesting point is that the transfer function of a chaotic signal from the transmitter to the receiver is almost constant for all the frequency range. On the other hand, the transfer function of a sinusoidal signal has a frequency dependence. Namely, a sinusoidal signal has a small response gain for lower frequency less than the relaxation oscillation, while a sinusoidal signal with higher frequency has a larger response gain than 0 dB. Thus, the difference for the response gains between chaotic and sinusoidal signals is the origin of the chaos-pass filtering effects. As another issue, the direct injection current modulation for the transmitter laser as a message encryption deteriorates the performance of chaos synchronization. Therefore, an external modulation for chaotic signals using an electro-optic (EO) modulator is frequently used. Uchida et al. (2003b) studied the effects of chaos-pass filtering in a chaos masking system using semiconductor lasers with optical feedback, and obtained similar results for the response between the transmitter and the receiver as discussed here. It is derived from the above discussion that chaotic carrier frequency, which is limited by the resonant oscillation frequency of semiconductor laser, must be much greater than a main message frequency to attain higher data-bit transmission in chaotic semiconductor laser systems. Thus the use of semiconductor lasers, which have high modulation bandwidth, is essential for massive chaotic secure communications.

13.3 Cryptographic Applications in Optical Injection Systems

We can apply the systems of semiconductor lasers subjected to optical injection discussed in Sect. 12.4 to chaotic communications. Figure 13.11 shows the schematic diagram of the chaotic communications. Message encoding and decoding schemes are the same as the systems of optical feedback in Figure 13.3. For CMA, CMO, and CSK, the equations for the transmitter laser

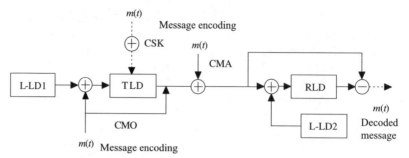

Fig. 13.11. Schematic diagram of chaotic communications in semiconductor lasers with optical injections. $m(t)$ is a message signal to be embedded. The position for each message encoding scheme is shown in the figure. The *solid lines* are optical connections and the *broken lines* are electronic connections

are written by

$$
\frac{dE_T(t)}{dt} = \frac{1}{2}(1 - i\alpha_T)G_{n,T}\{n_T(t) - n_{th,T}\}E_T(t)
$$
$$
+ \frac{\kappa_{inj,T}}{\tau_{in,T}}E_T(t)\exp(-i\Delta\omega_T t) + \eta_{CMO}m_{CMO}(t) \qquad (13.7)
$$
$$
\frac{dn_T(t)}{dt} = \frac{J_T}{ed}\{1 + \eta_{CSK}m_{CSK}(t)\} - \frac{n_R(t)}{\tau_{s,R}}
$$
$$
- G_{n,T}\{n_T(t) - n_{0,T}\}|E_T(t)|^2 \qquad (13.8)
$$

where $\Delta\omega = \omega_{inj,T} - \omega_{0,T}$ is the frequency detuning between the injection laser and the transmitter laser. Whereas the receiver driven by the transmitted signal can be described by

$$
\frac{dE_R(t)}{dt} = \frac{1}{2}(1 - i\alpha_R)G_{n,R}\{n_R(t) - n_{th,R}\}E_R(t)
$$
$$
+ \frac{\kappa_{inj,R}}{\tau_{in,R}}E_R(t)\exp(-i\Delta\omega_R t)
$$
$$
+ \frac{\kappa_{cp}}{\tau_{in,R}}E_T(t - \tau_c)\exp(i\omega_{0,T}\tau_c)
$$
$$
+ \eta_{CMA}m_{CMA}(t - \tau_c) + \eta_{CMO}m_{CMO}(t - \tau_c) \qquad (13.9)
$$
$$
\frac{dn_R(t)}{dt} = \frac{J_R}{ed} - \frac{n_R(t)}{\tau_{s,R}} - G_{n,R}\{n_R(t) - n_{0,R}\}|E_R(t)|^2 \qquad (13.10)
$$

Optical modulations of messages (CMA and CMO) are assumed to be applied to the complex fields, however electro-optic (EO) modulators are usually used as intensity modulations. We need some modifications of the above rate equations for the intensity modulations. However, (13.7)–(13.10) are again a good approximation for the optical injection systems as far as the modulation amplitude is small enough.

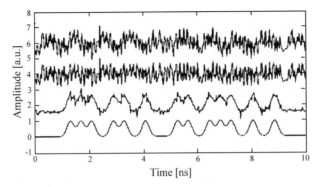

Fig. 13.12. Numerical example of message transmission in semiconductor lasers with optical injection in CMO. The system is an open-loop and the message is a 2.5 Gbps signal. From *top* to *bottom:* transmitted signal including message, receiver output, decoded signal by subtraction of receiver output from transmitted signal, and low-pass filtered signal for decoded message (after Liu JM, Chen HF, Tang S (2001a); ©2001 IEEE)

Figure 13.12 presents a numerical example of chaotic communications in a system of semiconductor lasers subjected to optical injection (Liu et al. 2001a). The figure is a message transmission of a 2.5 Gbps signal based on CMO. In the transmitter laser, a chaotic carrier is generated at the appropriate injection ratio and frequency detuning in the optical injection configuration. A binary message is encoded into the chaotic carrier. There are two schemes of chaotic modulations to the optical field in CMO; additive modulation and multiplicative modulation. This case is for additive modulation. After subtracting the receiver output from the transmitted signal and low-pass filtering it (bottom of the figure), they obtained a decoded message with good quality. The system of optical injection is also phase sensitive like the system of coherent optical feedback (Heil et al. 2003). We must pay attention to the optical phase to achieve good quality of communications.

13.4 Cryptographic Applications in Optoelectonic Systems

We here describe chaotic communications in semiconductor lasers with optoelectronic feedback systems. The system is incoherent coupling and the light from the laser is once detected by a photodetector. Then, the photocurrent is fed back into the bias injection current of the laser. Therefore, noises originating from photons are averaged out due to slow response of the carrier lifetime and high performance for synchronization between the transmitter and receiver lasers can be expected. A message is embedded into fast chaotic pulsation oscillations from sub-nano-second to pico-second and high-speed

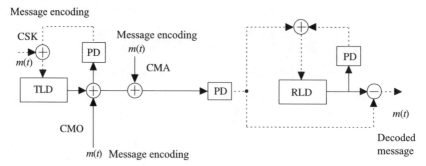

Fig. 13.13. Schematic diagram of chaotic communications in semiconductor lasers with optoelectronic feedback. $m(t)$ is a message signal to be embedded. The position for each message encoding scheme is shown in the figure. The *solid lines* are optical connections and the *broken lines* are electronic connections

chaotic data transmissions are also expected due to the availability of high speed electronic circuits. Figure 13.13 shows a model of chaotic communication systems in semiconductor lasers with optoelectronic feedback. The system can be described by only the photon number and the carrier density. A message is directly added to the laser intensity or the bias injection current, which is not the case for the systems of optical feedback and optical injection. The following formulation can be widely used in real optoelectronic systems without modification. For the transmitter, the rate equations read

$$\frac{\mathrm{d}S_T(t)}{\mathrm{d}t} = G_{n,T}\{n_T(t) - n_{th,T}\}E_T(t) + R_{sp,T} \tag{13.11}$$

$$\frac{\mathrm{d}n_T(t)}{\mathrm{d}t} = \frac{J_T}{ed}\{1 + \eta_{CSK}m_{CSK}(t)\}\{1 + s_T(t)\} - \frac{n_T(t)}{\tau_{s,T}}$$
$$- G_{n,T}\{n_T(t) - n_{0,T}\}S_T(t) \tag{13.12}$$

$$s_T(t) = \xi_T S_T(t - \tau_T) + \eta_{CMO}m_{CMO}(t - \tau_T) \tag{13.13}$$

When the system response is continuous, $s_T(t)$ is replaced by a continuous response function in the same manner as in (12.68) as

$$y_T(t) = \int_{-\infty}^{t} f(t' - t)s_T(t')\mathrm{d}t' \tag{13.14}$$

whereas the receiver driven by the transmitted signal can be described by

$$\frac{\mathrm{d}S_R(t)}{\mathrm{d}t} = G_{n,R}\{n_R(t) - n_{th,R}\}E_R(t) + R_{sp,R} \ , \tag{13.15}$$

$$\frac{\mathrm{d}n_R(t)}{\mathrm{d}t} = \frac{J_R}{ed}\{1 + s_R(t)\} - \frac{n_R(t)}{\tau_{s,R}} - G_{n,R}\{n_R(t) - n_{0,R}\}S_R(t) \tag{13.16}$$

$$s_R(t) = \xi_R S_R(t - \tau_R) + \xi_{cp}S_T(t - \tau_c) + \eta CMOm_{CMO}(t - \tau_c) \tag{13.17}$$

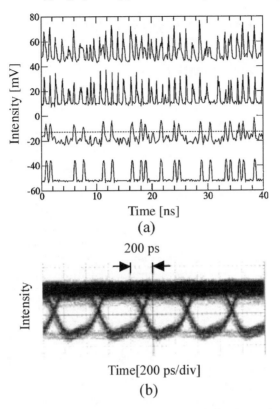

Fig. 13.14. Experimental example of message transmission in semiconductor lasers with optoelectronic feedback in CMO. The system is an open-loop using DFB lasers at the wavelength of 1.30 μm and the message is a 2.5 Gbps signal. **a** From *top* to *bottom:* transmitted signal including message, receiver output, decoded filtered signal, and original message. **b** Eye pattern of the decoded signal (after Tang S, Liu JM (2001a); © 2001 IEEE)

Similarly, $s_R(t)$ is replaced by a continuous response function for a finite response of the electronic feedback circuits.

Figure 13.14 shows the experimental results of chaotic communications in semiconductor lasers with optoelectronic feedback in CMO. The system is open-loop. A message is a pseudo-random signal of an NRZ pulse sequence. The message is embedded into the bias injection current with additive modulation. As discussed in Sect. 12.5, open-loop configuration shows better quality of chaos synchronization and modulation (Tang and Liu 2001a and Liu et al. 2002b). Figure 13.14a shows the plot of signals for data transmission and decoding. The decoded signal (the third signal from the top) well reproduces the original message above the threshold level (dotted line). From the time lag, it is recognized that the synchronization is a complete type, although the time lag is compensated to compare the waveforms. Since chaotic signals generated in semiconductor lasers with an optoelectronic feedback system are pulse-like irregular oscillations, additive modulation is suited for CMO rather than multiplicative modulation. Figure 13.14b shows the eye pattern for the decoded message. A good quality of opened eyes is obtained.

13.5 Performance of Chaotic Communications

We have discussed chaotic communications for three message encryption schemes (CMA, CMO, and CSK) in three different systems of semiconductor lasers (optical feedback, optical injection, and optoelectronic feedback systems). Each scheme in each system has merit and demerit for chaotic data transmissions. Liu et al. (2002b) numerically compared the performances of chaos communications for these methods. Figure 13.15 shows the results of the comparison of the performance for the three systems of (a) optical feedback, (b) optical injection, and (c) optoelectronic feedback. A message is a 10 Gbps pseudo-random pulse sequence. The relaxation oscillation frequency of the laser assumed is set to be 12 GHz at the operating bias injection current. Open-loop configurations are assumed for all the systems. For CMA and CMO in optical feedback and optical injection systems, the ratio of the amplitude for the encoding message to that of the chaotic amplitude is set at 0.05. For the CSK system, the injection current is modulated as an ON/OFF modulation and the corresponding modulation index to the optical field is taken to be the same as those in CMA and CMO. At the modulation index of 0.05, SNR of 30 dB has the channel noise at a level of an equivalent laser linewidth of $\Delta\nu = 0.66$ MHz. On the other hand, the modulation index is assumed to be 0.2 in the optoeletronic feedback system.

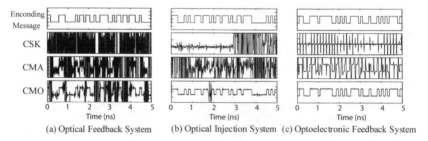

(a) Optical Feedback System   (b) Optical Injection System (c) Optoelectronic Feedback System

Fig. 13.15. Comparison of performance for chaotic communications in semiconductor laser systems at a data transmission rate of 10 Gbps. **a** Optical feedback system, **b** optical injection system, and **c** optoelectronic feedback system. All the systems are open-loop. From *top* to *bottom* in each figure: encrypted message, decoded signal in CSK, decoded signal in CMA, and decoded message in CMO (after Tang S, Liu JM (2001a); © 2001 IEEE)

A finite response time is required for chaos synchronization in a receiver system when each message is transmitted. In CSK, the chaotic state is always switched in accordance with the binary message and the receiver cannot follow the ON/OFF switching of the chaotic states. Thus, the performance of synchronization becomes worse and one cannot recover the message for the worst case. The basis of chaotic communications is that a message signal attached to the chaotic carrier, which is very small in comparison with the size of the trajectory as a chaotic attractor, will be averaged out and has almost no effect on the duplication of the chaotic trajectory. The whole chaotic carrier waveform can then be reproduced very precisely through all the local predictor functions. However, as already discussed, a message is essentially a perturbation for the chaotic attractor of a transmitter output in CMA and CSK even when it is small, thus the synchronization deviation increases for large message amplitude. On the other hand, the symmetry of the transmitter and receiver systems is preserved even if a message is embedded into the transmitter. Therefore, the system of CMO is robust for synchronization deviation compared with CMA and CSK. As a result, we can successfully recover the message with good quality for CMO in the system of optoelectronic feedback as shown in Fig. 13.15c. The synchronization is interrupted by synchronization deviation and bursts at the high data-transmission rate of 10 Gbps even for CMO in the systems of optical feedback and optical injection, and one cannot obtain the original messages. In all systems, the messages are not recovered in the CSK and CMA schemes in Fig. 13.15. For a higher data-transmission rate, a system of optoelectronic feedback with CMO is best suited for chaotic communications.

The reason why the system of optoelectronic feedback is better than those of optical feedback and optical injection for chaos synchronization has been discussed in Sect. 12.5.1. The carrier lifetime plays an important role in the system of optoelectronic feedback, while the photon lifetime is crucial for the other systems. The photon noises are averaged out due to a slow response of the carrier and, as a result, the system is robust for photon noises. As far as the response of the electronic circuits can follow the chaotic signal, we can expect good chaos synchronization and faithful message decoding.

The measure of the performance for a communication system is the bit rate error (BER) for the decoded message as a function of the signal-to-noise ratio (SNR) in the transmission channel. The SNR is defined as (Liu et al. 2002b)

$$\mathrm{SNR} = 10 \log \frac{S_\mathrm{m}}{\sigma_\mathrm{n}^2} \tag{13.18}$$

where S_m is the power of the transmitted message, and $\sigma_\mathrm{n}^2 = N_0/2T_\mathrm{b}$ is the variance of the channel noise with $N_0/2$ being the power spectral density of the channel noise and T_b being the bit duration. The channel SNR is a function of the channel noise, which is taken to additive white Gaussian noise, and the bit energy of the transmitted message, which depends on the modulation index of the message. BER arises from synchronization errors and bursts induced by channel errors and spontaneous emission noises in the lasers. The synchronization error σ_{error} has already been defined in (12.36).

Simply good quality of synchronization does not guarantee good retrieval of a message signal due to the sensitivity of the synchronized trace to any perturbation, including the perturbation caused by the intrinsic noise of the transmitter and that of the receiver. If some perturbation temporarily desynchronizes the synchronized transmitter and receiver for a period of time, the message signal within this period cannot be recovered. Therefore, the robustness of synchronization has to be considered in chaotic communications. Desynchronization could happen if the synchronized trace has any positive conditional Lyapunov exponent for a period of time while any perturbation is acting on this synchronized trace. It depends on the value of the positive conditional Lyapunov exponent and the strength of the perturbation during that period of time.

The synchronization errors are categorized into two origins of synchronization deviation, which is associated with the accuracy of synchronization and desynchronization bursts, which is related to the robustness of synchronization in the system. The correlation coefficient of chaos synchronization between the transmitter and receiver lasers is usually not close to unity, although the two chaotic signals are similar. The deterioration of the correlation coefficient corresponds to the synchronization error. On the other hand, the chaotic output in the receiver completely differs from the transmitter signal at the occurrence of desynchronization bursts and the transmitter and receiver outputs have no correlation. Desynchronization burst is observed at the marginal region of the allowed parameter mismatches for the synchronization. The occurrence of desynchronization burst depends on the combination of chaotic parameters in the systems. It is an essential phenomenon in chaos synchronization systems with parameter mismatches. Even when the parameters of the transmitter and receiver systems are equal, desynchronization burst may suddenly occur by noises in the electronic circuits. The bits error caused by synchronization deviation is measured by the concept of synchronized bit error rate (SBER)

$$\text{SBER} = \frac{\text{error bits caused by desynchronization bursts}}{\text{total number of bits tested}} \qquad (13.19)$$

The desynchronized bit error rate (DBER) induced by desynchronization bursts is defined by

$$\text{DBER} = \frac{\text{error bits caused by synchronization deviation}}{\text{total number of bits tested}} \qquad (13.20)$$

Then the total BER is simply defined by the sum of the SBER and the DBER as

$$\text{BER} = \text{DBER} + \text{SBER} \qquad (13.21)$$

Figure 13.16 is the result of BERs calculated by numerical simulations under the assumption of SNR = 30 dB (Liu et al 2002b). BER can be improved by applying a filter to the decoded messages. In this example, the testing filter used for examining the characteristics of error reduction is a digital Chbyshev Type I filter with the cutoff frequency equal to the bit rate, the sampling frequency equal to 20 times the cutoff frequency, 0.5 dB ripple on the passband and −30 dB attenuation on the stopband. The best performance of BER is obtained for the optoelectronic feedback system with CMO, which is a good coincidence with the result in Fig. 13.15. Figure 13.16b is the performance of BER obtained by changing the bit rate of the data transmission in CMO schemes. The BERs in optical feedback and optical injection systems stay constant for the change of the bit rate. However, we can expect

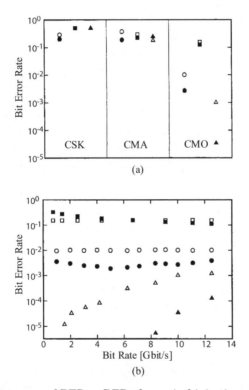

Fig. 13.16. Performance of BER. **a** BERs for optical injection system (marked as *circles*), optical feedback system (marked as *squares*), and optoelectronic feedback system (marked as *triangles*) of three different encryption schemes (after Tang S, Liu JM (2001a); © 2001 OSA). The *solid symbols* mark the BER after the filtering, and the *open symbols* mark the BER before the filtering. **b** BER versus bit rate for three systems under CMO. The meanings of the symbols are the same as those in **a** (after Tang S, Liu JM (2001a); © 2001 IEEE). The relaxation oscillation frequency of the lasers is assumed to be 12 GHz

a great improvement of BER in optoelectronic feedback systems for a lower data transmission rate.

13.6 Security of Chaotic Communications

The successful demonstrations of chaotic data transmissions in laser systems including semiconductor lasers, and solid state lasers, and fiber lasers have proved that these schemes are robust to some degree. However, there is still much to be done in terms of evaluating the robustness and practical trade-offs with the security based on parameter sensitivity. The security in chaotic communications is guaranteed by the coincidence of the system parameters between the transmitter and receiver including the device parameters and the operation conditions of the lasers. Namely, one cannot achieve chaos synchronization in the system without knowing parameters as a key for communications even if one can know the system configurations. Chaotic communications are essentially hardware-based techniques. However, we can make a virtual system of the chaotic communication on computer software when we know completely the system configurations and their mathematical descriptions. Even for such a case, it is still difficult to imitate the synchronous chaotic signal without knowing each system parameter value. Therefore, the tolerance for the parameter mismatch is quite important for the realization of secure chaotic communications and we require systems that have strict conditions for the parameter mismatches for communications. However, as noted, there is a tradeoff the difficulty for the range of synchronization in real systems.

One of the aspects of the security of messages hidden in chaotic waveforms is the difficulty of separating the message from the chaotic carrier by analyzing recorded waveforms. It has been demonstrated that systems with low dimensional chaos are not secure for data transmissions, in the sense that a low dimensional attractor is easily reconstructed from time series data, and system parameters are easily estimated from this attractor (Short 1994 and 1996). Decoding without knowing key parameters becomes more difficult with the increase of the dimension of the chaotic dynamics (Dachselt and Schwarz 2001). The chaotic signal from a semiconductor laser with optical feedback, as described by the theoretical model, is embedded in an infinite dimensional system due to the delay. However, the actual dimension of the dynamics is typically much lower, being restricted by the intrinsic response times in the laser. Quantitative analysis of the dimension of the synchronized chaos and the degree of security remains an important challenge for future study. Here, we limit our discussion to the issue of the matching of the receiver laser for message recovery by synchronization.

Another important issue is the noise problem. Not only optical and electronic circuits to generate chaotic signals but also optical channels include noises, however there is little study on the effects of noises on the performance of chaotic communications. The result of Fig. 13.16 is such an instance. It is a well-known result that chaos is sensitive to the initial conditions of the system. When the systems include noises, it may be considered that they disturb the systems and the receiver is not able to output synchronous oscillations even if the two nonlinear systems consist of the same components and have the same parameters. Contrary to the expectations, two nonlinear systems can synchronize with each other even they include noises. This fact is verified by numerical simulations and experiments, although the basis for the phenomenon has not been theoretically proved yet. Chaos induced by the nonlinearity of a system and statistical noises are completely separated and the system seems to discriminate them. Though noises are additive to chaotic signals, a chaotic evolution of a system is essentially determined by the pure initial conditions of the system without noises. For example, a chaotic attractor has slight deviations from the original one in the presence of noises, but the chaotic route and chaotic dynamics do not change, while, for example, the maxima and minima of the output in fixed and periodic states in the bifurcation are not points but have finite widths due to the external noises.

There are two types of chaos synchronization: one is complete chaos synchronization and the other is generalized chaos synchronization. In the following, we discuss the differences of the security in chaotic communications between the two types of synchronization. Take as an example optical feedback systems. As shown in Fig. 12.7, complete chaos synchronization is only realized at zero frequency detuning and a lower optical injection in the map of the unstable injection locking region. On the other hand, generalized chaos synchronization is attained in a wide range of frequency detuning and optical injection in the stable injection locking region. From the standpoint of security, the conditions for generalized chaos synchronization are loose compared with those for the complete case. We have investigated the effects of mismatches of the laser device parameters for chaos synchronization in Fig. 12.8. In that case, good synchronization is attained for a very small range of the parameter mismatches in complete chaos synchronization and the accuracy of synchronization rapidly becomes worse for the increase of the parameter mismatches. On the other hand, the tolerance for the parameter mismatches is rather large for the case of the generalized chaos synchronization. Strict conditions are imposed for the case of complete chaos synchronization and, as a result, the security is better than that of generalized chaos synchronization. As a whole, the scheme of complete chaos synchronization is suited for chaotic communications with respect to a high degree of security.

The methods of analog chaotic communications are based on the technique of embedding or hiding a message into a chaotic carrier as a secure code. The study of the security issue is still under way. Finally, we here briefly address other alternative techniques proposed at present. Among them, the method of code scrambling based on chaotic signal generations as discrete-sequence optical CDMA is used for chaos communications as digital techniques as discussed before (Kennedy et al. 2000). In recent years, the study has been carried out on developing chaotic algorithms, algorithms using the iteration of nonlinear functions, to efficiently generate random sequences with improved randomness and correlation properties for use in spread-spectrum, code-division multiplexing and error correction (Chen and Wornell 1998, Chen et al. 2001, and Uchida et al. 2003a). In any event, it is noted that the methods of analogue message encoding and decoding discussed here may not be the best ones for secure optical communications for chaotic data transmissions or for ultimate secure communication systems.

13.7 Chaotic Carrier and Bandwidth of Communications

The relaxation oscillation frequency is an important indicator of the maximum possible rate of data transmissions in chaotic systems of semiconductor lasers. The relaxation oscillation frequency ν_R of the solitary laser is given by $\nu_R = \sqrt{gS/\tau_{ph}}/2\pi$. The relaxation oscillation frequency of currently available semiconductor lasers is of the order of several GHz to 10 GHz. Chaotic variations in semiconductor lasers has a broad spectrum and the attainable maximum frequency is usually larger than the relaxation oscillation frequency of the solitary lasers. Therefore, we can transmit a signal which contains higher frequency components than the relaxation oscillation. For example, a message that contains more than the frequency components over 10 GHz was successfully transmitted in a chaotic communication system (Liu et al. 2001b). Also, over 100 Mbps messages were transmitted through a communication channel composed of systems of solid state lasers that had a relaxation oscillation frequency of less than several MHz (VanWiggeren and Roy 1998).

The semiconductor laser with a high frequency of chaotic carrier is desirable as a light source of chaotic communications to perform high data-rate transmission. Semiconductor laser with fast response is also essential for other applications such as ordinary optical communications and mass-data storage systems. By carefully choosing parameters of semiconductor materials and device structures, the effort for fabricating faster response semiconductor lasers is still ongoing. However, the attainable frequency for the relaxation oscillation is limited only by improving the materials and the device structures. On the other hand, the relaxation oscillation frequency can be greatly enhanced by strong optical injection from different lasers as discussed in Sect. 6.3.2. Wang et al. (1996) theoretically investigated the enhancement of the relaxation oscillation frequency of a semiconductor laser by using a small signal stability analysis of the laser rate equations at the stable injection locking steady-state and gave an example where the relaxation oscillation frequency of the injected laser at the solitary oscillation of 3 GHz was increased up to 12 GHz by strong optical injection. Thus, a semiconductor laser with enhanced modulation bandwidth by strong optical injection is effective as a light

source for chaotic communications with the capability of faster data transmissions (Takiguchi et al. 2003).

The receiver laser does not respond immediately after the transmitter signal is injected to the receiver laser and a finite transition time is required for
synchronization. The transition time depends on the device parameters and
the system configurations. We again consider a particular example. The model
is an optical feedback system and the scheme of CSK. The synchronization
recovery time affects the quality of synchronization especially in ON/OFF
CSK systems. The transition time is directly governed by the synchronization recovery time. So far, there have been few systematic studies of the recovery time for chaos synchronization. Vicente et al. (2002) investigated the
synchronization recovery time for open- and closed-loop systems of semiconductor lasers with optical feedback. Figure 13.17 shows the numerical results
of the synchronization time as a function of the external cavity round trip
time τ of the transmitter system, respectively. In Fig. 13.17a, the recovery
time is independent of the delay time in the open-loop system and the average
time required for the synchronization is very small, about 0.2 ns. The reason
for the constant delay time for the variation of the external cavity length is
that the receiver laser does not need to adjust the time of synchronization for
the parameters of the external cavity length, since the external cavity does
not exist in the receiver of the open-loop system. On the other hand, the recovery time of the closed-loop system increases for the increase of the external
cavity length (delay time). The time required in this case is much longer than

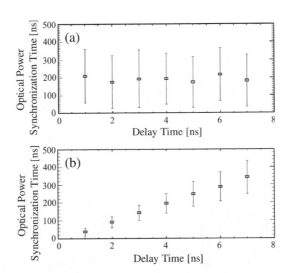

Fig. 13.17. Synchronization recovery time as a function of external-cavity round-
trip time in semiconductor lasers with optical feedback. The message encoding
scheme is CSK. **a** Open-loop and **b** closed-loop systems (after Vicente R, Pérez T,
Mirasso CR (2002); © 2002 IEEE)

that of the open-loop system. It is of the order of several tens of nano-seconds. In actual fact, there exist sudden bursts of synchronization even after chaos synchronization is achieved. This burst behavior also degrades the quality of chaos communications and the property of synchronization bursts remains an important problem.

13.8 Chaos Communications in the Real World

13.8.1 Chaos Masking Video Signal Transmissions

Many theoretical studies have been reported for secure communications using chaotic semiconductor lasers. Also, at the laboratory level, experimental chaotic communications have been demonstrated for data transmissions of sinusoidal signals and pseudo-random bit-sequence signals based on various data encryption methods. However, only a few studies have been reported on chaotic data transmission for real data. Larger et al. (2001) demonstrated data transmission of voice signals using chaotic semiconductor lasers with electro-optic feedback through wavelength filters, which is the same system as discussed in Sect. 7.4. In that system with a chaos modulation scheme, an AM voice signal encrypted into transmitter chaos is transmitted on a radio frequency and the signal is successfully decrypted in the receiver system. In the following, we show another example of real world data transmissions of video signals at 2.4 GHz side-band frequency embedded into chaotic carrier in semiconductor lasers with optical feedback.

Figure 13.18 shows a chaos masking system for video data transmission using a chaotic generator of semiconductor laser with optical feedback (Annovazzi-Lodi et al. 2005). The light source is a DFB semiconductor laser of an oscillation wavelength at 1.55 μm. To make the laser chaotic oscillations, the emitted light is fed back from a tip of a transmission fiber, which is located at 3 cm from the laser facet. The DFB lasers used were selected between first neighbors on the same wafer. As a chaos synchronization system, the system is an open-loop configuration and there is no optical feedback in the receiver system. The injection current is biased at $1.5I_{th}$ and the value of each parameter for the transmitter and receiver lasers, such as bias injection currents and temperatures, is finely tuned to coincide with each other. The levels both for the optical feedback in the transmitter and the transmission signal from the transmitter to receiver laser is set to be around 1% in the experiment. A message is obtained as a modulation of another DFB laser and is added to chaotic transmission signal (signal from Master laser) though 50/50 fiber coupler. The fiber length of the data transmission between transmitter and receiver is of about 1.2 km. After transmission, the signal is amplified through a semiconductor optical amplifier to increase the maximum injection level from the transmitter into the receiver. Then the signal passed through a birefringence controller to trim and adjust the injection level to the receiver

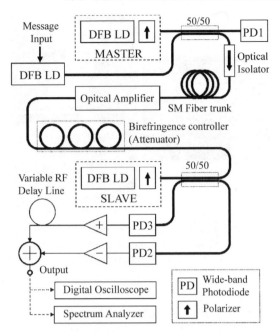

Fig. 13.18. Video signal encoding and decoding system using chaos masking method with 1.2 km fiber transmission (Annovazzi-Lodi V, Benedetti M, Merlo S, Norgia M, Provinzano B (2005); © IEEE 2005)

laser. Subtracting from the reproduced chaotic signal from the transmitted one, the decoded message is recovered. The type of chaos synchronization in this system is one for generalized synchronization (injection and amplification synchronization).

Figure 13.19 shows the results of data transmissions for a still TV pattern. In the system, a composite video signal with amplitude-modulated frequency at 2.4 GHz is used as a message. The quality of the received signal has been evaluated after synchronous detection and baseband filtering at the receiver output node. The signal amplitude has been adjusted as a compromise between efficient masking, low signal distortion and good quality of the recovered message. In Fig. 13.19, three photographs of the monitor screen are shown. Figure 13.19a is an original pattern to be transmitted without added chaos. Figure 13.19b shows the picture hidden within chaos and represents the message as it would be recovered by an eavesdropper. Figure 13.19c shows the extracted message after synchronization. The signal level has been adjusted as a trade-off between sufficient image masking by chaos and acceptable image quality after chaos cancellation. Figure 13.19b is obtained by setting the AM sideband level at about 4 dB over chaos. In these conditions, the signal-to-noise ratio of S/N = 16 ∼ 18 dB is obtained for the decoded message.

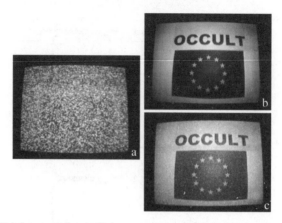

Fig. 13.19. TV frames of a still image transmitted by the setup of Fig. 13.14. **a** Original image to be send, **b** encoded pattern with chaotic signal, and **c** decoded pattern (Annovazzi-Lodi V, Benedetti M, Merlo S, Norgia M, Provinzano B (2005); © IEEE 2005)

13.8.2 Chaotic Signal Transmissions through Public Data Link

Arigyris et al. (2005) tested the effectiveness of chaotic data transmission in the existing public optical communication links. They employed two chaotic communication systems using semiconductor lasers as light sources; one is an electro-optic open-loop system and the other is all-optical open-loop system both based on chaos masking technique. Figure 13.20 shows the schematics of the systems. Figure 13.20a shows a chaotic electro-optic synchronization system. In this system, chaos is not generated from the nonlinearity of the semiconductor laser itself but from the nonlinear delayed response of light due to delayed opto-electronic hybrid feedback through electro-optic modulator (an integrated electro-optic Mach-Zehnder interferometer: MZ). The system has very high response over 10 GHz and is frequently used as a chaos generator (Gibbs 1985 and Davis 1990). The lasers used are DFB semiconductor lasers with an oscillation wavelength of 1.55 μm. A message to be sent, generated from another DFB laser, is simply added to chaotic signals through a 50/50 fiber coupler, which is an additive chaos masking scheme. The bandwidth of the system estimated is about 7 GHz.

Figure 13.20b shows the second case of all-optical system. The transmitter is a DFB semiconductor laser subjected to optical feedback from a digital variable reflector (R) located 6 m from the laser. The system is almost the same as one for the video signal transmission system in the previous section. A polarization controller (PC) is used within the cavity to adjust the polarization state of the light reflected back from the variable reflector. The message is added via a LiNbO$_3$ Mach-Zehnder modulator (MOD) at the transmitter's output. The scheme is multiplicative chaos masking. The bandwidth of the all-optical system is less than that of the electro-optic system and it is

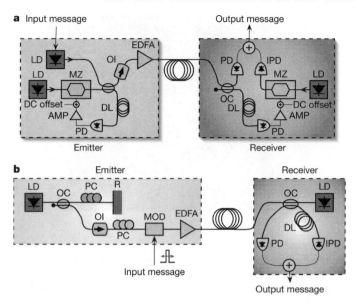

Fig. 13.20. Setups for two optical chaos communication systems. **a** Electro-optic open-loop system. **b** All-optical open-loop system (Argyris A, Syvridis D, Larger L, Annovazzi-Lodi V, Colet P, Fischer I, Garcia-Ojalvo J, Mirasso CR, Pesquera L, Shore AK (2005); © Nature Pub. 2005)

about 5 GHz. In both schemes, an erbium-doped fiber amplifier (EDFA) is used to compensate for the power lost upon transmission. Decoding is performed via subtraction of the transmitted signal from the signal filtered by the receiver. Operationally, the subtraction is performed by adding the photocurrents coming from an ordinary and a sign-inverting amplified photodiode (PD and IPD, respectively). Also the type of chaos synchronization in this system is one for generalized synchronization as is usually the case for real chaos communication systems.

Figure 13.21 shows the results for laboratory experiments of eye diagrams in the electro-optic setup after transmission of a binary message through single mode optical fiber of 50 km and dispersion compensation fiber of 6 km. The message is a pseudo-random bit sequence of $2^7 - 1$ bits and the transmission rate is 3 Gbps. The observed bit error rates (BERs) are of the order of 10^{-7}. As already discussed, the performance of data transmission for all-optical scheme is usually poor compared with that for electro-optic system. The transmission rate to attain the similar performance of BER as that in the electro-optic system is about 1 Gbps.

To test performance under 'real-world' conditions of chaotic communications, the chaos-based all-optical transmission system were implemented using an installed optical network infrastructure of single-mode fiber belonging to the metropolitan area network of Athens, Greece. The network has

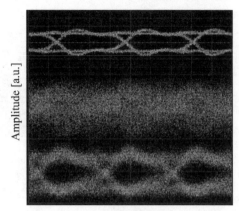

Fig. 13.21. Laboratory test of eye diagrams in the electro-optic system through 50 km fiber transmission. *Top trace*: test message, *middle trace*: encoded signal, *bottom trace*: decoded message (Argyris A, Syvridis D, Larger L, Annovazzi-Lodi V, Colet P, Fischer I, Garcia-Ojalvo J, Mirasso CR, Pesquera L, Shore AK (2005); © Nature Pub. 2005)

a total length of 120 km. The topology consists of three fiber rings, linked together at specific cross-connect points as shown in Fig. 13.22a. Through three cross-connect points, the transmission path follows the Ring-1 route, then the Ring-2 route, and finally the Ring-3 route. To cancel the chromatic dispersion that would be induced by the single-mode fiber transmission, a dispersion compensation fiber module is used in the link. Also, to compensate the optical losses and filtering of amplified spontaneous emission noise, erbium-doped fiber amplifiers and optical filters are used along the optical link. The pair of lasers of the transmitter and receiver is selected to exhibit parameter mismatches that are constrained below 3%. The mean optical power injected into the receiver has been limited to 0.8 mW, to avoid possible damage of the antireflective coating of the slave laser. Test messages are non-return-to-zero (NRZ) pseudorandom bit sequences applied by externally modulating the chaotic carrier by means of a modulator. The message amplitude is attenuated by 14 dB with respect to the carrier to maintain the message security in the communications. As a result, the BER of the transmitted signal after filtering is always larger than 6×10^{-2}, which is the instrumentation limit. A good synchronization performance of the transmitter-receiver setup leads to an efficient cancellation of the chaotic carrier and, thus, satisfactory decoded messages are obtained as shown in Fig. 13.22b. Figure 13.22c shows the performance of the chaotic transmission system for different message bit rates up to 2.4 Gbps for two different code lengths of $2^7 - 1$ and $2^{23} - 1$. All BER values have been measured after filtering the subtracted electric signal, by using low-pass filters with bandwidth adjusted each time to the

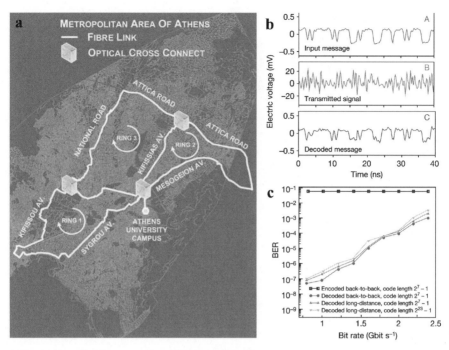

Fig. 13.22. Field experiment of fiber transmission. **a** Chaos-encoded data transmissions in the optical communication network of Athens, Greece. **b** Time trances of 1 Gbps message. *Trace A*: applied message of BER $< 10^{-12}$, *trace B*: Carrier with the encoded message of BER $\approx 6 \times 10^{-2}$, *trace C*: recovered message after 120 km transmission of BER $\approx 10^{-7}$. **c** The bit error rate (BER) performance. *Squares*: encoded signal, *circles*: back-to-back decoded message, *triangles*: decoded message after transmission for two different code lengths. LD: laser diode, MZ: electro-optic Mach-Zehnder interferometer, PD: photodiode, AMP: electronic amplifier, OI: optical isolator, DL: delay line, EDFA: erbium doped fiber amplifier, OC: optical fiber coupler, IPD: sign-inverting amplified photodiode, R: digital variable reflector, PC: polarization controller, MOD: modulator (Argyris A, Syvridis D, Larger L, Annovazzi-Lodi V, Colet P, Fischer I, Garcia-Ojalvo J, Mirasso CR, Pesquera L, Shore AK (2005); © Nature Pub. 2005)

message bit rate. For transmission rates in the gigabit per second range the recovered message exhibits BER values lower than 10^{-7}. For higher transmission rates, the corresponding BERs increase due to the fact of imperfect synchronization, as shown in Fig. 13.22c.

A Appendix: Chaos

About Chaos in Nonlinear Systems

Chaos is a phenomenon observed in a nonlinear system described by a certain set of differential equations and shows irregular oscillations for time or spatial evolutions (Lorenz 1963). We must distinguish chaos from the observations of random events, such as flipping of a coin. Namely, chaos is a disorder in a system written by deterministic equations. Chaos is observed in a wide variety of nonlinear systems, not only in physical and engineering systems, but also in biological systems and, even, in sociology and economy. Nonlinearity of a system is one of the important factors to observe chaos. Whether the system is continuous or discrete, it is a good candidate for a chaotic system, but the nonlinearity itself does not always guarantee chaotic oscillations in the system. Chaos occurs in a nonlinear system under appropriate parameter conditions and also certain parameter ranges. Chaos has a rigid definition for the irregular oscillations of the output from the system. Whether the output from a system is chaos or not is determined by the rigorous procedures by using mathematical tools, such as attractors, the Lyapunov exponent, and the Poincaré map of the output (Abarbanel 1996). In the following, we discuss what chaos is, what it looks like, and how chaotic data are analyzed. Also, as unique features of chaos, we show some examples of useful techniques for practical applications: control of chaos and chaos synchronization. This chapter does not aim at the rigid mathematical descriptions of chaos, but it intends to show the general aspects of chaos and give useful tools for the analyses of chaos for reading this book.

The following is a list of some additional reading for the general concept and mathematical treatments of chaos. The readers interested in physical and mathematical backgrounds of chaos and also the treatment of chaotic data may consult the following books.

Abarbanel HDI (1996) Analysis of observed chaotic data. Springer-Verlag, New York

Barker GL, Gollub JP (1990) Chaotic dynamics, an introduction. Cambridge University Press, Cambridge

Devaney RL (1989) An introduction to chaotic dynamical systems, Second edition. Addison-Wesley Pub.

Nagashima H, Baba Y, Nakahara M (1998) Introduction to chaos: physics and mathematics of chaotic phenomena. Inst of Physics Pub Inc.

Ott E (2002) Chaos in dynamical systems, Second Edn. Cambridge University Press, Cambridge

Jackson EA (1991) Perspectives of nonlinear dynamics, Vol.1, Vol.2. Cambridge University Press, Cambridge

Thompson JMT, Stewart HB (1986) Nonlinear dynamics and chaos, geometrical methods for engineers and scientists. John Wiley & Sons, New York

Tuckerman L (1986) Order within chaos. John Wiley & Sons, New York

A.1 Nonlinear Chaotic Systems

A.1.1 Discrete Systems

We consider the occurrence of an event that becomes the new input of the system and causes the next events. Namely, the system is described by discrete difference equations. We assume a nonlinear response of a discrete system and write its output as $x(n)$

$$x(n + 1) = f(x(n); \mu) \tag{A.1}$$

where n denotes the step of the occurrence, $f(x)$ is the nonlinear function, and μ is the system parameter. In general, $x(n)$ is a vector and the parameter $\boldsymbol{\mu}$ is also a vector.

Logistic mapping is the well-known mathematical relation for a discrete chaotic system. We consider the logistic map with a variable x and a parameter μ given by

$$x_{n+1} = \mu x_n(x_n - 1) \tag{A.2}$$

This is a simple mapping using a quadratic function. It is stable and its solution converges to a fixed point for any initial value of x as far as the parameter μ is roughly less than 3. The output x shows an oscillation for the evolution of n when the parameter μ has a value larger than 3. With a further increase of the parameter value, the system shows period-doubling oscillatory solutions and, finally, reaches irregular chaotic oscillations. Figure A.1 a shows such an example for the parameter value of $\mu = 3.8$. This irregularity is completely different from the random fluctuations observed in stochastic processes. The irregularity is simply derived from the deterministic difference equation in (A.2) and the irregularity is called "chaos." One of the unique features of chaos is the sensitivity for the initial condition. In the numerical calculation in (A.2), we can never obtain the same waveform for different initial values even the difference is extremely small. Fig. A.1b is the demonstration of the sensitivity for the initial condition.

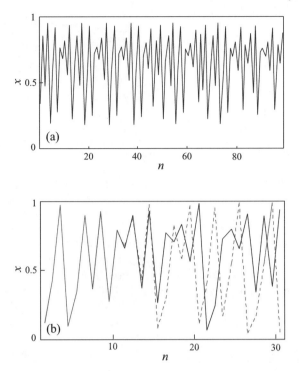

Fig. A.1. Chaotic oscillations in logistic mapping. **a** Chaotic oscillation at $\mu = 3.8$.
b Sensitivity for initial conditions. *Solid line:* $x_0 = 0.12000$ and *broken line:* $x_0 = 0.12001$

In the figure, we choose two initial conditions with a slight difference of $x_0 = 0.12000$ and $x_0 = 0.12001$. At first, x shows a quite similar orbit, however the difference between the two orbits deviate with each other and, finally, show completely different oscillations for the evolution of the event. This is a typical nature of the sensitivity for the initial conditions in a chaotic system.

From the plot of the maxima and minima of the variable x for the evolution of step n, we obtain a map for the parameter μ as shown in Fig. A.2. The map is called a bifurcation diagram or a chaotic bifurcation diagram. From this plot, we can see that a fixed point of the output x evolves to periodic oscillations and period-doubling oscillations occur for the increase of the parameter μ. Finally, the output x behaves with completely irregular oscillations at high values of the parameter μ, namely, chaotic oscillations. The type of bifurcation is called a period-doubling bifurcation. Period-doubling bifurcation is not the only chaotic route, but other routes for chaotic evolutions exist depending on the configuration of systems and chaos parameters, for example, quasi-period doubling bifurcations and intermittent chaotic bifurcations. A bifurcation diagram is important to know how the dynamics of

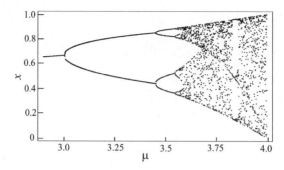

Fig. A.2. Bifurcation diagram for logistic mapping

a system change for the parameters. We can make similar plots of bifurcation diagrams for the change of the parameters for any chaotic systems and know the chaotic evolution route of the system output. The fact that such complex dynamics are induced from a simple relation in (A.2) is surprising. Indeed, we can observe complex dynamics in optical systems with discrete nonlinear properties.

A.1.2 Continuous Systems

Continuous systems have more or less nonlinear aspects. For engineering applications, linear response is usually assumed even when a system has non-linearity, or the linearization is frequently applied for the nonlinear system within suitable ranges of parameters, because nonlinear systems are too complex to be modeled and analyzed. They show unwanted irregular oscillations for a wide range of the change of variables. Therefore, only linear parts of the systems are a good model for engineering applications. Until recently, analysis for nonlinear systems was only of fundamental interest as concerns the dynamic behaviors of the systems. However, many phenomena of nonlinear behaviors have been observed in engineering models. For example, strange relaxation oscillations were observed in a triode used for electric circuits, which is a chaotic oscillation and is now known as a van der Pol oscillator (van der Pol 1926). Chaos, which is irregular oscillations induced in deterministic nonlinear systems, became common understanding after the work done by Lorenz in 1963, who investigated the behaviors of convective flow for the atmospheric model. He obtained three differential equations for the model by appropriately scaling the three variables. In spite of the simple forms of the three equations, one cannot obtain analytical solutions and needs numerical calculations. Later, it was proved that lasers were described by the same Lorenz equations with three variables (Haken 1975).

Here, we show an example of chaotic oscillations in a continuous nonlinear system. We consider a system of damping oscillation driven by an external force known as the Duffing spring model. The Duffing model is described by

$$\frac{d^2x}{dt^2} + k\frac{dx}{dt} + x^3 = B\cos\omega t \tag{A.3}$$

where x is the variable for the displacement of the spring, ω is the angular frequency of the external driving force, B is the amplitude of the external force, and k is the coefficient of the friction. Using the notation $y = dx/dt$, the Duffing equation is written by the following two differential equations:

$$\frac{dx}{dt} = y \tag{A.4}$$

$$\frac{dy}{dt} = -ky - x^3 + B\cos\omega t \tag{A.5}$$

Namely, a nonlinear chaotic system is generally described by a vector variable $\boldsymbol{u}$ and a nonlinear function f as

$$\frac{d\boldsymbol{u}}{dt} = f(\boldsymbol{u}; \boldsymbol{\mu}) \tag{A.6}$$

This equation is quite similar to the discrete nonlinear system described in (A.1). $\boldsymbol{\mu}$ is the vector parameter of the system. The Duffing model is essentially a damping oscillation system, but it shows various stable and unstable oscillations depending on the parameter.

Figure A.3a plots the trajectory of the variables x and y for the parameter values of $k = 0.225$, $B = 0.3$, and $\omega = 1$. The output of the system converges to a periodic orbit (periodic attractor) for the time evolution. The diagram in Fig. A.3a is called a "phase space" and the attraction in the phase space is called "attractor." When $k = 0.5$, $B = 0.3$ $\omega = 1$, the orbit shows a complicated trajectory within a limited compact space as shown in Fig. A.3b. The attractor is so strange that it is called a "strange attractor." A strange attractor is a typical feature of chaotic dynamics. Complex dynamics appear to be such a simple differential equation. Chaos generated from various differential systems for physical and engineering models has been extensively studied after the discovery of chaos by Lorenz. The energy from a physical or engineering system is generally dispersed for the time evolution and such a system is called a dissipative system. In a dissipative system, the volume of the trajectory generally shrinks in the phase space in accordance with the dispersion of energy. However, for dissipative chaotic systems, the trajectory stays within a finite volume in the phase space for the time evolution like in Fig. A.3b due to the presence of an external drive force and it shows a strange attractor. On the other hand, for a system that holds the conservation of energy, the attractor is always fixed within a finite area in the phase space.

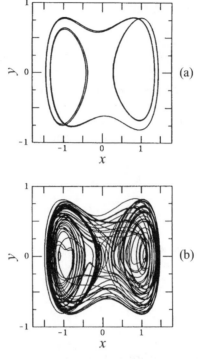

Fig. A.3. Attractors of the Duffing system. **a** Periodic attractor for $k = 0.225$, $B = 0.3$, and $\omega = 1$. **b** Chaotic attractor for $k = 0.5$, $B = 0.3$ and $\omega = 1$

A.1.3 Delay Differential Systems

The system described by delay differential equations is a nonlinear system as it is. For the control of a system, a fraction of the output is usually fed back into the system with a finite delay time. When the delay cannot be ignored in order to consider the system response, the system must be described by delay differential equations. In this book, we treat various delayed feedback systems in semiconductor lasers. The general form of a delay differential system is given by the following equation:

$$\frac{\mathrm{d}\boldsymbol{u}(t)}{\mathrm{d}t} = f(\boldsymbol{u}(t - \tau); \boldsymbol{\mu}) \tag{A.7}$$

where τ is the delay time in the system. As demonstrated in this book, we can observe a rich variety of chaotic dynamics in delay differential systems. The complexity of a system described by delay differential equations is much stronger than that of simple differential systems, since the mapping of a de-layed system is of continuous nature. The chaotic dimension (discussed in Sect. A.2.3) of differential systems is low, while it is high for delay differen-tial systems. Therefore, a delay differential system is sometimes refered to as a high-dimensional chaotic system.

A.2 Analysis and Characteristic Descriptions for Chaotic Data

A.2.1 Phase Space, Attractor, and Poincaré Map

We have already shown some tools for the analysis of chaotic oscillations in nonlinear systems. We discussed the phase space for chaotic systems in Sect. A.1.2. We here first go into details of the explanation for the phase space. The phase space of a dynamical system is a mathematical space with orthogonal coordinate directions representing each of the variables needed to specify the instantaneous state of the system. A phase map may be constructed in several ways. For the Lorenz system, for example, the state of the system can be described by three variables x, y, and z and parameters σ, r, and b as

$$\frac{\mathrm{d}x}{\mathrm{d}t} = -\sigma(x - y) \tag{A.8}$$

$$\frac{\mathrm{d}y}{\mathrm{d}t} = -xz - rx - y \tag{A.9}$$

$$\frac{\mathrm{d}z}{\mathrm{d}t} = xy - bz \tag{A.10}$$

Figure A.4 shows the time evolution of Lorenz equations with a diagram in a three dimensional phase space. The diagram is called the phase portrait and the orbit is called the trajectory in the phase space. The Lorenz system gives an excellent instance of what we called a strange attractor. We see a beautiful butterfly-shaped attractor on which a trajectory starting from any initial point will come back to the near neighborhood of the initial point but never precisely repeat it. An important feature of the phase portrait is that

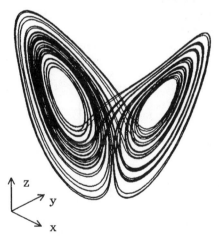

Fig. A.4. Lorenz attractor for parameter values of $\sigma = 10$, $r = 28$, and $b = 8/3$

two trajectories will never cross each other. This non-crossing property derives from the fact that both past and future states of a deterministic system are uniquely determined by the system state at a given time. A crossing of trajectories inevitably introduces ambiguity into past and future states and contradicts the assumed uniqueness of the trajectory. However, a projection of a higher dimensional space onto a plane might show apparent crossings which do not represent actual interactions. The phase portraits for the periodic, quasi-periodic, and chaotic time variations in general appear to be a limit cycle, a torus, and a strange attractor, respectively, as shown in Appendix A.2.2.

Poincaré map

The Poincaré map introduced by H. Poincaré is a classical technique for analyzing a dynamical system. It is obtained by viewing the trajectory stroboscopically. It is also a useful tool for understanding the characteristics of nonlinear systems. For an n-dimensional trajectory Γ as shown in Fig. A.5, take an $(n-1)$-dimensional hyper plane Σ transverse to the trajectory at X_0. The trajectory emanating from X_0 will hit Σ at $X_1, X_2, \ldots$ at the following transversings. The Poincaré map P is defined as

$$X_{k+1} = \mathrm{P}(X_k) = \mathrm{P}(\mathrm{P}(X_{k-1})) = \mathrm{P}^2(X_{k-1}) = \cdots \qquad (A.11)$$

where k is an integer. The Poincaré map replaces the continuous dynamical system into a discrete map, which is much easier to deal with mathematically. With the Poincaré map, one can dramatically reduce the data number that is especially necessary in the experiments. As demonstrated in Appendix A.2.2, one can easily distinguish among the periodic, quasi-periodic, and chaotic variations from the appropriate Poincaré maps. In particular, the Poincaré map for chaotic systems exhibits remarkable features. The map does not result in a simple geometrical structure and, when it is magnified, the fine

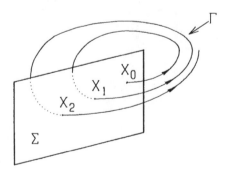

Fig. A.5. Poincaré map of an n-th order autonomous system. Γ: n-dimensional trajectory, Σ: $(n-1)$-dimensional hyper plane

structure resembles the gross one. Namely, the system trajectory has a fractal structure.

A.2.2 Steady State Behaviors

We can classify dynamical systems in terms of their steady-state solutions and describe some typical examples for particular cases. The steady-state refers to the asymptotic behavior of the system as time t approaches infinity. Typically four different types of steady-state behaviors can be identified to chaotic system. Each steady-state will be described from the following different points of view: in the time domain, in the frequency domain, in the phase space, and in the Poincaré map. We here mainly focus on the Poincaré maps for chaotic systems.

Equilibrium points

An equilibrium point x^0 of an autonomous system is a constant solution in (A.6). In most cases, this means $F(x^0; \mu) = 0$ or $x^0 = f(x^0; \mu)$. The phase portrait for an equilibrium point is the point itself.

Periodic solutions

A time dependent variable x(t) is periodic if $x(t + T) = x(t)$ for all t and some minimal period T. For a model of Duffing's equations in (A.4) and (A.5), a periodic solution was plotted in Fig. A.3a. Such closed curve is called a limit-cycle for it can be regarded as a diffeomorphic copy of a cycle. The corresponding Poincaré map is displayed in Fig. A.6a by sampling the trajectory with the driving period. The Poincaré map for a periodic solution in general consists of finite points and the number of points depends on the sampling period.

Quasi-periodic solutions

A quasi-periodic solution is the sum of periodic solutions each of whose frequency is a linear coupling of a finite set of base frequencies. The simplest case is the one with two frequencies. We here consider the van der Pol equations to demonstrate how quasi-periodic solutions arise in dynamical systems. The van der Pol equations read

$$\frac{dx}{dt} = y \tag{A.12}$$

$$\frac{dy}{dt} = (1 - x^2)y - x + \gamma \cos \omega t \tag{A.13}$$

In the van der Pol equations, there exist two periodic components: one is the driving signal with the period $T_1 = 2\pi/\omega$ and the other is the intrinsic

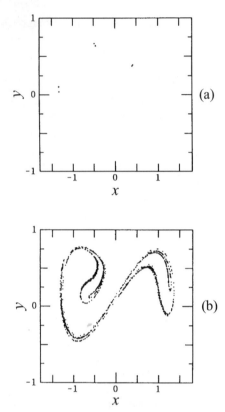

Fig. A.6. Poincaré map for Duffing oscillation. **a** Map for periodic oscillation for Fig. A.3a. **b** Map for chaotic oscillation for Fig. A.3b

oscillation whose period T_2 depends on the system parameters. When T_2 is synchronized with T_1, there appears a periodic oscillation. However, if T_1 and T_2 are incommensurate and neither of them can dominate the output, the system has a quasi-periodic behavior. An example of the quasi-periodic solutions in the van der Pol model is shown in Fig. A.7. Figure A.7a shows a time series for the quasi-periodic solution. The spectrum (Fig. A.7b) consists of main spectral peaks corresponding to the intrinsic oscillation frequencies of the system and the tightly spaced side bands due to the driving signal. The difference between the intrinsic frequency and the driving frequency can be measured from the spacing of the harmonics within the sideband. The corresponding attractor is plotted in Fig. A.7c. In the phase space, the trajectory lies on a diffeomorphic copy of the two-torus as shown in Fig. A.8. Since the trajectory is a curve and two-torus is a surface, not every point on the torus lies on the trajectory, however it can be shown that the trajectory repeatedly passes closely every point on the torus in an arbitrary manner. Depending on the sampling period, the Poincaré map for the quasi-periodic solutions ap-

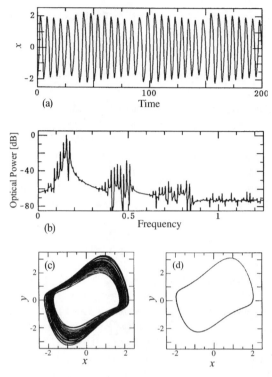

Fig. A.7. Quasi-periodic solution for the van der Pol oscillator at $\gamma = 0.5$ and $\omega = 1.1$. **a** Chaotic oscillation, **b** power spectrum, **c** attractor, and **d** Poincaré map

Fig. A.8. Torus with two periodic components denoted as S_1 and S_2

pears as various limit cycles. In the van der Pol model, we obtain a Poincaré map by sampling the trajectory with the driving period as in Fig. A.7d.

Chaos

Chaos is different from any one of the above three solutions. Once more we recall Duffing's equations to generate the chaotic attractor in Fig. A.3b. It behaves with more irregular oscillations than those of periodic and quasi-periodic states. In general, the spectrum becomes broadened, and no distinct spectral peak is visible, but has a continuous broadband spectral structure.

The Poincaré map corresponding to Fig. A.3b shows a very interesting structure and is displayed in Fig. A.6b.

A.2.3 Fractal Dimension and Correlation Dimension

We here discuss some statistical measures of chaotic attractors by introducing the concept of the fractal dimension. A chaotic trajectory has generally a fractal structure as discussed in Sect. A.2.1. We first start from the capacity dimension. For simplicity, a geometric structure, i.e., a limit cycle, a torus, or a chaotic attractor, is regarded as a set. Consider a set located in an m-dimensional phase space. Imagine we cover the attractor by $N(\varepsilon)m$-dimensional boxes of ε on each side. The capacity dimension d_c is defined as

$$d_c = \lim_{\varepsilon \to 0} \frac{\log N(\varepsilon)}{\log(1/\varepsilon)} \tag{A.14}$$

For a dissipative dynamical system such as a delay differential model, the attractor is located in an m-dimensional phase space (embedding space), but the fractal dimension is in general less than the embedding dimension m. Furthermore, chaotic attractors generally have non-integer dimension.

There are several kinds of definitions about the fractal dimension. For an experimental data or a model of a high-dimensional dynamical system, another type of dimension is more efficient in data calculation than the capacity dimension. The method is the correlation dimension d_g. Suppose that many points are scattered over of a set and consider the correlation between any two points. The typical number of neighbors of a given point will vary more rapidly with the distance from that point if the set has higher dimension than otherwise. The correlation dimension may be computed from the correlation function $C(\rho)$ defined by

$$C(\rho) = \lim_{N \to \infty} \frac{1}{N^2} \sum_{i,j=1}^{N} \theta(\rho - |x_i - x_j|) \tag{A.15}$$

where x_i and x_j are the points on the attaractor, $\theta(y)$ is the Heaviside function ($\theta(y) = 1$ if $y > 0$ and 0 otherwise), and N is the number of the points randomly chosen from the entire data set. The Heaviside function simply counts the number of the points within a radius ρ of the point x_i, and $C(\rho)$ gives the average function of the points within ρ. The correlation dimension is defined by the variation of $C(\rho)$ with ρ approaching zero as

$$C(\rho) \approx \rho^{d_g} \ (\rho \to 0) \tag{A.16}$$

Therefore, the correlation dimension is obtained from the slope of the graph of $\log C(\rho)$ versus $\log \rho$.

A.2.4 Lyapunov Exponent

The fractal dimension provides a qualitative measure of the singularity of the chaotic attractor. However, it is a static measure and, from the fractal dimension, we know nothing about how the trajectory varies in the phase space as the time evolution. Moreover, this kind of measure has no explicit relation with the most important characteristics of chaos, i.e. the sensitivity to the initial condition. Therefore, we have to resort to some dynamical measure – the method is known as the Lyapunov exponent.

The basic idea of the Lyapunov exponent is to measure the average rate of the divergence for the neighboring trajectories on the attractor. The direction of the maximum divergence or convergence locally changes on the attractor. The motion must be monitored at each point along the trajectory. Therefore, a small sphere is defined, whose center is a given point on the attractor and whose surface consists of phase points from nearby trajectories. As the center of the sphere and its surface points evolve in time, the sphere becomes an ellipsoid with the principal axes in the directions of the contraction and the expansion. The average rates of the expansion or the contraction along the principal axes are the Lyapunov exponents. For the i-th principal axis, the corresponding exponent is defined by

$$\lambda_i = \lim_{t \to \infty} \left\{ \frac{1}{t} \frac{L_i(t)}{L_i(0)} \right\} \tag{A.17}$$

where $L_i(t)$ is the radius of the ellipsoid along the i-th principal axis at time t. In this expression, the growth rate is always measured along the i-th principal axis, but the absolute orientation in the phase space of that axis is not fixed. It is impractical to perform the actual computation in the way suggested in the definition, because the initially close phase points would soon diverge from each other by distances approaching the size of the attractor, and the computation would then fail to capture the local rates of the divergence and the contraction. Therefore, vectors connecting the surface of the ellipsoid to the center must be shrunk periodically or renormalized to ensure that the size of the ellipsoid remains small and that its surface points correspond to trajectories near that of the center point. The calculation algorithm for the Lyapunov exponent from time series can be found in the reference (Abarbanel 1996).

There are some important points related to the Lyapunov exponents as discussed in the followings.

(1) For a chaotic system, at least one of the Lyapunov exponents must be positive to allow the sensitive dependence on the initial conditions.
(2) According to the definition of the Lyapunov exponents, a small volume V in the phase space will change in time as

$$V(t) = V_0 \exp \left(\sum_{i=1}^{n} \lambda_i t \right) \tag{A.18}$$

and hence the rate of the change of the volume $V(t)$ is simply

$$\frac{dV(t)}{dt} = \sum_{i=1}^{n} \lambda_i V(t) \tag{A.19}$$

Therefore, for a dissipative chaotic system, the sum of all the Lyapunov exponents must be negative, i.e., $\sum_i \lambda_i < 0$.

(3) If we order λ_i $(i = 1, 2, ..., n)$ as $\lambda_1 > \lambda_2 > ... > \lambda_n$, then the Lypapunov dimension, d_1, is defined as

$$d_1 = j + \frac{\lambda_1 + \lambda_2 + \cdots + \lambda_j}{|\lambda_{j+1}|} \tag{A.20}$$

where j is the number of the Lyapunov exponents, which gives a positive sum, but adding λ_{j+1} would make the sum negative, i.e., $\sum_{i=1}^{j} \lambda_i > 0$ but $\sum_{i=1}^{j+1} \lambda_i < 0$. The relationship among the capacity dimension d_c, the correlation dimension d_g, and the Lyapunov dimension d_1 is $d_c \leq d_g \leq d_1$. The equality holds only for the case that the points on the fractal are approximately uniformly distributed (Aberbanel 1996).

A.3 Chaos Control

Controlling chaos

Real physical and engineering systems are more or less chaotic systems. Stable control for nonlinear systems is essential for practical applications. However, a system sometimes seems to be out of control for any parameters when it operates under chaotic oscillations. The state of the system is surrounded by chaotic sea and the system could not be stabilized by any parameter changes as far as the change is small. However, there always exist stable islands scatteredly located not far from the operating point of the system (they are usually not exact stable points but unstable periodic orbits). By appropriately choosing a perturbation to the system and attracting the state to such a stable orbit, one might successfully stabilize the system even when the original state is a chaotic oscillation. This kind of stabilization is called "chaos control." Chaos control is different from the ordinary technique of forced control in respect to a small perturbation for the system. In ordinary forced control, the power imposed to the system is so large that the original state of the system is completely changed after the control. On the other hand, the applied signal in chaos control is small enough not to change the original state and the system with chaotic oscillation is attracted to a periodic or even a fixed state by the small perturbation. The dynamics of the original state is not changed after the control is achieved. This is the idea of chaos control. Of course, a chaotic state may also be stabilized to a stable oscillation by a large control signal. This is different from chaos control, but it is categorized into ordinary forced control. The principle of chaos control was first proposed by Ott, Grebogi, and Yorke in 1990. They gave the mathematical proof of chaos control for chaotic systems and the method is now called OGY method or OGY algorithm after their names. In the following, we present a brief proof of the OGY algorithm. However, the application of the OGY method requires the full mathematical description of a nonlinear model. We need the attractors or the Poincaré map in advance to analyze and control the system. It is sometimes difficult to apply the OGY method for the control of practical systems. As an alternative method (but it still has an essence of the OGY algorithm), we also introduce the continuous chaos control methods, which are applicable to real experimental systems.

OGY algorithm

Take as an example a single-input two-dimensional discrete nonlinear system and consider chaos control in the system. The system is written by

$$x_{n+1} = f(x_n, \mu) \tag{A.21}$$

where f is again the nonlinear function and μ is the control parameter. We assume that the system has a fixed point at $\mu = \mu_0$, namely

$$x_f = f(x_f, \mu_0) \tag{A.22}$$

The operating point for the system parameter μ is assumed to be close to the parameter μ_0 with the accompanying output x_f. We define the difference of the outputs x_n and x_f as

$$\xi_n = x_n - x_f \tag{A.23}$$

Since the difference ξ is small, we linearize the difference and set the control signal μ_n so as to force the original state x_n to the stable state x_f. The relation is given by

$$\xi_{n+1} = M x_n + a \mu_n \tag{A.24}$$

where M and a are Jacobian matrices for the variable and the parameter at the fixed point. The point we consider is a type of saddle node unstable point. Then, one of the eigenvalues of M must be $|\lambda_u| < 1$ and the other one is $|\lambda_s| < 1$. We write the eigenvectors for respective eigenstates as e_u and e_s, and define the accompanying contravariant basis vectors v_u and v_s. Using the vector relations of $v_u \cdot e_u = v_s \cdot e_s = 1$ and $v_s \cdot e_u = v_u \cdot e_s = 0$, the matrix M is written by

$$M = \begin{bmatrix} e_u & e_s \end{bmatrix} \begin{bmatrix} \lambda_u & 0 \\ 0 & \lambda_s \end{bmatrix} \begin{bmatrix} v_u \\ v_s \end{bmatrix} = \lambda_u e_u \cdot v_u + \lambda_s e_s \cdot v_s \tag{A.25}$$

When a small control signal is applied to put the difference ξ_{n+1} on the stable orbit in accordance with (A.24), the difference ξ_{n+1} is parallel to the vector e_s. Then, we obtain the condition

$$v_u \cdot \xi_{n+1} = 0 \qquad (A.26)$$

Taking into account the relations in (A.24) and the condition of $v_n \cdot a \neq 0$ from (A.24), we obtain the following signal suitable for the control:

$$\mu_n = -\frac{\lambda_u v_n \cdot x_n}{v_n \cdot a} \qquad (A.27)$$

When the state of the system is close enough to the point x_f and the signal given by (A.27) is applied to the system, the state falls down to the fixed point x_f. Figure A.9 schematically shows the OGY control algorithm. In Fig. A.9a,

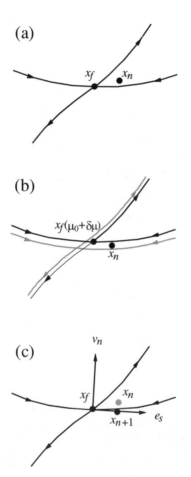

Fig. A.9. OGY control algorithm. **a** Initial state, **b** application of small perturbation to the parameter, and **c** controlled state

the initial state x_n is close enough to the fixed point x_f. A small perturbation is applied to the control parameter according to (A.27) as shown in Fig. A.9b. Then, the next state is shifted onto the stable orbit in Fig. A.9c and, finally, the state is attracted to the fixed point. The model here is for a low-dimensional system, but the procedure is also applied to high-dimensional chaotic systems. For a continuous system, it must be approximated to a discrete form in order to apply the OGY algorithm. Further, we must in advance know all the parameter values of the system to perform the control. Therefore, the method has a limitation for application to real nonlinear systems. For this reason, various control algorithms have been proposed without losing the essence of the OGY algorithm.

Continuous control

We here describe the continuous chaos control proposed by Pyragas (1993) as an alternative method of chaos control for practical use. The system configuration is shown in Fig. 9.1. The control signal is the difference between the present and time delayed outputs. The delay time introduced is almost equal to the response time of the system (but may not be exactly equal to the response time). Namely, the delay τ corresponds to the typical frequency of the chaotic oscillation. When we can obtain exactly the mathematical description of the system, we can calculate the delay time from the equations. Otherwise, the typical frequency of chaotic oscillations in the experiment would give a good estimate for the delay time. In the continuous control, the system under a certain chaotic oscillation is synchronized to a periodic state by feeding back the difference signal.

Using one of the variables, $y(t)$, in a chaotic system and writing the other variables as a vector representation, $x(t)$, the equations for the system are given by

$$\frac{dy}{dt} = f(y, x) \tag{A.28}$$

$$\frac{dx}{dt} = g(y, x) \tag{A.29}$$

where f and g are the nonlinear functions. The feedback signal in the continuous control is proportional to the difference of the present output $y(t)$ and the delayed output $y(t - \tau)$ defined by

$$u(t) = K\{y(t - \tau) - y(t)\} \tag{A.30}$$

For the control, the equation (A.28) is modified as

$$\frac{dy}{dt} = f(y, x) + u(t) \tag{A.31}$$

where K is the strength of the feedback and the effective control is performed by changing this value.

The control signal in (A.30) becomes zero after the success of the control, i.e., $u(t - \tau) = u(t)$ and the output shows periodic oscillation with a fundamental period of τ. The target of the periodic orbit to be controlled must be close to the initial state of the chaotic system and the delay time must be accurately estimated to some degree in advance either theoretically or experimentally. One cannot tell to which state the system is attracted among the possible orbits, when the system includes several unstable periodic orbits close to the initial state. The continuous control does not require the calculation of the Poincaré map such as in the OGY method and it is robust for noises. The continuous control is an extension of the OGY method for continuous differential systems. The control is easily realized for practical engineering systems, when the response time is not so fast to implement by using electronic circuits.

We show an example of the continuous control for the Rössler system shown in Fig. A.10. The equations of the system are given by

$$\frac{\mathrm{d}x}{\mathrm{d}t} = -y - z \tag{A.32}$$

$$\frac{\mathrm{d}y}{\mathrm{d}t} = x + ay + u(t) \tag{A.33}$$

$$\frac{\mathrm{d}x}{\mathrm{d}t} = b + z(x - c) \tag{A.34}$$

$u(t)$ in (A.30) is the control term. In the figure, the quasi-periodic or weak chaotic state is controlled to period-3 oscillation (Pyragas 1992). As can be easily seen from the figure, the control signal diminishes after the success of the control.

Another example of the method of chaos control is the occasional proportional feedback (OPF). The OPF control is discussed in Sect. 9.2.2. The control in the OPF method is also a periodic perturbation to the system and the system with chaotic state is attracted to an unstable periodic orbit near the initial state. A small periodic control signal is produced by processing

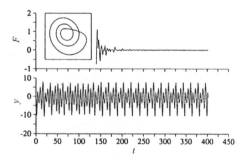

Fig. A.10. Continuous chaos control for the Rössler system. The chaotic oscillation is controlled to a period-3 state at the parameter values of $a = 0.2$, $b = 0.2$, $c = 5.7$, $\tau = 17.5$, and $K = 0.2$ (after Pyragas K (1992); © 1992 Elsevier)

from the system's output in a digital electronic circuit with an appropriate synchronous reference. Different from the continuous control, the feedback control circuit in the OPF system continually outputs the control signal even after the success of the control. However, the control signal is usually small enough not to disturb the original chaotic attractor and does not change the dynamics of the system.

A.4 Chaos Synchronization

Synchronization between two nonlinear chaotic systems is not self-evident and was first demonstrated by Pecora and Carroll in 1990. After the proposal of the method of chaos synchronization, it has been demonstrated in various nonlinear systems including lasers. We prepare almost the same two nonlinear systems with the same parameter values. A fraction of the output from one of the chaotic systems (chaotic transmitter) is sent to the other system (chaotic receiver). Then, the receiver output synchronizes with the transmitter signal under appropriate parameter conditions. The mathematical basis for chaos synchronization is given for discrete and differential systems, however the mathematical explanation for chaos synchronization in delay differential systems is not easy and the proof has not been given yet. However, numerical and experimental demonstrations of chaos synchronization in discrete, differential, and delay differential systems have been reported.

We explain the general idea of chaos synchronization using a simple differential model. Consider a chaotic system with a set of vector variables u and v. We divide the transmitter system into two subsystems and describe the nonlinearities for the respective subsystems as $f(u, v)$ and $g(u, v)$. Then, the transmitter system is characterized by

$$\frac{du}{dt} = f(u, v) \tag{A.35}$$

$$\frac{dv}{dt} = g(u, v) \tag{A.36}$$

The subsystems are mutually coupled and the outputs of u and v are assumed to be chaotic. We prepare the receiver system for chaos synchronization. The receiver system consists of only one of the subsystems and it has the same form as (A.35). Namely, the receiver is written by

$$\frac{dw}{dt} = f(w, v) \tag{A.37}$$

Without signal transmission from the transmitter, the variable v is treated as a constant vector. The subsystem is described by the nonlinear function, however the output w may be either a chaotic or a stable oscillation when the receiver receives no transmitted signal. Even if the output is chaotic, the output w would never show the same chaotic oscillations as the output u in the transmitter since chaos has strict sensitivity for initial conditions. However, the output w shows completely the same chaotic oscillation as the transmitter output u when a fraction of the transmitter output is sent to the receiver under appropriate conditions. In a mathematical sense, there is a condition for having a negative value of the maximum conditional Lyapunov exponent for the difference of the outputs u and w.

Take an example of a Lorenz system for the demonstration of chaos synchronization. The differential equations with the variables x, y, and z, and the transmitter system is divided into two subsystems as

subsystem 1

$$\left.\begin{aligned} \frac{\mathrm{d}y}{\mathrm{d}t} &= -xz + rx - y \\ \frac{\mathrm{d}z}{\mathrm{d}t} &= xy - bz \end{aligned}\right\} \tag{A.38}$$

subsystem 2

$$\frac{\mathrm{d}x}{\mathrm{d}t} = \sigma(y - x) \tag{A.39}$$

The receiver system is assumed as a copy of subsystem 1 of the transmitter and it is given by

$$\left.\begin{aligned} \frac{\mathrm{d}y'}{\mathrm{d}t} &= -xz' + rx - y' \\ \frac{\mathrm{d}z'}{\mathrm{d}t} &= xy' - bz' \end{aligned}\right\} \tag{A.40}$$

As already noted, the variable x in the receiver is treated as a constant without the transmission of a signal from the transmitter. The outputs x, y, and z of the transmitter are chaotic at the parameter values of $\sigma = 16$, $b = 4$, and $r = 45.92$. On the other hand, it is proved that the maximum Lyapunov exponents of the outputs y' and z' of the receiver are negative. Therefore, the receiver exhibits stable outputs without receiving any chaotic signal from the transmitter.

Figure A.11 shows the transients for the chaos synchronization in a Lorenz system. The output z' from the receiver is pulled into the transmitter output z due to the presence of the transmission signal and shows a chaotic oscillation. For a sufficiently elapsed time, the two outputs show the same chaotic oscillation as shown in Fig. A.11a. Figs. A.11b and c show the plots of the trajectories for the $y - z$ and $y' - z'$ planes, respectively. The starting points are the time when the transmitted signal is received at the receiver. For the elapse of time, the distance between the two trajectories is reduced and the trajectories finally overlap with each other. Thus, chaos synchronization is

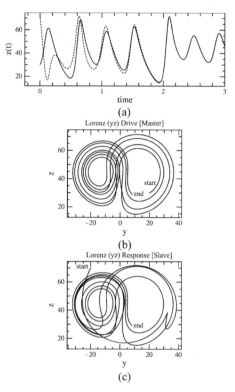

Fig. A.11. Transients of chaos synchronization in the Lorenz system. The values of the parameters are $\sigma = 16$, $b = 4$, and $r = 45.92$. **a** Time series of outputs z (*solid curve*) and z' (*broken gray line*), **b** trajectory at $y - z$ plane, and **c** trajectory at $y' - z'$ plane. The start point is the time at which the receiver receives the transmitted signal

achieved. Chaos synchronization is only attained under certain conditions. We must take care when choosing a set of subsystems and parameter conditions for successful chaos synchronization. To make suitable subsystems for chaos synchronization, the receiver system must be coincident with the transmitter system when one or more of the transmitter outputs are sent to the receiver. In this instance of the Lorenz system, the receiver conforms the transmitter system with transmission of signal x. In laser systems, however, the method is not straightforwardly applicable, since we cannot divide the dynamics of laser variables into subsystems. As an alternative technique, two same sets of laser systems are prepared as a transmitter and a receiver and a small portion of the laser output in the transmitter laser is sent to the receiver. Successful chaos synchronization is achieved under appropriate conditions of the system parameters and signal transmission as discussed in Chap. 12.

References

1. Abarbanel HDI (1996) Analysis of observed chaotic data. Springer-Verlag, New York
2. Abarbanel HDI, Kennel MB, Illing L, Tang S, Chen HF, Liu JM (2001) Synchronization and communication using semiconductor lasers with opto-electronic feedback. IEEE J Quantum Electron 37:1301–1311
3. Abraham NB, Mandel P, Narducci LM (1988) Dynamical instabilities and pulsations in lasers. In: Wolf E (ed) Progress in optics Chap. 1 Vol. 25. North-Holland, Amsterdam
4. Adachihara H, Hess O, Abraham E (1993) Spatiotemporal chaos in broad-area semiconductor lasers. J Opt Soc Am B 10:658–665
5. Acket GA, Lenstra D, den Boef J, Verbeek BH (1984) The influence of feedback intensity on longitudinal mode properties and optical noise in index-guided semiconductor lasers. IEEE J Quantum Electron 20:1163–1169
6. Agrawal GP (1984) Line narrowing in a single mode injection laser due to external optical feedback. IEEE J Quantum Electron 20:468–471
7. Agrawal GP (1985) Chirp minimization and optimum biasing for current-modulated coupled-cavity semiconductor lasers. Opt Lett 1:10–12
8. Agrawal GP, Klaus JT (1991) Effect of phase-conjugate feedback on semiconductor laser dynamics. Opt Lett 16:1325–1327
9. Agrawal GP, Gray GR (1992) Effect of phase-conjugate feedback on the noise characteristics of semiconductor lasers. Phys Rev A 46:5890–5898
10. Agrawal GP, Dutta NK (1993) Semiconductor lasers, 2nd edn. Van Nostrand Reinhold, New York
11. Ahamed M, Yamada M (2002) Influence of instantaneous mode competition on the dynamics of semiconductor lasers. IEEE Quantum Electron 38:682–693
12. Ahlers V, Parlitz U, Lauterborn W (1998) Hyperchaotic dynamics and synchronization of external-cavity semiconductor lasers. Phys Rev E 58:7208–7213
13. Aida T, Davis P (1992) Oscillation modes of diode-laser pumped hybrid bistable system with large delay and applications to dynamical memory. IEEE J Quantum Electron 28:686–699
14. Altés JB, Gatare I, Panajotov K, Thienpont H, Sciamanna M (2006) Mapping of the dynamics induced by orthogonal optical injection in vertical-cavity surface-emitting lasers. IEEE J Quantum Electron 42:198–207
15. Anderson OK, Fischer APA, Lane LC, Louvergneaux E, Stolte S, Lenstra D (1999) Experimental stability diagram of a diode laser subject to weak phase-conjugate feedback from a Rubidium vapor cell. IEEE J Quantum Electron 35:577–582

16. Annovazzi-Lodi V, Donati S, Manna M (1994) Chaos and locking in a semi-conductor laser due to external injection. IEEE J Quantum Electron 30:1537–1541

17. Annovazzi-Lodi V, Donati S, Scirè A (1996) Synchronization of chaotic injected-laser systems and its application to optical cryptography. IEEE J Quantum Electron 32:953–959

18. Annovazzi-Lodi V, Donati S, Scirè A (1997) Synchronization of chaotic lasers by optical feedback for cryptographic applications. IEEE J Quantum Electron 33:1449–1454

19. Annovazzi-Lodi V, Benedetti M, Merlo S, Norgia M, Provinzano B (2005) Optical chaos masking of video signals. IEEE Photon Technol Lett 17:1995–1997

20. Aoshima T, Ohtsubo J (1992) Two-dimensional vector LDV using laser diode frequency change and self-mixing effect. Opt Commun 92:219–224

21. Arakawa Y, Yariv A (1985) Theory of gain, modulation response, and spectral linewidth in AlGaAs quantum well lasers. IEEE J Quantum Electron 21:1666–1674

22. Arecchi FT, Meucci R, Puccioni GP, Tredicce JR (1982) Experimental evidence of subharmonic bifurcations, multistability, and turbulence in a Q-switched gas laser. Phys Rev Lett 49:1217–1220

23. Arecchi FT, Lippi GL, Puccioni GP, Tredicce JR (1984a) Deterministic chaos in laser with injected signal. Opt Commun 51:308–314

24. Arecchi FT, Lippi GL, Pucchioni GP, Tredicce JR (1984b) In: Eberly JH, Mandel L, Wolf E (eds) Coherence and quantum optics. Plenum, New York

25. Argyris A, Syvridis D, Larger L, Annovazzi-Lodi V, Colet P, Fischer I, Garcia-Ojalvo J, Mirasso CR, Pesquera L, Shore AK (2005) Chaos-based communications at high bit rates using commercial fibre-optic links. Nature 438:343–346

26. Arteaga MA, Unold HJ, Ostermann JM, Michalzik R, Thienpont H, Panajotov K (2006) Investigation of polarization properties of VCSELs subject to optical feedback from an extremely short external cavity part II: experiments. IEEE J Quantum Electron 42:102–107

27. Arteaga MA, López-Amo M, Thienpont H, Panajotov K (2007) Role of external cavity reflectivity for achieving polarization control and stabilization of vertical cavity surface emitting laser. Appl Phys Lett 90:031117–1–3

28. Arimoto A, Ojima M (1984) Diode laser noise at control frequency in optical videodisc players. Appl Opt 23:2913–2920

29. Arimoto A, Ojima M, Chinone N, Oishi A, Gotoh T, Ohnuki N (1986) Optimum conditions for the high frequency noise reduction method in optical videodisk system. Appl Opt 25:1398–1403

30. Arnold G, Russer P, Peterman K (1982) Modulation of laser diodes. In: Kressel H (ed) Semiconductor devices for optical communication, Chap. 7. Springer-Verlag, Berlin

31. Asatuma T, Takiguchi Y, Frederico S, Furukawa A, Hirata S (2006) Successive phase change and stability of near-field patterns for broad-area laser diodes. SPIE Proc 6104:61040C

32. Barchanski A, Gensty T, Degen C, Fischer I, Elsäßer W (2003) Picosecond emission dynamics of vertical-cavity surface-emitting lasers: spatial, spectral, and polarization-resolved characterization. IEEE J Quantum Electron 39:850–858

33. Bennett S, Snowden CM, Iezekiel S (1997) Nonlinear dynamics in directly modulated multiple-quantum-well laser diodes. IEEE J Quantum Electron 33:2076–2083
34. Besnard P, Robert F, Charès ML, Stéphan GM (1997) Theoretical modeling of vertical-cavity surface-emitting lasers with polarized optical feedback. Phys Rev A 56:3191–3205
35. Besnard P, Charès ML, Stéphan G, Robert F (1999) Switching between polarized modes of a vertical-cavity surface-emitting laser by isotropic optical feedback. J Opt Soc Am B 16:1059–1064
36. Binder JO, Cormack D, and Somani A (1990) Intermodal tuning characteristics of an INGaAsP laser with optical feedback from an external grating reflector. IEEE J Quantum Electron 26:1191–1199
37. Bloch F (1946) Nuclear induction. Phys Rev 70:460–474
38. Bochove E (1997) Theory of a semiconductor laser with phase-conjugate optical feedback. Phys Rev A 55:3891–3899
39. Boers PM, Vlaardinerbrek MT (1975) Dynamic behavior of semiconductor lasers. Electron Lett 11:206–208
40. Bosch T, Servagent N, Donati S (2001) Optical feedback interferometry for sensing application. Opt Eng 40:20–27
41. Botez D (1981) InGaAsP/InP double-heterosturcture lasers: simple expressions for wave confinement, beamwidth, and threshold current over wide range in wavelength $(1.1 - 1.65 \ \mu m)$. IEEE J Quantum Electron 17:178–186
42. Burkhard T, Ziegler MO, Fischer I, Elsäßer W (1999) Spatio-temporal dynamics of broad area semiconductor lasers and its characterization. Chaos, Solitons & Fractals 10:845–850
43. Buss J, Adams MJ (1979) Phase and group indices for double heterostructure lasers. Solid-State Electron Dev 3:189–195
44. Cao H (2003) Lasing in disordered media. In: Wolf E (ed) Progress in optics, chap. 6, vol. 45. North-Holland, Amsterdam
45. Carr TW, Erneux T (2001) Dimensionless rate equations and simple conditions for self-pulsing in laser diodes. IEEE J Quantum Electron 37:1171–1177
46. Casperson LW (1978) Spontaneous coherent pulsations in laser oscillators. IEEE J Quantum Electron 14:756–761
47. Champagne Y, Mailhot S, McCarthy N (1995) Numerical procedure for the lateral-mode analysis of broad-area semiconductor lasers with an external cavity. IEEE J Quantum Electron 31:795–810
48. Chen YC, Winful HG, Liu JM (1985) Subharmonic bifurcations and irregular pulsing behavior of modulated semiconductor lasers. Appl Phys Lett 47:208–210
49. Chen B, Wornell G (1998) Analog error-correcting codes based on chaotic dynamical systems. IEEE Trans Commun 46:881–890
50. Chen HF, Liu JM, Simpson TB (2000) Response characteristics of direct current modulation on a bandwidth-enhanced semiconductor laser under strong injection locking. Opt Commun 173:349–355
51. Chen HF, Liu JM (2000) Open-loop chaotic synchronization of injection-locked semiconductor lasers with gigahertz range modulation, IEEE J Quantum Electron 36:27–34
52. Chen CC, Yao K, Umeno K, Biglieri E (2001) Design of spread-spectrum sequences using chaotic dynamical systems and ergodic theory. IEEE Trans Circuits Syst. I 48:1498–1509

53. Chen NS, Yu SF (2004) Transient response of ARROW VCSELs under external optical feedback. IEEE Photon Technol Lett 16:1610–1612

54. Cheng DL, Yen TC, Chang JW, Tsai JK (2003) Generation of high-speed single-wavelength optical pulses in semiconductor lasers with orthogonal-polarization optical feedback. Opt Commun 222:363–369

55. Chern JL, Otsuka K, Ishiyama F (1993) Coexistence of two attractors in lasers with delayed incoherent optical feedback. Opt Commun 96:259–266

56. Chi MJ, Thestrup B, Mortensen JL, Nielsen ME, Petersen PM (2003) Improvement of the beam quality of a diode laser with two active broad-area segments. J Opt A 5: S338-S341

57. Chi M, Thestrup B, Petersen PM (2005) Self-injection locking of an extraordinarily wide broad-area diode laser with a 1000-μm-wide emitter. Opt Lett 30:1147–1149

58. Chiang MC, Chen HF, Liu JM (2005) Experimental synchronization of mutually coupled semiconductor lasers with optoelectronic feedback. IEEE J Quantum Electron 41:1333–1340

59. Chiang M, Chen HF, Liu JM (2006) Synchronization of mutually coupled systems. Opt Commun 261:86–90

60. Chinone N, Aiki K, Nakamura M, Ito R (1978) Effects of lateral mode and carrier density profile on dynamic behaviors of semiconductor lasers. IEEE J Quantum Electron 14:625

61. Cho Y, Umeda M, (1986) Observation of chaos in a semiconductor laser with delayed feedback. Opt Commun 59:131–136

62. Chow WW, Koch SW, Sargent III M (1993) Semiconductor laser physics. Springer-Verlag, New York

63. Chraplyvy AR, Liou KY, Tkach RW, Eisenstein G, Jhee YK, Koch TL, Anthony PJ, Chakrabarti UK (1986) Simple narrow-linewidth 1.5 mm InGaAsP DFB external-cavity laser. Electron Lett 22:88–90

64. Cohen JS, Dentine RR, Verbeek BH (1988) The effect of optical feedback on the relaxation oscillation in semiconductor lasers. IEEE J Quantum Electron 24:1989–1995

65. Cohen J, Wittgrefe F, Hoogerland MD, Woerdman JP (1990) Optical spectra of a semiconductor laser with incoherent optical feedback. IEEE J Quantum Electron 26:982–990

66. Colet P, Roy R (1994) Digital communication with synchronized chaotic lasers. Opt Lett 19:2056–2058

67. Cook DD, Nash FR (1975) Gain-induced guiding and astigmatic output beam of GaAs lasers. J Appl Phys 46:1660–1672

68. Cuomo KM, Oppenheim AV, Stogatz SH (1993) Synchronization of Lorenz-based chaotic circuits with applications to communications. IEEE Trans Circuits Syst. II 40:626–633

69. Dachselt F, Schwarz W (2001) Chaos and cryptography. IEEE Trans Circuits Sys I 48:1498–1500

70. Dahmani B, Hollberg L, Drullinger R (1987) Frequency stabilization of semiconductor lasers by resonant optical feedback. Opt Lett 12:876–878

71. Damen TC, Duguay MA (1980) Optoelectronic regenerative pulser. Electron Lett 16:166–167

72. Danckaerta J, Naglera B, Alberta J, Panajotova K, Veretennicoff I, Erneux T (2002) Minimal rate equations describing polarization switching in vertical-cavity surface-emitting lasers. Opt Commun 201:129–137

73. Davis P (1990) Application of optical chaos to temporal pattern search in a nonlinear optical resonator. Jpn J Appl Phys 29:L1238–1240
74. Davis P, Liu Y, Aida T (2001) Chaotic wavelength hopping device for multi-wavelength optical communications. IEEE Trans Circuits Syst I 48:1523–1527
75. Debernardi P, Ostermann JM, Feneberg M, Jalics C, Michalzik R (2005) Reliable polarization control of VCSELs through monolithically integrated surface gratings: A comparative theoretical and experimental study. IEEE J Select Topics Quantum Electron 11:107–116
76. Degen C, Fischer I, Elsäßer W (1999) Transverse modes in oxide confined VCSELs: influence of pump profile, spatial hole burning, and thermal effects. Opt Exp 5:38–47
77. De Jagher PC, van der Graaf WA, Lenstra D (1996) Relaxation-oscillation phenomena in an injection-locked semiconductor laser. Quantum Semiclass Opt 8:805–822
78. DeTienne DH, Gray GR, Agrawal GP, Lenstra D (1997) Semiconductor laser dynamics for feedback from a finite-penetration-depth phase-conjugate mirror. IEEE J Quantum Electron 33:838–844
79. Detoma E, Tromborg B, Montrosset I (2005) The complex way to laser diode spectra: example of an external cavity laser strong optical feedback. IEEE J Quantum Electron 41:171–182
80. Diehl R (2000) High-power diode laser, fundamentals, technology, applications. Springer-Verlag, Berlin
81. Donati S, Giuliani G, Merlo S (1995) Laser diode feedback interferometer for measurement of displacements without ambiguity. IEEE J Quantum Electron 31:113–119
82. Donati S, Falzoni L, Merlo S (1996) A PC-interfaced, compact laser-diode feedback interferometer for displacement measurements. IEEE Trans Instrum Meas 45:942–947
83. Donati S, Merlo S (1998) Applications of diode laser feedback interferometry. J Opt 29:156–161
84. Dutta NK, Olsson NA, Koszi LA, Besomi P, Wilson RB (1984) Frequency chirp under current modulation in InGaAsP injection lasers. J Appl Phys 56:2167–2169
85. Eguia MC, Mindlin GB, Giudici M (1998) Low-frequency fluctuations in semiconductor lasers with optical feedback are induced with noise. Phys Rev E 58:2636–2639
86. Elsäßer W, Göbel EO (1984) Spectral linewidth of gain- and index-guided InGaAs semiconductor lasers. Appl Phys Lett 45:353–355
87. Elsäßer W, Göbel EO (1985) Multimode effects in the spectral linewidth of semiconductor lasers. IEEE J Quantum Electron 21:687–692
88. Eriksson S, Lindberg AM (2001) Periodic oscillation within the chaotic region in a semiconductor laser subjected to external optical injection. Opt Lett 26:142–144
89. Erneux T, Kovanis V, Gavrielides A, Alsing PM (1996) Mechanism for period-doubling bifurcation in a semiconductor laser subject to optical injection. Phys Rev A 53:4372–4380
90. Erzgräber H, Krauskoph B, Lenstra D, Fischer APA, Vemuri G (2006) Frequency versus relaxation oscillations in a semiconductor laser with coherent filtered optical feedback. Phys Rev E 73:055201-1–4

91. Ewald G, Knaak KM, Otte SG, Wendt KDA, Kluge HJ (2005) Development of narrow-linewidth diode lasers by use of volume holographic transmission gratings. Appl Phys B 80:483–487

92. Fan L, Fallahi M, Murray JT, Bedford R, Kaneda Y, Zakharian AR, Hader J, Moloney JV, Stolz W, Koch SW (2006) Tunable high-power high-brightness linearly polarized vertical-external-cavity surface-emitting lasers. Appl Phys Lett 88:021105–13

93. Feature section on optical chaos and applications to cryptography (2002) IEEE J Quantum Electron 38 (9)

94. Fischer I, van Tartwijk GHM, Levine AM, Elsäßer W, Gobel EO, Lenstra D (1996a) Fast pulsing and chaotic itinerancy with a drift in the coherence collapse of semiconductor lasers. Phys Rev Lett 76:220–223

95. Fischer I, Hess O, Elsäßer W, Göbel E (1996b) Complex spatio-temporal dynamics in the near-field of broad-area semiconductor laser. EuroPhys Lett 35:579–584

96. Fischer I, Liu Y, Davis P (2000a) Synchronization of chaotic semiconductor laser dynamics on subnanosecond time scales and its potential for chaos communication. Phys Rev A 62:011801–1–4

97. Fischer APA, Andersen OK, Yousefi M, Stolte S, Lenstra D (2000b) Experimental and theoretical study of filtered optical feedback in a semiconductor laser. IEEE J Quantum Electron 36:375–384

98. Fischer APA, Yousefi M, Lenstra D, Carter MW, Vemuri G (2004a) Filtered optical feedback induced frequency dynamics in semiconductor lasers. Phys Rev Lett 92:023901–1–4

99. Fischer APA, Yousefi M, Lenstra D, Carter MW, Vemuri G (2004b) Experimental and theoretical study of semiconductor laser dynamics due to filtered optical feedback. IEEE J Select. Topics Quantum Electron 10:944–954

100. Fleming MW, Mooradian A (1981a) Fundamental line broadening of single-mode (AaAl)As diode lasers. Appl Phys Lett 38:511–513

101. Fleming MW, Mooradian A (1981b) Spectral characteristics of external-cavity controlled semiconductor lasers. IEEE J Quantum Electron 17:44–59

102. Fravre F (1987) Theoretical analysis of external optical feedback on DFB semiconductor lasers. IEEE J Quantum Electron 23:81–88

103. Fujino H, Ohtsubo J (2000) Experimental synchronization of chaotic oscillations in external cavity semiconductor lasers. Opt Lett 25:625–627

104. Fujino H, Ohtsubo J (2001) Synchronization of chaotic oscillations in mutually coupled semiconductor lasers. Opt Rev 8:351–357

105. Fujita Y, Ohtsubo J (2005) Optical-feedback-induced stability and instability in broad-area semiconductor lasers. Appl Phys Lett 87:031112–13

106. Fujiwara M, Kubota K, Lang R (1981) Low-frequency intensity fluctuation in laser diode with external optical feedback. Appl Phys Lett 38:217–220

107. Fujiwara N, Takiguchi Y, Ohtsubo J (2003a) Observation of low-frequency fluctuations in vertical-cavity surface-emitting lasers. Opt Lett 28:896–898

108. Fujiwara N, Takiguchi Y, Ohtsubo J (2003b) Observation of the synchronization of chaos in mutually injected vertical-cavity surface-emitting semiconductor lasers. Opt Lett 28:1677–1679

109. Fujiwara N, Ohtsubo J (2004) Synchronization in chaotic vertical-cavity surface-emitting semiconductor lasers. SPIE Proc 5349:282–289

110. Fukushima T (2000) Analysis of resonator eigenmodes in symmetric quasi-stadium laser laser diodes. J Lightwave Technol 18:2208–2216

111. Fukushima T, Miyazaki H, Ando T, Tanaka T, Sakamoto T (2002) Nonlinear dynamics in directly modulated self-pulsating laser diodes with a highly doped saturable absorption layer. Jpn J Appl Phys 41 Part 1:117–124

112. Fukuchi S, Ye SY, Ohtsubo J (1999) Relaxation oscillation enhancement and coherence collapse in semiconductor lasers with optical feedback. Opt Rev 6:365–371

113. Furukawa A, Sasaki S, Hoshi M, Matsuzono A, Moritoh K, Baba T (2004) High-power single-mode vertical-cavity surface-emitting lasers with triangular holey structure. Appl Phys Lett 85:5161–5163

114. Furuya K, Suematsu Y, Sakakibara Y, Yamada M (1979) Influence of intraband electronic relaxation on relaxation oscillation of injection lasers. Trans IECE Japan E-62:241–245

115. Gallagher DFG, White IH, Carroll JE, Plumb RG (1987) Gigabit pulse position bistability in semiconductor lasers. J Lightwave Technol LT-5:1391–1398

116. Garcia-Ojalvo J, Casademont J, Mirasso CR, Torrent MC, Sancho JM (1999) Coherence and synchronization in diode-laser arrays with delayed global coupling. Int J Bifurcation Chaos 9:2225–2229

117. Gatare I, Sciamanna M, Buesa J, Thienpont H, Panajotov K (2006) Nonlinear dynamics accompanying polarization switching in vertical-cavity surface-emitting lasers with orthogonal optical injection. Appl Phys Lett 88:101106–1–3

118. Gavrielides A, Kovanis V, Erneux T (1997) Analytical stability boundaries for a semiconductor laser subject to optical injection. Opt Commun 136:253–256

119. Grigorieva EV, Haken H, S.A. Kaschenko SA (1999) Theory of quasiperiodicity in model of lasers with delayed optoelectronic feedback. Opt Commun 165:279–292

120. Gehrig E, Hess O (2000) In: Diehl R (ed) Dynamics of high-power diode lasers, in high-power diode lasers – fundamentals, technology, applications. Springer-Verlag, Berlin

121. Gehrig E, Hess O (2002) Mesoscopic spatiotemporal theory for quantum-dot lasers. Phys Rev A 65:033804–1-16

122. Gehrig E, Hess O (2003) Spatio-temporal dynamics and quantum fluctuations in semiconductor lasers. Springer-Verlag, Heidelberg

123. Gehrig E, Hess O, Ribbat C, Sellin RL, Bimberg D (2004) Dynamic filamentation and beam quality of quantum-dot lasers. Appl Phys Lett 84:1650–1652

124. Genty G, Gröhn A, Talvitie H, Kaivola M, Ludvigsen H (2000) Analysis of the linewidth of a grating-feedback GaAlAs laser. IEEE J Quantum Electron 36:1193–1198

125. Giacomelli G, Calzavara M, Arecchi FT (1989) Instabilities in a semiconductor laser with delayed optoelectronic feedback. Opt Commun 74:97–101

126. Gibbs HM (1985) Optical bistability: controlling light with light. Academic Press, New York

127. Gilbert T, Gammon RW (2000) Stable oscillations and devil's staircase in the van der Pol oscillator. J Bifurcation Chaos 10:155–164

128. Giudici M, Balle S, Ackemann T, Barland S, Tredicce JR (1999) Polarization dynamics in vertical-cavity surface-emitting lasers with optical feedback: experiment and model. J Opt Soc Am B 16:2114–2123

129. Giuliani G, Norgia M (2000) Laser diode linewidth measurement by means of self-mixing interferometry. IEEE Photon Technol Lett 12:1028–1030

130. Giuliani G, Donati S, Passerini M, Bosch T (2001) Angle measurement by injection detection in a laser diode. Opt Eng 40:95–99

131. Giuliani G, Norgia M, Donati S, Bosch T (2002) Laser diode self-mixing technique for sensing applications. J Opt A: Pure Appl Opt 4: S283–S294

132. Giuliani G, Bozzi-Pietra S, Donati S (2003) Self-mixing laser diode vibrometer. Meas Sci Technol 14:24–32

133. Giuliani G, Donati S (2005) Laser interferometry. In: Kane D, Shore KA (eds) Unlocking dynamical diversity optical feedback effects on semiconductor lasers, Chap. 7. Wiley, Chichester

134. Glas P, Müller R, Klehr A (1983) Bistability, self-sustained oscillations, and irregular operation of a GaAs laser coupled to an external resonator. Opt Commun 47:297–301

135. Goedgebuer JP, Larger L, Porte H, Delorme F (1998a) Chaos in wavelength with a feedback tunable laser diode. Phys Rev E 57:2795–2798

136. Goedgebuer JP, Larger L, Porte H (1998b) Optical cryptosystem based on synchronization of hyperchaos generated by a delayed feedback tunable laser diode. Phys Rev Lett 80:2249–2252

137. Goedgebuer JP, Levy P, Larger L, Chen CC, Rhodes WT (2002) Optical communication with synchronized hyperchaos generated electrooptically. IEEE J Quantum Electron 38:1178–1183

138. Goldberg L, Taylor HF, Dandrige A, Weller JF, Miles RO (1982) Coherence collapse in single mode semiconductor lasers due to optical feedback. IEEE J Quantum Electron 18:555–679

139. Gray GR, Ryan AT, Agrawal GP, Gage EC (1993) Control of optical-feedback-induced laser intensity noise in optical data recording. Opt Eng 32:739–745

140. Gray GR, Huang D, Agrawal GP (1994) Chaotic dynamics of semiconductor lasers with phase-conjugate feedback. Phys Rev A 49:2096–2105

141. Gray GR, DeTienne Dh, Agrawal GP (1995) Mode locking in semiconductor lasers by phase-conjugate optical feedback. Opt Lett 20:1295–1297

142. Green K, Krauskopf B (2006) Mode structure of a semiconductor laser subject to filtered optical feedback. Opt Commun 258:243–255

143. Groot PJ, Gaillatin GM (1989) Backscatter-modulation velocimetry with an external-cavity laser diode. Opt Lett 14:165–167

144. Grigorieva EV, Haken H, Kaschenko SA (1999) Theory of quasiperiodicity in model of lasers with delayed optoelectronic feedback. Opt Commun 165:279–292

145. Haken H (1975) Analogy between higher instabilities in fluids and lasers. Phys Lett 53A:77–78

146. Haken H (1985) Light, vol 2. North-Holland, Amsterdam

147. Harth W (1973) Large signal direct modulation of injection lasers. Electron Lett 9:532–533

148. Hashimoto M, Ohtsu M (1987) Experiments on a semiconductor laser pumped rubidium atomic clock. IEEE J Quantum Electron 23:446–451

149. Haug H (1969) Quantum-mechanical rate equations for semiconductor lasers. Phys Rev 184:338–348

150. Hegarty SP, Huyet G, Porta P, McInerney JG (1998) Analysis of the fast recovery dynamics of a semiconductor laser with feedback in the low-frequency fluctuation regime. Opt Lett 23:1206–1208

151. Heil T, Fischer I, Elsäßer W (1998) Coexistence of low-frequency fluctuations and stable emission on a single high-gain mode in semiconductor lasers with external optical feedback. Phys Rev A 58: R2672–2675

152. Heil T, Fischer I, Elsäßer W (1999) Influence of amplitude-phase coupling on the dynamics of semiconductor lasers subject to optical feedback. Phys Rev A 60:634–641

153. Heil T, Fischer I, Elsäßer W, Gavrielides A (2001a) Dynamics of semiconductor lasers subject to delayed optical feedback: the short cavity regime. Phys Rev Lett 87:243901-1–4

154. Heil T, Fischer I, Elsäßer W, Mulet J, Mirasso CR (2001b) Chaos synchronization and spontaneous symmetry breaking in symmetrical delay-coupled semiconductor lasers. Phys Rev Lett 86:795–798

155. Heil H, Mulet J, Fischer I, Mirasso CR, Peil M, Colet P, Elsäßer W (2003a) ON/OFF phase shift keying for chaos-encrypted communication using external-cavity semiconductor lasers. IEEE J Quantum Electron 38:1162–1169

156. Heil T, Fischer I, Elsäßer W, Krauskopf B, Green K, Gavrielides A (2003b) Delay dynamics of semiconductor lasers with short external cavities: bifurcation scenarios and mechanisms. Phys Rev E 67:066214-1–11

157. Heil T, Uchida A, Davis P, Aida T (2003c) TE-TM dynamics in a semiconductor laser subject to polarization-rotated optical feedback. Phys Rev A 68:033811-1–8

158. Helms J, Petermann K (1990) A simple analytic expression for the stable operation range of laser diode with optical feedback. IEEE J Quantum Electron 26:833–836

159. Hemery H, Chusseau L, Lourtioz JM (1990) Dynamic behaviors of semiconductor lasers under strong sinusoidal current modulation: modeling and experiments at 1.3 µm. IEEE J Quantum Electron 26:633–641

160. Henry CH (1982) Theory of the linewidth of semiconductor lasers. IEEE J Quantum Electron 18:259–264

161. Henry CH (1983) Theory of phase noise and power spectrum of a single mode injection laser. IEEE J Quantum Electron 19:1391–1397

162. Henry CH (1986) Theory of spontaneous emission noise in open resonators and its applications to lasers and optical amplifiers. J Lightwave Technol LT-4:288–297

163. Henry CH, Kazarinov RF (1986) Instability of semiconductor lasers due to optical feedback from distant reflectors. IEEE J Quantum Electron 22:294–301

164. Hess O, Koch SW, Moloney JV (1995) Filamentation and beam propagation in broad-area semiconductor lasers. IEEE J Quantum Electron 31:35–43

165. Hess O, Kuhn T (1996a) Maxwell-Bloch equations for spatially inhomogeneous semiconductor lasers. I. theoretical formulation. Phys Rev A 54:3347–3359

166. Hess O, Kuhn T (1996b) Maxwell-Bloch equations for spatially inhomogeneous semiconductor lasers. II. spatiotemporal dynamics. Phys Rev A 54:3360–3368

167. Hodgson N, Weber H (1997) Optical resonators. Springer, London

168. Hogenboom EHM, Klische W, Weiss CO, Godone A (1985) Instabilities of a homogeneously broadened laser. Phys Rev Lett 55:2571–2574

169. Hohl A, Gavrielides A, Erneux T, Kovanis V (1997) Localized synchronization in two coupled nonidentical semiconductor lasers. Phys Rev Lett 78:4745–4748

170. Hohl A, Gavrielides A (1998) Experimental control of a chaotic semiconductor laser. Opt Lett 23:1606–1608

450   References

171. Hohl A, Gavrielides A, Erneux T, Kovanis V (1999) Quasiperiodic synchronization for two delay-coupled semiconductor lasers. Phys Rev A 59:3941–3949
172. Holm MA, D. Burns D, Ferguson AI, M. D. Dawson MD (1999) Actively stabilized single-frequency vertical-external-cavity AlGaAs laser. IEEE Photon Technol Lett 11:1551–1553
173. Hori Y, Serizawa H, Sato H (1988) Chaos in directly modulated semiconductor lasers. J Opt Soc Am B 5:1128–1133
174. Huffaker DL, Deng H, Deppe DG (1998) 1.15-µm wavelength oxide-confined quantum-dot vertical-cavity surface-emitting laser. IEEE Photon Technol Lett 10:185–187
175. Hülsewede R, Sebastian J, Wenzel H, Beister G, Kanauer A, Erbert G (2001) Beam quality of high power 800 nm broad-area laser diodes with 1 and 2 mm large optical cavity structure. Opt Commun 192:69–75
176. Hunt ER (1991) Stabilizing high-period orbits in a chaotic system: the diode resonator. Phys Rev Lett 67:1953–1955
177. Hwang SK, Liu JM (2000) Dynamical characteristics of an optically injected semiconductor laser. Opt Commun 183:195–205
178. Hwang SK, Liu JM, White JK (2004) 35-GHz intrinsic bandwidth for direct modulation in 1.3-µm semiconductor lasers subject to strong injection locking. IEEE Photon Technol Lett 16:972–974
179. Iannelli JM, Shevy Y, Kitching SJ, Yariv A (1993) Linewidth reduction and frequency stabilization of semiconductor lasers using dispersive losses in an atomic vapor. IEEE J Quantum Electron 29:1253–1261
180. Ikeda K (1979) Multiple-valued stationary state and its instability of the transmitted light by a ring cavity system. Opt Commun 30:257–261
181. Ikegami T, Suematsu Y (1967) Resonance-like characteristics of the direct modulation of a junction laser. Proc IEEE 55:122–133
182. Ikegami T, Suematsu Y (1968) Carrier lifetime measurement of a junction laser using direct modulation. IEEE J Quantum Electron 4:148–151
183. Ikuma Y, Ohtsubo J (1998) Dynamics in a compound cavity semiconductor laser induced by small external-cavity-length change. IEEE J Quantum Electron 34:1240–1246
184. Inaba H (1982) In: Suematsu Y (ed) Optical devices and fibers. North-Holland, Amsterdam
185. Ishii Y (2004) Laser-diode interferometry. In: Wolf E (ed) Progress in optics, chap. 3, vol.46. North-Holland, Amsterdam
186. Ishiyama F (1999) Bistability of quasi periodicity and period doubling in a delay-induced system. J Opt Soc Am B 16:2202–2206
187. Ito H, Yokoyama. H, Murata S, Inaba H (1981) Generation of picosecond optical pulses with highly RF modulated AlGaAs DH laser. IEEE J Quantum Electron 17:663–670
188. Jiang S, Pan Z, M. Dagenais M, Morgan RA, and K. Kojima K (1993) High-frequency polarization self-modulation in vertical-cavity surface-emitting lasers. Appl Phys Lett 63:3545–3547
189. Jones RJ, Sivaprakasam S, Shore KA (2000) Integrity of semiconductor laser chaotic communications to naïve eavesdroppers. Opt Lett 22:1663–1665
190. Jones RJ, Rees P, Spencer PS, Shore KA (2001) Chaos and synchronization of self-pulsating laser diodes. J Opt Soc Am B 18:166–172
191. Juang C, Chen MR, Juang J (1999) Nonlinear dynamics of self-pulsating laser diodes under external drive. Opt Lett 24:1346–1348

192. Juang C, Hwang TM, Juang J, Lin WW (2000) A synchronization scheme using self-pulsating laser diodes in optical chaotic communication. IEEE J Quantum Electron 36:300–304

193. Kakiuchida H, Ohtsubo J (1994) Characteristics of a semiconductor laser with external feedback. IEEE J Quantum Electron 30:2087–2097

194. Kane MD, Shore KA (2005) Unlocking dynamical diversity-optical feedback effects on semiconductor lasers. John Wiley & Sons, Chichester

195. Kao YH, Tsai CH, Wang CS (1992) Observation of bifurcation structure in a rf driven semiconductor laser using an electronic simulator. Rev Sci Instrum 63:75–79

196. Kao YH, Lin HT (1993) Virtual Hoph precursor of period-doubling route in directly modulated semiconductor lasers. IEEE J Quantum Electron 29:1617–1623

197. Katagiri Y, Hara S (1994) Increased spatial frequency in interferential undulations of coupled cavity lasers. Appl Opt 33:5564–5570

198. Kawaguchi H (1994) Bistabilities and nonlinearities in laser diodes. Artech House, London

199. Kazarinov RF, Henry CH (1987) The relation of line narrowing and chirp reduction resulting from coupling of a semiconductor laser to a passive resonator. IEEE J Quantum Electron 23:1401–1409

200. Kennedy MP, Rovatti R, Setti G (2000) Chaotic electronics in telecommunications. CRC Press, Boca Raton

201. Kikuchi K, Okoshi T (1982) Simple formula giving spectrum-narrowing ratio of semiconductor-laser output obtained by optical feedback. Electron Lett 18:10–12

202. Kikuchi N, Liu Y, Ohtsubo J (1997) Chaos control and noise suppression in external-cavity semiconductor lasers. IEEE J Quantum Electron 33:56–65

203. Kirkby PA, Goodwin AR, Thompson GHB, Selway PR (1977) Observations of self-focusing in stripe geometry semiconductor lasers and the development of a comprehensive model of their operation. IEEE J Quantum Electron 13:705–719

204. Kittel A, Pyragas K, Richter R (1994) Prerecorded history of a system as an experimental tool to control chaos. Phys Rev E 50:262–268

205. Klische W, Telle HR, Weiss CO (1984) Chaos in a solid-state laser with a periodically modulated pump. Opt Lett 9:561–563

206. Koelink MH, Slot M, de Mul FFM, Grever R, Graaff AC, Dasse M, Aaroudse JG (1992) Laser Doppler velocimeter based on the self-mixing effect in a fiber-coupled semiconductor laser: theory. Appl Opt 31:3401–3408

207. Koshio H, Ohtsubo J (2004) Instability and dynamic characteristics of self-pulsating semiconductor lasers with optical feedback. SPIE Proc 5349:385–365

208. Kourogi M, Ohtsu M (1995) Phase noise and its control in semiconductor lasers. In: Agrawal GP (ed) Semiconductor lasers past, present, and future, chap. 3. AIP Press, Woodbury

209. Kovanis V, Gavrielides A, Simpson TB, Liu JM (1995) Instabilities and chaos in optically injected semiconductor lasers. Appl Phys Lett 67:2780–2782

210. Kressel H, Butler JK (1977) Semiconductor lasers and heterojunction LEDs. Academic Press, New York

211. Kürz P, Mukai T (1996) Frequency stabilization of semiconductor laser by external phase-conjugate feedback. Opt Lett 21:1369–1371

212. Kürz P, Nagar R, Mukai T (1996) Highly efficient phase conjugation using spatially nondegenerate four-wave mixing in a broad-area laser diode. Appl Phys Lett 68:1180–1182

213. Kusumoto K, Ohtsubo J (2002) 1.5-GHz message transmission based on synchronization of chaos in semiconductor lasers. Opt Lett 27:989–991

214. Kusumoto K, Ohtsubo J (2003) Anticipating chaos synchronization at high optical feedback rate in compound cavity semiconductor lasers. IEEE J Quantum Electron 39:1531–1536

215. Kuznetsov M, Tsang DZ, Walpole JN, Liau ZL, Ippen EP (1986) Chaotic pulsation of semiconductor lasers with proton-bombarded segment. In: Boyd RW, Raymer MG, Narducci LM (eds) Optical instabilities, vol 4. Cambridge University Press, Cambridge, UK

216. Lang R (1979) Lateral transverse mode instability and its stabilization in stripe geometry injection laser. IEEE J Quantum Electron 15:718–726

217. Lang R, Kobayashi K (1980) External optical feedback effects on semiconductor injection properties. IEEE J Quantum Electron 16:347–355

218. Langley LN, Shore KA (1994) Intensity noise and linewidth characteristics of laser diodes with phase conjugate optical feedback. IEE Proc OptoElectron 141:103–108

219. Larger L, Goedgebuer JP, Merolla JM (1998a) Chaotic oscillator in wavelength: a new setup for investigating differential difference equations describing nonlinear dynamics. IEEE J Quantum Electron 34:594–601

220. Larger L, Goedgebuer JP, Delorme F (1998b) Optical encryption system using hyperchaos generated by an optoelectronic wavelength oscillator. Phys Rev E 57:6618–6624

221. Larger L, Goedgebuer JP, Udaltsov VS, Rhodes WT (2001) Radio-transmission system using high dimensional chaotic oscillator. Electron Lett 37:594–595

222. Lau KY, Yariv A (1985) Ultra-high speed semiconductor lasers. IEEE J Quantum Electron 21:121–138

223. Laurent PH, Clairon A, Bréant CH (1989) Frequency noise analysis of optically self-locked diode lasers. IEEE J Quantum Electron 25:1131–1142

224. Law JY, van Tartwijk GHM, Agrawal GP (1997) Effects of transverse-mode competition on the injection dynamics of vertical-cavity surface-emitting lasers. Quantum Semiclass Opt 9:737–747

225. Law JY, Agrawal GP (1997a) Effects of spatial hole burning on gain switching in vertical-cavity surface-emitting lasers. IEEE J Quantum Electron 33:462–468

226. Law JY, Agrawal GP (1997b) Mode-partition noise in vertical-cavity surface-emitting lasers. IEEE Photon Technol Lett 9:437–439

227. Law JY, Agrawal GP (1998) Feedback-induced chaos and intensity-noise enhancement in vertical-cavity surface-emitting lasers. J Opt Soc Am B 15:562–569

228. Lawrence JS, Kane DM (1999) Injection locking suppression of coherence collapse in a diode laser with optical feedback. Opt Commun 167:273–282

229. Lawrence JS, Kane DM (2002a) Nonlinear dynamics of a laser diode with optical feedback systems subject to modulation. IEEE J Quantum Electron 38:185–192

230. Lawrence JS, Kane DM (2002b) Broad-area diode lasers with plane-mirror and phase-conjugate feedback. J Lightwave Technol 20:100–104

231. Lax M (1960) Fluctuations from the nonequilibrium steady state. Rev Mod Phys 32:25–64
232. Lax M, Louisell WH (1969) Quantum noise, XII. Density-operator treatment of field and population fluctuations. Phys Rev 185:568–591
233. Lee CH, Yoon TH, Shin Y (1985) Period doubling and chaos in a directly modulated laser diode. Appl Phys Lett 46:95–97
234. Lee CH, Shin SY, Lee SY (1988) Optical short-pulse generation using diode laser with negative optoelectronic feedback. Opt Lett 13:464–466
235. Lee CH, Shin SY (1993) Self-pulsing spectral bistability, and chaos in a semiconductor laser diode with optoelectronc feedback. Appl Phys Lett 62:922–924
236. Lee EK, Pang HS, Park JD, Lee H (1993) Bistability and chaos in an injection-locked semiconductor laser. Phys Rev A 47:736–739
237. Lee SB, Lee JH, Chang JS (2002) Observation of scarred modes in asymmetrically deformed microcylinder lasers. Phys Rev Lett 88:033903-1–4
238. Lee KH, Baek JH, Hwang IK, Lee YH, Lee GH, Ser JH, Kim HD, Shin HE (2004) Square-lattice photonic-crystal vertical-cavity surface-emitting lasers. Opt Exp 17:4136–4143
239. Lenstra D, van Vaalen M, Jaskorzynska B (1984) On the theory of a single-mode laser with weak optical feedback. Physica 125C:255–264
240. Lenstra L, Verbeek BH, den Boef AJ (1985) Coherence collapse in single-mode semiconductor lasers due to optical feedback. IEEE J Quantum Electron QE-21:674–679
241. Lenstra D (1991) Statistical theory of the multistable external-feedback laser. Opt Commun 81:209–214
242. Lenstra D, van Tartwijk GHM, van der Graaf WA, De Jagher PC (1993) Multiwave-mixing dynamics in a diode laser. SPIE Proc 2039:11–22
243. Lenstra D, Vemuri G, Yousefi M (2005) Generalized optical feedback. In: Kane DM, Shore KA (eds) Unlocking dynamical diversity. Wiley, Chichester
244. Levine AM, van Tartwijk GHM, Lenstra D, Erneux T (1995) Diode lasers with optical feedback: stability of the maximum gain mode. Phys Rev A 52: R3436–3439
245. Levy G, Hardy AA (1997) Chaotic effects in flared lasers: a numerical analysis. IEEE J Quantum Electron 33:26–32
246. Li H, Ye J, McInerney JG (1993) Detailed analysis of coherence collapse in semiconductor lasers. IEEE J Quantum Electron 29:2421–2432
247. Li L (1994a) Static and dynamic properties of injection-locked semiconductor lasers. IEEE J Quantum Electron 30:1701–1708
248. Li L (1994b) A unified description of semiconductor lasers with external light injection and its application to optical bistability. IEEE J Quantum Electron 30:1723–1731
249. Li H, Lucas TL, McInerney JG, Wright MW, Morgan RA (1996) Injection locking dynamics of vertical cavity semiconductor lasers using conventional and phase conjugate injection. IEEE J Quantum Electron 32:22–235
250. Li H, Iga K (2002) Vertical-cavity surface-emitting laser devices. Springer-Verlag, Berlin
251. Liby BW, Statman D (1996) Controlling the linewidth of a semiconductor laser with photorefractive phase conjugate feedback. IEEE J Quantum Electron 32:835–838

252. Lin C, Burrus Jr. CA, Coldren LA (1984) Characteristics of single-longitudinal mode selection in short-coupled-cavity (SCC) injection lasers. J Lightwave Technol 2:544–549

253. Lin FY, Liu JM (2003a) Nonlinear dynamics of a semiconductor laser with delayed negative optoelectronic feedback. IEEE J Quantum Electron 39:562–568

254. Lin FY, Liu JM (2003b) Nonlinear dynamical characteristics of an optically injected semiconductor laser subject to optoelectronic feedback. Opt Commun 221:173–180

255. Linke AR(1985) Modulation induced transient chirping in single frequency lasers. IEEE J Quantum Electron 21:593–597

256. Liu Y, Ohtsubo J (1991) Observation of higher-harmonic bifurcations in a chaotic system using a laser diode active interferometer . Opt Commun 85:457–461

257. Liu Y, Ohtsubo J (1992a) Chaos in an active interferometer. J Opt Soc Am B 9:261–265

258. Liu Y, Ohtsubo J (1992b) Period three-cycle in a chaotic system using a laser diode active interferometer. Opt Commun 93:311–317

259. Liu HF, Ngai WF (1993) Nonlinear dynamics of a directly modulated 1.55 μm InGaAsP distributed feedback semiconductor laser. IEEE J Quantum Electron 29:1668–1675

260. Liu Y, Ohtsubo J (1993) Regeneration spiking oscillation in semiconductor laser with a nonlinear delayed feedback. Phys Rev A 47:4392–4399

261. Liu Y, Ohtsubo J (1994a) Experimental control of chaos in a laser-diode interferometer with delayed feedback. Opt Lett 19:448–450

262. Liu Y, Ohtsubo J (1994b) Controlling chaos of a delayed optical bistable system. Opt Rev 1:91–93

263. Liu Y, Ohtsubo J, Shoji Y (1994) Accessing of high mode oscillations in a delayed optical bistable system. Opt Commun 105:193–198

264. Liu Y (1994) Study of chaos in a delay-differential system with a laser diode active interferometer. PhD Thesis, Shizuoka University

265. Liu JM, Simpson TB (1994) Four-wave mixing and optical modulation in a semiconductor laser. IEEE J Quantum Electron 30:957–966

266. Liu Y, Kikuchi N, Ohtsubo J (1995) Controlling of dynamical behavior of semiconductor lasers with external optical feedback. Phys Rev E 51: R2697–2700

267. Liu Y, Ohtusbo J, Ye SY (1996) Dynamical and coherent characteristics of semiconductor lasers with external optical feedback. SPIE Proc 2886:120–127

268. Liu Y, Ohtsubo J (1997) Dynamics and chaos stabilization of semiconductor lasers with optical feedback from an interferometer. IEEE J Quantum Electron 33:1163–1169

269. Liu JM, Chen HF, Meng XJ, Simpson TB (1997) Modulation bandwidth noise, and stability of semiconductor laser subjected to strong injection locking. IEEE Photon Technol Lett 9:1325–1327

270. Liu Y, Davis P (2000) Dual synchronization of chaos. Phys Rev E 61: R2176–2179

271. Liu Y, Davis P (2001) Synchronized chaotic mode hopping in DBR lasers with delayed opto-electric feedback. IEEE J Quantum Electron 37:337–352

272. Liu JM, Chen HF, Tang S (2001a) Optical-communication systems based on chaos in semiconductor lasers. IEEE Trans Circuits Syst I 48:1475–1483

273. Liu JM, Chen HF, Tang S (2001b) Synchronization of chaos in semiconductor lasers. Nonlin Anal 47:5741–5751

274. Liu Y, Chen HF, Liu JM, Davis P, Aida T (2001c) Communication using synchronization of optical-feedback-induced chaos in semiconductor lasers. IEEE Tans Circuits Syst I 48:1484–1490

275. Liu Y, Takiguchi Y, Davis P, Aida T, Saito S, Liu JM (2002a) Experimental observation of complete chaos synchronization in semiconductor lasers. Appl Phys Lett 80:4306–4308

276. Liu JM, Chen HF, Tang S (2002b) Synchronized chaotic optical communications at high bit rates. I EEE J Quantum Electron 38:1184–1196

277. Liu H, Yan M, Shum P, Ghafouri-Shiraz H, Liu D (2004) Design and analysis of anti-resonant reflecting photonic crystal VCSEL lasers. Opt Exp 18:4269–4274

278. Loiko NA, Samson AM (1992) Possible regimes of generation of a semiconductor laser with a delayed optoelectronic feedback. Opt Commun 93:66–72

279. Lorenz EN (1963) Deterministic nonperiodic flow. J Atmos Sci 20:130–148

280. Lucero AJ, Tkach RW, Derosier RM (1988) Distortion of the frequency modulation spectra of semiconductor lasers by weak optical feedback. Electron Lett 24:337–339

281. Luo LG, Chu PL, Liu HF (2000) 1-GHz Optical communication system using chaos in erbium-doped fiber lasers. IEEE Photon Tech Lett 12:269–271

282. Mandel P, Zeghlache H (1983) Stability of a detuned single mode homogeneously broadened ring laser. Opt Commun 47:146–150

283. Mandre SK, Fischer I, Elsäßer W (2003) Control of the spatiotemporal emission of a broad-area semiconductor laser by spatially filtered feedback. Opt Lett 28:1135–1137

284. Mandre SK, Fischer I, Elsäßer W (2005) Spatiotemporal emission dynamics of a broad-area semiconductor laser in an external cavity: stabilization and feedback-induced instabilities. Opt Commun 244:355–365

285. Marciante JR, Agrawal GP (1998) Spatio-temporal characteristics of filamentation in broad-area semiconductor lasers: experimental results. IEEE Photon Technol Lett 10:54–56

286. Marcuse D, Lee TP (1983) On approximate analytical solutions of rate equations for studying transient spectra of injection lasers. IEEE J Quantum Electron 19:1397–1406

287. Marino F, Barland S, Balle S (2003) Single-mode operation and transverse-mode control in VCSELs induced by frequency-selective feedback. IEEE Photon Technol Lett 15:789–791

288. Martín-Regalado J, van Tartwijk GHM, Balle S, San Miguel M (1996a) Mode control and pattern stabilization in broad-area lasers by optical feedback. Phys Rev A 54:5386–5393

289. Martín-Regaldo J, Balle S, Abraham NB (1996b) Modeling spatio-temporal dynamics of gain-guided multistripe and broad-area lasers. IEE Proc Opto Electron 143:17–23

290. Martín-Regalado, Prati JF, San Miguel M, Abraham NB (1997) Polarization properties of vertical-cavity surface-emitting lasers. IEEE J Quantum Electron 33:765–783

291. Masoller C, Abraham NB (1998) Stability and dynamical properties of the coexisting attractors of an external-cavity semiconductor laser. Phys Rev A 57:1313–1322

456   References

292. Masoller C, Abraham NB (1999a) Low-frequency fluctuations in vertical-cavity surface-emitting semiconductor lasers with optical feedback. Phys Rev A 59:3021–3031
293. Masoller C, Abraham NB (1999b) Polarization dynamics in vertical-cavity surface-emitting lasers with optical feedback through a quarter-wave plate Appl Phys Lett 74:1078–1080
294. Masoller C (2001) Anticipation in the synchronization of chaotic semiconductor lasers with optical feedback. Phys Rev Lett 86:2782–2785
295. Masoller C, Torre MS, Mandel P (2006) Influence of the injection current sweep rate on the polarization switching of vertical-cavity surface-emitting lasers. J Appl Phys 99:026108-1-3
296. McCall SL, Platzman PM (1985) An optimized $\pi/2$ distributed feedback laser. IEEE J Quantum Electron 21:1899–1894
297. McCmber DE (1966) Intensity fluctuations in the output of cw laser oscillators I. Phys Rev 141:306–322
298. Merbach D, Hess O, Herzel H, Schöll E (1995) Injection induced bifurcations of transverse spatiotemporal patterns in semiconductor arrays. Phys Rev E 52:1571–1578
299. Merlo S, Donati S (1997) Reconstruction of displacement waveforms with a single-channel laser-diode feedback interferometer. IEEE J Quantum Electron 33:527–531
300. Milloni PW, Eberly JH (1988) Laser. Wiley, New York
301. Mils RO, Dandrigdge A, Tveten B, Taylor HF, Tgiallorenzi TG (1980) Feedback induced line broadening in cw channel-substrate planar laser diodes. Appl Phys Lett 37:990–992
302. Miltyeni E, Ziegler MO, Hofmann M, Sacher J, Elsäßer W, Göbel EO, MacFarlane DL (1995) Long-term stable mode locking of a visible diode laser with phase-conjugate feedback. Opt Lett 20:734–736
303. Mirasso CR, Colet P, Garcia-Fernandez P (1996) Synchronization of chaotic semiconductor lasers: application to encoded communications. IEEE Photon Tech Lett 8:299–301
304. Mirasso CR, van Tartwijk GHM, Hernández-García E, Lenstra D, Lynch S, Landais P, Phelan P, O'Gorman J, San Miguel M, Elsäßer W (1999) Self-pulsating semiconductor lasers: theory and experiment. IEEE J Quantum Electron 35:764–770
305. Mirasso CR, Kolesik M, Matus M, White JK, Moloney JV (2002a) Synchronization and multimode dynamics of mutually coupled semiconductor lasers. Phys Rev A 65:013805-1-4
306. Mirasso CR, Mulet J, Masoller C (2002b) Chaos shift-keying encryption in chaotic external-cavity semiconductor lasers using a single-receiver scheme. IEEE Photon Technol Lett 14:456–458
307. Mogensen F, Olesen H, Jacobsen G (1985) Locking conditions and stability properties for a semiconductor laser with external light injection. IEEE J Quantum Electron 21:784–793
308. Mørk J, Tromborg B, Christiansen PL (1988) Bistability and low-frequency fluctuations in semiconductor lasers with optical feedback: a theoretical analysis. IEEE J Quantum Electron 24:123–133
309. Mørk J, Mark J, Tromborg B (1990a) Route to chaos and competition between relaxation oscillations for a semiconductor laser with optical feedback. Phys Rev Lett 65:1999–2002

310. Mørk J, Semkow M, Tromborg B (1990b) Measurement and theory of mode hopping in external cavity laser. Electron Lett 26:284–285

311. Mørk J, Tromborg B (1990) The mechanism of mode selection for an external cavity laser. IEEE Photon Technol Lett 2:21–23

312. Mørk J, Tromborg B, Mark J (1992) Chaos in semiconductor lasers with optical feedback: theory and experiment. IEEE J Quantum Electron 28:93–108

313. Mørk J, Sabbatier H, Sørensen MP, Tromborg B (1999) Return-map for low-frequency fluctuations in semiconductor lasers with optical feedback. Opt Commun 171:93–97

314. Mourat G, Servagent N, Bosch T (2000) Distance measurements using the self-mixing effect in a 3-electrode DBR laser diode. Opt Eng 39:738–743

315. Mulet J, Balle S (2002) Spatio-temporal modeling of the optical properties of VCSELs in the presence of polarization effects. IEEE J Quantum Electron 38:291–305

316. Mulet J, Masoller C, Mirasso CR (2002) Modeling bidirectionally coupled single-mode semiconductor lasers. Phys Rev A 65:063815–1–12

317. Münkel M, Kaiser F, Hess O (1996) Spatio-temporal dynamics of multi-stripe semiconductor lasers with delayed optical feedback. Phys Lett A 222:67–75

318. Münkel M, Kaiser F, Hess O (1997) Stabilization of spatiotemporally chaotic semiconductor laser arrays by means of delayed optical feedback. Phys Rev E 56:3868–3875

319. Murakami A, Ohtsubo J, Liu Y (1997) Stability analysis of semiconductor laser with phase-conjugate feedback. IEEE J Quantum Electron 33:1825–1831

320. Murakami A, Ohtsubo J (1998) Dynamics and linear stability analysis in semiconductor lasers with phase-conjugate feedback. IEEE J Quantum Electron 34:1979–1986

321. Murakami A, Ohtsubo J (1999) Dynamics of semiconductor lasers with optical feedback from photorefractive phase conjugate mirror. Opt Rev 6:350–364

322. Murakami A (1999) Dynamics of semiconductor lasers with phase-conjugate optical feedback. PhD Thesis, Shizuoka University

323. Murakami A, Ohtsubo J (2002) Synchronization of feedback-induced chaos in semiconductor lasers by optical injection. Phys Rev A 65:033826–1–7

324. Murakami A, Kawashima K, Atsuki K (2003) Cavity resonance shift and bandwidth enhancement in semiconductor lasers with strong light injection. IEEE J Quantum Electron 39:1196–1194

325. Murakami A, Shore KA (2005) Chaos-pass filtering in injection-locked semiconductor lasers. Phys Rev A 72:053810–1–8

326. Murakami A, Shore KA (2006) Analogy between optically driven injection-locked laser diodes and driven damped linear oscillators. Phys Rev A 73:043804–1–9

327. Nakamura M, Aiki K, Chinone N, Ito R, Umeda U (1978) Longitudinal-mode behaviors of mode stabilized $Al_xGa_{1-x}As$ injection lasers. J Appl Phys 49:4644–4548

328. Naumenko AV, Loiko NA, Turovets SI, Spencer PS, Shore KA (1998) Controlling dynamics in external-cavity laser diodes with electronic impulsive delayed feedback. J Opt Soc Am B 15:551–561

329. Naumenko A, Besnard P, Loiko N, Ughetto G, Bertreux JC (2003a) Characteristics of a semiconductor laser coupled with a fiber Bragg grating with arbitrary amount of feedback. IEEE J Quantum Electron 39:1216–1228

458   References

330. Naumenko AV, Loiko NA, Sondermann M, Ackemann T (2003b) Description and analysis of low-frequency fluctuations in vertical-cavity surface-emitting lasers with isotropic optical feedback by a distant reflector. Phys Rev A 68:033805-1–16

331. Ning C, Haken H (1990) Detuned lasers and the complex Lorenz equations: subcritical and supercritical Hopf bifurcations. Phys Rev A 41:3826–3837

332. O'Gorman J, Hawdon BJ, Hegarty J (1989) Frequency locking and quasiperiodicity in a modulated external cavity injection lasers. J Appl Phys 66:57–60

333. Ohtsu M, Kotajima S (1985) Linewidth reduction of a semiconductor laser by electrical feedback. IEEE J Quantum Electron 21:1905–1912

334. Ohtsu M, Murata M, Kourogi M (1990) FM noise reduction and subkilohertz linewidth of an AlGaAs laser by negative electronic feedback. IEEE J Quantum Electron 26:231–241

335. Ohtsu M (1996) Frequency control of semiconductor lasers. Wiley-Interscience, New York

336. Ohtsubo J, Liu Y (1990) Optical bistability and multistability in active interferometer. Opt Lett 15:731–733

337. Ohtsubo J (1999) Feedback induced instability and chaos in semiconductor lasers and their applications. Opt Rev 6:1–15

338. Ohtsubo J (2002a) Chaotic dynamics in semiconductor lasers with optical feedback. In: Wolf E (ed) Progress in optics, chap.1, vol. 44. North-Holland, Amsterdam

339. Ohtsubo J (2002b) Chaos synchronization and chaotic signal masking in semiconductor lasers with optical feedback. IEEE J Quantum Electron 38:1141–1154

340. Ohtsubo J, Davis P (2005) Chaotic optical communication. In: Kane D, Shore KA (eds) Unlocking dynamical diversity optical feedback effects on semiconductor lasers, chap. 10. Wiley, Chichester

341. Okoshi T, Kikuchi K, Nakayama A (1980) Novel method for high resolution measurement of laser output spectrum. Electron Lett 16:630–631

342. Olesen H, Saito S, Mukai T, Saitoh T, Mikami O (1983) Solitary spectral linewidth and its reduction with external grating feedback from a 1.55 mm InGaAsP BH laser. Jpn J Appl Phys 22: L664-L666

343. Olesen H, Osmundsen JH, Tromborg B (1986) Nonlinear dynamics and spectral behavior for an external cavity laser. IEEE J Quantum Electron 22:726–773

344. Osinski M , Buss J (1987) Linewidth broadening factor in semiconductor lasers – an overview. IEEE J Quantum Electron 23:9–29

345. Osmundsen JH, Gade N (1983) Influence of optical feedback on laser frequency spectrum and threshold conditions. IEEE J Quantum Electron 19:465–469

346. Otsuka K, Chern JL (1991) High-speed picosecond pulse generation in semiconductor lasers with incoherent optical feedback. Opt Lett 16:1759–1761

347. Otsuka K (1999) Nonlinear dynamics in optical complex systems. KTK Scientific Publishers, Tokyo

348. Ott E, Grebogi C, Yorke JA (1990) Controlling chaos. Phys Rev Lett 64:1196–1199

349. Özdemir SK, Shinohara S, Takamiya S, Yoshida H (2000) Noninvasive blood flow measurement using speckle signals from a self-mixing laser diode: in vitro and in vivo experiments. Opt Eng 39:2574–2580

350. Pan MW, Shi BP, Gray GR (1997) Semiconductor laser dynamics subject to strong optical feedback. Opt Lett 22:166–168

351. Paoli TL, Ripper JE (1970) Direct modulation of semiconductor lasers. Proc IEEE 58:1457–1465

352. Paoli TL (1981) Optical response of a stripe-geometry junction laser to sinusoidal current modulation at 1.2 GHz. IEEE J Quantum Electron 17:675–680

353. Papoulis A (1984) Probability, random variables, and stochastic processes. McGraw-Hill, New York

354. Patzak E, Sugimura A, Satio S, Mukai T, Olesen H (1983) Semiconductor laser linewidth in optical feedback configurations. Electron Lett 19:1026–1027

355. Paul J, Lee MW, Shore KA (2004) Effects of chaos pass filtering on message decoding quality using chaotic external-cavity laser diodes. Opt Lett 29:2497–2499

356. Paulus P, Langenhorst R, Jäger D (1987) Stable pulsations of semiconductor lasers by optoelectronic feedback with avalanche photodiodes. Electron Lett 23:471–472

357. Pecora LM, Carroll TL (1990) Synchronization in chaotic systems. Phys Rev Lett 64:821–824

358. Pecora LM, Carroll TL (1991a) Driving systems with chaotic signals. Phys Rev Lett 44:2374–2384

359. Carroll TL, Pecora ML (1991b) Synchronizing in chaotic systems. IEEE Trans Circuits Syst. 38:453–456

360. Petermann K (1979) Calculated spontaneous emission factor for double-heterostructure injection lasers with gain-induced waveguiding. IEEE J Quantum Electron 15:566–570

361. Petermann K, Arnold G (1982) Noise and distortion characteristics of semiconductor lasers in optical fiber communication systems. IEEE J Quantum Electron 18:543–555

362. Petermann K (1988) Laser diode modulation and noise. Kluwer Academic, Dordrecht

363. Peters-Flynn S, Spencer PS, Sivaprakasam S, Pierce I, Shore KA (2006) Identification of the optimum time-delay for chaos synchronization regimes of semiconductor lasers. IEEE J Quantum Electron 42:427–434

364. Phillips MW, Gong H, Ferguson AI, Hanna DC (1987) Optical chaos and hysteresis in a laser-diode pumped Nd-doped fibre laser. Opt Commun 61:215–218

365. Piazzolia S, Spano P, Tamburrini T, (1986) Small signal analysis of frequency chirping in injection-locked semiconductor lasers. IEEE J Quantum Electron 22:2219–2223

366. Pieroux D, Erneux T, Otsuka K (1994) Minimal model of class-B laser with delayed feedback: cascading branching of periodic solutions and period-doubling bifurcation. Phys Rev A 50:1822–1829

367. Pittoni F, Gioannini M, Montrosset I (2001) Time-domain analysis of fiber grating semiconductor laser operation in active mode-locking regime. IEEE J Select Topics Quantum Electron 7:280–286

368. Pochi Y (1993) Introduction to photorefractive nonlinear optics. Wiley, New York

369. Poincaré H (1913) The foundation of science: science and method. Science Press, Lancaster

370. Press WH, Flannery BP, Teukolsky SA, Vetterling WT (1986) Numerical recipes: the art of scientific computing. Cambridge University Press, Cambridge
371. Pyragas K (1992) Continuous control of chaos by self-controlling feedback. Phys Lett A 170:421–428
372. Pyragas K (1993) Predictable chaos in slightly perturbed unpredictable chaotic systems. Phys Lett A 181:203–210
373. Pyragas K (2001) Control of chaos via an unstable delayed feedback controller. Phys Rev Lett 86:2265–2268
374. Raab V, Menzel R (2002) External resonator design for high-power laser diodes that yields 400 mW of TEM_{00} power. Opt Lett 27:167–169
375. Rahman L, Winful H (1994) Nonlinear dynamics of semiconductor arrays: a mean field model. IEEE J Quantum Electron 30:1405–1406
376. Ribbat Ch, Sellin RL, Kaiander I, Hopfer F, Ledentsov NN, Bimberg D, Kovsh AR, Ustinov VM, Zhukov AE, Maximov MV (2003) Complete suppression of filamentation and superior beam quality in quantum-dot lasers. Appl Phys Lett 82:952–954
377. Risch C, Voumard C (1977) Self-pulsation in the output intensity and spectrum of GaAs-AlGaAs cw diode lasers coupled to a frequency-selective external optical cavity. J Appl Phys 48:2083–2085
378. Risken H (1996) The Fokker-Planck equation: methods of solution and applications, 2nd edn. Springer-Verlag, Berlin
379. Ritter A, Haug H (1993a) Theory of laser diodes with weak optical feedback: 1. small-signal analysis and side-mode spectra. J Opt Soc Am B 10:130–114
380. Ritter A, Haug H (1993b) Theory of laser diodes with weak optical feedback: 2. limit-cycle behavior, quasi-periodicity, frequency locking, and route to chaos. J Opt Soc Am B 10:145–154
381. Rosigter F, Mégret P, Deparis O, Blondel M, Erneux T (1999) Suppression of low-frequency fluctuations and stabilization of a semiconductor laser subjected to optical feedback from a double cavity: theoretical results. Opt Lett 24:1218–1220
382. Rogister F, Sukow DW, Gavrielides A, Mégret P, Deparis O, Blondel M (2000) Experimental demonstration of suppression of low-frequency fluctuations and stabilization of an external-cavity laser diode. Opt Lett 25:808–810
383. Rogister F, Locquet A, Pieroux D, Sciamanna M, Deparis O, Megret P, Blondel M (2001) Secure communication scheme using chaotic laser diodes subject to incoherent optical feedback and incoherent optical injection. Opt Lett 26:1466–1469
384. Romanelli M, Hermier JP, Giacobino E, Bramati A (2005) Demonstration of single-mode operation of a vertical-cavity surface-emitting laser with optical feedback: the intensity-noise-measurement approach. J Opt Soc Am B 22:2596–2600
385. Roy R, Murphy TW, Maier TD, Gills Z, Hunt ER (1992) Dynamical control of a chaotic laser: Experimental stabilization of a globally coupled system. Phys Rev Lett 68:1259–1262
386. Roy R, Thornburg Jr. KS (1994) Experimental synchronization of chaotic lasers. Phys Rev Lett 72:2009–2015
387. Ruiz-Oliveras FR, Pisarchik AN (2006) Phase-locking phenomenon in a semiconductor laser with external cavities. Opt Express 14:12859–12866

388. Ryan A, Agrawal GP, Gray GR, Gage EC (1994) Optical feedback-induced chaos and its control in multimode semiconductor lasers. IEEE J Quantum Electron 30:668–679

389. Ryvkinn BS, Panajotov K, Avrutin EA, Veretennicoff I, Thienpont H (2004) Optical-injection-induced polarization switching in polarization-bistable vertical-cavity surface-emitting lasers. J Appl Phys 96:6002–6007

390. Sacher J, Elsäßer W, Göbel EO (1989) Intermittency in the coherence collapse of a semiconductor laser with external feedback. Phys Rev Lett 63:2224–2227

391. Sacher J, Baums D, Panknin P, Elsäßer W, Göbel EO (1992) Intensity instabilities of semiconductor lasers under current modulation, external light injection, and delayed feedback. Phys Rev A 45:1893–1905

392. Saleh B (1978) Photoelectron statistics. Springer-Verlag, Berlin

393. San Miguel M, Feng Q, Moloney JV (1995) Light polarization dynamics in surface-emitting semiconductor lasers. Phys Rev A 52:1729–1740

394. Sánches-Díaz A, Mirasso CR, Colt P, García-Fernández P (1999) Encoded Gbit/s digital communications with synchronized chaotic semiconductor lasers. IEEE J Quantum Electron 35:292–296

395. Sano T (1994) Antimode dynamics and chaotic itinerancy in the coherence collapse of semiconductor lasers with optical feedback. Phys Rev A 50:2719–2726

396. Sauer M, Kaiser F (1998) On-off intermittency and bubbling in the synchronization break-down of coupled lasers. Phys Lett A 243:38–46

397. Scalise L (2002) Self-mixing feedback laser Doppler vibrometry. SPIE Proc 4827:374–384

398. Schremer A, Fujita T, Lin CF, Tang CL (1988) Instability threshold resonances in directly modulated external-cavity semiconductor lasers. Appl Phys Lett 52:263–265

399. Schunk N, Petermann K (1988) Numerical analysis of the feedback regimes for a single-mode semiconductor laser with external feedback. IEEE J Quantum Electron 24:1242–1247

400. Schunk N, Petermann K (1989) Stability analysis for laser diodes with short external cavities. IEEE Photon Technol Lett 1:49–51

401. Sciamanna M, Erneux T, Rogister F, Deparis O, Mégret P, Blondel M (2002a) Bifurcation bridges between external-cavity modes lead to polarization self-modulation in vertical-cavity surface-emitting lasers. Phys Rev A 65:041801-1–4

402. Sciamanna M, Rogister F, Deparis O, Mégret P, Blondel M, Erneux T (2002b) Bifurcation to polarization self-modulation in vertical-cavity surface-emitting lasers. Opt Lett 27:261–263

403. Sciamanna M, Masoller C, Abraham NB, Rogister F, Mégret P, Blondel M (2003a) Different regimes of low-frequency fluctuations in vertical-cavity surface-emitting lasers. J Opt Soc Am B 20:37–39

404. Sciamanna M, Masoller C, Rogister F, Mégret P, Abraham NB, Blondel M (2003b) Fast pulsing dynamics of a vertical-cavity surface-emitting laser operating in the low-frequency fluctuation regime. Phys Rev A:015805-14

405. Sciamanna M, Valle A, Mégret P, Blondel M, Panajotov K (2003c) Nonlinear polarization dynamics in directly modulated vertical-cavity surface-emitting lasers. Phys Rev E 68:016207-1–4

406. Sciamanna M, Panajotov K (2005) Two-mode injection locking in vertical-cavity surface-emitting lasers. Opt Lett 30:2903–2905

407. Sciamanna M, Panajotov K (2006) Route to polarization switching induced by optical injection in vertical-cavity surface-emitting lasers. Phys Rev A 73:023811–1–17

408. Servagent N, Gouaux F, Bosch T (1998) Measurements of displacement using the self-mixing interference in a laser diode. J Optics 29:168–173

409. Shibasaki N, Uchida A, Yoshimori Y, Davis P (2006) Characteristics of chaos synchronization in semiconductor lasers subject to polarization-rotated optical feedback. IEEE J Quantum Electron 42:342–350

410. Shinohara S, Naito H, Yoshida H, Ikeda H, Sumi M (1989) Compact and versatile self-mixing type semiconductor laser doppler velocimeters with direction discrimination circuit. IEEE Trans Instrum Meas 38:674–577

411. Short KM (1994) Step toward unmasking secure communications. Int J Bifurcation Chaos 4:959–979

412. Short KM (1996) Unmasking a modulated chaotic communication scheme. Int J Bifurcation Chaos 6:367–375

413. Siemsen D (1978) Observation of inherent oscillations and subharmonic resonances in the light output GaAs DH lasers. Int J Electron 45:63–67

414. Simmendinger C, Preißer D, Hess O (1999a) Stabilization of chaotic spatiotemporal filamentation in large broad area lasers by spatially structured optical feedback. Opt Exp 5:48–54

415. Simmendinger C, Münkel MM, Hess O (1999b) Controlling complex temporal and spatio-temporal dynamics in semiconductor lasers. Chaos, Solitons Fractals 10:851–864

416. Simpson TB, Liu JM, Gavrielides A, Kovanis V, Alsing PM (1994) Period-doubling route to chaos in a semiconductor laser subject to optical injection. Appl Phys Lett 64:3539–3541

417. Simpson TB, Liu JM, Gavrielides A, Kovanis V, Alsing PM (1995a) Period-doubling cascades and chaos in a semiconductor laser with optical injection. Phys Rev A 51:418–4185

418. Simpson TB, Liu JM, Gavrielides A (1995b) Bandwidth enhancement and broadband noise reduction in injection-locked semiconductor lasers. IEEE Photon Technol Lett 7:709–911

419. Simpson TB, Liu JM, Gaverielides A (1996) Small-signal analysis of modulation characteristics in semiconductor laser subject to strong optical injection. IEEE J Quantum Electron 32:1456–1468

420. Simpson TB, Liu JM, Huang KF, Tai K (1997) Nonlinear dynamics induced by external optical injection in semiconductor lasers. Quantum Semiclass Opt 9:765–784

421. Simpson TB, Liu JM (1997) Enhanced modulation bandwidth in injection-locked semiconductor lasers. IEEE Photon Technol Lett 9:1322–1324

422. Simpson TB, Doft F, Strzelecka E, Liu JJ, Member, Chang W, Simonis GJ (2001) Gain saturation and the linewidth enhancement factor in semiconductor lasers. IEEE Photon Technol Lett 13:776–778

423. Simpson TB (2003) Mapping the nonlinear dynamics of a distributed feedback semiconductor laser subjected to external optical injection. Opt Commun 215:135–151

424. Sivaprakasam S, Shore KA (1999) Signal masking for chaotic optical communication using external-cavity diode lasers. Opt Lett 24:1200–1202

425. Sivaprakasam S, Shahverdiev EM, Shore KA (2000a) Experimental verification of the synchronization condition for chaotic external cavity diode lasers. Phys Rev E 62:7505–7507

426. Sivaprakasam S, Shore KA (2000b) Critical signal strength for effective decoding in diode laser chaotic optical communications. Phys Rev E 61:5997–5999

427. Sivaprakasam S, Shore KA (2000c) Message encoding and decoding using chaotic external-cavity diode lasers. IEEE J Quantum Electron 36:35–39

428. Sivaprakasam S, Shahverdiev EM, Spencer PS, Shore KA (2001) Experimental demonstration of anticipating synchronization in chaotic semiconductor lasers with optical feedback. Phys Rev Lett 87:154101–1–3

429. Smowton PM, Pearce EJ, Schneider HC, Chow WW, Hopkinson M (2002) Filamentation and linewidth enhancement factor in InGaAs quantum dot lasers. Appl Phys Lett 81:3251–3253

430. Sondermann M, Bohnet H, Ackemann T (2003) Low-frequency fluctuations and polarization dynamics in vertical-cavity surface-emitting lasers with isotropic feedback. Phys Rev A 67:021802–1–4

431. Special issue, Instability in active optical media (1985) J Opt Soc Am B 2 (1)

432. Special issue on applications of chaos in modern communication systems (2001) IEEE Tans. Circuits Syst I 48 (12)

433. Spencer PS, Mirasso CR, Colet P, Shore KA (1998) Modeling of optical synchronization of chaotic external-cavity VCSEL's . IEEE J Quantum Electron 34:1673–1679

434. Spencer PS, Kane DM, Shore KA (1999) Coupled-cavity effects in FM semiconductor lasers. J Lightwave Technol 17:1072–1078

435. Spencer PS, Mirassso CR (1999) Analysis of optical chaos synchronization in frequency-detuned external-cavity VCSELs. IEEE J Quantum Electron 35:803–809

436. Stoehr H, Mensing F, Helmcke J, Sterr (2006) Diode laser with 1 Hz linewidth. Opt Lett 31:736–738

437. Sugawara T, Tachikawa M, Tsukamoto T, Shimizu T (1994) Observation of synchronization in laser chaos. Phys Rev Lett 72:3502–3506

438. Sukow DW, Heil T, Fischer I, Gavrielides A, Hohl-AbiChedid A, Elsäßer W (1999) Picosecond intensity statistics of semiconductor lasers operating in the low-frequency fluctuation regime. Phys Rev A 60:667–673

439. Sukow DW, Blackburn KL, Spain AR, Babcock KJ, Bennett JV, Gavrielides A (2004) Experimental synchronization of chaos in diode lasers with polarization-rotated feedback and injection. Opt Lett 29:2393–2395

440. Sukow DW, Gavrielides A, Erneux T, Baracco MJ, Parmenter ZA, Blackburn KL (2005) Two-field description of chaos synchronization in diode lasers with incoherent optical feedback and injection. Phys Rev A 72:043818–1–6

441. Sukow DW, Gavrielides A, McLachlan T, Burner G, Amonette J, Miller J (2006) Identity synchronization in diode lasers with unidirectional feedback and injection of rotated optical fields. Phys Rev 74:023812–1–8

442. Tabaka A, Panajotov K, Veretennicoff I, Sciamanna M (2004) Bifurcation study of regular pulse packages in laser diodes subject to optical feedback. Phys Rev E 70:036211–1–9

443. Tabaka A, Peil M, Sciamanna M, Fischer I, Elsasser W, Thienpont H, Veretennicoff I, Panajotov K (2006) Dynamics of vertical-cavity surface-emitting lasers in the short external cavity regime: pulse packages and polarization mode competition. Phys Rev A 73:013810–1–14

444. Takiguchi Y, Liu Y, Ohtsubo J (1998) Low-frequency fluctuation induced by injection-current modulation in semiconductor lasers with optical feedback. Opt Lett 23:1369–1371

445. Takiguchi Y, Liu Y, Ohtsubo J (1999a) Low-frequency fluctuations and frequency-locking in semiconductor lasers with optical feedback. Opt Rev 6:339–401

446. Takiguchi Y, Liu Y, Ohtsubo J (1999b) Low frequency fluctuation in semiconductor lasers with long external cavity feedback and its control. Opt Rev 6:424–432

447. Takiguchi Y, Fujino H, J Ohtsubo J (1999c) Experimental synchronization of chaotic oscillations in external cavity semiconductor lasers in low-frequency fluctuation regime. Opt Lett 24:1570–1572

448. Takiguchi Y (2002) Chaotic oscillations in semiconductor lasers with optical feedback: control and applications. PhD Thesis, Shizuoka University

449. Takiguchi Y, Kan H, Ohtsubo J (2002) Modulation induced low-frequency fluctuations in semiconductor lasers with optical feedback and their suppression by synchronous modulation. Opt Rev 9:234–237

450. Takiguchi Y, Ohyagi K, Ohtsubo J (2003) Bandwidth-enhanced chaos synchronization in strongly injection-locked semiconductor lasers with optical feedback. Opt Lett 28:319–321

451. Takiguchi Y, Asatsuma T, Hrata S (2006) Effect of the threshold reduction on a catastrophic optical mirror damage in broad-area semiconductor lasers with optical feedback. SPIE Proc 6104:61040X

452. Tamburrini M, Spano P, Piazolla S (1983) Influence of an external cavity on semiconductor laser phase noise. Appl Phys Lett 43:410–412

453. Tang S, Liu JM (2001a) Message encoding/decoding at 2.5 Gbits/s through synchronization of chaotic pulsing semiconductor lasers. Opt Lett 26:1843–1845

454. Tang S, Liu JM (2001b) Synchronization of high-frequency chaotic optical pulses. Opt Lett 26:596–598

455. Tang S, Liu JM (2001c) Chaotic pulsing and quasi-periodic route to chaos in a semiconductor laser with delayed opto-electronic feedback. IEEE J Quantum Electron 37:329–336

456. Tang S, Chen HF, Liu JM (2001) Stable route-tracking synchronization between two chaotically pulsing semiconductor lasers. Opt Lett 26:1489–1491

457. Tang S, Liu JM (2003) Chaos synchronization in semiconductor lasers with optoelectronic feedback. IEEE J Quantum Electron 39:708–715

458. Tang S, Vicente R, Chiang MC, Mirasso CR, Liu JM (2004) Nonlinear dynamics of semiconductor lasers with mutual optoelectronic coupling. IEEE J Select Topics Quantum Electron 10:936–943

459. Tang X, van der Ziel JP, Chang B, Johnson R, Tatum JA (1997) Observation of bistability in GaAs quantum-well vertical-cavity surface-emitting lasers. IEEE J Quantum Electron 33:927–932

460. Temkin H, Olsson NA, Abeles TH, Logan RA, Panish MB (1986) Reflection noise in index-guided InGaAsP lasers. IEEEE J Quantum Electron QE-22:286–293

461. Thompson GHB (1980) Physics of semiconductor laser devices. Wiley, Chichester

462. Tkach RW, Chraplyvy AR, (1985) Line broadening and mode splitting due to weak feedback in single frequency 1.5 μ m lasers. Electron Lett 21:1081–1083

463. Tkach RW and Chraplyvy AR (1986) Regimes of feedback effects in 1.5 μm distributed feedback lasers. J Lightwave Technol 4:1655–1661

464. Tredicce JR, Arecchi FT, Lippi GL, Puccioni P (1985) Instabilities in lasers with an injected signal. J Opt Soc Am B2:173–183

465. Tromborg B, Osmundsen JH, Olesen H (1984) Stability analysis for a semiconductor laser in an external cavity. IEEE J Quantum Electron 20:1023–1032

466. Tromborg B, Olssen H, Pan X, Saito S (1987) Transmission line description of optical feedback and injection locking for Fabry-Perot and DFB lasers. IEEE J Quantum Electron 23:1875–1889

467. Tromborg B, Mørk J, Valichansky V (1997) On mode coupling and low-frequency fluctuations in external-cavity laser diodes. J Opt B 9:831–851

468. Tronciu VZ, Yamada M, Ohno T, Ito S, Kawakami T, Taneya M (2003) Self-pulsation in an InGaN laser – theory and experiment. IEEE J Quantum Electron 39:1509–1514

469. Tucker RS (1985) High-speed modulation of semiconductor lasers. J Lightwave Technol 3:1180–1192

470. Turovets SI, Dellunde J, Shore KA (1997) Nonlinear dynamics of a laser diode subjected to both optical and electronic feedback. J Opt Soc Am B 14:200–208

471. Uchida A, Sato T, Ogawa T, Kannari F (1999) Characteristics of transients among periodic attractors controlled by high-frequency injection in a chaotic laser diode. IEEE J Quantum Electron 35:1374–1371

472. Uchida A, Davis P, Itaya S (2003a) Generation of information theoretic secure keys using a chaotic semiconductor laser. Appl Phys Lett 83:3213–3215

473. Uchida A, Liu Y, Davis P (2003b) Characteristics of chaotic masking in synchronized semiconductor lasers. IEEE J Quantum Electron 39:963–970

474. Uomi K, Chinone N, Ohtoshi T, Kajimura T(1985) High relaxation oscillation frequency (beyond 10 GHz) of GaAlAs multiquantum well lasers. Jpn J Appl Phys 24: L539–541

475. Vahala K, Yariv A (1983a) Semiclassical theory of noise in semiconductor lasers – part I. IEEE J Quantum Electron 19:1096–1101

476. Vahala K, Yariv A (1983b) Semiclassical theory of noise in semiconductor lasers – part II. IEEE J Quantum Electron 19:1102–1109

477. Vahala K, Harder C, Yariv A (1983) Observation of relaxation resonance effects in the field spectrum of semiconductor lasers. Appl Phys Lett 42:211–213

478. Vahala K, Kyuma K, Yariv A, Kwong SK, Cronin-Golomb M, Lau KY (1986) Narrow linewidth, single frequency semiconductor laser with a phase conjugate external cavity mirror. Appl Phys Lett 49:1563–1565

479. Vainio M (2006) Phase-conjugate external cavity diode laser. IEEE Photon Technol Lett 18:2047–2049

480. Valle A, Sarma J, Shore KA (1995a) Spatial holeburning effects on the dynamics of vertical-cavity surface-emitting laser diodes. IEEE J Quantum Electron 31:1423–1431

481. Valle A, Sarma J, Shore KA (1995b) Dynamics of transverse competition in vertical cavity surface emitting laser diodes. Opt Commun 115:297–302

482. van der Graaf WA, Pesquera L, Lenstra D (1998) Stability of a diode laser with phase-conjugate feedback. Opt Lett 23:256–258

483. van der Graaf WA, Pesquera L, Lenstra D (2001) Stability and noise properties of diode lasers with phase-conjugate feedback. IEEE J Quantum Electron 37:562–573

484. van der Pol B (1926) On relaxation-oscillations. Phil Mag 7:978–992
485. van der Ziel JP (1985) In: Tsang WT (ed) Semiconductors and semimetals, 22 Part B Academic, Orlando
486. van Tartwijk GHM, van der Linden HJC, Lenstra D (1992) Theory of a diode laser with phase-conjugate feedback. Opt Lett 17:1590–1592
487. van Tartwijk GHM, Lenstra D (1994) Nonlocal potential for class-B lasers with external optical feedback. Phys Rev A 50: R2837–2840
488. van Tartwijk GHM, Lenstra D (1995) Semiconductor lasers with optical injection and feedback. Quantum Semiclass Opt 7:87–148
489. van Tartwijk GHM, Levine AM, Lenstra D (1995) Sisyphus effect in semiconductor lasers with optical feedback. IEEE J Selected Topic Quantum Electron 1:466–472
490. van Tartwijk GHM, San Miguel M (1996) Optical feedback on self-pulsating semiconductor lasers. IEEE J Quantum Electron 32:1191–1202
491. van Tartwijk GHM, Agrawal GP (1998) Laser instabilities: a modern perspective. Prog. Quantum Electron 22:43–122
492. van Voorst PD, Offerhaus HL, Boller KJ (2006) Single-frequency operation of a broad-area laser diode by injection locking of a complex spatial mode via a double phase conjugate mirror. Opt Lett 31:1061–1063
493. VanWiggeren GD, Roy R (1998a) Communication with chaotic lasers. Science 279:1198–1200
494. VanWiggeren GD, Roy R (1998b) Optical communication with chaotic waveforms. Phys Rev Lett 81:3547–3550
495. Vicente R, Pérez T, Mirasso CR (2002) Open- versus closed-loop performance of synchronized chaotic external-cavity semiconductor lasers. IEEE J Quantum Electron 38:1197–1204
496. Viktorov EA, Mandel P (2000) Low frequency fluctuations in a multimode semiconductor laser with optical feedback. Phys Rev Lett 85:3157–3160
497. Von Lehmen AC, Florez LT, Stoffel NG (1991) Dynamics, polarization, and transverse mode characteristics of vertical cavity surface emitting lasers. IEEE J Quantum Electron 27:1402–1409
498. Voss HU (2000) Anticipating chaotic synchronization. Phys Rev E 61:5115–5119
499. Voumard C (1977) External-cavity-controlled 32 MHz narrow-band cw GaAlAs-diode lasers. Opt Lett 1:61–63
500. Wang J, Haldar MK, Li L, Mendis VC (1996) Enhancement of modulation bandwidth of laser diodes by injection locking. IEEE Photon Technol Lett 8:34–36
501. Weiss CO, King H (1982) Oscillation period doubling chaos in a laser. Opt Commun 44:59–61
502. Weiss CO, Godone A, Olafsson A (1983) Routes to chaotic emission in a cw He-Ne laser. Phys Rev A 28:892–895
503. Weiss CO, Klische W, Ering PS, Cooper M (1985) Instabilities and chaos of a single mode NH_3 ring laser. Opt Commun 44:405–408
504. Weiss CO, Brock J (1986) Evidence for Lorenz-type chaos in a laser. Phys Rev Lett 57:2804–2806
505. White JK, Moloney JV (1999) Multichannel communication using an infinite dimensional spatiotemporal chaotic systems. Phys Rev A 59:2422–2426

506. Wieczorek S, Bernd Krauskopf B, Lenstra D (1999) A unifying view of bifurcations in a semiconductor laser subject to optical injection. Opt Commun 172:279–295

507. Wieczorek S, Krauskopf B, Lenstra D (2000) Mechanisms for multistability in a semiconductor laser with optical injection. Opt Commun 183:215–226

508. Wieczorek S, Krauskopf B, Lenstra D (2001a) Unnested islands of period doublings in an injected semiconductor laser. Phys Rev E 64:056204–1–9

509. Wieczorek S, Krauskopf B, Lenstra D (2001b) Sudden chaotic transitions in an optically injected semiconductor laser. Opt Lett 11:816–818

510. Wieczorek S, Krauskopf B, Lenstra D (2001c) Bifurcation transitions in an optically injected diode laser: Theory and experiment. Opt Commun 215:125–134

511. Wieczorek S, Simpson TB, Krauskopf B, Lenstra D (2002) Global quantitative predictions of complex laser dynamics. Phys Rev E 65: R045207–1-4

512. Wieland J, Mirasso CR, Lenstra D (1997), Prevention of coherence collapse in diode lasers by dynamic targeting. Opt Lett 22:469–471

513. Winful HG, Chen YC, Liu JM (1986) Frequency locking, quasiperiodicity, and chaos in modulated self-pulsating semiconductor lasers. Appl Phys Lett 48:616–618

514. Winful HG, Rahman L (1990) Synchronized chaos and spatiotemporal chaos in arrays of coupled lasers. Phys Rev Lett 65:1575–1578

515. Winful HG (1992) Instability threshold for an array of coupled semiconductor lasers. Phys Rev A 46:6093–6094

516. Wolff S, Fouckhardt H (2000) Intracavity stabilization of broad area lasers by structured delayed optical feedback. Opt Exp 7:222–227

517. Wolff S, Rodionov A, Sherstobitov VE, Fouckhardt H (2003) Fourier-optical transverse mode selection in external-cavity broad-area lasers: experimental and numerical results. IEEE J Quantum Electron 39:448–458

518. Wyatt R, Devlin WJ (1983) 10 kHz linewidth 1.5 μ m InGaAsP external cavity laser with 55 nm tuning range. Electron Lett 19:110–112

519. Wyatt R (1985) Spectral linewidth of external cavity semiconductor lasers with strong, frequency selective feedback. Electron Lett 21:658–659

520. Yabre G, (1996) Effect of relatively strong light injection on the chirp-to-power ratio and the 3 dB bandwidth of directly modulated semiconductor lasers. J Lightwave Technol 14:2367–2373

521. Yamada M (1993) A theoretical analysis of self-sustained pulsation phenomena in narrow-stripe semiconductor lasers. IEEE J Quantum Electron 29:1330–1336

522. Yamada M (1996) Theoretical analysis of noise-reduction effects in semiconductor lasers with help of self-sustained pulsation phenomena. J Appl Phys 79:61–71

523. Yamada M (1998a) Computer simulation of feedback induced noise in semiconductor lasers operating with self-sustained pulsation. IEICE Trans Electron E81-C:768–780

524. Yamada M (1998b) A theoretical analysis of quantum noise in semiconductor leasers operating with self-sustained pulsation. IEICE Trans Electron E81-C:290–298

525. Yamamoto Y(1983) AM and FM quantum noise in semiconductor lasers – part I: theoretical analysis. IEEE J Quantum Electron QE-19:34–46

526. Yariv A (1997) Optical electronics in modern communications. Oxford University Press, Oxford

527. Ye J, Li H, McInerney JG (1993) Period-doubling route to chaos in a semiconductor laser with weak optical feedback. Phys Rev A 47:2249–2252

528. Ye SY, Ohtsubo J (1998) Experimental investigation of stability enhancement in semiconductor lasers with optical feedback. Opt Rev 5:280–284

529. Yen TC, Chang JW, Lin JM, Chen RJ (1998) High-frequency optical generation in a semiconductor laser by incoherent optical feedback. Opt Commun 150:158–162

530. Yoon TH, Lee CH, Shin SY (1989) Perturbation analysis of bistable and period doubling bifurcation in directly-modulated laser diodes. IEEE J Quantum Electron 25:1993–2000

531. Yoshino T, Nara M, Mnatzakanian S, Lee BS, Strand TC (1987) Laser diode feedback interferometer for stabilization and displacement measurements. Appl Opt 26:892–897

532. Yousefi M, Lenstra D (1999) Dynamical behavior of semiconductor laser with filtered external optical feedback. IEEE J Quantum Electron 35:970–976

533. Yousefi M, Lenstra D (2003) Nonlinear dynamics of a semiconductor laser with filtered optical feedback and the influence of noise. Phys Rev E 67:046213-1–11

534. Yu SF (1999) Nonlinear dynamics of vertical-cavity surface-emitting lasers. IEEE J Quantum Electron 35:332–331

535. Yu HC, Wang JS, Su YK, Chang SJ, Lai FI, Chang YH, Kuo HC, Sung CP, Yang HPD, Lin KF, Wang JM, Chi JY, Hsiao RS, Mikhrin S (2006) 1.3 μm InAs-InGaAs quantum-dot vertical-cavity surface-emitting laser with fully doped DBRs grown by MBE. IEEE Photon Technol Lett 18:418–420

536. Yu Y, Giuliani G, Donati S (2004) Measurement of the linewidth enhancement factor of semiconductor lasers based on the optical feedback self-mixing effect. IEEE Photon. Technol. Lett. 16:990–992

537. Zah CE, Osinski JS, Menocal SG, Tabatabaie N, Lee TP, Dentai AG, Burrus CA (1987) Wide-bandwidth and high-power 1.3 mm InGaAsP buried crescent lasers with semi-insulating Fe-doped InP current blocking layers. Electron Lett 23:52–53

538. Zeghlache H, Mandel P (1985) Influence of detuning on the properties of laser equations. J Opt Soc Am B 2:18–22

539. Zhu H, Ruset IC, Hersman FW (2005) Spectrally narrowed external-cavity high-power stack of laser diode arrays. Opt Lett 30:1342–1344

540. Zorabedian P, Trutna WR, Cutler LS (1987) Bistability in grating tuned external-cavity semiconductor laser. IEEE J Quantum Electron 23:1855–1860

Index

Springer Series in
OPTICAL SCIENCES

Springer Series in
OPTICAL SCIENCES